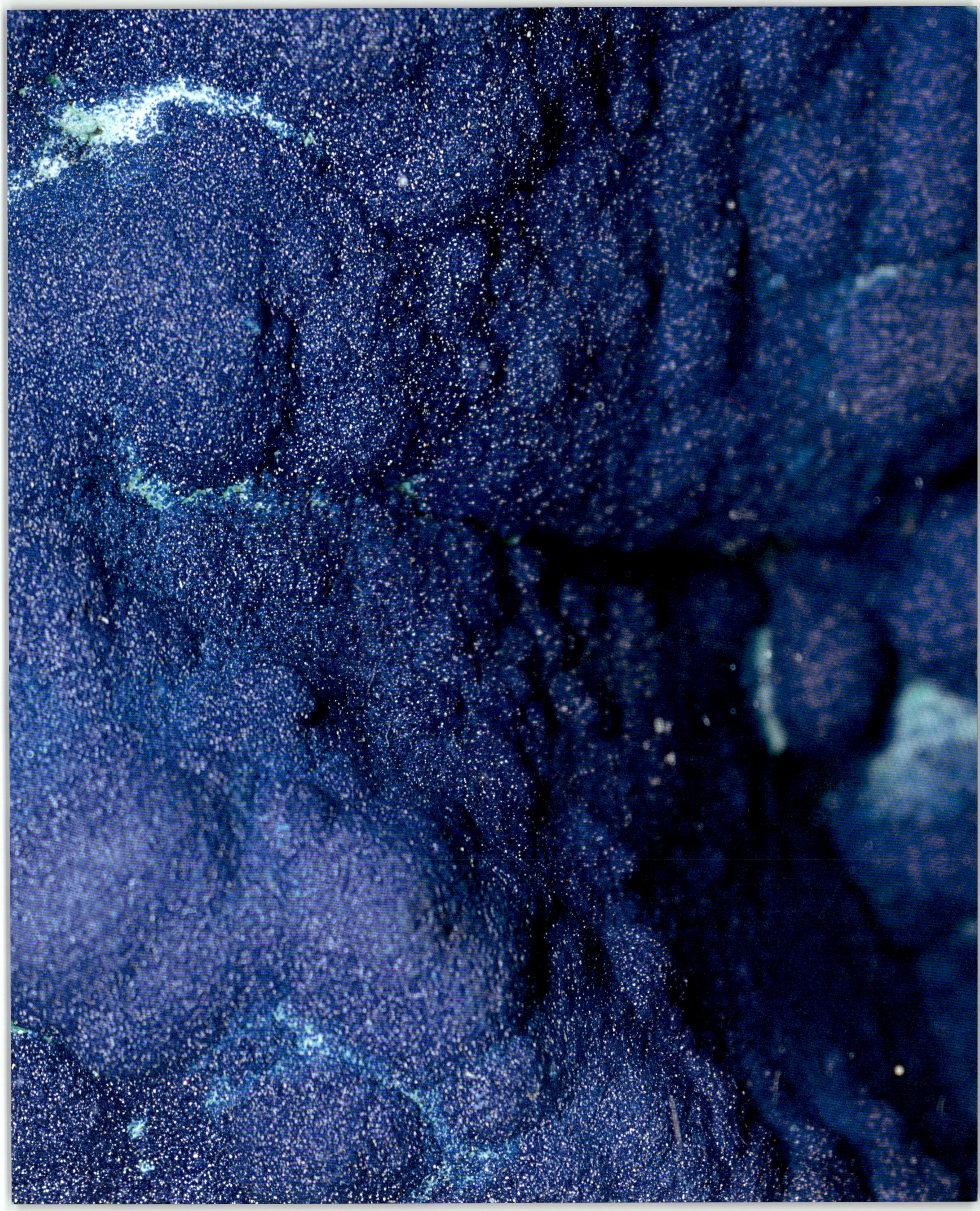

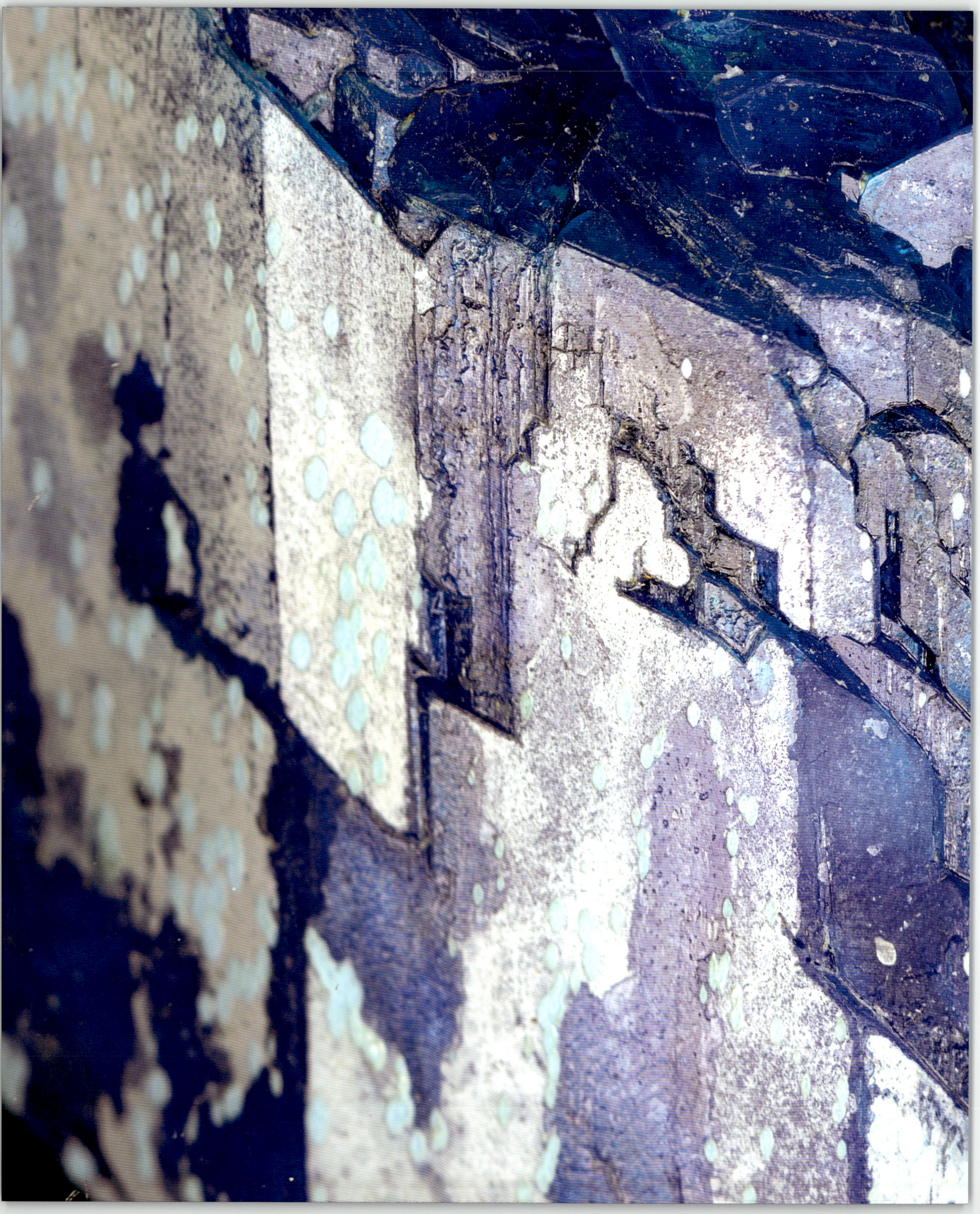

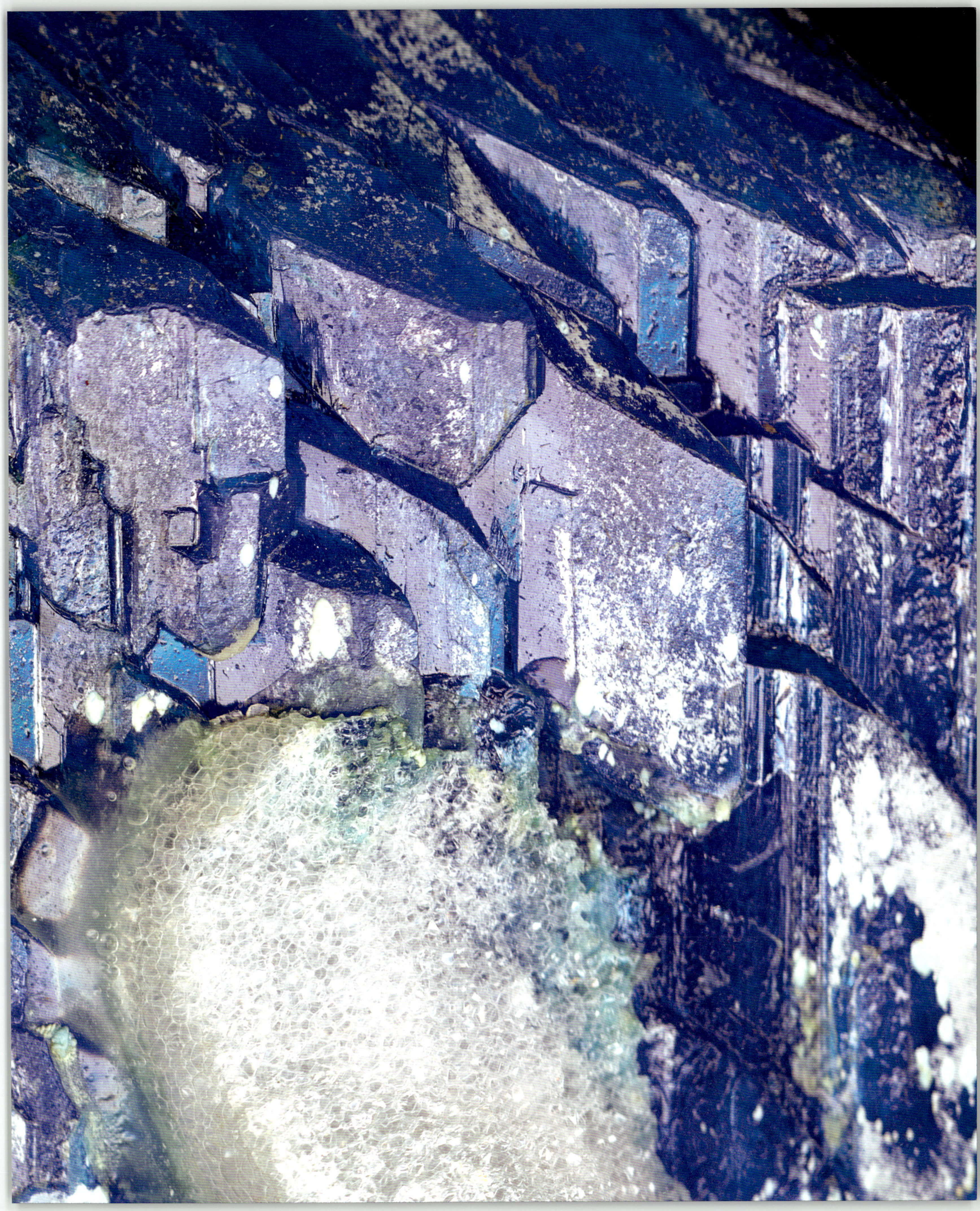

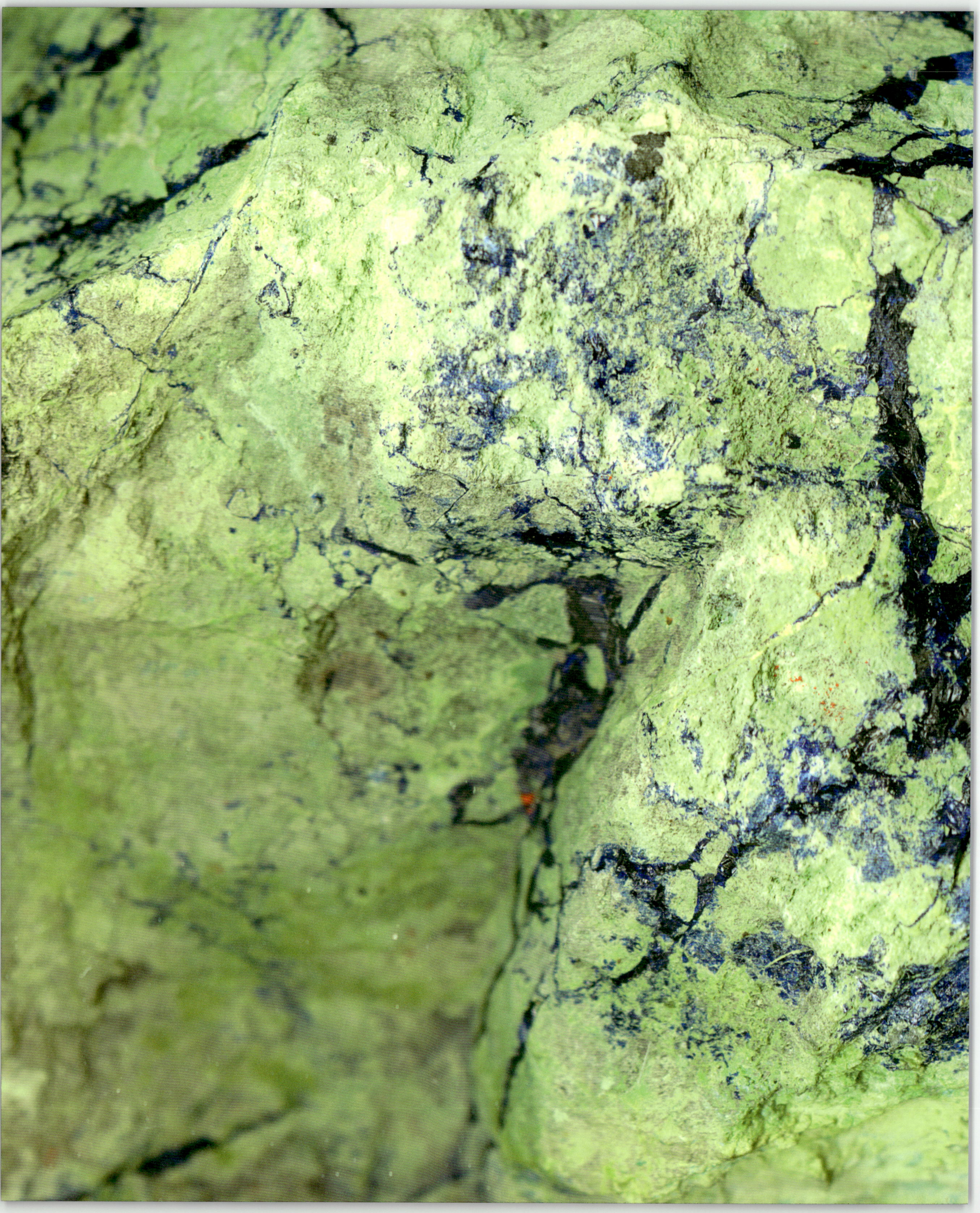

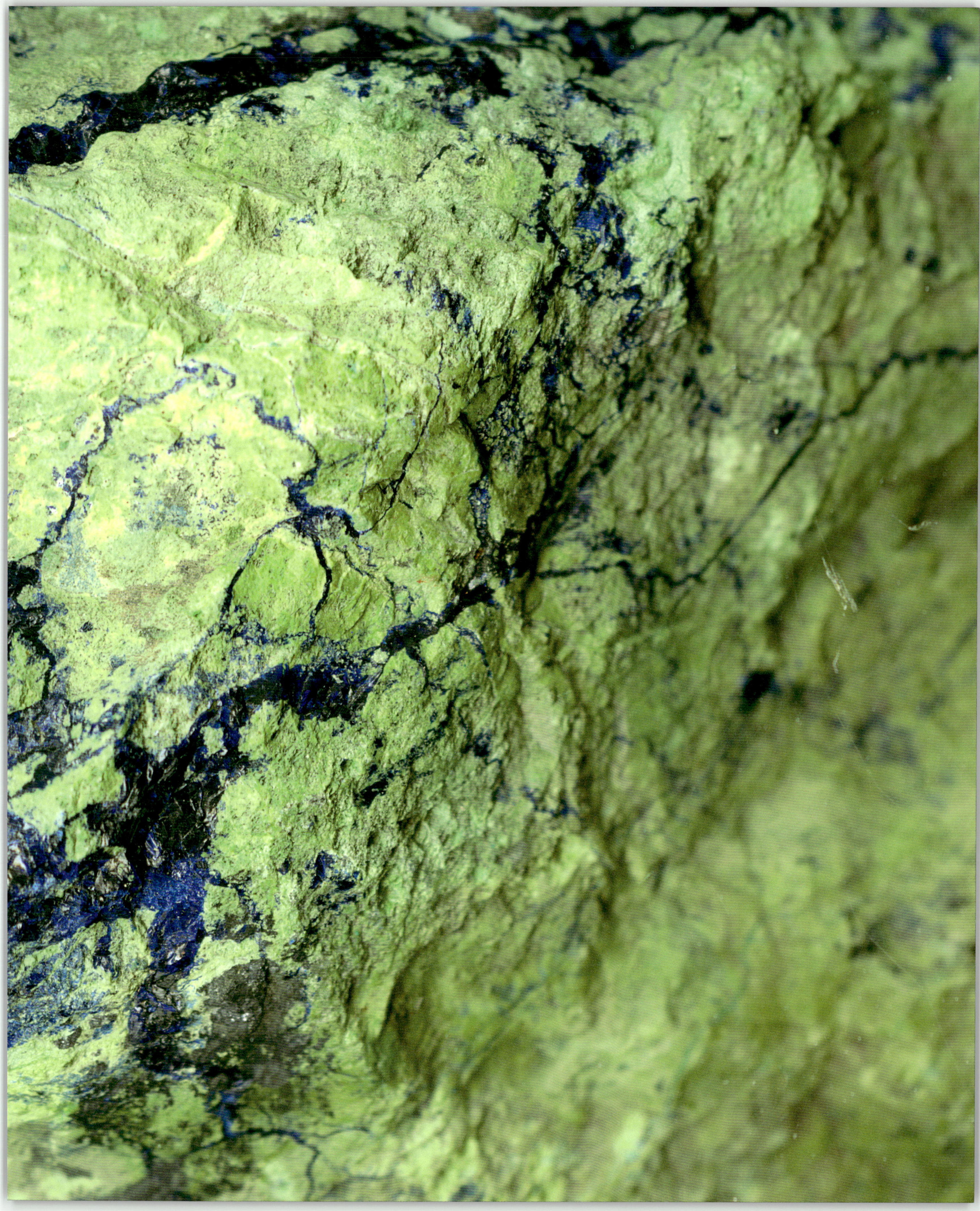

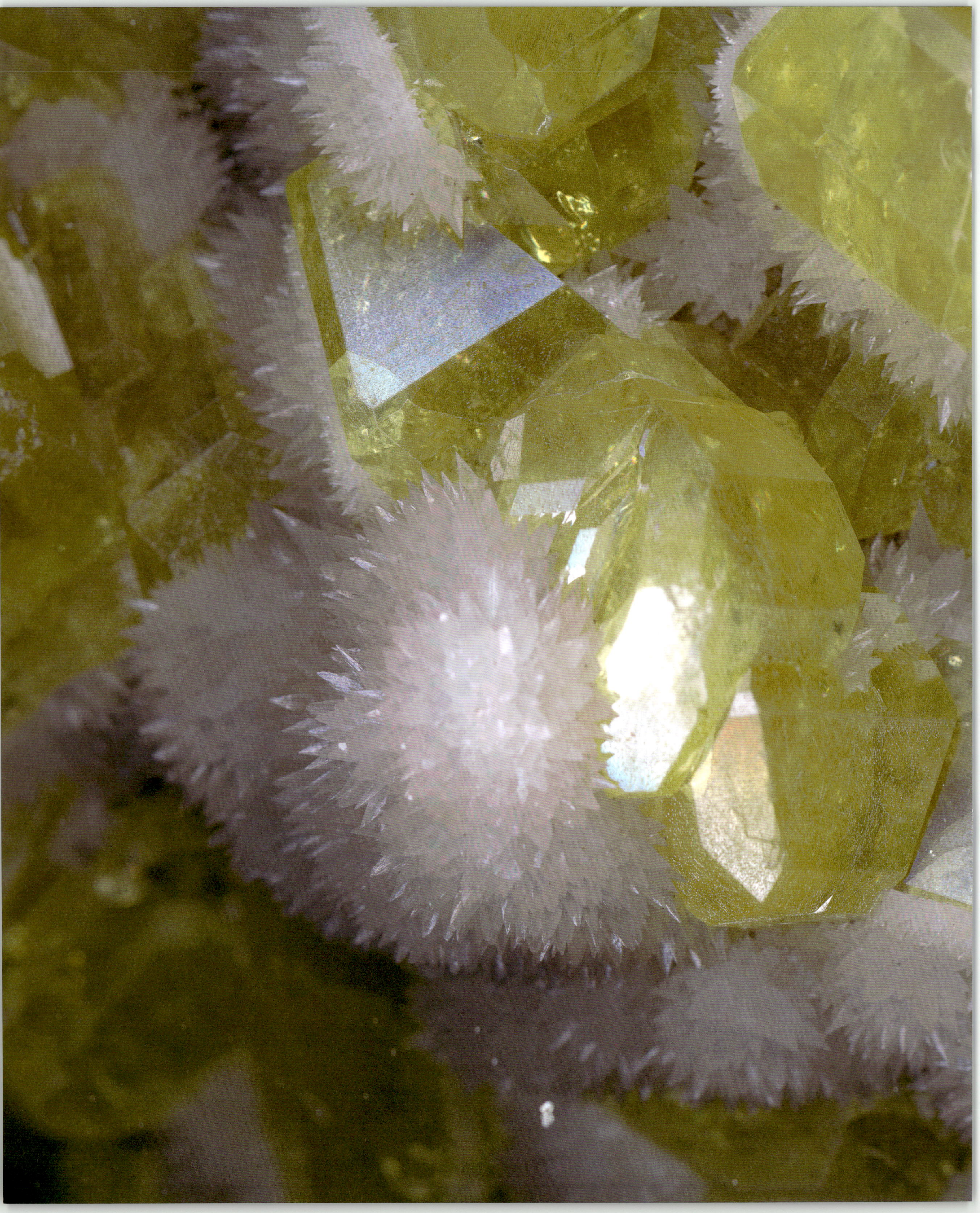

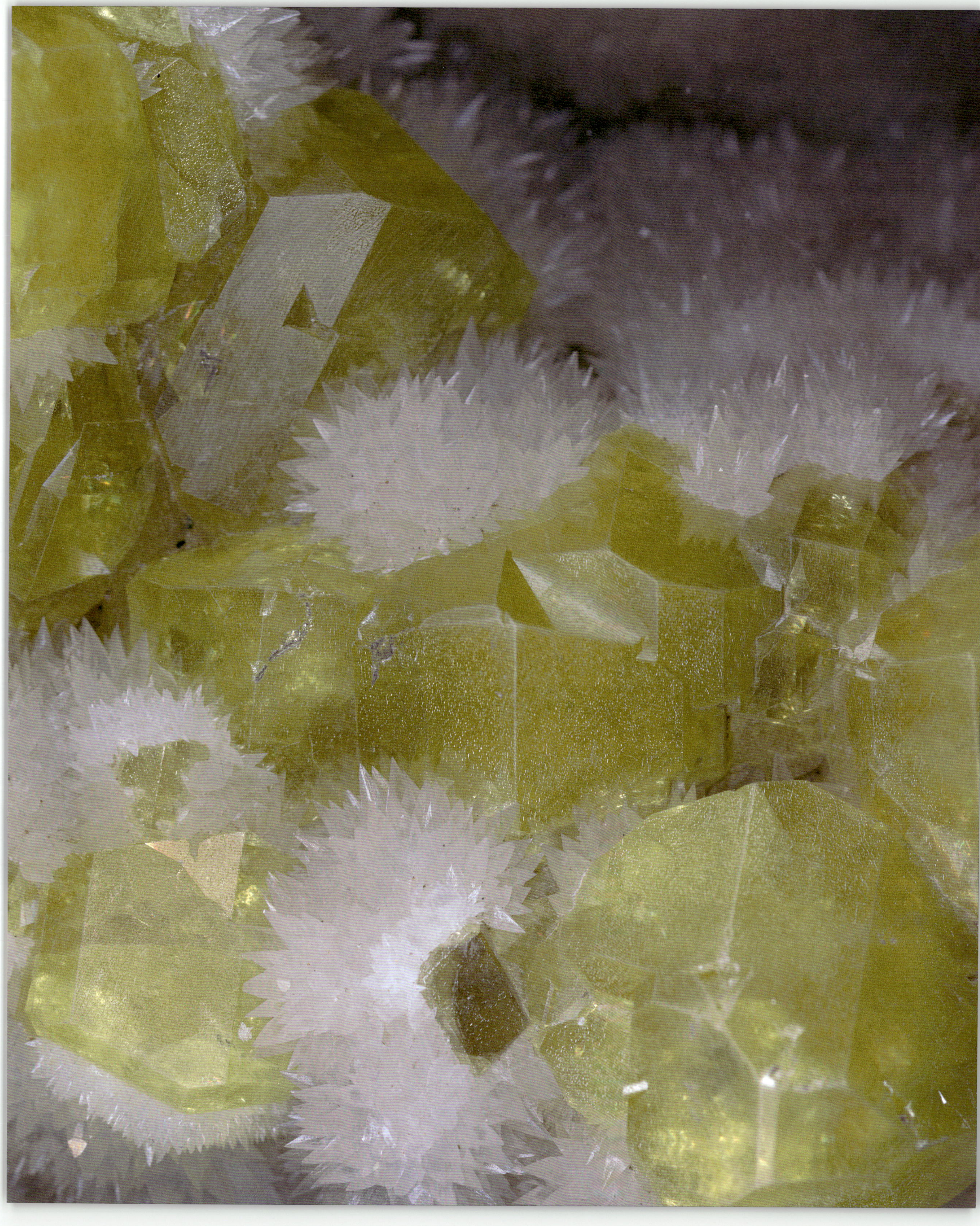

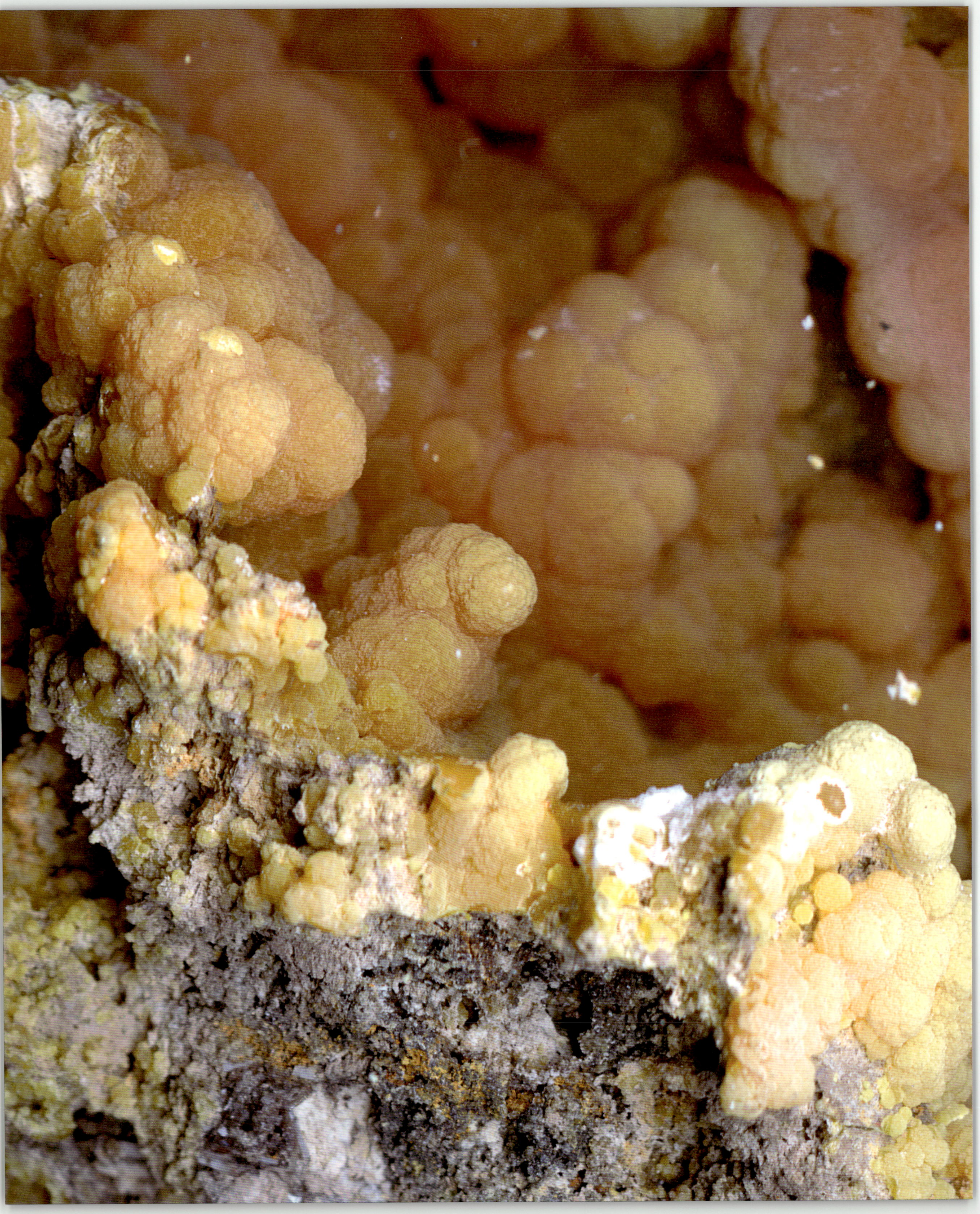

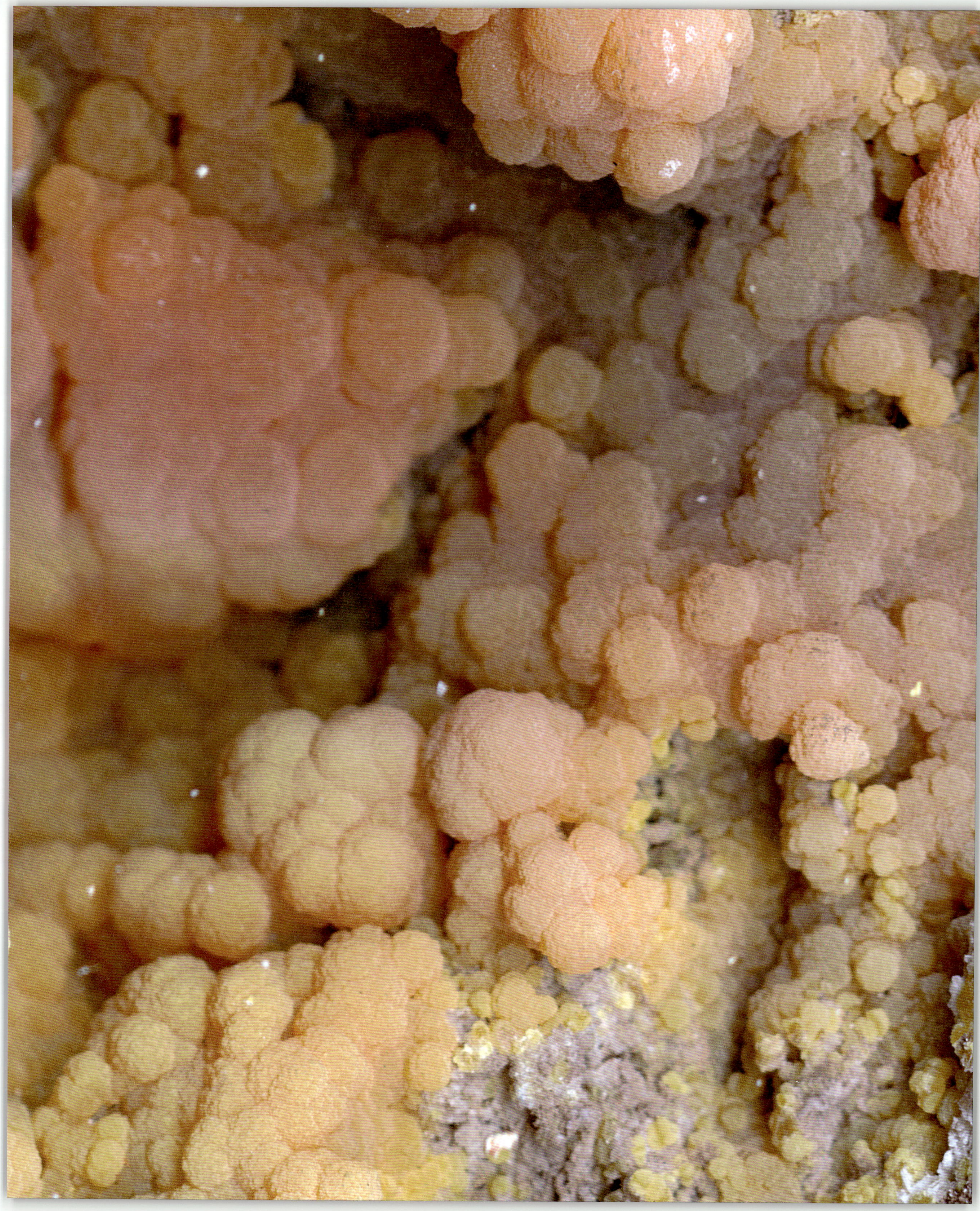

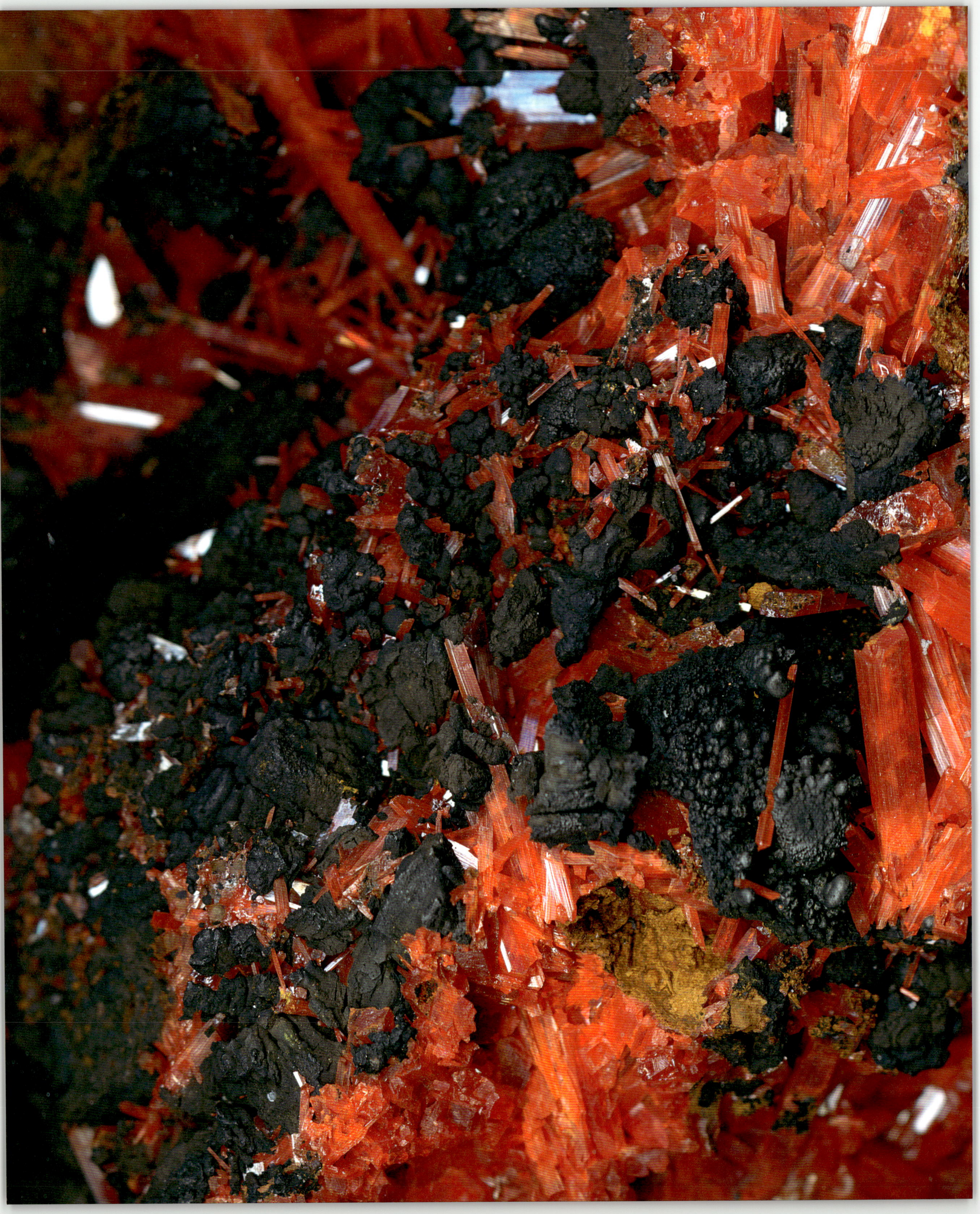

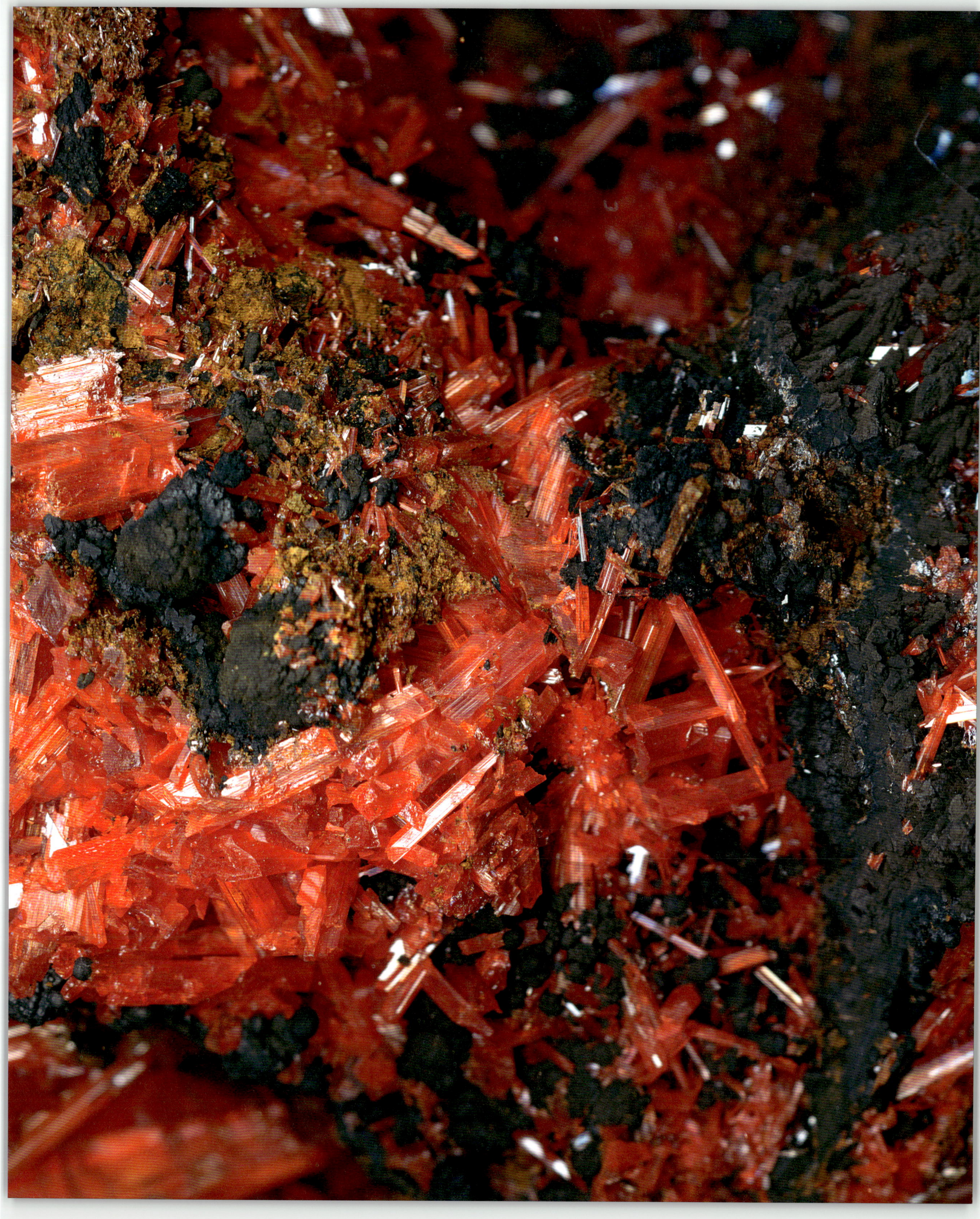

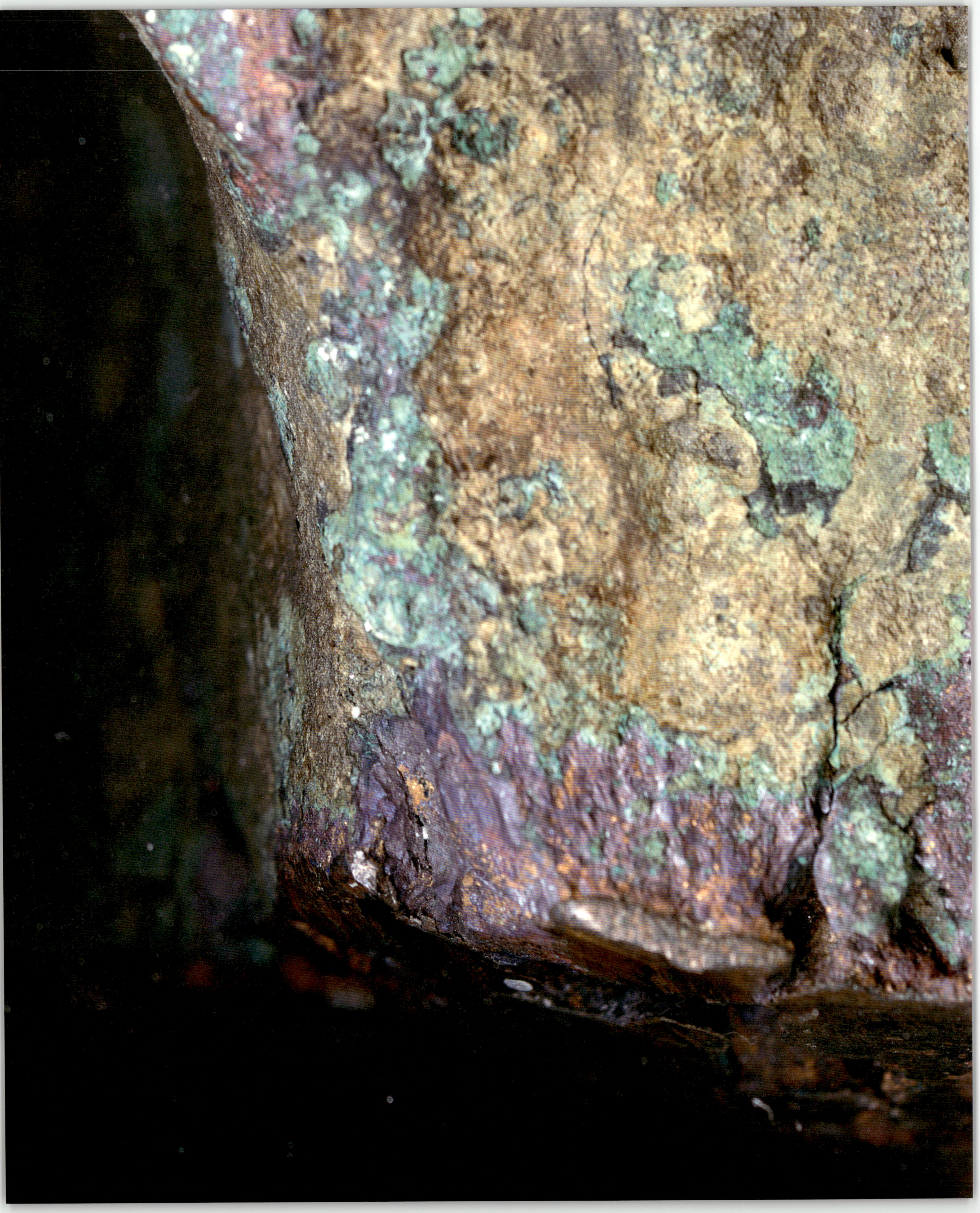

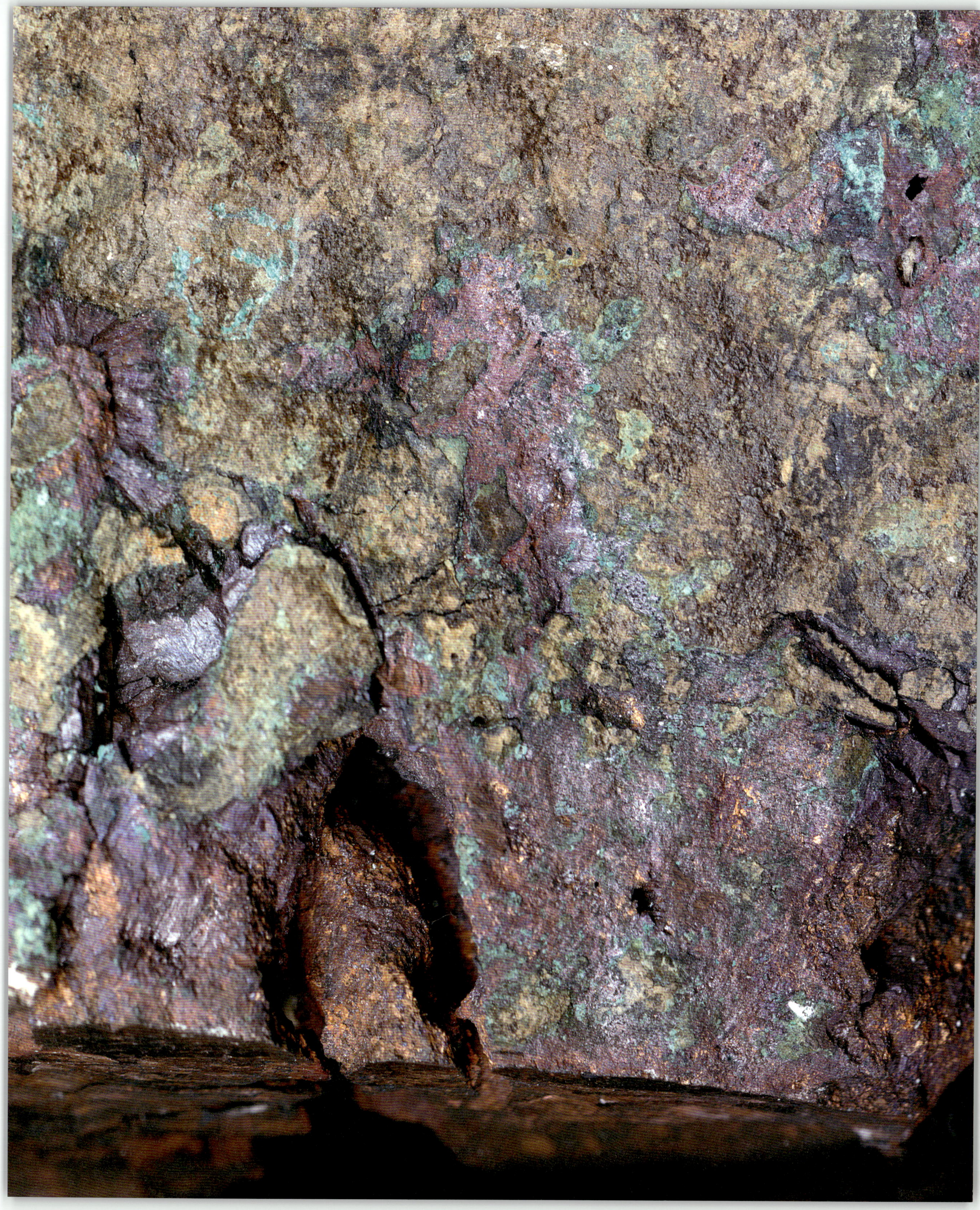

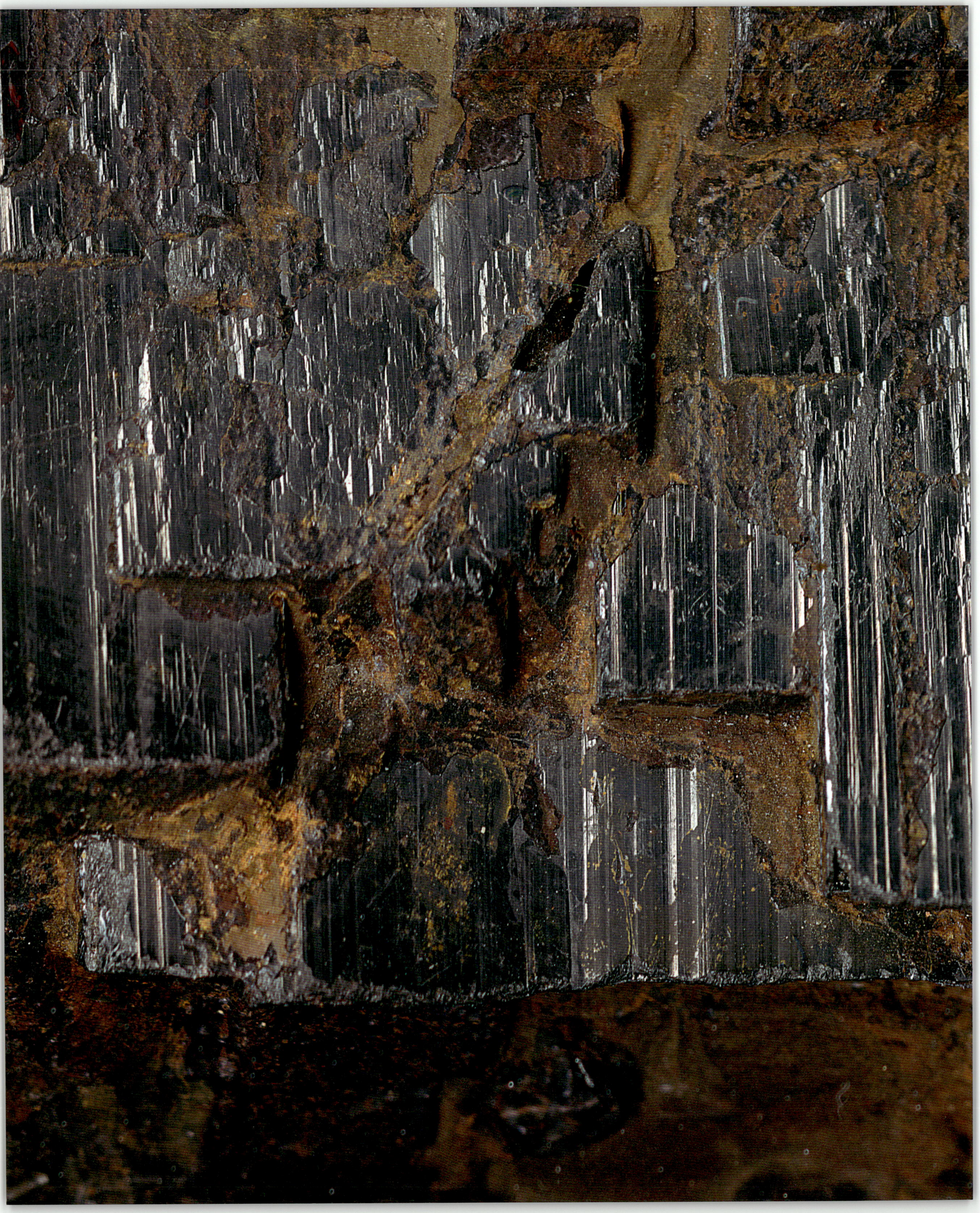

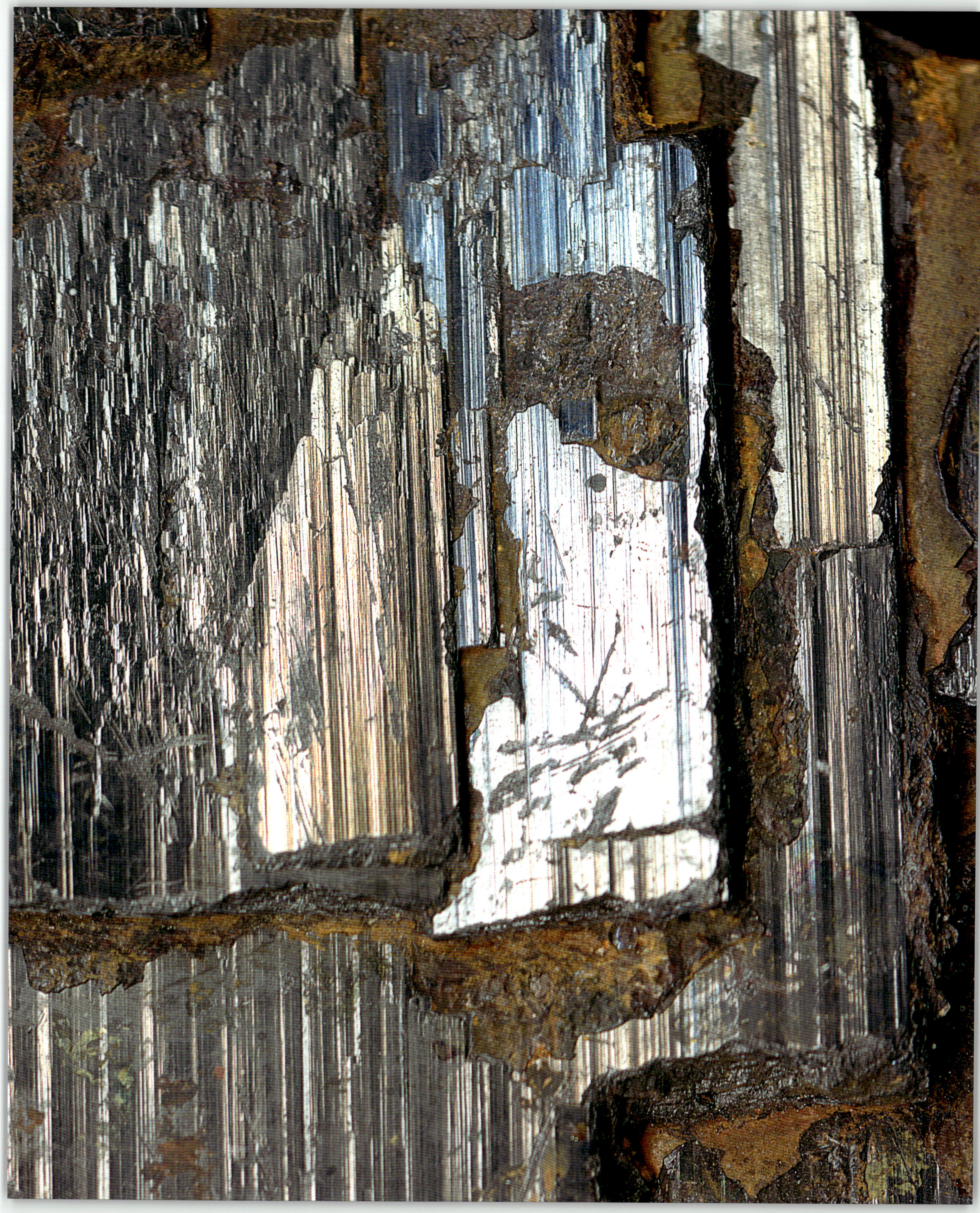

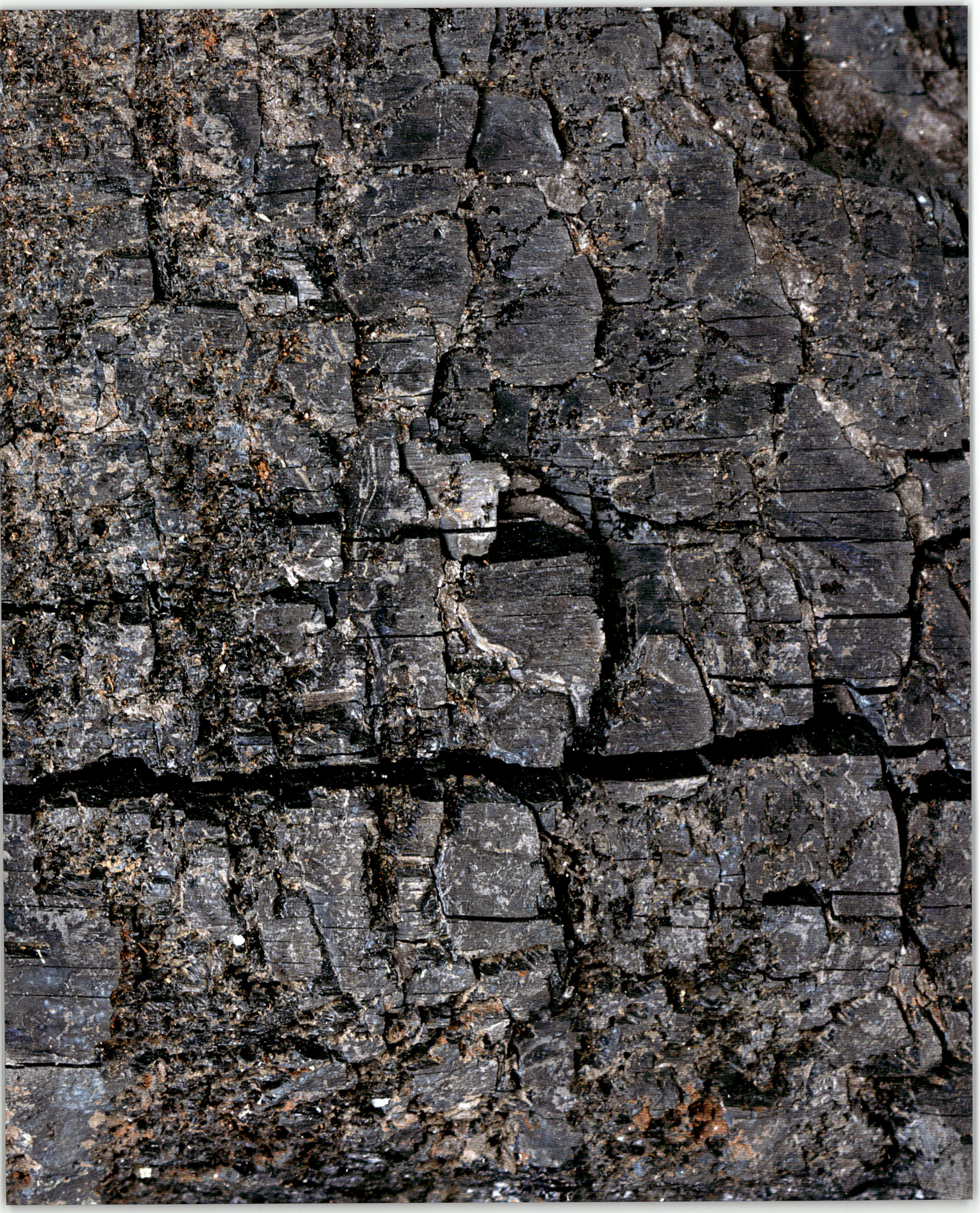

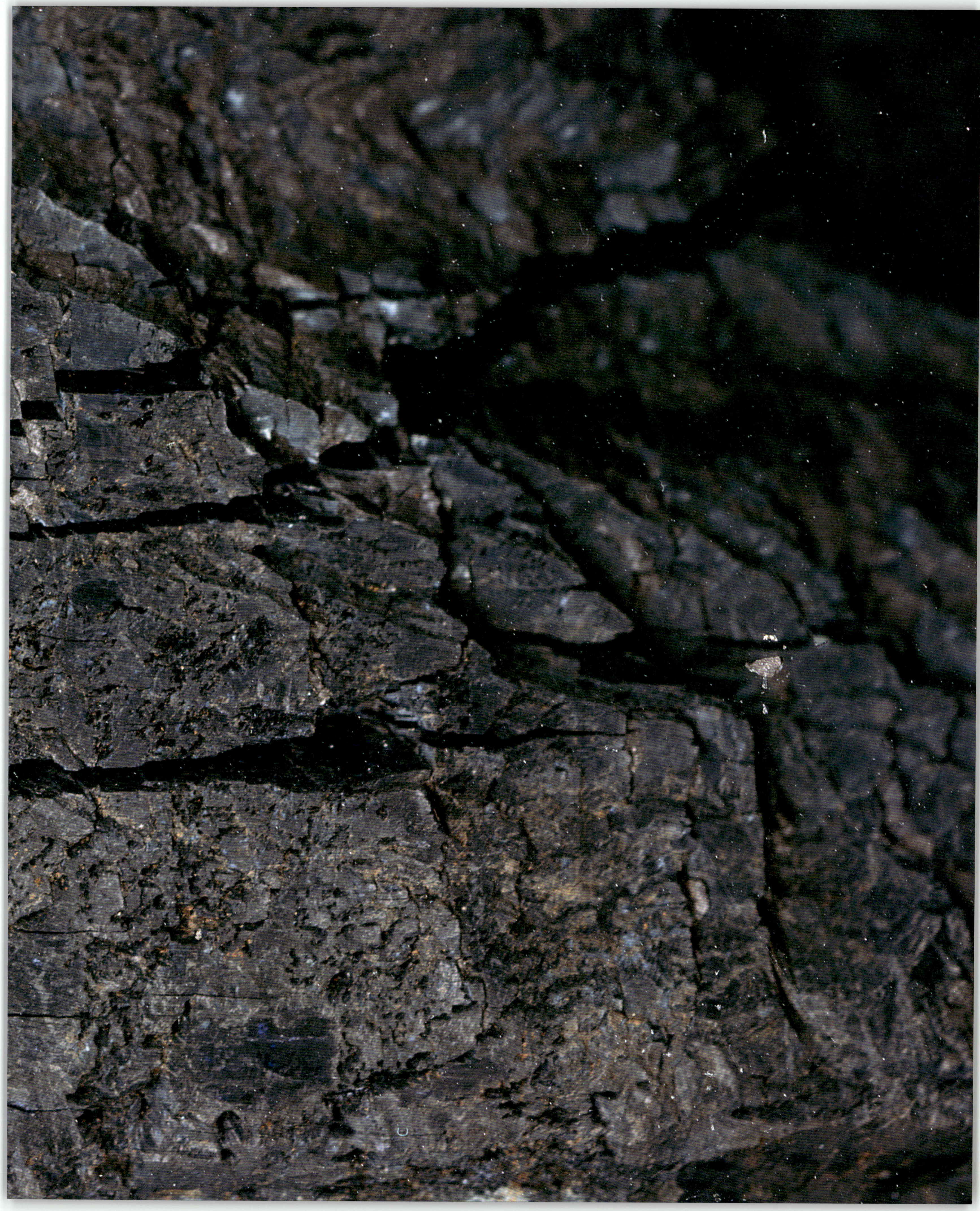

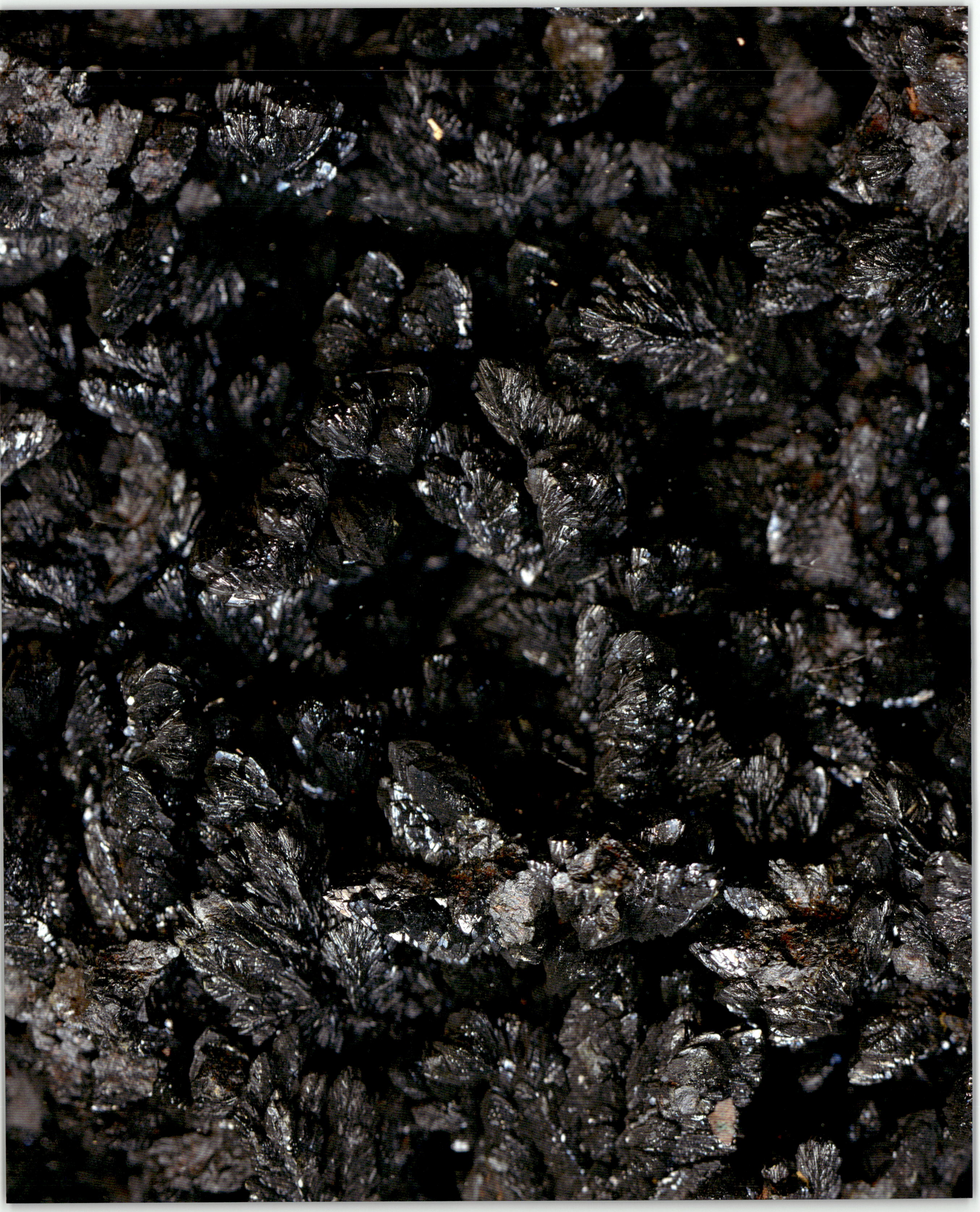

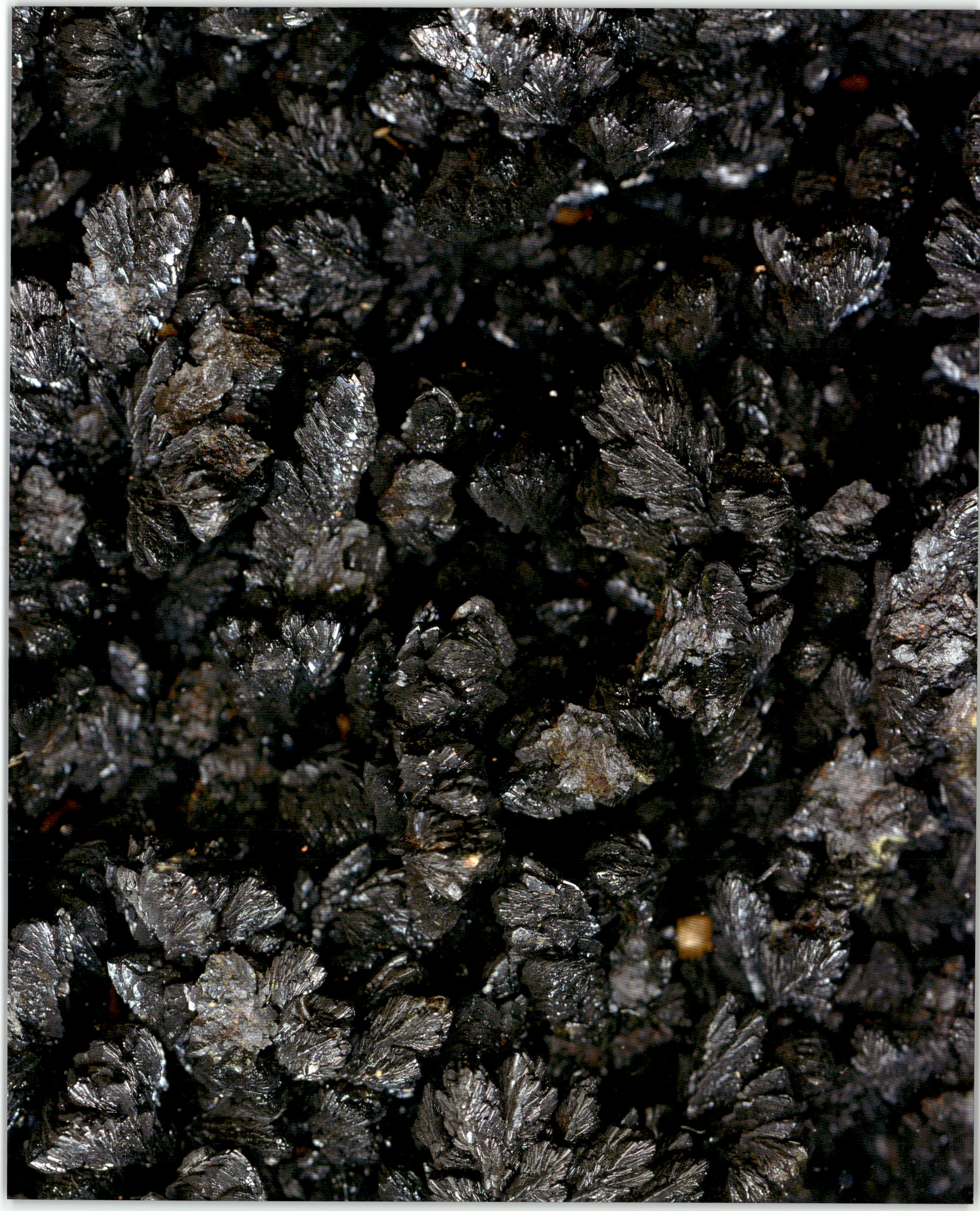

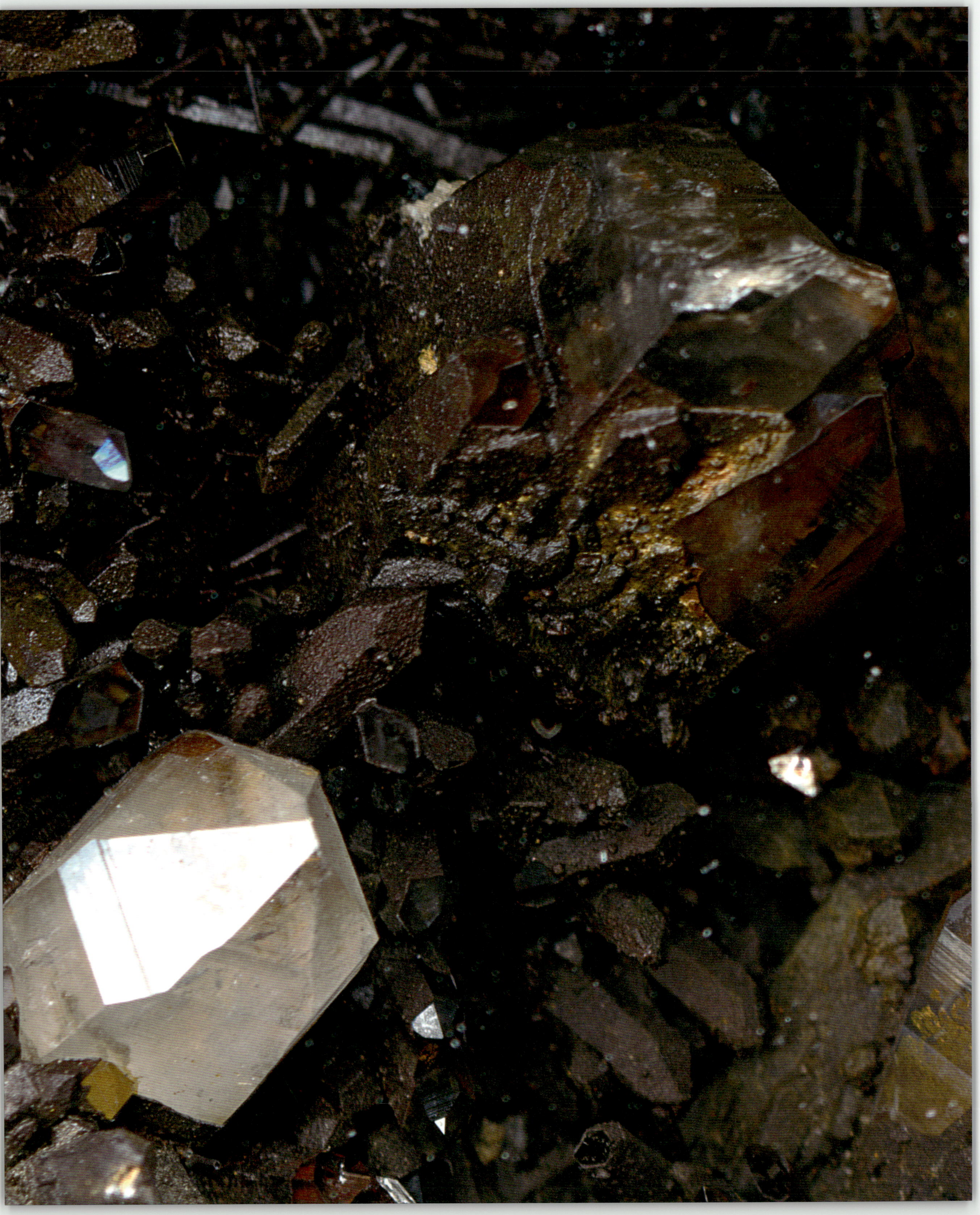

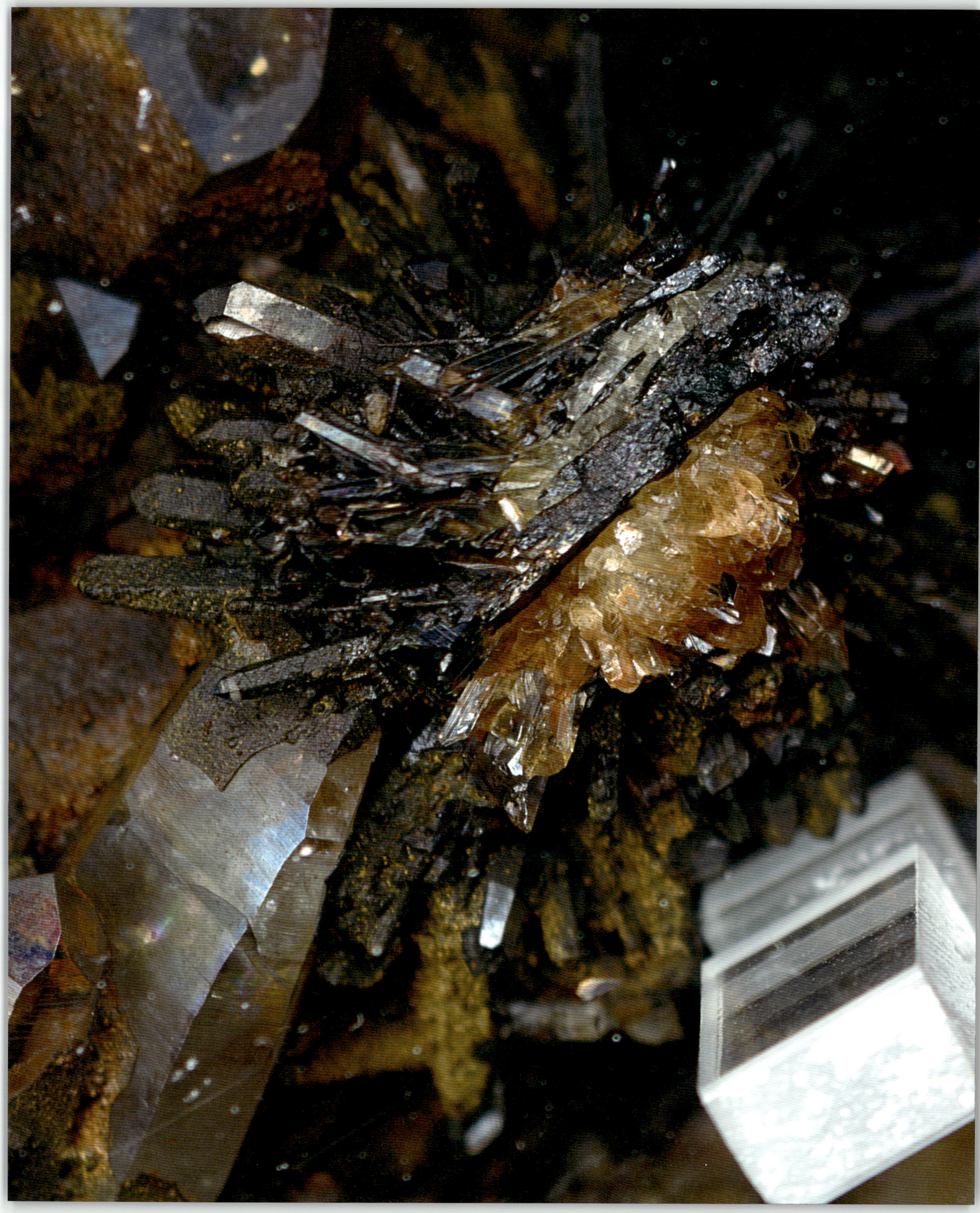

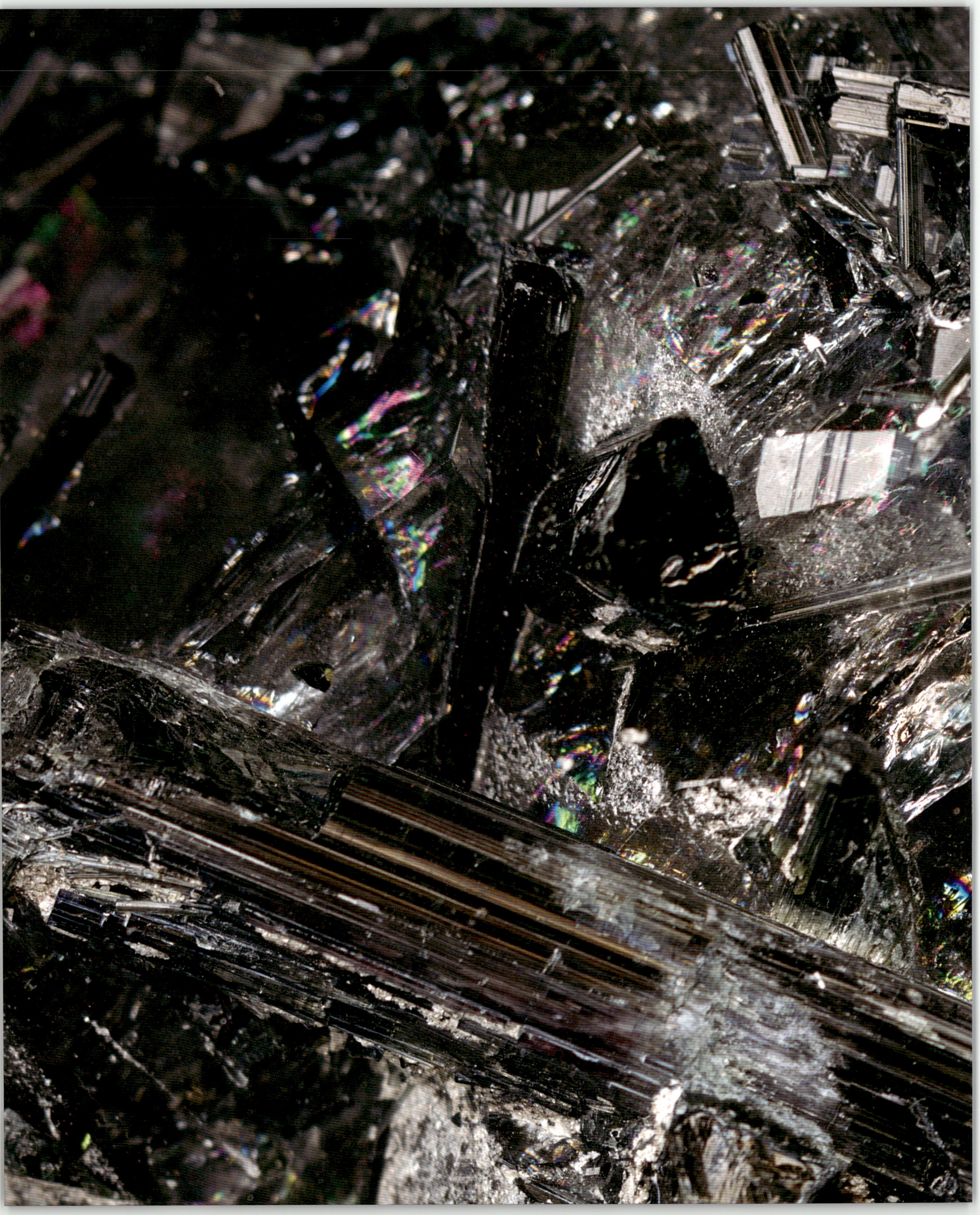

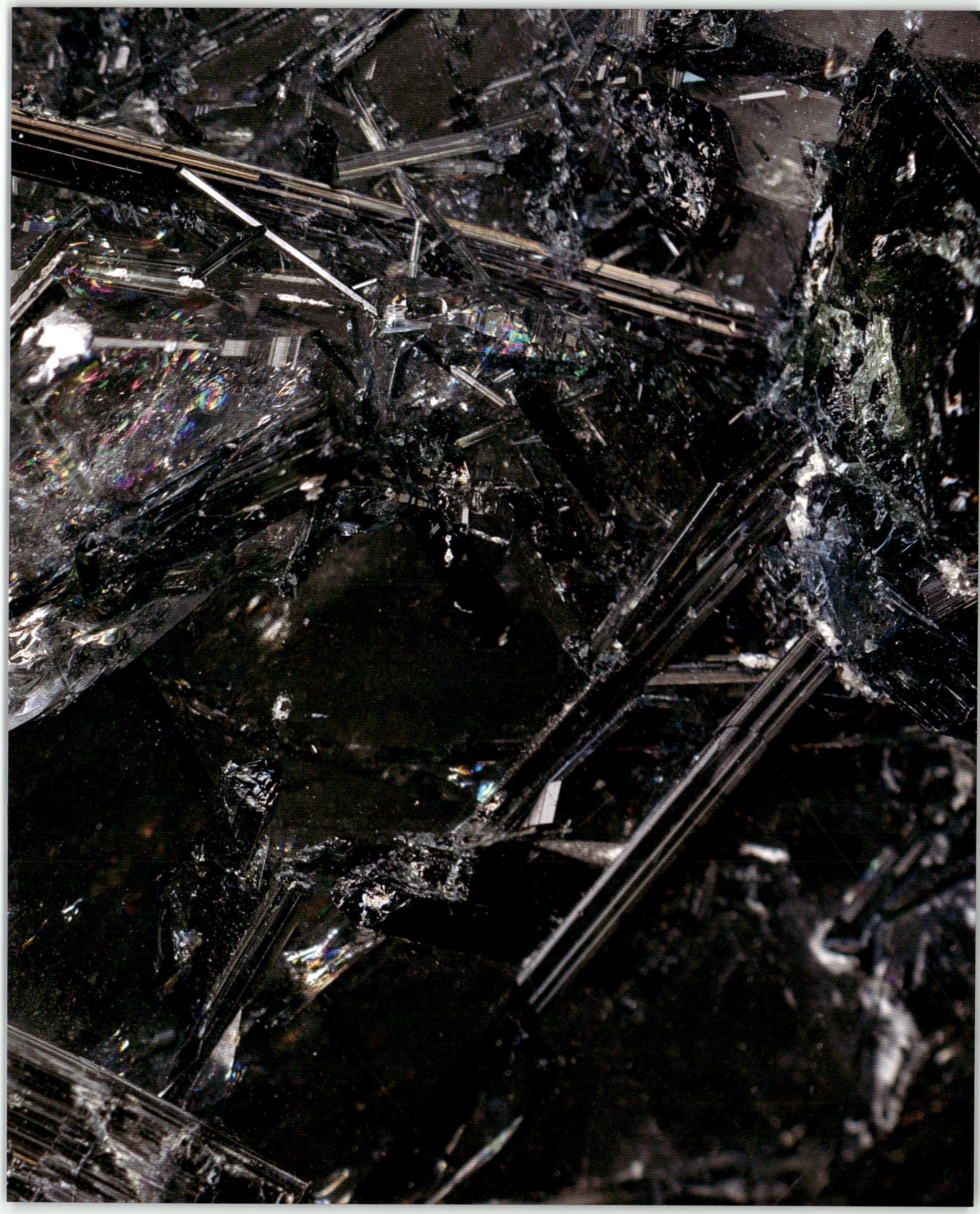

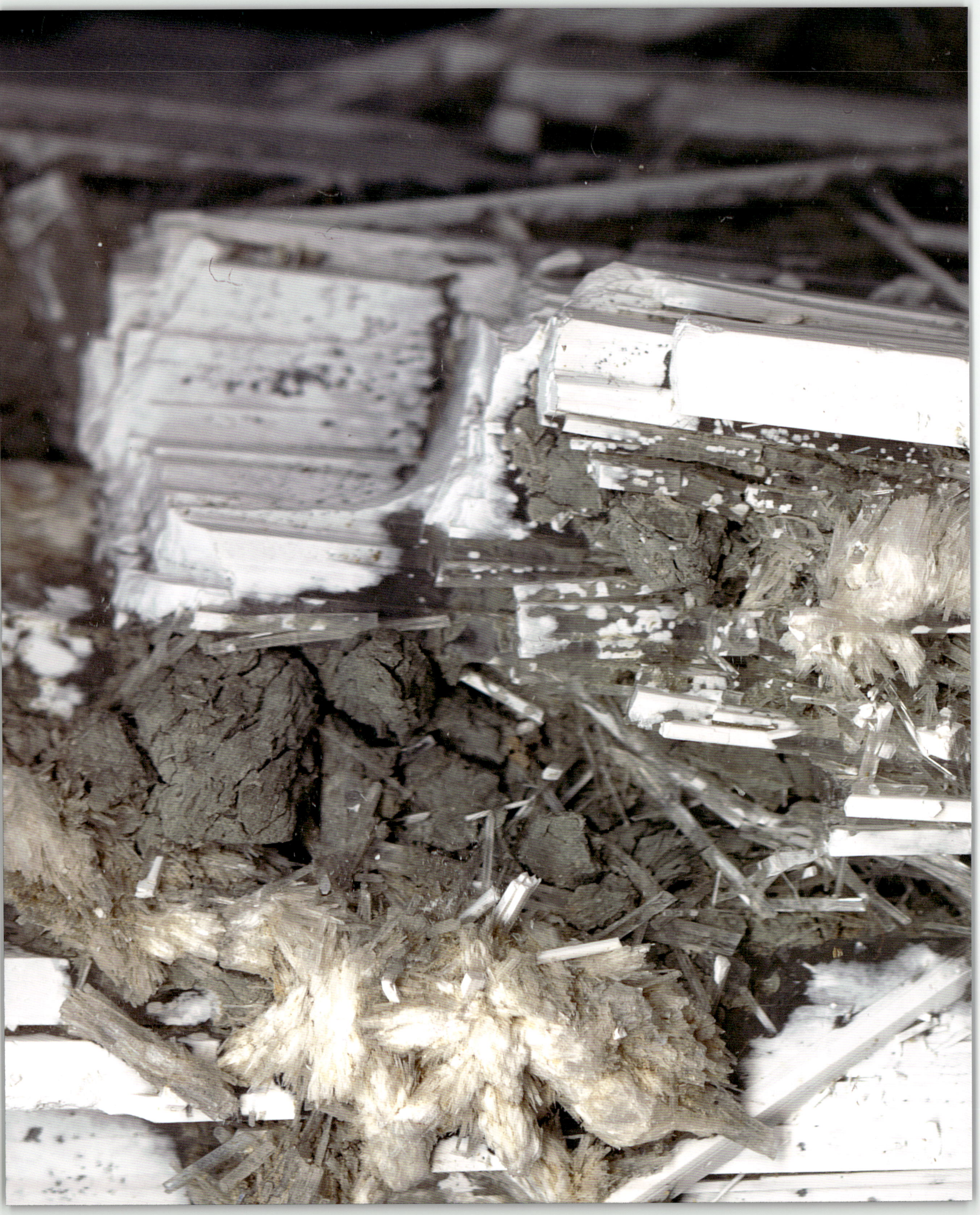

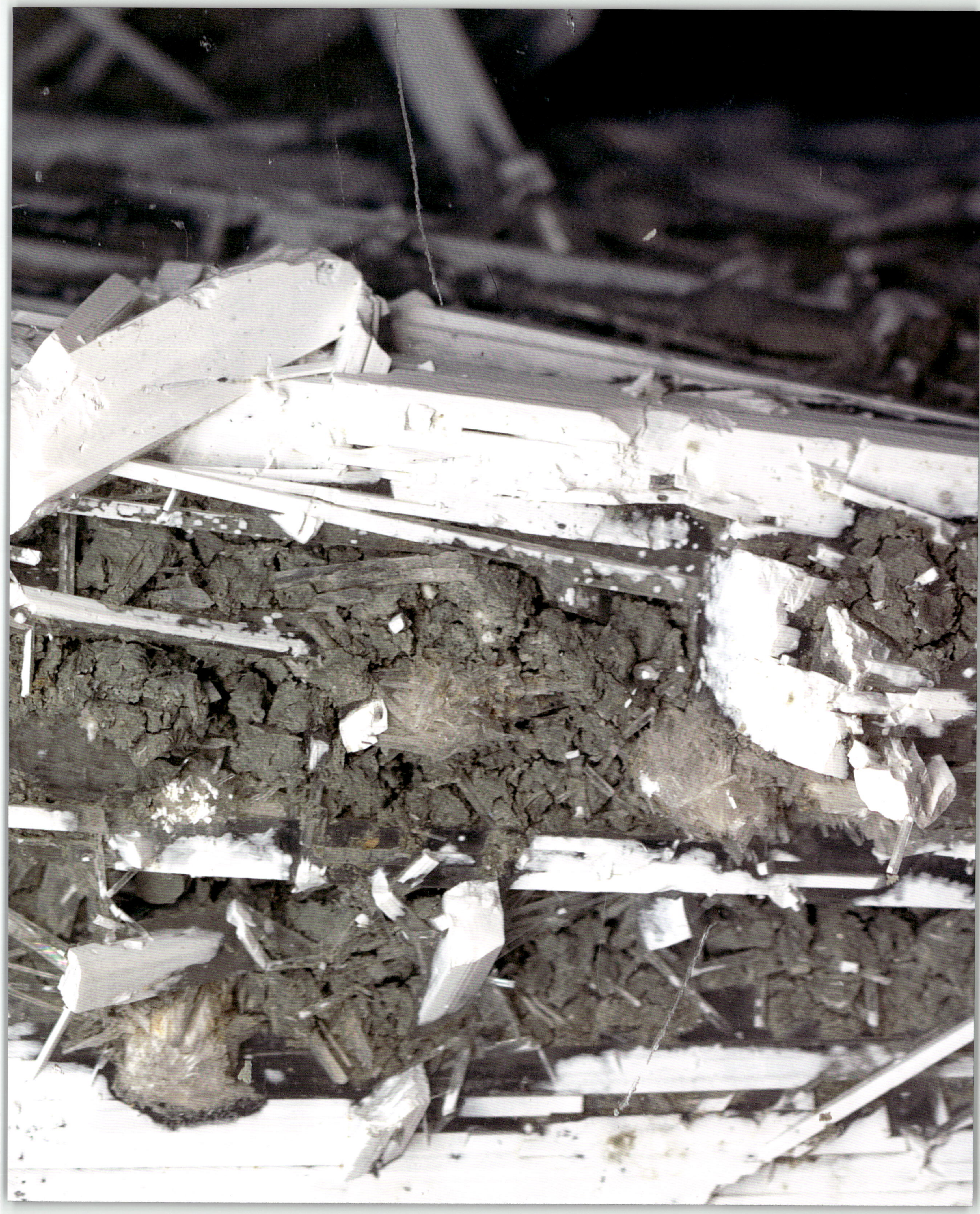

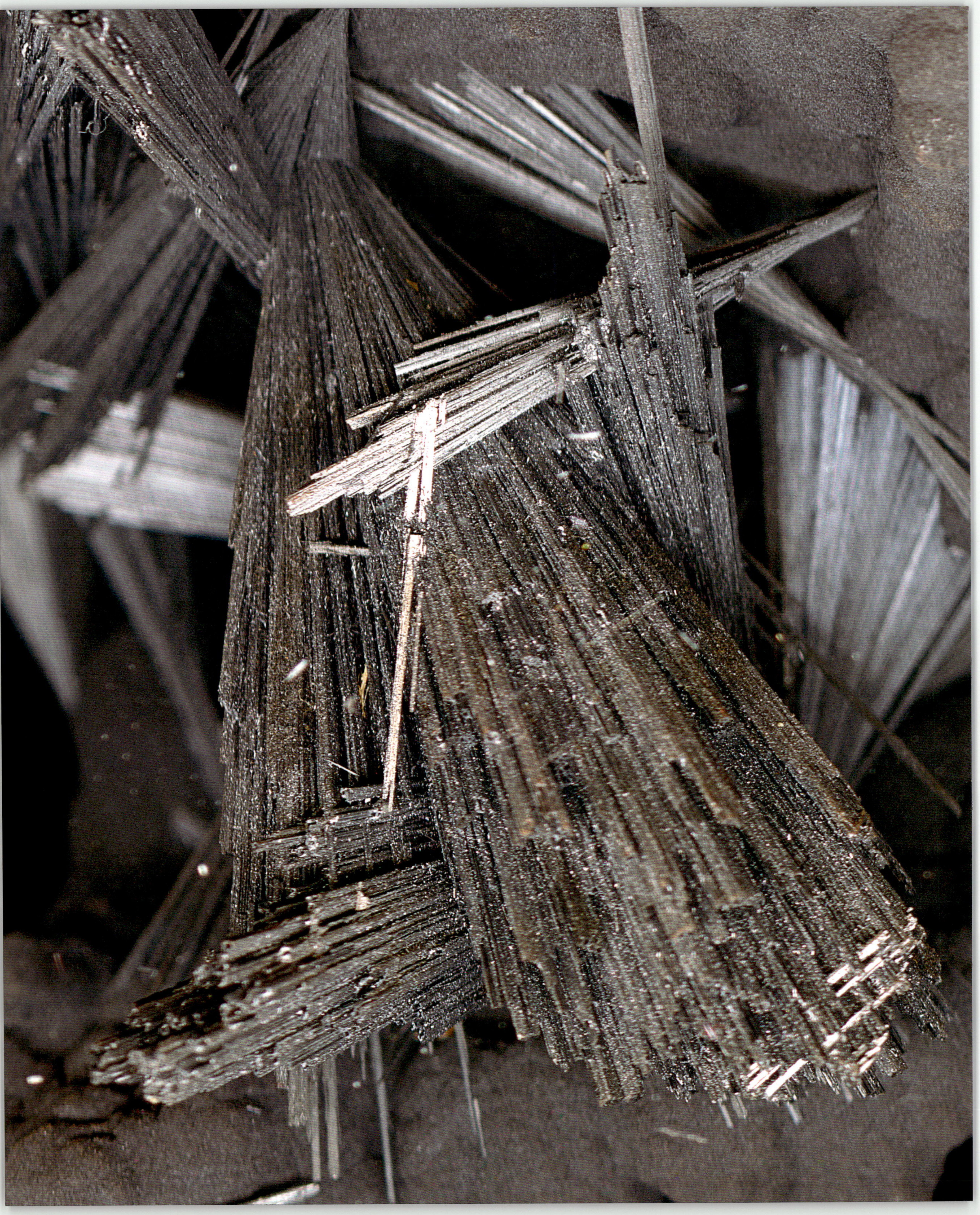

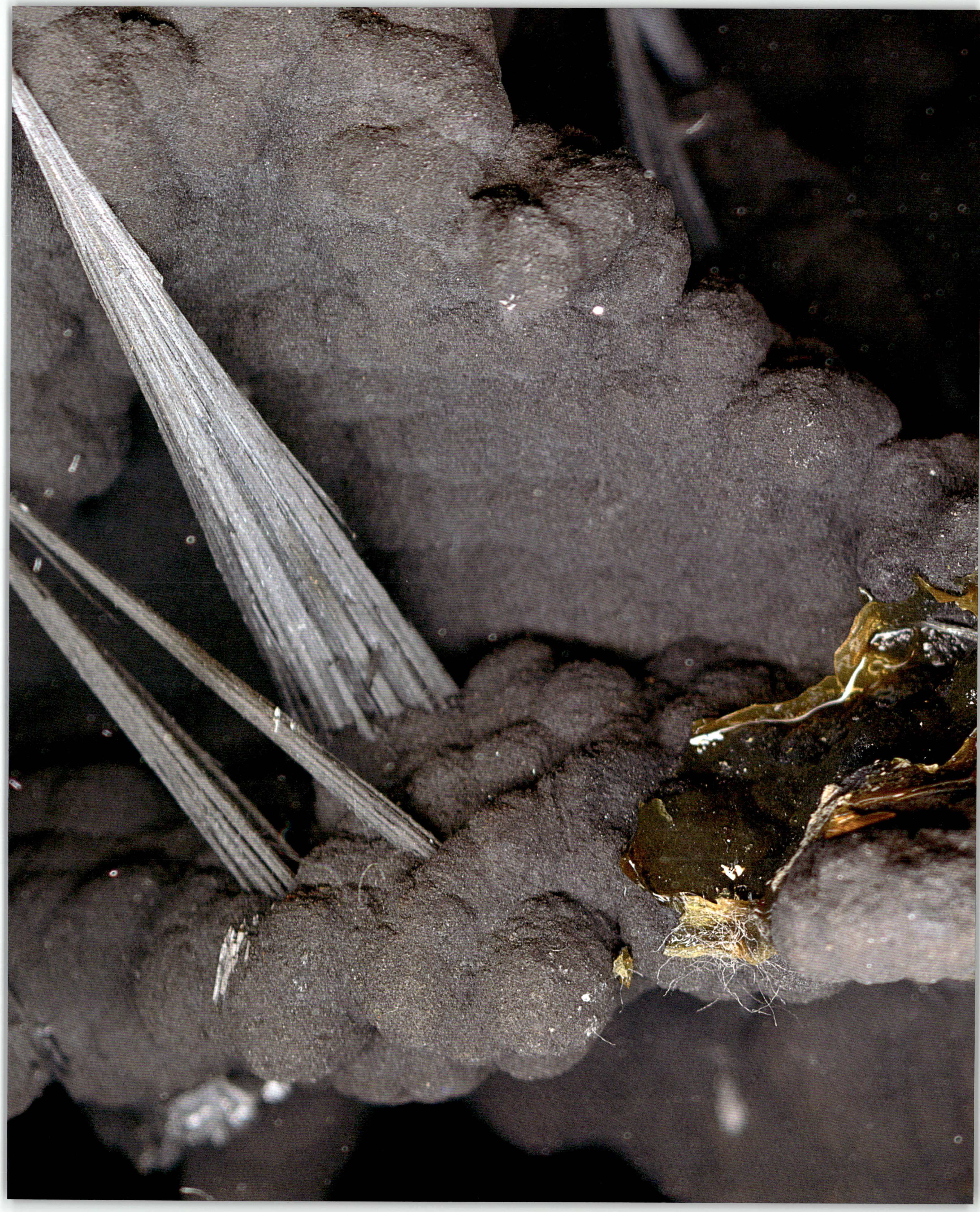

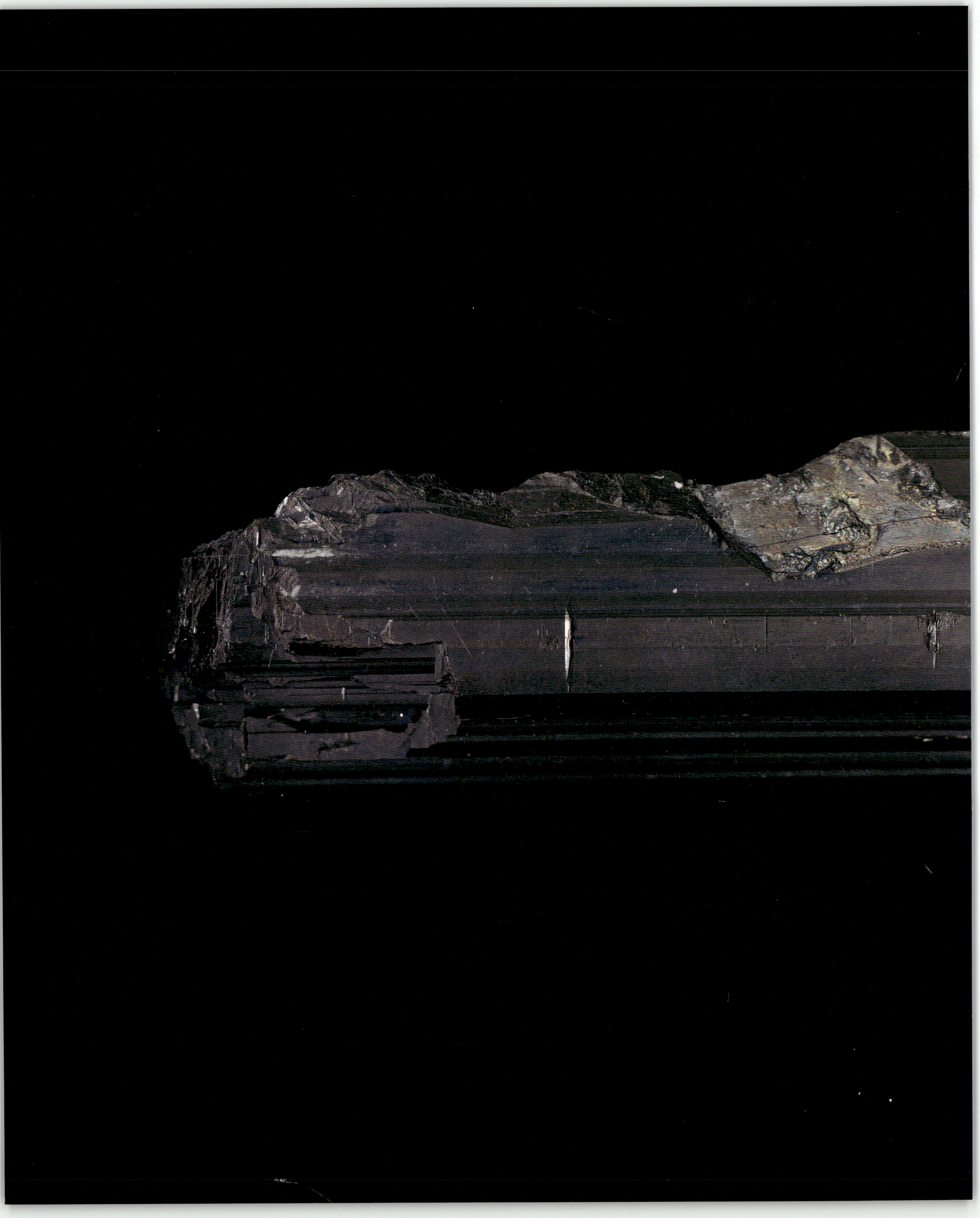

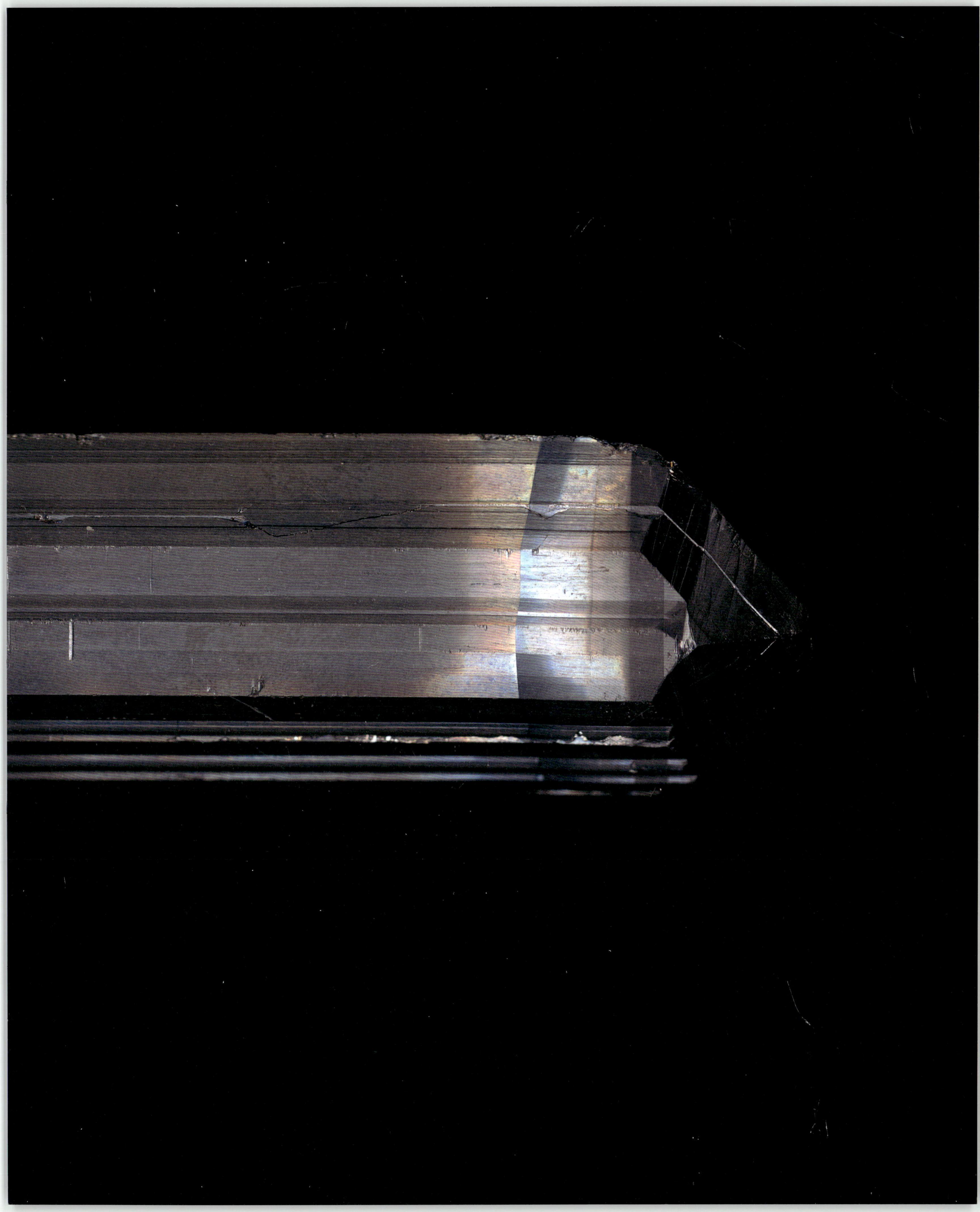

Fig. 1–15 [pp.1–31]

TRANSPOSITIONAL GEOLOGIES

SPECTRES OF COLONIALITY

Kerber | Berlin 2025

SASCHA MIKLOWEIT

This publication is part of the *Transpositional Geologies* project:
www.transpositional-geologies.org

Editor
Sascha Mikloweit

Design
Irakli Kiziria, Sascha Mikloweit

Photography Transpositional Geologies
Manuel Pandalis, Amanda Holmes

Texts
Paul Basu
Charmaine//Gamxamûs
R. Johannes Giebel
Elizabeth Grosz
Maria-Oo Haihambo
Selby Hearth
Bastian Herbst
Chris Hill
Herbert Jauch
Juuso Ndinolune David Kaluvi
Veripuami Nandee Kangumine
Malina Lauterbach
Helmut Maier
Prince Kamaazengi Marenga I
Sascha Mikloweit
Hidipo Nangolo
Saima Nakuti Ndahangwapo
Paul O'Kane
Jermaine Solunga
Peripatetic Thinking
Ellison Tjirera
Kuhepa Tjondu
Samo Tomšič
Kathryn Yusoff

Copyediting
Brendon Hooper

Proofreading
Sally Parker, Wessex Editorial

Index
Helen Taylor Indexing

Lithography
Sascha Mikloweit, Manuel Pandalis

Production Kerber
Jens Bartneck

Project management Kerber
Verena Simon

Print
Gutenberg Beuys Feindruckerei

Paper
Fedrigoni Splendorlux Mirror Argento, 320g
Fedrigoni Symbol Tatami, 150g
Munken Print White 15, 115g

Printed in Germany

Kerber Verlag
Rudi-Dutschke-Str. 26
10969 Berlin
Germany
+49 30 259 28 280
info@kerberverlag.com
kerberverlag.com

Kerber publications are distributed worldwide:

ACC Art Books
Sandy Lane
Old Martlesham
Woodbridge, IP12 4SD
UK
+44 1394 38 99 50
+44 1394 38 99 99 (F)
uksales@accartbooks.com
accartbooks.com

Artbook | D.A.P.
75 Broad Street, Suite 630
New York, NY 10004
USA
+1 (212) 627-1999
+1 (212) 627-9484 (F)
orders@dapinc.com
artbook.com

AVA Verlagsauslieferung AG
Industrie Nord 9
5634 Merenschwand
Switzerland
+41 44 762 42 00
+41 44 762 42 10 (F)
avainfo@ava.ch

Zeitfracht Medien GmbH
Distribution
Germany +49 711 7860 2254
service.zeitfracht.de

The German National Library (Deutsche Nationalbibliothek) lists this publication in the Deutsche Nationalbibliografie: dnb.de.

www.kerberverlag.com

Publication funding
Kunststiftung NRW
(Arts Foundation of North Rhine-Westphalia)
Landschaftsverband Rheinland (LVR)
Mineralogical Museum of the University of Bonn
Embassy of the Federal Republic of Germany
in Windhoek
Goethe-Institut Namibia

Transpositional Geologies 2019–2024

Associated institutions
Tsumeb Municipality, Namibia
Mineralogical Museum of the University of Bonn
Mineralogical Collections of the Technical
University of Berlin
Mineralogical Collection of the Ministry of
Mining and Energy Namibia
Embassy of the Federal Republic of Germany
in Windhoek
Goethe-Institut Namibia
Förderverein Filmkultur Bonn e.V.
Kunstmuseum Bonn

Project funding
Kunststiftung NRW
(Arts Foundation of North Rhine-Westphalia)
Institut für Auslandsbeziehungen (ifa)
BBK Modul C Neustart Kultur
Mineralogical Museum of the University of Bonn
Landschaftsverband Rheinland (LVR)
Stiftung Kunstfonds
Senate Department for Culture and Europe Berlin
Sparkassen-Kulturstiftung Rheinland
Stiftung Kunst der Sparkasse Bonn
Embassy of the Federal Republic of Germany
in Windhoek
Goethe-Institut Namibia
Bonn University Foundation
Kulturamt Bonn
TRA Global Heritage Lab University of Bonn

Technical Support
LaserSoft Imaging
711rent Hamburg

Limited Edition
Transpositional Geologies—Spectres of Coloniality is also available in a limited edition of 25 signed copies. This special edition features an image-only cover with no typography, three-side, four-colour edge printing, and is accompanied by two edition prints, all housed in a custom-made box.

Contact Transpositional Geologies
tg-t075@transpositional-geologies.org

ISBN 978-3-7356-0971-7

KERBER

Table of Contents

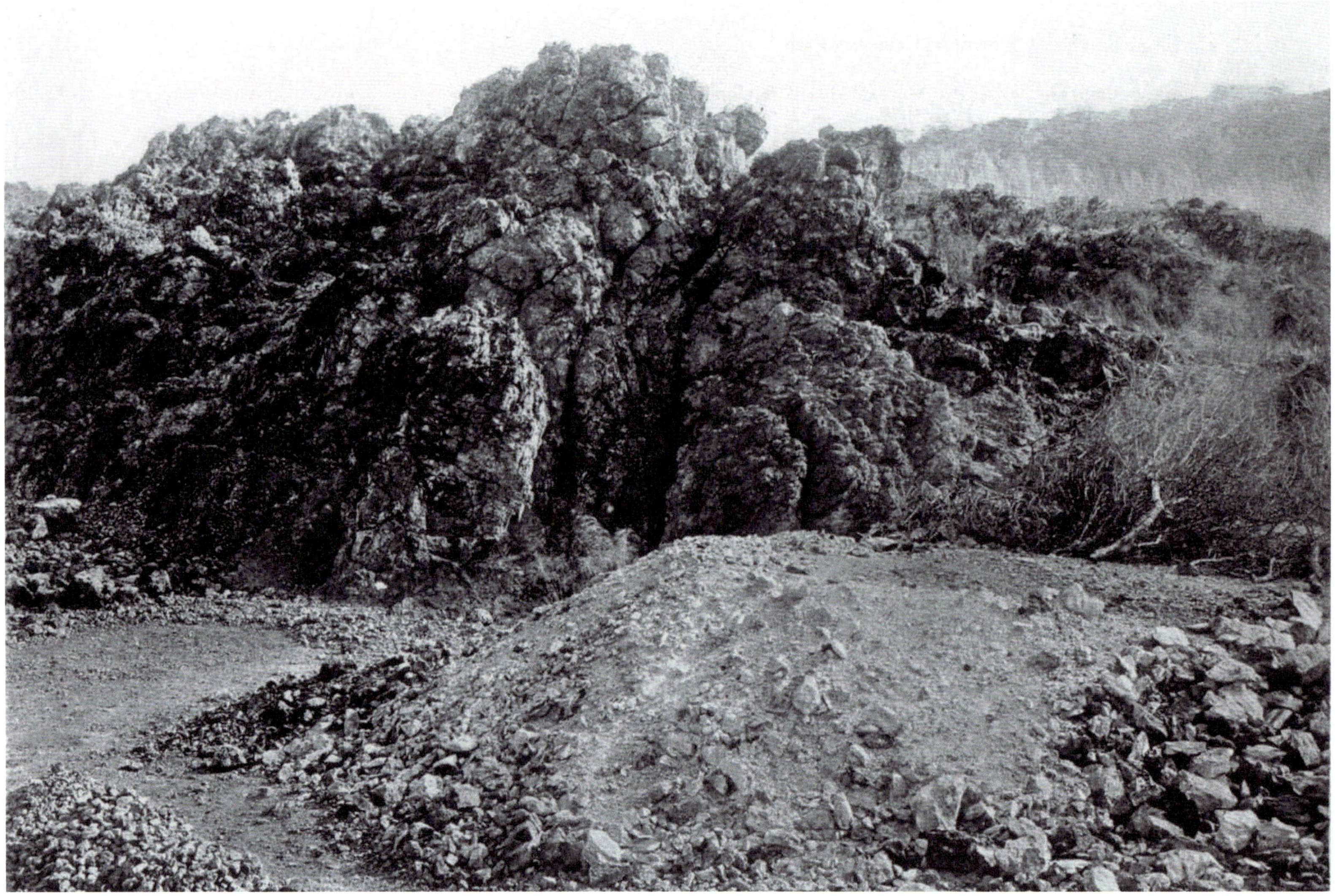

Preface

Sascha Mikloweit

If you have ever visited an institutional mineralogical collection, like me, you would have typically encountered glass cabinets of minerals organised by classification systems according to their material properties. Yet, each mineral also carries with it a history of extraction, destruction, (dis) possession, and global relations. These histories are often obscured by practices of classification and display, which present minerals as 'neutral' scientific specimens divorced from their socio-political contexts.

Transpositional Geologies: Spectres of Coloniality undertakes a rigorous and multidisciplinary exploration of the entangled legacies of geology, colonialism, and transhistorical extractive economies. In locating institutional geoscientific collections as indices to the afterlife of colonial and imperial power structures, *Transpositional Geologies* reads mineral specimens as objects of 'culture' rather than merely of 'nature'—a dichotomy deeply rooted in modern Western thought. This binary mode of thinking has not only shaped scientific disciplines such as geology and mineralogy, but has also obscured the inherently political nature of their practices. Industrial-scale mining, as a quintessential example of human capitalist interaction with the Earth, embodies the transhistorical violence of extractive economies with their legacies of colonial exploitation. By examining these legacies, this volume re-entangles the natural and cultural dimensions of geology, proposing we think in terms of an evolving 'political geology'.

The genesis of this project lies in the institutional geoscientific collection of the Mineralogical Museum of the University of Bonn and its archived objects. My reflections began with these specimens, leading to an emerging question as to what had been omitted from their complex socio-political histories by the subject-specific gaze of the geosciences. This realisation underscored the necessity of a multidirectional reflection that led to a first research field trip to Namibia in 2022, where Tsumeb—with its intricate and layered histories—further emerged as a focal point and model for this inquiry.

Fig. 16

Capturing my five-year engagement of artistic collaboration with scientists, historians, cultural and municipality workers in Namibia and Germany, in this volume I bring together international voices from fields including anthropology, critical theory, geology, history, museum studies, philosophy, poetry, fine art, public administration—and those of minerals such as boltwoodite, cerussite, and smithsonite. Rock by rock, this volume invites you to engage with a progressively nuanced reading of geology's history: its epistemic violence, omissions, and racial regimes, and how the lasting residues of its colonial legacies continue to shape our present-day extractive realities.

Through scholarly essays, artistic interventions, and poetic reflections, this volume presents a dialogue on the complexities of these relationships. Each contribution in this book is marked by a title system unique to the project: *transposition* followed by a randomised three-digit number. This choice alludes to the inherent gaps and incompleteness of all thinking, a nod to the always fragmented nature of knowledge production—regardless of its claims of universality. Incrementally, the contributions of this volume form a map, or perhaps they behave like cross sections of sediment—a dynamic and open-ended thinking process that still resists closure and universality.

Methodological Evolution and Artistic Collaborations

At the heart of this volume lies the methodological framework established through the *Transpositional Geologies* project's exhibitions. These were held at the Mineralogical Museum of the University of Bonn, Germany and in Tsumeb, Namibia, in 2023, serving as both a precursor and a conceptual primer for this book. Site-specific—and partially immersive—installations, process sculptures, performance art, public lectures with invited international speakers, experimental film and collaborative projects interrogated the mineralogical collection and its colonial legacies.

Artistic collaboration was integral to this endeavour, beginning for example with *transposition 001* (*Happy Banality of Everything*), a cinematic and performative exploration of mineral objects from the Mineralogical Museum of the University of Bonn's collection by Mikloweit in collaboration with musicians and graphic designers Irakli Kiziria and Holger Risse. Premiered at the Kunstmuseum Bonn as a site-specific performance, this work critiqued the neo-colonial minerality embedded in museum architectures and their broader power structures. In Namibia and Germany, *transposition 015* (*re—existence*) involved the Tsumeb Municipality, the Mineralogical Collections of the Technical University of Berlin, and the Mineralogical Museum of the University of Bonn, constructing the transnational journey of sediment and a pulverised colonial museum specimen of the Bonn collection through performative acts of re-contextualisation, destruction, and restitution.

This volume further documents collaborations reflecting on colonial binaries and shared histories, including with Tsumeb-based jewellery designer Richardlee Shoombe, local artist August Kapolo, German-Greek photographer Manuel Pandalis, and local businesses such as Steinbach Beergarden, Medala's Car Wash, and the former TCL Mine Recreation Club. *transposition 368* (*The Future Sound of Destruction*), developed with Berlin-based Paul Philipp Heinze, engaged the sensorial dimensions of destruction through a soundscape installed at the Mineralogical Museum of the University of Bonn.

From Political Geology to Institutional Critique

The opening sequence of *Transpositional Geologies: Spectres of Coloniality* introduces a series of Sascha Mikloweit's Object-Images based on geoscientific collection specimens, and in them the aesthetic tussles with the political. These double-paged images juxtapose the aesthetic qualities of geoscientific specimens with contextualising figure texts, re-linking them with broader socio-political and historical frameworks. Together, they foreground the complex histories of coloniality, extraction, and material culture, setting the stage for the multidirectional inquiry that follows.

Paul Basu's foundational essay, *Transpositional Geologies: Presencing the Politics of Mineral Specimens*, critiques the assumed neutrality of geoscientific objects in museum collections. Basu introduces the concept of 'presence' to critique the silences and absences in museum narratives, advocating for a transpositional approach to recovering marginalised histories.

Sascha Mikloweit then deepens this analysis in *Thinking Through Minerals: We Build Our Languages Out of Rocks*, reframing mineral specimens as cultural artefacts. Using the Mineralogical Museum of the University of Bonn as a case study, Mikloweit exposes the colonial entanglements of geoscientific collections and proposes a decolonial reimagination of their epistemic frameworks.

Colonial Histories and Extractive Economies

Selby Hearth's *Contested Land and the Extraction of Geologic Knowledge: Tsumeb and the Otavi Mountains as Reconstructed through White (pre-)colonial Reporting* examines early geological mapping in Namibia, centring the intersections of omitted Indigenous knowledge, extraction, and colonial scientific practices. Helmut Maier's *Colonial Geology, the Prussian State Geological Institute, and the South West African Copper (1850–1919)* highlights the role of institutional geology in facilitating colonial resource extraction and legitimising imperial power.

Juuso Ndinolune David Kaluvi's testimony, presented in *We Toiled for Zilch*, offers a personal narrative of labour exploitation in Tsumeb's mine during the apartheid system, complemented by Herbert Jauch's contextual analysis and Hidipo Nangolo's introduction. Christopher R. Hill's *Out of Space and Place: Tales of Namibian Uranium through Mineral Objects* expands this discourse, tracing uranium's scientific, symbolic, and geopolitical dimensions, while Saima Nakuti Ndahangwapo's *'Rössing' Uranium Mine: The German Link to a British Enterprise* foregrounds colonial legacies of industrial mining. Ellison Tjirera's *Land Inequality as Structural Violence: Vignettes of Namibia's Enduring 'Land Question'* critiques Namibia's persisting socio-economic disparities, linking colonial land dispossession to contemporary inequities.

Bastian Herbst's *Extracts from the Colonial Archives* is based on materials from Namibia's National Archives, obtained during the *Transpositional Geologies* research field trip in 2022. Through them he examines the socio-political and environmental history of mining in Tsumeb, rendering visible the intersections of labour exploitation, colonial corporate power, and the violence of extraction economies, including links to the Herero and Nama genocide.

Poetic Resonances

Then, a sequence of seven Object-Images, all based on Tsumeb minerals from German collections, are placed in conversation with poetic contributions from contemporary Namibia. Poems by practitioners who intimately experience—and comment on—the current reality of Black life in Namibia, engaging with themes of resilience, ecological loss, and colonial extraction: Veripuami Nandeekua Kangumine's *The Colour of Ripe Mangos*, Maria-Oo Haihambo's *Pink Salt*, Charmaine//Gamxamus' *Desert Mother,* Kuhepa Tjondu's *Wounded Earth,* Prince Kamaazengi Marenga I's *TSUMEB—A Poem,* and Jermaine Solunga's *History! My Human Geography*.

Theoretical Reflections and Artistic Practice

Philosopher Samo Tomšič's *On the Impurity of Epistemic Objects* bridges philosophical and scientific discourses, interrogating how knowledge production inherently reflects social and political biases and the lingering epistemological exceptionalism of European modernity. This contribution critiques the evolution of science into a regime of knowledge accumulation that often disguises its socio-political entanglements, from racialisation to extractivist metaphysics.

In an interdisciplinary dialogue, Kathryn Yusoff and Elizabeth Grosz's *Art and the Earth* extend this discourse, arguing how extractive economies not only exploit human and natural resources but also impose a racialised and gendered hierarchy of labour. They examine how creative practices can challenge these structures by imagining alternative values and geologies, positioning the earth as both a site of struggle and a medium for radical reimagination.

In examining *Geoscientific Collections as Spaces of Sedimented Histories*, Malina Lauterbach, R. Johannes Giebel, and Sascha Mikloweit reframe institutional geoscientific collections as repositories of cultural and historical memory. Extending his reflection on memory culture, Sascha Mikloweit's *Down the Cemetery Path* explores the bureaucratic and societal challenges of engaging with colonial memory in Bonn, Germany, during his *Transpositional Geologies* project. His contribution critically reflects on the limitations and potential of public remembrance in addressing historical accountability. Throughout the second half of the volume the documentation of Mikloweit's experimental artistic contributions disrupt the logic of conventional memory culture and classification systems, emphasising the relational and impermanent nature of geologic knowledge production.

Malina Lauterbach's *On the Impermanence of Sascha Mikloweit's transposition 015* and Peripatetic Thinking's *A Synopsis* reflects on the fluidity of Mikloweit's artistic practice. Paul O'Kane concludes the volume with his reading of *A Non-Violent Explosion in the Taxonomic Schloss (Towards a Neo-Baroque)* situates Mikloweit's work, unfolding his reflections from initial theme of 'explosion', and followed by a last set of Mikloweit's Object-Images.

Multidirectional Thinking

Together, these contributions critique, reinterpret, and expand the discourse on geology, coloniality, cultural memory, and transhistorical extractive economies. The interstitial spaces between essays, poems, artistic interventions, and images invite you to engage with the complexities of the material world, offering pathways for reflection and transformative inquiry.

Transpositional Geologies: Spectres of Coloniality is thus not merely a book but a dynamic network of relations, grounded on, seeking out, and celebrating the collective input of its contributors and collaborators. At the outset of this work, it was an email to the Director of the Mineralogical Museum of the University of Bonn proposing my project; then my small team of existing collaborators grew through introductions of colleagues, chance encounters in archives with fellow researchers, dedicated invitations to leading thinkers, and embracing self-initiated expressions of interest of others to cooperate. This community includes the federal and state funding institutions and local and national administrations who supported this work and are detailed in the Acknowledgments. This open-ended approach reflects the essence of my artistic practice as a thinking device, and was also essential to bringing the project into fruition.

As readers navigate this multidirectional thinking, the hope is not only to unearth new understandings but also to continue this dialogue. The past is not the past, but continues to shape the present—not only in Namibia and Germany. This volume invites us to reflect on this enduring truth, encouraging us to see geology not only as a science of the Earth, but as a deeply human and political endeavour.

Transpositional Geologies
Presencing the Politics of Mineral Specimens

Paul Basu

Abstract: As an introduction to *Transpositional Geologies*, this essay considers what histories and knowledges are explicitly or implicitly 'presenced' by mineral specimens encountered in geology museums. It explores how disciplines and disciplinary museum institutions understand specimens differently. The geological museum's continued presentation of specimens as natural, scientific objects, devoid of politics, is contrasted with the increasing politicisation of cultural objects in the ethnographic museum. The essay argues that both disciplines are implicated in histories of colonial violence and extractivism, and that insights gained within anthropology as it has sought to acknowledge and address its colonial entanglements could be productively applied in geology. Museum objects, whether geological or anthropological, are conceptualised as fragments, which stand as metonyms for absent contexts. The essay discusses how these absences are presenced explicitly through museum interpretation, but also more elliptically through other strategies, including the various 'transpositional' interventions documented and enacted in the book.

Keywords: political geology, mineral specimens, epistemic objects, presence, metonymy, material culture, museums, extractivism, Broken Hill/Kabwe, Nature-Culture

The perceived separation between Nature and Culture is one of the constitutive characteristics of modern Western thought. Yet, as Bruno Latour has argued, there is nothing 'natural' about such a separation; rather it must be brought into being through what Latour terms 'the work of purification' (Latour, 1993, p.30). Through such work, ideas relating to nature and the nonhuman, including the natural sciences, are separated from ideas relating to human culture and society. One consequence of this separation is that the 'science of things' may become disentangled from the 'politics of men' (Latour, 1993, p.30).[1] No matter how impossible this modernist project of purification, it has been pursued relentlessly, and nowhere more brutally—or so profitably—than in contexts of colonial and neo-colonial extractivism.[2]

This volume seeks to re-entangle Nature and Culture, nonhuman and human, science and politics, to produce what Latour might have called a 'quasi-object' (an object-like thing embodying an impure mix of nature-culture, human-nonhuman). It is a book that acknowledges the inextricability of Nature and Culture, and which recognises that the natural sciences—including geology and mineralogy—are necessarily also political sciences. *Transpositional Geologies* is, then, a work of 'political geology'; a critical engagement, which, in the words of Kathryn Yusoff, 'undermines the *givenness* of geology as an innocent or natural description of the world', and considers it instead as 'a relation of power' that continues to constitute relations of power through its 'toxic legacies' (Yusoff, 2018, p.10).

There are many sites at which one might begin exploring the politics of geology. One might start by tracing the impact of mining operations on communities and environments. One might follow the transformation of rocks into materials and into manufactured consumer products. One might examine how wealth is created through such transformations and consider who benefits—and who, or what, does not. Mining companies and their beneficiaries are all too aware of how politicised these sites of extraction and transformation are, making this a notoriously difficult sector to research critically. It is a sector replete with toxicities, both literal and metaphorical.

The starting point for *Transpositional Geologies* (one of them, at least) is, at first sight, both less obvious and less politicised. It begins with a fragment of greenish rock, a scientific specimen among thousands of similar specimens kept in drawers and display cases in the rarely-visited mineralogical museum at Bonn University. It is a piece of smithsonite, accessioned with the number 50215; sourced, so the museum's records tell us, in the early twentieth century from a mine in Tsumeb in what was then colonial German South-West Africa, now Namibia.[3] While research and teaching methods in what we now call the Earth Sciences have moved away from collecting, assembling and working with such specimens, they were, as Samo Tomšič's contribution to the volume discusses, once important epistemic and pedagogical objects—'containers of knowledge' (Tomšič, this volume, p.129; see also Chalk, 2012). The question that this volume revolves around is, precisely what knowledges do these fragments of rock 'contain'?

If the answer to this question seemed self-evident to generations of geologists, mineralogists and visitors to natural history museums (the '*givenness* of geology' that Yusoff refers to), Sascha Mikloweit's magnified 'Image-Objects', which punctuate this book, invite us to look again at these seemingly innocuous specimens, and, in so doing, confront our understanding of them. Are they merely samples of different types of rocks and minerals, containers only of geological knowledge? Despite being 'frag-ments of nature', as scientific specimens these epistemic objects are, of course, also culturally-constructed things. They too are quasi-objects: an impure mix of the natural and the cultural. What they are—and what knowledges they contain—is more elusive and shifts according to one's positionality, one's perspective and the knowledges that one brings to them. Like all museum objects, they resist the imposed fixity of classificatory schema. Despite their apparent solidity, these rocks are mutable things.[4]

[1] I consciously retain Latour's gendering of this politics.

[2] https://as.nyu.edu/departments/french/people/Faculty/bruno-latour.html

[3] This fragment of rock, and the epistemic regimes it was subjected to, were powerfully represented in Tsumeb in Sascha Mikloweit's *transposition 086* and *transposition 995*, documented in this volume, pp.196–201, pp.206–207.

[4] This mutability is most literally expressed in Sascha Mikloweit's pulverisation of another Tsumeb mineral specimen from the University of Bonn Mineralogical Museum collection (see this volume pp.228–245), but, of course, it is ubiquitous (though often invisible) in the metal and mineral components of consumer products that surround us (see, for example, Acosta, 2018).

Museums, Collections and Disciplinary Knowledges

As well separating Nature from Culture, the sciences from the arts and humanities, the work of purification has also produced a division of epistemic labour even within the sciences. There has been a proliferation of discrete disciplines and subdisciplines, each determining its own specialist field and professing its distinct theories, schools of thought and institutional structures. While, in theory, we celebrate the possibilities of interdisciplinarity, in practice, entrenched disciplinary conventions, methodologies, prejudices and forms of expertise often make it difficult to speak across the divides that we ourselves have created. It is worth reminding ourselves that things could have been otherwise—perhaps they still can.

Let us reflect briefly on the history of museum collections and their relation to disciplinary knowledge formation. Founded in 1683, the Ashmolean Museum at my own institution, the University of Oxford, is one of the world's oldest surviving public museums. Its original building, which now houses the University's History of Science Museum, once incorporated laboratories and lecture rooms as well as displays of Elias Ashmole's remarkable collections. Under one roof, the Museum exhibited all manner of curiosities, many of which were inherited from the even earlier Museum Tradescantium, embracing both *naturalia* and *artificialia*, Nature and Culture.

Among hundreds of objects, here was to be found: 'two ribs of a whale', 'all kinds of foreign plants', 'a large magnet stone', 'bright coloured birds from India', 'some very light wood from Africa', 'the robe of the King of Virginia', 'a toad fish', 'a human bone weighing 42 lbs.', 'the hand of a mummy', and 'a stone found in the West Indies in the water, whereupon are graven Jesus, Mary and Joseph'.[5] While the work of European colonial extractivism is already in evidence in this pre-modern, early Enlightenment miscellany, the work of disciplinary purification had yet to be effected. This would, however, occur gradually over the following two centuries, and, by the late-nineteenth century, the Ashmolean had narrowed its scope to become a museum dedicated to European art and archaeology, while new, separate museums and departments had been established at Oxford to house and study botanical, zoological, anatomical, geological, anthropological and ethnological collections. This increasing institutional specialisation was typical of museums and universities elsewhere.

Once collections had been separated, it is interesting to reflect on the relations between these disciplinary assemblages. Oxford's new University Museum had been founded in 1860, with distinct sections for human and comparative anatomy, zoology, palaeontology, geology and mineralogy. When a dedicated anthropological museum—the Pitt Rivers Museum—opened 24 years later as an extension to the University Museum, its ethnographical collections were understood and displayed as if they too were natural history specimens. Individual specimens were not valued as unique artworks or artefacts, but as exemplars of particular types of things. Each specimen indexed something beyond itself: a particular category of thing, which itself indexed a particular 'ethnos' or 'people', who, in turn, represented something more than themselves in anthropological theory-making.

Under this natural science paradigm, anthropologists assembled vast collections of specimens of 'primitive technologies' with which to conduct comparative analysis. Indeed, the mania for collecting extended to include the anatomical remains of human beings themselves, as well as 'physical type' photographs which reduced living people into the status of specimens, all to be collated, catalogued, compared and classified. Profoundly influenced by the natural sciences, the ethnographic museum thus became a laboratory for theorising human cultural as well as physical evolution: artefacts and anatomy became indices by which one could measure civilisational progress. However misguided, having the semblance, authority and supposed neutrality of 'scientific method', such theorising supported ideas of racialised hierarchies of human development, from the most primitive to the most civilised, which, in turn, justified the European colonisation of large parts of the world as well as the seizure of its extractable resources.

At the same time, the status of anthropology *as a science* remained contested. In words attributed to the American cultural anthropologist, Alfred Kroeber: anthropology was regarded as 'both the most humanistic of the sciences and the most scientific of the humanities'—quasi-science. By the 1960s, few anthropologists were comfortable with the idea that people or their arts and artefacts could be studied as if they were natural history specimens; cultural evolutionist theories were long-since discredited; and the historical entanglement of anthropology and colonialism elicited increasing self-critique within the discipline. Influenced by postcolonial critique, the 1980s saw a period of sustained critical reflexivity within anthropology, with anthropologists taking it upon themselves to interrogate the mechanisms through which they constructed their object of study (the primitive 'Other' to Euro-America's supposedly civilised self) and how the authority of anthropological knowledge claims were established, mainly through literary devices used in 'writing culture', but also in ethnographic filmmaking and in museum display (see, for example, Fabian, 1983; Clifford and Marcus, 1986; Karp and Lavine, 1991; Loizos, 1993).

By the 1980s, the majority of academic anthropologists had long-since distanced themselves from the ethnographic museum, and many ethnographic displays were dismantled, packed away and consigned to a dubious disciplinary history that many would sooner forget. With the revival of interest in anthropological material culture studies in the 1990s, however, and alongside the emergence of critical museology, the ethnographic museum became a new, politically-charged space for thinking about the legacies of colonialism and its forms of knowledge production. Questions were posed concerning the future of such museums and the rubrics under which they might be redeemed (e.g. Harris and O'Hanlon, 2013). Thus ethnographic museums were rebranded and euphemised as museums of 'world cultures' or 'world art', while other critics heralded the 'post-ethnographic' museum as a laboratory for addressing the epistemic violences of the colonial past, particularly through collaboration with

[5] Excerpts from a description of the Tradescant Collection by Georg Christoph Stirn, 1638 (quoted in MacGregor, 1983, p.21).

contemporary artists from previously colonised regions (Deliss, 2020). More recently still, even curators within the field have asserted that there is no future for such 'warehouses of disaster capitalism' other than to repurpose them as 'sites of remembrance and conscience for the human lives, environments and cultures destroyed by European colonialism' (Hicks, 2020, pp.15, 228).

Given that these fraught, but productive, debates have been taking place for decades now in the ethnographic museum, it is surprising how little of this has spilled over into other disciplinary museum spaces (particularly botanical gardens, natural history and geology museums), as if these museums, too, were not part of the same capitalist-colonial knowledge complex. Indeed, while there has been a powerful drive to recognise and reject extractivist modes of knowledge production within anthropology, it could be argued that the most shockingly literal and overt traces of colonial resource extraction in the museum sector are not to be found in the ethnographic museum, but in the geological museum. Colonialism, after all, was not motivated by the desire to classify indigenous peoples and their material culture, but by the prospect of seizing and exploiting the resources, and especially the mineral wealth, that could be extracted from the lands of others. While ethnographic and history museums have been mined metaphorically for lessons in social justice and postcolonial ethical practice (Corrin, 1993), those museums most directly implicated in colonial mining enterprises have remained strangely neutral, apolitical spaces. In the geological museum, the 'science of things' has yet to be re-entangled with the 'politics of men'.

Mineral Specimens as Material Culture

Given the transformation of the ethnographic museum from a supposedly neutral, scientific institution, in which a 'natural history' of humankind was prosecuted, into a highly contested space, which has, in many cases, acknowledged the politics of its colonial entanglement, what lessons might this provide for the geological museum? What if, for example, we approached geological collections not as specimens of natural materials but as material culture?

In a well-known essay, 'Objects of Ethnography', the anthropologist and museologist Barbara Kirschenblatt-Gimblett asks what makes an object 'ethnographic'? Kirschenblatt-Gimblett contends that there is nothing intrinsically ethnographic about a given artefact; rather 'objects become ethnographic by virtue of being defined, segmented, detached, and carried away by ethnographers' (1991, p.387). 'Disciplines make their objects', Kirschenblatt-Gimblett argues, 'and in the process make themselves' (ibid.). The same is, of course, also true of the making of geological specimens. And since there is nothing intrinsically ethnographical or geological about these things, we might subject a geological specimen to anthropological scrutiny and in so doing 'remake' it as an ethnographic object.
Actually, the 'material turn' of the 1990s, though led by anthropologists and archaeologists, was self-consciously 'undisciplined' in character, drawing on a wide range of theoretical and empirical influences. Central to material culture studies, however, is the relationship between people and things: understanding things, that is, in their differing social, cultural, economic and political, as well as spatial and temporal contexts, and considering how things materialise, mediate and move between these different contexts (Miller and Tilley, 1996, p.5; Basu, 2017).

Here, I am particularly interested in how material things may be said to contain, presence or narrate particular contexts. This is, of course, essential to the way in which museums communicate knowledge through collections. Thus, Neil MacGregor, then director of the British Museum, opens his book *A History of the World in 100 Objects* with the seemingly straight-forward statement: 'Telling history through things is what museums are for' (MacGregor, 2010, p.xiii). In museums, some histories (or other narratives or forms of knowledge) are made explicit through object labels, contextualising images, catalogue entries or, as in the case of *A History of the World in 100 Objects*, even radio broadcasts. And here the curatorial power to narrate, to shape and control the knowledge communicated, becomes all-important. Whose history is told? From what perspective? Whose voices are present? Whose are silenced? (Trouillot, 1995). The history of the world told in MacGregor's book and radio series was criticised, unsurprisingly, because it represented the authoritative voice of the British Museum, narrating objects that had been appropriated, sometimes in contexts of violence, from throughout the world in the age of empire.

One decolonial riposte was a project which seeks to pluralise not only the histories that objects can tell, including 'formerly subaltern stories' that museums have often ignored, but also the 'worlds'—the ontological contexts—in which those histories emerge and make sense. In an introductory statement on the project website, *100 Histories of 100 Worlds in 1 Object*, the project team note how the British Museum radio series and book 'reinstate the idea of the museum as a "view from nowhere" and everywhere at the same time. It was presented as a place to see the world; yet without reflection on how the institution itself obtained and reframed the objects in order to create its own seemingly universal narrative'. As well as empowering 'previously excluded voices' to tell their own histories about museum objects taken from their ancestors, the project seeks to 'show that one museum object can have 100 histories and exist in 100 worlds'.[6] It is clear that the fragment of smithsonite accessioned under number 50215 at Bonn University's Mineralogical Museum would not only have *meant* something different, but was, indeed, something *essentially* different, to the German industrial chemist, Carl Duisberg, who initially acquired it for his collection, and the unknown African mineworker who pick-axed it from an ore seam for the Otavi Mining and Railway Company in an era of colonial violence, which culminated in the Herero and Nama genocide. Aside from a museum's explicit narrations, whether articulated by powerful institutions or marginalised community members, the majority of the histories or other forms of knowledge that an object might speak to remain implicit or latent. MacGregor's *History of the World* begins with a fragment of rock not so dissimilar to the specimen from Tsumeb. This object, accessioned at the British

[6] [...]100 histories and exist in 100 worlds' (https://100histories100worlds.org).

Museum under number 1934,1214.1, is a basalt pebble found by the paleoanthropologist and archaeologist Louis Leakey in 1931 in the Olduvai Gorge in what was then colonial Tanganyika, now Tanzania. What places this stone in the collection of the Prehistory Department of the British Museum, rather than the Earth Sciences Department of the Natural History Museum—why it is constituted as an archaeological object and not a geological object—is evidence of it being intentionally chipped to form a cutting edge, and, crucially, its excavation by Leakey from what he interpreted as an archaeological, rather than geological, context, next to fossilised butchered animal bones, that could be dated to between 1.8 and 2 million years ago. Were it not for this contextual knowledge, which cannot be discerned from the object itself, rather than being one of the oldest known humanly-made things in the world (MacGregor, 2010, p.9), it is doubtful that this otherwise unremarkable lump of basalt would arouse the interest of any museum. But what other histories, knowledges or worlds might this chipped stone invoke?

Material Culture, Metonymy and Presencing Absences

Rather than ethnographic (or archaeological, geological, etc.) *objects*, Kirschenblatt-Gimblett proposes that we consider these things as *fragments*. Her rationale is that they are made through excision or detachment from a whole. Detachment refers partly to 'the physical act of producing fragments' (and we might reflect on how different disciplines *extract* their specimens from their contexts), but also to what Kirschenblatt-Gimblett describes as 'the detached attitude that makes that fragmentation and its appreciation possible' (1991, p.388). She writes of a 'poetics' of the fragment. This is primarily about the perceived relationship between the detached fragment, perhaps accessioned into a museum collection, and the 'absent whole' that it remains part of. This is a metonymic relationship, a relationship of parts and the absent wholes that they in some way stand for. The historian Eelco Runia describes metonymy as 'the trope [...] of "presence in absence"' (2006, p.6). Kirschenblatt-Gimblett's essay is mainly concerned with examining the implications of different strategies that museums adopt in 'exhibiting the fragment' and thereby explicitly presencing particular absences. She contrasts approaches that place the object 'in-situ', mimetically evoking the whole that was 'left behind', with those that set an object 'in context'. Regarding the latter, Kirschenblatt-Gimblett acknowledges that 'there are as many contexts for an object as there are interpretative strategies' (1991, p.390), and this is dependent on the argument being made, or the histories, knowledges or questions being explored.

Outside the explicit contextualisations of museum display, however, what an object/fragment might presence or stand for beyond itself often remains latent: something present but not necessarily visible. Perhaps the most important contexts that give meaning and significance to an object remain unarticulated in these explicit narrations. Indeed, this is an important insight from anthropological material culture studies: that material things have a language of their own. They are expressive, but in tacit, non-linguistic ways. As Christopher Tilley notes, 'through its silent "speech" and "written" presence', an artefact 'speaks what cannot be spoken, writes what cannot be written, and articulates that which remains conceptually separated in social practice' (1999, p.103). In this way, museum objects such as the fragments of rock from Tsumeb and Olduvai Gorge presence other histories and other knowledges without explicitly narrating them; indeed, their materiality may resist or contradict the dominant narrations imposed upon them. With their 'silent speech', these fragments express their own counter-narratives. Here, perhaps, we find consonance with Édouard Glissant's declaration, 'I build my language out of rocks', which inspired the title of Sascha Mikloweit's exhibition in Bonn (this volume, pp.53–60) and draws attention to the representational limitations of language in the face of 'the world's proliferating opacity' (Dash, 2017, p.357). Things—fragments of rock included—speak for themselves, even if they do so opaquely.

As part of the work of purification, science has been granted a monopoly on the production of knowledge. Yet, in other cultural contexts (and Latour would argue in our own too), knowledge and wisdom is found in what science has dismissed as irrational. Anthropology has been one of the few disciplines that has insisted on the rationality of other ways of knowing and being in the world, and one motivation for the collection of ethnographic objects has been precisely to illustrate the logic of what Claude Lévi-Strauss famously termed 'the science of the concrete'—the rendering of 'the sensible world in sensible terms', rather than through abstract language and concepts (1966, p.16). Such ways of understanding the world, which Lévi-Strauss argued were no less 'scientific' than understandings derived through the natural sciences (ibid.), share much with how knowledge and affect are produced in art. This again challenges the privilege given to spoken or written language, and embraces the possibilities of non-linguistic forms of communication via objects, images, sound and performance. As evidenced in the various transpositional interventions documented in this book, such approaches posit alternative—often indeterminate—knowledges (making present, while not making explicit, absent histories), while destabilising what we think we know (Schwab, 2023).

Presencing Broken Hill

While *Transpositional Geologies* is focused on histories and legacies of German colonial extractivism in Namibia, including its material traces in the mineralogical collections of the University of Bonn and Technical University of Berlin, I should like to reflect on a context which is 'adjacent' both in terms of European colonial power and colonised African territory. Walking between the rows of antique vitrines in the Mineral Gallery of London's Natural History Museum, one encounters thousands of systematically-ordered mineral specimens. As with the majority of such displays, written on a small label beside each specimen is a brief description of the specimen itself and the location from which it was sourced. Though by no means the original, hand-written labels, many of the locations on the labels in the Gallery still refer to colonial territories and use colonial-era place names. In one case, for example, is a cluster of specimens from 'Tsumeb, South West Africa'. 'South West Africa' was the name given to Namibia after the First World War, when it became a mandate territory of South Africa, until its independence in 1990. In another case, an impressive specimen is labelled 'green prisms on limonite, Broken Hill, Northern Rhodesia'. 'Northern Rhodesia' was, of course, named after the imperialist Cecil Rhodes, whose British South Africa Company had been granted mineral rights over vast territories in Southern Africa, as well as the means to administer the territories for its own profit. Declared a British protectorate in 1924, it became Zambia on its independence from Britain in 1964, while Broken Hill reverted to its Bemba toponym, Kabwe, soon after.

The ongoing coloniality of museum labelling, including the use of now offensive terms and colonial-era placenames, has been the subject of considerable debate and a number of projects within ethnographic museums.[7] In the Natural History Museum's Mineral Gallery, the anachronistic dissonance of the labels at least presences a brutal colonial context that is otherwise absent. Indeed, as with the museums in Bonn and Berlin, into the few, pared words of these labels are elided histories of colonial conquest, extraction and exploitation that have global reach. The placename 'Broken Hill' itself evokes a landscape mutilated through the violence of mining and 'over-writes' indigenous place-knowledge with Eurocentric sensibilities. It was Thomas Davey, an Australian mining engineer working for the Rhodesia Copper Company, who 'discovered' a rich outcrop of lead and zinc at a place called Kabwe-ka Mukuba in 1902 and who renamed it Broken Hill after the mining town of the same name in New South Wales. Kabwe-ka Mukuba is a Bemba toponym, meaning 'place of ore smelting', alluding to the long precolonial history of (small-scale and sustainable) mining and metal production in the area (Musambachime, 2016).

We might note that both the outcrop after which the Australian Broken Hill was named in 1844, and its Zambian namesake, have long-since been mined away (as, indeed, has the 'Green Hill' after which Tsumeb is named). Fragments of both Broken Hills are juxtaposed, though without comment, in the same Natural History Museum display case (Fig. 17). They presence the empire-wide reach of colonial resource extraction, but, perhaps even more poignantly, they presence absent places in the landscape: places that have literally been scoured from the face of the earth and can never be recovered.

It is quite remarkable that as recently as 2019 a scholarly article could be published on the 'Minerals of Kabwe Mine', describing how the Broken Hill/Kabwe lead-zinc deposit 'was fortuitously discovered' by Davey, how its naming was 'an optimistic invitation of comparison with the extraordinarily rich mineral discovery' previously made in Australia, and that Davey and his assistant 'pegged claims on six *kopjes* (hills)', all without reference to the exploitative colonial contexts (in both Zambia and Australia) in which this took place (Southwood, Cairncross and Rumsey, 2019, p.115). As the authors of the article, who include the principal curator responsible for mineral collections at the Natural History Museum, note, full-scale mining operations commenced at Broken Hill/Kabwe in 1904, and in the following eighty

[7] See, for example, the 'Labelling Matters' project at the Pitt Rivers Museum (https://prm.ox.ac.uk/labelling-matters) and the 'Words Matter' project at the National Museum of World Cultures, Leiden (https://materialculture.nl/en/publications/words-matter).

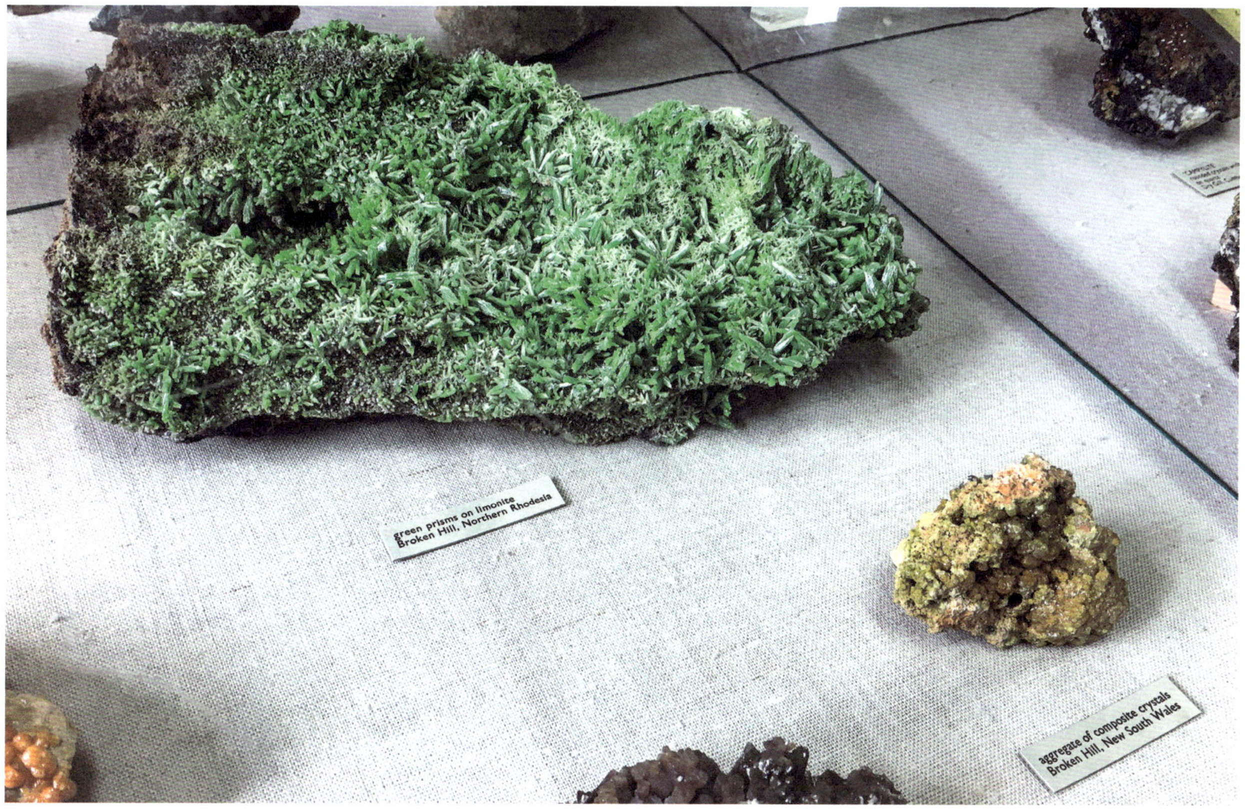

Fig. 17

years, more than 1.8 million tonnes of zinc; 0.8 million tonnes of lead; 7,816 tonnes of vanadium oxide; 80,000 kilograms of silver; and 235,000 kilograms of cadmium were extracted from the mine (ibid.). The authors do acknowledge, in passing, that 'a century of mining and metallurgical treatment [at Broken Hill/Kabwe] has left a legacy of severe pollution', and, 'in 2013, Kabwe was listed as the sixth most toxic site in the world' (ibid., p.116).

With the commencement of industrial scale mining, the British South Africa Company extended its railway from Bulawayo to Broken Hill/Kabwe, enabling the region's mineral wealth to be extracted, initially, via the port of Beira on the coast of Mozambique. Broken Hill fast developed as a major urban and administrative centre, and also the base for the further exploration of minerals and metal ore deposits in what became known as the Copperbelt. As Fergus Macpherson writes in *Anatomy of a Conquest: British Occupation of Zambia*, with the expansion of mining interests in the area, the British South Africa Company put into operation a land policy it had developed over the preceding years, 'which involved the claim to mineral rights wherever minerals were found; the ignoring and indeed annulling of the land ownership of the local rulers and communities; and the extensive eviction of villages in the vicinity of the mining operations and European townships which developed in such areas' (1981, p.222). Furthermore, mining companies were also able to extract native labour at 'phenomenally low cost' from subjugated populations through a process of 'semi-conscription', which was deemed necessary because mining work was so unpopular due to its dangerous nature and the harsh treatment of workers at the hands of European managers (ibid.).

To return to the work of purification: if the European colonial annexation of African territories, the dispossession and exploitation of its people, and the extraction of its wealth seems to us very much about the 'politics of men', the case of Zambia, like that of Namibia, reminds us that this was achieved through the separation of this politics from the 'science of things'. In an article examining the relationship between science and the extension of colonial power in the Zambian Copperbelt in the 1920s and '30s, Tomas Frederiksen argues that, through the use of new 'scientific' prospecting techniques, geologists were able to 'see' the Northern Rhodesian landscape—and, more particularly, what lay *under* the Northern Rhodesian landscape—in new ways (Frederiksen, 2013).

Crucially, it was through the production of such scientific knowledge, which was consumed in distant financial centres through the abstract representations of charts, diagrams, maps, tables, statistics, and through the circulation of specimens, that enabled 'elements of nature' to be disembedded from their social and cultural, as well as physical, contexts, and moved 'into the logics of capital as commodities which could be priced and exchanged on global markets' (Frederiksen, 2013, p.272). Frederiksen characterises this as a process of 'geologising the Copperbelt'. Emptied of people and of culture, the Northern Rhodesian 'subsurface' was thereby 'produced as a "resource" and made amenable to specific forms of analysis and understanding' (ibid.). In the case of Northern Rhodesia, then, Frederiksen concludes that 'scientific endeavour fundamentally changed socio-natural relationships', and 'the extension of colonial presence and power rested on the activities of commercially-funded scientists', who 'perhaps unwittingly [also] became the architects of African dispossession' (ibid., p.280).

There are other histories—and histories of science—presenced by the Broken Hill specimen in London. During open-cast mining operations at Kabwe in 1921, for example, mineworkers uncovered the near-complete fossilised skull of an ancient human, which, when it came to the attention of palaeontologists, was initially identified as a new species of human ancestor and named *Homo rhodesiensis*, or 'Rhodesian Man'.[8] This fragment of an ancient human-turned-stone was also transformed into a specimen—indeed, a 'type specimen'—and deposited at the Natural History Museum in London, where it was described at the time as constituting 'one of the scientific treasures of that Institution' (Hrdlička, 1926, p.173). The discovery was sufficiently sensational to warrant an eight-page feature in the *Illustrated London News*. This included detailed drawings and photographs of the disfigured hill (*kopje*) and mine workings where it was found; photographs of the skull arranged alongside other primate and human skulls for comparison; and a sketch imaginatively reconstructing 'Broken Hill Man as he lived in South Africa'. Today, the find spot, the so-called 'Bone Cave', would be inscribed on the UNESCO World Heritage List—if it existed. Within months of the discovery, however, the *Illustrated London News* article makes clear the shocking scale of the erasure of a whole landscape resulting from the continuing mining operations:

> Before mining operations commenced, there stood at this spot a kopje, or hill, 50 to 60 feet high, with a slight depression in the centre. Mining operations have demolished the hill, and have excavated to the depth of over 90 feet below ground level where the hill stood, and it was at this depth that the skull was discovered. The entrance to the 'Bone Cave' was at ground level (Harris, 1921, p.680).

Given the huge wealth that has been created outside Zambia through the extraction and exploitation of its 'natural' resources, there is an irony in the fact that it is the petrified skull of Rhodesian Man—understood as a 'cultural' resource—that is at the centre of contemporary Zambian restitution debates. Arguing for the return of the skull, the Zambian archaeologist Francis Musonda points to the colonial context in which it was discovered, the illegality that surrounded its removal to Britain (contravening the colonial government's own legislation), and the consequences of this 'cultural loss' to the people of Zambia. Musonda argues that, since the history of the skull is 'enmeshed in colonial domination', the Natural History Museum 'should endeavour to correct the historical injustice and disengage itself from the colonial legacy' by returning the skull to Zambia (2013, pp.206, 217). Welcome though the return of the 'Broken Hill skull' would be, no act of restitution can, of course, repair the devastated landscape left in the wake of the colonial mining operations, or compensate generations of Kabwe's citizens for the chronic health problems caused by its toxic waste.

[8] The skull has subsequently been classified as one of the best preserved fossils of another extinct human species called *Homo heidelbergensis*, thought to have first appeared approximately 600,000 years ago.

If the palaeontological remains of *Homo rhodesiensis* were of primary interest to *physical* anthropologists concerned with the science of human evolution, the history of British *social* anthropology is also entangled in the story of colonial Broken Hill and the wider Northern Rhodesian Copperbelt. By the 1930s, a new generation of sociologically-minded anthropologists had gathered around Bronislaw Malinowski at the London School of Economics. They were adamant in their rejection of the older ethnological tradition with its evolutionary assumptions and diffusionist theories 'born of the dust and welter' of the ethnographic museum (Basu, 2013).

A new 'problem-oriented' anthropology emerged in Northern Rhodesia, pioneered by anthropologists such as Audrey Richards, Godfrey Wilson and Monica Hunter, and pursued by others into the 1960s at the Rhodes-Livingstone Institute, the first social science research institute to be established in Africa (Schumaker, 2001). The focus of their work was specifically to investigate the impact of colonial development on local populations, including researching 'cultural contact', 'detribalisation', 'urbanisation', and new social dynamics as local people migrated to work in the mines or in other colonial infrastructure projects, such as the railways or road building. Wilson, who was appointed as the first director of the Rhodes-Livingstone Institute in 1938, initially planned to conduct research among African mineworkers in the Copperbelt. Here, however, he met with resistance from mining companies and colonial administrators, who, in the aftermath of violent strikes in 1935, were concerned that his inquiries would 'stir up discontent' in the labour force (Brown, 1973, p.190). Consequently, Wilson shifted the location of his study of migrant labour and urbanisation to the mining town of Broken Hill, where he conducted research in 1939 and 1940 (Wilson, 1941, 1942; Hansen, 2015). (The Copperbelt became the primary focus of the Rhodes-Livingstone Institute's research over the following 20 years.)

The Rhodes-Livingstone Institute was funded through the Colonial Development and Welfare Fund. Thus, even though their position was more critical, social scientists such as Wilson were also part of the colonial infrastructure, and their work reflects 'the fundamental ambiguity which lay at the heart of the relationship between anthropology and colonial rule' (Brown, 1973, p.174). At the same time, by focusing on the social and cultural implications of colonial resource extraction, they re-humanised the geologised Copperbelt. Reintroducing the 'politics of men' into the 'science of things' that facilitated the commoditisation of a continent's natural (and human) resources, their policy-focused work also provides a bridge to the present and the future, reminding us that the specimens in the mineralogical museum not only presence pasts, but also presence the ongoing struggles that people must contend with to overcome the most literally toxic of colonial legacies. This includes the struggle of campaign groups, such as Children of Kabwe, to hold to account the global mining companies that benefitted from extracting Kabwe's mineral resources for the consequent contamination of the local environment and associated health implications caused.[9]

[9] [...] health implications caused. (https://childrenofkabwe.com).

Microliths as Monuments?

In this introductory essay to *Transpositional Geologies*, I have sought to explore how absent histories, knowledges and worlds are presenced, both explicitly and implicitly, by mineralogical specimens in geological museum collections. Even though mineralogical museums typically perpetuate an ideological separation between Nature and Culture in their explicit presentation and narration of what are constructed as 'scientific specimens' void of politics, their discursive control of these epistemic objects may be resisted—or, one might say, 'refused'—by these fragmentary things themselves. Thinking through some of the transpositional interventions in this book relating to the specimens from Tsumeb, and through the adjacent context of Kabwe, it is clear that these fragments contain within them the potential to presence histories and knowledges that are suppressed, even denied, in dominant geological narrations. *Transpositional Geologies* documents and enacts various approaches to making these absent narratives present: through scholarly essays, but also through photography, poetry, exhibition, film, (post-)conceptual art, and other unashamedly 'unscientific' ways of knowing and telling. There will, however, always be an excess of meanings 'contained' in these fragments; meanings as yet unrepresented and perhaps unrepresentable.

In his historiographical discussion of the concept of 'presence', defined as 'the unrepresented way in which the past is present in the present', Eelco Runia (2006) contrasts the operation of metaphor and metonymy in relation to monuments. He describes 'metaphorical monuments', which are typically more representational in form, as 'primarily engaged in a transfer of meaning'. 'Metonymic monuments', on the other hand—and he uses Eisenman's Memorial to the Murdered Jews of Europe in Berlin as an example—'concentrate on a transfer of presence'. Such monuments, he argues, entail, firstly, 'a metonymical transposition of "substance" [...] from one place to another', and, secondly, 'in an act of artistic transubstantiation, this metonymical *Fremdkörper* [a 'foreign body' or thing that is out of place] is metaphorically associated with what is to be commemorated' (ibid., p.17). 'Whereas metaphor "gives" meaning', Runia writes, 'metonymy insinuates that there is an urgent *need* for meaning' (ibid., p.19).

Runia's reflections on metonymy, monuments and meaning are lenses through which we might look upon the geological specimens in the mineralogical museum. Like their megalithic counterparts, might these fragments of rock, these microliths, not also be regarded as monuments? Despite the museum's insistent claims on them, they are without doubt matter out of place. And, as with the statues of imperialists that have seemingly stood inert and ignored for decades, has the time not come to question the 'geological givenness' of these specimens and recognise the other knowledges that they presence? Knowledges of injustices that are no longer—indeed, never have been—tolerable. And just as the campaigners demand that the statues of Rhodes must fall, so might we discern a kind of 'fallism' in Mikloweit's pulverisation of the Tsumeb specimen in Bonn, its return as dust to the crater in Nambia where a mountain once stood?

References

Acosta, I. 2018. *Copper Geographies*. Barcelona: RM.
Basu, P., 2013. Material Culture: Ancestries and Trajectories in Material Culture Studies. In: J. G. Carrier and D. B. Gewertz, eds. 2013. *The Handbook of Sociocultural Anthropology*. London: Routledge, pp.370–390.
____ed., 2017. *The Inbetweenness of Things: Materializing Mediation and Movement between Worlds*. London: Bloomsbury.
Brown, R., 1973. Anthropology and Colonial Rule: The Case of Godfrey Wilson and the Rhodes-Livingstone Institute, Northern Rhodesia. In: T. Asad, ed., 1973. *Anthropology and the Colonial Encounter*. London: Ithaca Press, pp.173–197.
Chalk, H.-L., 2012. *The Uses, Meanings and Values of Natural Objects: University Earth Science Objects and Collections as Material Culture*. Unpublished PhD Thesis. University of Manchester.
Clifford, J. and Marcus, G. E., eds., 1986. *Writing Culture: The Poetics and Politics of Ethnography*. Berkeley: University of California Press.
Corrin, L. G., 1993. Mining the Museum: An Installation Confronting History. *Curator*, 36, pp.302–313.
Dash, J. M., 2017. The Stranger by the Shore: The Archipelization of Caliban in Antillean Theatre. In: B. R. Roberts, M. A. Stephens, eds., 2017. *Archipelagic American Studies*. Durham, NC: Duke University Press, pp.356–370.
Deliss, C., 2020. *The Metabolic Museum*. Stuttgart: Hatje Cantz.
Fabian, J., 1983. *Time and the Other: How Anthropology Makes Its Object*. New York: Columbia University Press.
Frederiksen, T., 2013. Seeing the Copperbelt: Science, Mining and Colonial Power in Northern Rhodesia. *Geoforum*, 44, pp.271–281.
Hansen, K. T., 2015. Urban Research in a Hostile Setting: Godfrey Wilson in Broken Hill, Northern Rhodesia, 1938–1940. *Kronos*, 41, pp.193–214.
Harris, C. and O'Hanlon, M., 2013. The Future of the Ethnographic Museum. *Anthropology Today*, 29, pp.8–12.
Harris, W. E., 1921. The Finding of the Broken Hill Skull: The Mystery of the Great Bone Cave. *Illustrated London News*, 19 November, p.680.
Hicks, D., 2020. *The Brutish Museums: The Benin Bronzes, Colonial Violence and Cultural Restitution*. London: Pluto.
Hrdlička, A., 1926. The Rhodesian Man. *American Journal of Physical Anthropology*, 9(2), pp.173–204.
Karp, I. and Lavine, S. D. eds., 1991. *Exhibiting Cultures: The Poetics and Politics of Museum Display*. Washington, DC: Smithsonian Institution Press.
Kirschenblatt-Gimblett, B., 1991. Objects of Ethnography. In: I. Karp and S. D. Lavine, eds., 1991. *Exhibiting Cultures: The Poetics and Politics of Museum Display*. Washington, DC: Smithsonian Institution Press, pp.386–443.
Latour, B., 1993. *We Have Never Been Modern*. Cambridge, MA: Harvard University Press.
Lévi-Strauss, C., 1966. *The Savage Mind*. Chicago: University of Chicago Press.
Loizos, P. 1993. *Innovation in Ethnographic Film: From Innocence to Self-Consciousness*. Chicago: University of Chicago Press.
MacGregor, A., 1983. The Tradescants as Collectors of Rarities. In: A. MacGregor, ed., 1983. *Tradescant's Rarities: Essays on the Foundation of the Ashmolean Museum*. Oxford: Clarendon Press, pp.17–23.
MacGregor, N., 2010. *A History of the World in 100 Objects*. London: Allen Lane.
Macpherson, F., 1981. *Anatomy of a Conquest: The British Occupation of Zambia, 1884-1924*. Harlow: Longman.
Miller, D. and Tilley, C. 1996. Editorial. *Journal of Material Culture*, 1, pp.5–14.
Musambachime, M., 2016. *Wealth from the Rocks: Mining and Smelting of Metals in Pre-Colonial Zambia*. Lexington: Xlibris.
Musonda, F. B., 2013. Decolonising the Broken Hill Skull: Cultural Loss and a Pathway to Zambian Archaeological Sovereignty. *African Archaeological Review*, 30, pp.195–220.
Runia, E., 2006. Presence. *History and Theory*, 45, pp.1–29.
Schumaker, L., 2001. *Africanizing Anthropology: Fieldwork, Networks and the Making of Cultural Knowledge in Colonial Africa*. Durham, NC: Duke University Press.
Schwab, M., 2023. Editorial. *JAR: Journal for Artistic Research*, 28. [online] Available at <http://www.jar-online.net/en/issues/28> [Accessed 1 September 2024].
Southwood, M. Cairncross, B. and Rumsey, M., 2019. Minerals of the Kabwe ('Broken Hill') Mine, Central Province, Zambia. *Rocks & Minerals*, 94, pp.114–149.
Tilley, C., 1999. *Metaphor and Material Culture*. Oxford: Blackwell.
Trouillot, M.-R., 1995. *Silencing the Past: Power and the Production of History*. Boston, MA: Beacon.
Wilson, G., 1941. *An Essay on the Economics of Detribalization in Northern Rhodesia, Part I*. Livingstone: Rhodes-Livingstone Institute.
____1942. *An Essay on the Economics of Detribalization in Northern Rhodesia, Part II*. Livingstone: Rhodes-Livingstone Institute.
Yusoff, K., 2018. *A Billion Black Anthropocenes or None*. Minnesota: University of Minnesota Press.

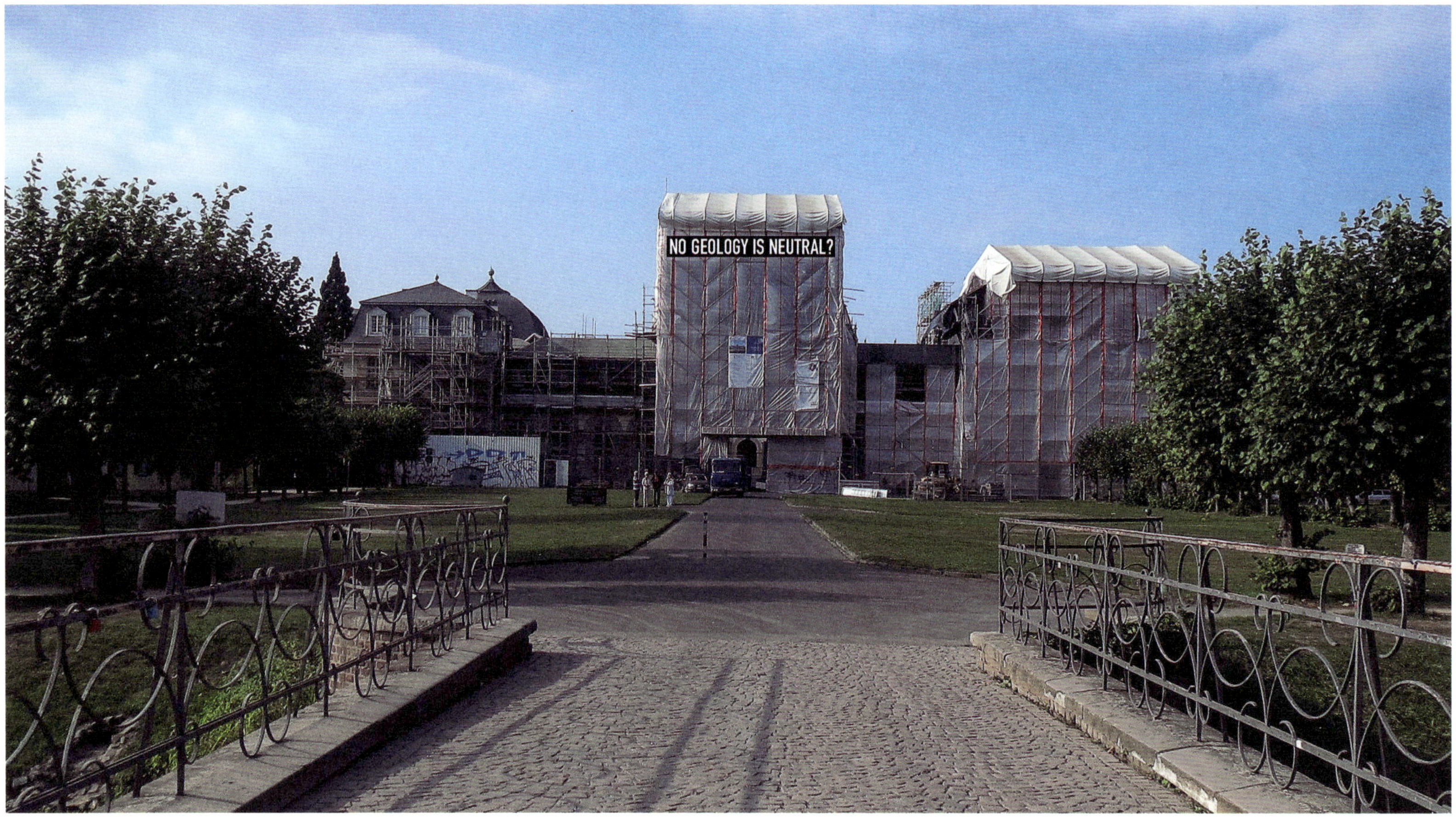
NO GEOLOGY IS NEUTRAL?

Thinking Through Minerals: We Build Our Languages Out Of Rocks

Sascha Mikloweit[1]

Abstract: This study critically examines the intricate entanglements of geology, epistemic violence, and colonial history through an exploration of the Mineralogical Collection at the University of Bonn, with a specific emphasis on specimens from Tsumeb, Namibia and their relation to German colonial history. By reinterpreting mineral specimens as 'objects of culture' rather than simply 'objects of nature', this analysis highlights the potential of such specimens to serve as indices to racialised socio-political power structures. Re-contextualising the mineral collection, this project challenges traditional reductive classification systems, interrogating the contemporary significance of colonial legacies embedded in such archives. Reframing how mineral collections are understood and displayed, this study critiques conventional curatorial practices and emphasises the need for an expanded classification system that re-connects geoscientific objects to their broader cultural and historical contexts. In contributing to the decolonisation of knowledge in the geosciences, this approach offers a model for rethinking institutional mineral collections globally.

Keywords: Geology, geoscientific collections, critical museum studies, decolonising scientific knowledge, colonialism, minerals, political geology

No geology is neutral.
(Kathryn Yusoff, 2018, p.111)

Nature is only the raw material of culture, appropriated, preserved, enslaved, exalted, or otherwise made flexible for disposal by culture in the logic of capitalist colonialism.
(Donna Haraway, 1988, p.592)

[…] Mankind [sic], once bewildered by sheer facts, finally dominated them through reflection, observation, and experiment. Henceforth mankind [sic] […] knows how to utilize the world. […] The natural sciences classify, but the quid proprium of things eludes them. […] In short, scientific knowledge enumerates, measures, classifies, and kills. […] To acquire the impersonality of scientific knowledge mankind [sic] depersonalized itself, deindividualized itself. An impoverished knowledge […]
(Aimé Césaire, 1973, p.xIii)

I build my language out of rocks.
(Édouard Glissant, 1969, p.43)

Towards a 21st century multidisciplinary classification system for minerals in institutional geoscientific collections

If you enter a mineral collection in Germany you are most likely to come across glass cabinets in which objects are organised based on the Nickel-Strunz classification system, structured according to the object's chemical compounds: sulfides, oxides, phosphates, etc. (Strunz and Nickel, 2001). Only small labels tell you about the origin of the objects on display with a bare minimum of information. Context beyond naming the place of extraction appears to be rather secondary in these settings—and provenance is mainly of interest if it comes to the object as commodity, since mineral specimens with unknown origins are commercially unattractive in the collector's market (Fritscher, 2012).[2]

Yet, from which contexts do these minerals originate? What are the legacies of the extraction sites, communities and histories that they were literally unearthed from? In fact, the sparse indication of a region of origin on the archive labels provides a single point of departure: thousands of objects—not simply neutral chalcopyrites, andradites or goethites—but the keys to thousands of histories, from the very moment the mineral was extracted, possessed and traded, marked with a number and certified with an archival label and index card. From this point onward, these objects of geology are bound to histories shaped by *man*. In a different register from their *deep time* existence, minerals become dormant indices to personal and collective biographies, power relations and political dynamics of individuals and societies involved in—or subject to—extraction:

> Geology is a mode of accumulation, on the one hand, and a mode of dispossession, on the other, depending on which side of the geologic colour line you end up on.
> (Yusoff, 2018, p.3)[3]

Geological science constitutes 'extractivist violence'[4] (Yusoff, 2021a, 2021b) since its inception; its findings and knowledge productions have been proliferating rapidly since the 16th century. In fact, the science of geology can be regarded as a substructure to modern life, as key to conceptualising earth as *raw material* that keeps on feeding contemporary materialities of consumption (Klare, 2012; Hecht, 2012; Parikka, 2015; Yusoff, 2020; Crawford, 2021). Just as much as geological science is an instrument of power (Yusoff, 2018), also representation in institutional collections of mineralogy

[1] This is an edited version of a text originally written in March 2023 and published as part of the exhibition *We Build Our Languages Out Of Rocks—Transpositional Geologies*, Mineralogical Museum of the University of Bonn, Germany (Mikloweit, 2023a). An archive in progress of the *Transpositional Geologies* project is accessible online: www.transpositional-geologies.org

[2] The same applies to scientific collections, where the knowledge of an object's origin is a crucial requirement.

[3] The concept of 'race' as a scientifically established fact, when in reality it is a mere scientific imaginary, gave birth to a 'colour line', which has led to the global division of humanity based on skin colour. This division has been identified by W.E.B. Du Bois as 'the problem of the twentieth century' (1903, p.15) and has impacted all aspects of human existence (ibid., pp.68, 113, 121). Today, the 'colour line' continues to exist as a means of subordination, serving to perpetuate the myth of endless economic growth that primarily benefits a privileged few, while exploiting a marginalised majority through the use of cheap labour (Yusoff, 2018).

[4] Extractivist violence refers to harm caused by extractive industries, including the displacement of communities, the destruction of ecosystems, and economic inequality. It is perpetuated by a global economic system—historically constituted by aggressive colonial operations and juridical frameworks and their subsequent evolution—that prioritises the financial interests of actors who benefit from extractive economies, ranging from mining companies and their suppliers to raw material traders, producers, and end-users, over the wellbeing of the environment, and the interests of the societies, communities, or individuals affected by extraction operations. Marginalised demographics are disproportionately impacted (Acosta, 2013; Veltmeyer and Petras, 2014).

Fig. 18

is (Hearth and Robbins, 2022). *Transpositional Geologies* reads these institutions as the material traces of a long journey of knowledge production—an epistemic history that has, in some aspects, resulted in profoundly negative consequences.

Since the late 15th century, European systems of knowledge production proliferated into the sciences,[5] which in turn became the major instrument in a fabrication of difference: the invention of 'the other', be it *nature, gender* or *race*, was—and still is—the constitutive epistemic element of 'modern life' (Eder, 1988; Haraway, 1988; Spivak, 1988; Wynter, 2003, 2006; Descola, 2005; Brassier, 2007; Quijano 2007). The proliferating natural sciences formed institutions—such as botanical, zoological and geological collections[6]—in which their ways of thinking 'mattered': what at an earlier stage of mineral collecting and display was driven by the desire of an affluent class to demonstrate sophistication and social status to their peers, developed later into effective institutionalised tools to teach academics, scientists and engineers in the field of (colonial) geology. Eventually these systems of thought, along with their imaginaries, were transmitted[7] to a specific public audience—the modern European citizen—in the format of the museum of mineralogy (Hearth and Robbins, 2022; Yusoff, 2021a; Maier, 2020).

Among collections of geological, botanical and natural history specimens the institutional 'mineralogical collections, still presented as "neutral" scientific specimens, are arguably the most politically charged collections of all' (Basu, 2023a). Approaching institutional collections of mineral objects today, we might ask ourselves what the (epistemic) violence[8] of these collections is and how it rests within the common classification systems for minerals that are used to structure and present these collections.[9] There are three common classification systems by which the approximately 1,500 institutional mineralogical collections maintained worldwide are currently organised:[10]

(i) the Dana classification system, developed in the 19th century by American geologist, mineralogist, volcanologist and zoologist James Dwight Dana (1818–1895) (Dana, 1837; Gaines et al., 1997);
(ii) the Strunz classification, developed by German mineralogist Hugo Strunz (1910–2006) (Strunz,1941);[11]
(iii) the Nickel-(Tennyson-)Strunz classification system[12], as an update of the Strunz classification, developed in the late 20th century by Canadian, Australian mineralogist Ernst Heinrich Nickel (1925–2009). (Strunz and Klein, 1977; Strunz and Tennyson, 1982; Nickel and Mandarino, 1987; Nickel and Grice, 1998; Mandarino, 1998; Strunz and Nickel, 2001).

All three systems base their organisation of mineral objects on chemical composition and crystal structure. Such classification systems themselves serve as vivid examples of how a science, driven by particular field-specific interests, produces a specialised and narrowed reading of its object of inquiry. This epistemic perspective in turn shapes the physical formal museal presentation, as it can often be observed with the display of systemically ordered mineral specimens presented in glass vitrines.[13] This specialised gaze aims to instil a setting of scientific 'objectivity', but such teleological forms of display omit cause and effect: the socio-political context of objects presented in this manner is not accessible, neither to a casual observer, nor to most scholars. In this sense, the field-specific specialisation decouples: the compartmentalisation of knowledge isolates the mineral objects from their complex world-relations in favour of subjective scientific interests.[14]

With field-specific, decoupled readings, questions of ethics are seldomly raised. What were and are the socio- and geo-political histories of these mining places (Hearth, 2021)? What were the efforts in international relations and politics that gravitated around such mining operations (Hill, 2023; Hill and Ashipala, 2024)? Who was involved in these power structures (Drechsler, 1966; Klare, 2012)? Who benefitted (Bond, 2006)? Who brought these objects to the light of day (Kaluvi, Nangolo and Jauch, 2024), and in which conditions did they live (ibid., 2024)? What were the impacts on their own and their families' health caused by the mining operations? What is the fallout from industrial mining operations in these places (Imalwa, Kaluwapa and Gramlich, 2023), and what are the consequences to the land surrounding mining operations in regards to toxicity and destruction? Which trade routes did mineral objects share with 'illegitimate' human specimens (Alberti, 2005, p.563),[15] cultural artefacts, and zoological and botanical specimens?

There are many more questions that remain to be investigated by historians of science, by historians of law, and by other researchers from a diverse range of fields who are yet to discover the unexpected relations between geoscientific collections and their research interests.

Geology and its practices have inadvertently created a *History*[16] of mineral objects divorced from their socio-political realities and histories by a specialised and narrowed reading of their epistemic objects. In educating students of geology today, mineralogical collections mainly ceased to function as teaching tools and fell dormant, became spaces of a forgetful past; in some cases, however, they still operate under highly charged and problematic ideological paradigms that are—one way or another—part of ongoing violent regimes of knowledge production (Hearth and Robbins, 2022; Manson and Sayner, 2019).

Expanding traditional scientific classification systems of mineral objects[17] in institutional collections to include information such as socio-political and cultural-historical contexts could lead toward a multidisciplinary classification system for mineral objects. This could unlock a vast epistemic potential, activating the field of geo-sciences as a gateway

[5] It is not the aim of *Transpositional Geologies* (TG) as a project to provide a general critique of geoscientific knowledge: there have been countless achievements that were borne out of geosciences at large and will continue to result from its cultural practices of rigour, enquiry and exploration in the future. The focus of this longterm project is rather to investigate and question the conditions of its knowledge production: its ideological transhistorical underpinnings, its racial regimes and power dynamics, its prolonging epistemological violence. Such excluding roots of respective discourse fields are often well sedimented (i.e. Lloyd, 2019). In our approach we revisit the epistemic violence that remains omitted in telling the geosciences' History (regarding history with capital H see fn.16). This omitted violence continuously mutates and replicates in our present globalised world system, resulting in a prolongation of injustices. The unfinished project of 'mono-humanism' (Wynter and McKittrick, 2007) needs to be nurtured further towards a humanism that is open to all forms of existence, ways of life and cosmologies (Scott, 2000; Descola, 2005), allowing for 'everyone's right to opacity' (Glissant, 1994a, p.45) where it is due.

[6] In places such as here in the Poppelsdorfer Schloss in Bonn, Germany or the Royal Museum for Central Africa, Terveuren, Belgium, all three categories are united in an architectural proximity under one roof.

[7] An imaginary refers to the shared cultural beliefs, values, and myths that shape our understanding of the world and our place in it. It is socially constructed and subject to change over time, and dominant imaginaries can serve to exclude and marginalise certain groups, while reinforcing power structures and systems of oppression (Wynter, 2006).

[8] Here the project is rooted in Elizabeth Grosz's reflections on the concept of violence in Jacques Derrida, and his notion of the arch-violence and the arch-writing (Grosz, 1999). Derrida's (1967; 1972; 1989; 1991) 'arch-violence' refers to the inherent violence and domination within power structures, often hidden or masked. "Arch-writing" is Derrida's term for the process of deconstruction, which challenges these structures by revealing their hidden biases and opening up new possibilities for interpretation and understanding. Together, these concepts highlight the intimate connection between language, power, and ideology, emphasising the need for critical analysis and radical change in social, political, and cultural systems.

[9] In question here is not the usefulness of such systems for the geo-sciences themselves within their own logic of inquiry, but the standalone application of such systems to structure institutional mineralogical collections and their public display today. The standalone application as is common practice now creates topical silences (Mason and Sayner, 2019)—it divorces the mineral objects from their socio-political contexts and histories, which are relevant to unearth the historical relations of the geo-sciences to larger structures of power.

[10] This is the case if they use one of these official systems. A fair amount of collections, especially older ones, are likely to use their individual classification system. These most likely are, in a majority of cases, organised by chemical class. (With thanks to Selby Hearth for pointing this out.) The Mineralogical Museum University Bonn collection uses the Strunz classification, as do the collections of Technical University of Berlin. It is important to briefly mention the 'Hey Systematic', or 'Hey's Mineral Index', developed by the British mineralogist Max H. Hey (1904–1984) first published in 1950, 2nd edition in 1955, this was reprinted with corrections in 1962, appendices were published in 1963 and 1973, and a last updated 3rd edition in 1993 (Clark, 1993). The system Hey developed at the Natural History Museum, London is limited to a perspective of mineral chemistry and classifies a total of 32 mineral groups. Interestingly, even though it was developed at the Natural History Museum, it was never applied to its mineralogical collections: "[...] although it could have been applied to the collections here, my understanding is that it was not applied because the practicality of reorganising such a large collection into a new systematic order would have been very difficult—as it would have required closing the museum to the public, for an extended period of time." (Rumsey, 2024).

[11] Strunz started working on his classification system in 1937 at the Mineralogical Museum Berlin (part of the Natural History Museum), where he was assistant to Paul Ramdohr (1890–1985). From his appointment as professor at the TU Berlin in 1951 onwards he developed a complete and corrected version of his system for the first time. This was achieved in collaboration with Prof Dr Christel Tennyson with whom he collaborated from the 1950s to the late 1990s. Out of this joint working the 4th edition (1966), 5th edition (1970), 6th edition (1977), 7th edition (1978) and 8th edition (1982) of the *Strunz Mineralogical Tables* saw the light of day. As a physical manifestation and legacy Strunz conceptualised, planned and oversaw the building works of the TU show collection from about 1955 onwards until 1969. He then used this collection as a device to teach 'his' classification system to students. The original state of this collection is preserved at the Mineralogical Collection of the TU Berlin to the present day.

to what it once separated. Such enhanced classification systems could transcend disciplinary boundaries[18], re-establish connections, and bridge knowledge divides created by the specialisation of geo-sciences towards an epistemic decolonisation. Is it not the ultimate form of restitution to destabilise and eventually disrupt these universalising forms of knowledge production that dominate today and perpetuate inequality?

Hence, the often overlooked format of the 'Institutional Mineralogical Collection' raises fundamental questions:

(i) What purposes may the ~1,500 institutional mineralogical collections maintained around the globe fulfil today?
(ii) What can these collections subsequently become in the future?
(iii) What can specialists and the general public learn from their archives if looked at from multidirectional perspectives?
(iv) In which way can we employ these institutional collections to advance our understanding of other concepts further than compounds, structures and properties?

Developing effective ways to address these questions may disrupt and transform the systemic perpetuation of epistemic violence—both past and present—in which geo-science and its acolytes play a role. Such violence ultimately enables health, environmental, social, and economic inequalities that negatively impact the well-being and livelihood of populations.

These collections should not be storage facilities of dead matter, but rather *laboratories of thought* where the individual mineral specimen becomes a *thinking device* to unearth power structures that facilitate historical and current inequalities. They could enrich reductive logics of scientific thought, leave behind its compartmentalisation, and instead *re-link, re-connect*, and *relay* them to their multiple dormant histories.[19]

We Build Our Languages Out Of Rocks—Transpositional Geologies, an artistic-curatorial format, engages with Glissant's concept of 'relays' (1990, pp.169–179) and explores geo-poetical experimental thought (Glissant, 1969, p.33; Last, 2015), navigating between historical and contemporary 'periphery' and 'centre' contexts.

Neither *Transpositional Geologies* as an extended journey through space and time, nor the experimental framework of this exhibition specifically, will produce direct answers to all the questions that emerge. Rather, it aims at multiplying the questions posed here until their accumulating polyphony unfolds within the silences of the museum (Mason and Sayner, 2019) and becomes perceptible to a general public.
The Mineralogical Collection of the University of Bonn thus becomes an alternate epistemic space, an exemplary laboratory, a model: realities that were neutralised, omitted and suppressed in its historical organisational forms in the course of a natural science approach, are re-forming relations.

The methods and rigours of classification systems could be employed to 'explode' them into a future that avoids mindlessly grooming modernity's barbaric flip-side (Glissant, 1981; Dussel, 1996; Quijano and Ennis, 2000; Quijano, 2007; Bond, 2006; regarding the notion of 'explode' see O'Kane, 2023).

Key elements of my methodology relevant to this exhibition

We Build Our Languages Out Of Rocks—Transpositional Geologies explores the inextricable and tangible links between the science of geology, its *epistēmē*, mineral extraction, and racism in German colonial history with a focus on aesthetics and aisthesis (Rancière, 2011). Through working with the archive of the Mineralogical Museum of the University of Bonn, *Transpositional Geologies* acknowledges, understands and handles mineral objects as keys to German colonial history: in this exhibition the particular attention is with mineral specimens from Tsumeb, Otavi district, Namibia.

Why Tsumeb? Because it is unique: from 1884 to 1915, Namibia was under the colonial rule of the German Empire who named it 'German South West Africa' (Wallace, 2013, pp. 97–102).[20] Within this context of German colonial rule, the Otavi Mining and Railways Company (OMEG)[21] was founded in 1900 in Berlin, Germany (Schnorbus, 2000, pp. 23–30). This operation was unique because of its economic interests—the construction of a railway line from Swakopmund to Tsumeb, completed in 1906, and the commencement of Tsumeb's first industrial-scale mining operations in 1907 (Drechsler, 1966, pp. 193–221; Mindat.org, n.d.). These activities disrupted a long tradition of local, small-scale mining operations and their complex ecologies (Hearth, 2021). Quintessentially, these factors contributed to the first genocide of the 20th century: the Herero and Nama genocide by the German Empire (Ibid., 225; Dedering, 1993). The Tsumeb mine itself, from a geo-scientific point of view, was to become one of the world's most prolific mineralogical sites before it stopped operating in 1996 (Wilson, et al., 1977; Hearth, 2021. p.433). Its specimens are held in significant mineralogical collections internationally. 'Tsumeb is one of the most mineralogically diverse sites in the world' (Mindat.org, 2020), (see Selby Hearth, in this publication). Consequently, Tsumeb minerals are likely present in nearly all of the approximately 1,500 institutional mineralogical collections worldwide. This widespread presence makes the case of Tsumeb an ideal subject for analysis, offering insights and producing learnings that are transferable to a majority of institutional

mineralogical collections.
This exhibition is based on a two-fold, site-specific approach:[22]

(i) An extensive visual and contextual study of minerals archived in the collection of the Mineralogical Museum of the University of Bonn, and;
(ii) Archive work and an explorative field trip to Windhoek, Swakopmund and Tsumeb, Namibia that took place from October to November 2022.[23]

The visual and initial contextual study has been conducted since September 2019 at the museum, extended during the field trip to Namibia in 2022 and will continue at the Mineralogical Collections of the Technical University of Berlin.[24] Both collection locales represent current and former seats of the German government and in case of the latter—the place in which Hugo Strunz[25] developed his Strunz Classification System.[26]

In collaboration with on-site staff, I identified sub-collections of minerals related to *geologies of race* [27] (Yusoff, 2018, p.6; 2021, pp.664, 666). They were then examined using an approach that I have been developing and refining continuously since 2009. While my earlier focus was on objects that were exposed to immense forces of physical violence in institutional contexts—2009–2010: Luggage left unattended and destroyed on suspicion of terrorism in a controlled explosion by the British Transport Police at Waterloo Station, London (Mikloweit, 2010; 2011); 2015–2017: Remnants of a spacecraft explosion of a catastrophic Ariane rocket launch failure at the spaceport in French Guiana, 1996 (Mikloweit, 2015a; 2015b; 2017)—, I now apply this approach to the tangent of the physical force of extraction and epistemic substrata of colonial oppression. From 2019–2023 I further developed and formalised my methodology, working with the mineralogical collection of MMdUB (Mikloweit, 2019; 2023a; 2023b; 2024).

At the core of this approach lies, as method, a transformative image production process in which the physical object of the archive (*O*) becomes *Object-Image* (*O–I*) by means of employing an off-the-shelf high-resolution scanner with modified hard- and software, which leads to the generation of a digital high-resolution image file library, an archive of mobile *O–Is*. These re-mobilised objects, transposed into *O–Is*, become modular devices throughout the further working process, and re-materialise as *Image-Objects* (*I–O*) eventually. [28] These are deployed in three ways:

(i) As devices to communicate and engage institutions of interest for the TG process;
(ii) As devices to communicate and engage actors within these milieus;
(iii) As aesthetic raw material, the *O–I* allows the original archive objects to be transposed into differing media and contexts: The resulting *I–Os* are central in bringing to life subject specific bodies of work in which the different entities relate to each other in a way that had been theorised by the scholar Michael Schwab (2018) as 'Transpositions'[29]

By scrutinising the mineral objects through the gaze of the technical apparatus of the scanner, the resulting visual characteristics of the images are of voyeuristic proximity, allowing for the object under investigation to be seen anew.[30] By altering the scale of the object, a relationship between viewer and *Image-Object* (*I—O*) is established that is both alienating and intimate at the same time, challenging pre-formatted positions of the spectators.

In combination, these different strands form an approach that allows a transpositional engagement through and around the mineral object and its housing institution: from deep geological time to human histories, from a chemical compound to the institutional realities of a mine, and the reverberations it has today.

The exhibition in Bonn (20 April–11 June 2023), the former capital of West Germany, stretched over four locations—the Mineralogical Museum University Bonn, its Exhibition Hall I & II; the University Bonn's Lecture Hall of the Institute of Geosciences, Department of Geochemistry/Petrology; and the historical lecture hall H1 situated in the University's main building, the Electoral Palace Bonn, and the grave of Lothar von Trotha (1848–1920) at Poppelsdorf Cemetery. Throughout the four locations 18 transpositions were exhibited, comprising experimental film screenings, process sculptures, image objects, an immersive installation including acoustic installation, curatorial and sculptural interventions into institutional and public spaces, lecture interventions, performances, and texts. Each transposition can stand alone for individual consideration, yet in relation they generated interchanging flows of information,[31] prompting the spectator to continually reevaluate their perspective on the subject matter: the coloniality[32] of mineral collections.

By showing a way to shift perspectives on the role of the object on a meta-level, particularly in the German (post-)colonial context, this exhibition—creating a multilogue between different branches of knowledge production, curatorial practice, and artistic formats—offered a contribution to the topical discourse on institutional approaches to contemporary mineralogical collections, their display, and their legacies.

[12] Here I thank Dr R. Johannes Giebel, curator of the Mineralogical Collections of the Technical University of Berlin, who pointed out that: 'Actually, this is the classification according to Strunz. After a long break following the publication of the previous (8th) edition, Dr Ernst Heinrich Nickel has finally contributed to the latest (9th) edition with an updated extension of the existing classification, which includes numerous data on new minerals and new data on minerals already listed. However, nothing fundamental has changed in the classification, it is still based on the same general principles as in the earlier editions. Some references, however, subsequently refer to the entire systematic as the Nickel-Strunz classification. Personally, I find this questionable because Prof Dr Christel Tennyson (1925–2010)—a colleague of Strunz at TU Berlin—worked intensively on the classification in previous editions (4th-8th editions), but there is no sufficient recognition of her, such as the "Strunz-Tennyson Classification". Prof Tennyson contributed significantly more than Dr. Nickel, yet Nickel is mentioned alongside Strunz in various references, without acknowledging Tennyson's contributions. Therefore, by using the term "Nickel-Strunz Classification", Tennyson's contribution is indirectly devalued.' A classic *Matilda Effect*.

[13] This for example is the case in the exhibition hall 1 of the Mineralogical Museum of the University of Bonn, Germany and the display of the Mineralogical Collections of the Technical University of Berlin, Germany.

[14] This in itself would of course be an interesting subject matter for analysis by historians of science.

[15] 'Illegitimate' human specimens refer to those that ended up in medical collections such as the Herero skulls in the Charité in Berlin, Germany (Bandle, et. al., 2013)—as well as other ethnological, taxidermic or botanical objects (Alberti, 2005, p.563). Accumulated colonial objects were transported on respectively established trade routes. Alberti remarks in his reflection on human anatomy museums that: 'Human remains collected from elsewhere moved along the same acquisition routes as plants, animals, and books, in the process shifting from subject to object.' (ibid., p.563). That mineral objects intended for collections shared the same routes is likely.

[16] I am using history with and without a capital H in Glissant's understanding: '"History [with a capital H] ends where the histories of those peoples once reputed to be without history come together." [continues p.58]

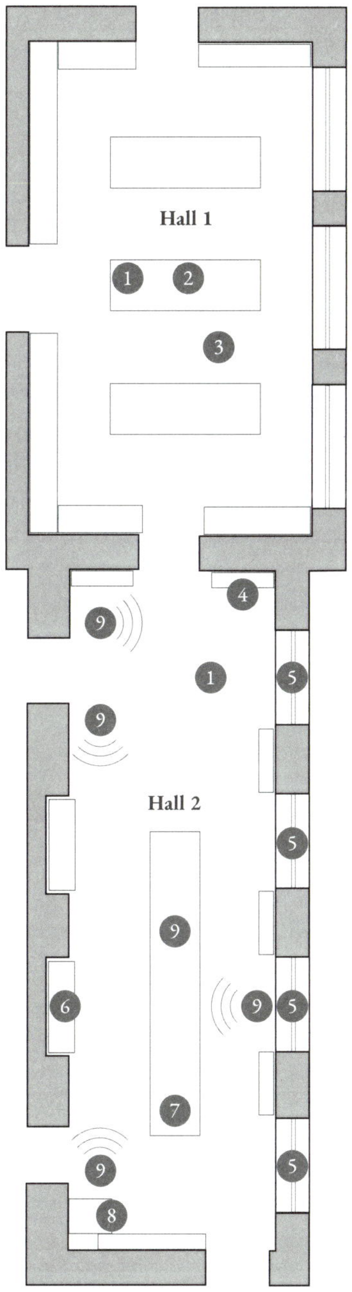

Central Exhibition:
Mineralogical Museum of the University of Bonn Exhibition Hall 1 (Systemic Collection) and Exhibition Hall 2 (Special Exhibitions), Poppelsdorfer Schloss

1 transposition 015 (this volume, pp.228–245)
2 transposition 441 (this volume, pp.192–195)
3 transposition 904 (this volume, pp.160–175)
4 transposition 504 (this volume, pp.204–205)
5 transposition 756 (this volume, pp.202–203)
6 transposition 741 (this volume, pp.208–211)
7 transposition 064 (this volume, pp.156–159)
8 transposition 665 (this volume, pp.216–217)
9 transposition 368 (this volume, pp.212–215)

Lecture Interventions:
Institute of Geosciences
Department of Geochemistry/ Petrology, Lecture Hall, Poppelsdorfer Schloss

transposition 135
transposition 633
transposition 495
transposition 512
(all four lecture interventions are accessible at:
http://transpositional-geologies.org/archive)

Film screenings:
Electoral Palace
(University of Bonn main building), Lecture Hall H1

transposition 001 (this volume pp.252–265)

Performances:
Bonn Poppelsdorf Cemetery

transposition 904 / 846 (this volume pp.160–191)

Fig. 19

Bibliography

Achinte, A. A., 2012. Epistemes "Otras": ¿Epistemes Disruptivas?. *KULA: Antropólogos del Atlántico Sur*, 6, pp.22–34.

Acosta, A., 2013. Extractivism and Neoextractivism: Two Sides of the Same Curse. In: B. D. Theobald, ed., 2013. *Beyond Development: Alternative Visions from Latin America*. Quito, Ecuador: Rosa Luxemburg Foundation, pp.61–86.

Alberti, S., 2005. Objects and the Museum. *Isis,* 96(4), pp.559–571.

Alli, S. and Azoulay, A. A., 2020. It Is Not Possible to Decolonize the Museum without Decolonizing the World. *Guernica*, [online] 12 March. Available at: <https://www.guernicamag.com/miscellaneous-files-ariella-aisha-azoulay> [Accessed 03 April 22].

Azoulay, A. A., 2019. *Potential History: Unlearning Imperialism*. London: Verso.

Bandle, A. L., et. al., 2013. *Case 20 Herero and Nam Skulls – Namibia and Charité Universitätsmedizin Berlin, Germany*. Geneva: ArThemis Art-Law Centre University of Geneva. [online] Available at: <https://plone.unige.ch/art-adr/cases-affaires/20-skulls-of-colonial-victims-2013-namibia-and-charite-universitatsmedizin-berlin/case-note-2013-20-herero-and-nama-skulls-2013-namibia-and-charite-berlin/view> [Accessed 12 March 2023].

Barad, K., 2007. *Meeting the Universe Halfway: Quantum Physics and the Entanglement of Matter and Meaning*. Durham, London: Duke University Press.

____2008. Posthumanist Performativity: Toward an Understanding of how Matter Comes to Matter. In: S. Alaimo, S. Hekman, eds., 2008. *Material Feminisms*. Bloomington: Indiana University Press.

____2009. Matter feels, converses, suffers, desires, yearns and remembers—Interview with Karen Barad. In: R. Dolphijn and I. Van der Tuin, eds., 2012. *New Materialism: Interviews & Cartographies*. Ann Arbor: Open Humanities Press., pp. 48–70.

Basu, P., 2023a. *Coloniality and Collecting: Extractivism and the Production of Knowledge – Materials and Material Culture* (Abstract). [online] Available at: <http://transpositional-geologies.org/archive/tg-t512> [Accessed 25 May 2024].

____2023b. *transposition 512 (Prof. Dr. Paul Basu (UK/DE): Coloniality and Collecting: Extractivism and the Production of Knowledge—Materials and Material Culture).* [video online] Available at: <http://transpositional-geologies.org/archive/tg-t512> [Accessed 25 May 2024].

Beech, A. 2016. Culture Without Mirrors–Restructuring Creative-Cognitive Power. *Glass Bead: Site 0. Castalia: the Game of Ends and Means*, 2016(1).

____2020. Art Beyond Identity: Constructive Identifications for Real Worlds. In: A. Beech and R. Mackay, eds., 2020. *Construction Site for Possible Worlds*. Falmouth: Urbanomic, pp.119–143.

Bond, P., 2006. *Looting Africa: The Economics of Exploitation*. London: Zed Books.

Brand, D., 2011. *A Map to the Door of No Return*. Toronto: Vintage Canada.

Brassier, R., 2007. The Thanatosis of Enlightenment. In: R. Brassier and C. Kerslake, eds., 2007. *Origins and Ends of the Mind: Philosophical Essays on Psychoanalysis*. Leuven: Leuven University Press, pp.135–148.

Brown, N., 2021. *Assembly*. London: Penguin Books.

Césaire, A., 1973. Poetry and Knowlege. In: *Lyric and Dramatic Poetry 1946–82*. Translated from French by Clayton Eshleman, Annette Smith, 1990. Charlottesville: The University Press of Virginia.

Césaire, S., 2009. *The Great Camouflage—Writings of Dissent (1941–1945): Suzanne Césaire*. Translated from French by K.L. Walker, 2012. Middletown: Wesleyan University Press.

Chorover, S. L., 1979. *From Genesis to Genocide: The Meaning of Human Nature and the Power of Behavior Control*. Cambridge, London: MIT Press.

Clark, A. M ., 1993. *Hey's Mineral Index. Mineral Species, Varieties and Synonyms*, 3rd ed. London, Glasgow, New York, Tokyo, Melbourne, Madras: Chapman & Hall.

Crawford, K., 2021. *Atlas of AI: Power, Politics, and the Planetary Costs of Artificial Intelligence*. New Haven, London: Yale University Press.

Crimp, D., 1993. *On The Museum's Ruins*. Cambridge: MIT Press.

Dana, J. D., 1837. *System of Mineralogy*. New York: John Wiley & Sons.

D'Avignon, R., 2022. *A Ritual Geology – Gold and Subterranean Knowledge in Savanna West Africa*. Durham, London: Duke University Press.

Dedering, T., 1993. The German-Herero War of 1904: Revisionism of Genocide or Imaginary Historiography? *Journal of Southern African Studies*, 19(1), pp.80–88.

Derrida, J., 1967. *Of Grammatology*. Translated from French by C.G. Spivak, 1997. Baltimore: The Johns Hopkins University Press.

____1972. *Margins of Philosophy*. Translated from French by A. Bass, 1982. Sussex: The Harvester Press, University of Chicago, pp.1–27.

____1989. Force of Law: The "Mystical Foundation of Authority". Translated from French by M. Quaintance. *Cardozo Law Review 11*, pp.920–1045.

____1991. *Given Time*. Translated from French by P. Kamuf, 1992. Chicago: University of Chicago Press.

Descola, P., 2005. Par-delà nature et culture. Paris: Éditions Gallimard. Translated from French by J. Lloyd, 2013. *Beyond nature and culture*. Chicago: University of Chicago Press.

Drechsler, H., 1966. *Südwestafrika unter deutscher Kolonialherrschaft: Die großen Landgesellschaften und Minengesellschaften (1885–1914)*. Stuttgart: Franz Steiner Verlag.

D'Souza, R., 2018. *What's Wrong with Rights?: Social Movements, Law and Liberal Imaginations*. London: Pluto Press.

Du Bois, W. E. B., 2007 (1903). *The Souls of Black Folk*. New York: Oxford University Press.

Dussel, E. D., 1992. *The Invention of the Americas: Eclipse of 'the Other' and the Myth of Modernity*. Translated by M. D. Barber, 1995. New York: Continuum.

____1993. Eurocentrism and Modernity (Introduction to the Frankfurt Lectures). *boundary 2*, 20(3): pp.65–76.

____1996. *The Underside of Modernity: Apel, Ricoeur, Rorty, Taylor, and the Philosophy of Liberation*. Translated by E. Mendieta. New Jersey: Humanities Press International.

____1997. The Architectonic of the Ethics of Liberation: On Material Ethics and Formal Moralities. *Philosophy & Social Criticism,* 23(3), pp.1–35.

____2000. Europe, Modernity, and Eurocentrism. *Nepantla: Views from South,* 1(3), pp.465–478.

____2002. World-System and 'Trans'-Modernity. *Nepantla: Views from South,* 3(2), pp.221–244.

Eder, K., 1988. *The Social Construction of Nature: A Sociology of Ecological Enlightenment*. Translated from German by M. Ritter, 1996. London: Sage.

Everett, P., 2021. *The Trees*. Minneapolis, Minnesota: Graywolf Press.

Fritscher, B., 2012. Making Objects Move: On Minerals and Their Dealers in 19th Century Germany. *Journal of History of Science and Technology*, 5(1), pp.84–105.

Gaines, R. V., Skinner H. C., Foord, E. E., Mason, B., and Rosenzweig, A., 1997. Dana's New Mineralogy: *The System of Mineralogy of James Dwight Dana and Edward Salisbury Dana*, 8th Edition. New York: John Wiley & Sons.

Glissant, È., 1958. *The Ripening*. Translated from French by M. Dash, 1985. London: Heinemann Educational Books.

____1969. *Poetic Intention*. Translated from French by N. Stephens and A. Malena, 2010. Callicoon: Nightboat Books.

____1981. *Caribbean Discourse: Selected Essays*. Translated from French by J. M. Dash, 1999. Charlottesville: University of Virginia Press.

____1990. *Poetics of Relation*. Translated by B. Wing, 1997. Ann Abor: The University of Michigan Press.

____1994a. Culture and Identity. In: *Introduction to a Poetics of Diversity*. Translated from French by C. Britton, 2020. Liverpool: Liverpool University Press.

____1994b. *The Collected Poems of Èdouard Glissant*. Translated from French by J. Humphries and M. Manolas, 2005. Minneapolis, London: University of Minnesota Press.

____1997. *Mahagony*. Translated from French by B. Wing, 2020. Lincoln: University of Nebraska Press.

Glissant, E., and Diawara, M., 2011. One World in Relation: Édouard Glissant in Conversation with Manthia Diawara. *Journal of Contemporary African Art*, 28(1), pp.5–18.

Grosz, E., 1999. The Time of Violence; Deconstruction and Value. *College Literature*, 26(1). pp.8–18.

____2008 (2020). *Chaos, Territory, Art—Deleuze and the Framing of the Earth*. New York: Columbia University Press.

____2011. *Becoming Undone: Darwinian Reflections on Life, Politics, and Art*. Durham, London: Duke University Press.

Haraway, D., 1988. Situated Knowledges: The Science Question in Feminism and the Privilege of Partial Perspective. *Feminist Studies,*14 (3), pp. 575–599.

Hearth, S., 2021. The 'World's Greatest Mineral Location': Hai||om, Ndonga, Herero, and the Early Colonial Histories of Tsumeb, Namibia. *Earth Sciences History*, 40(2), pp.433–460.

Hearth, S. and Robbins, C. 2022. Mineral displays as embodiments of geological thought and colonial invisibility. *Journal of Natural Science Collections,* 10, pp.3–17.

History is a highly functional fantasy of the West, originating at precisely the time when it alone "made" the history of the World. If Hegel relegated African peoples to the ahistorical, Amerindian [sic] peoples to the prehistorical, in order to reserve History for European peoples exclusively, it appears that it is not because these African or American peoples "have entered History" that we can conclude today that such a hierarchical conception of "the march of History" is no longer relevant.' (Glissant, 1981, p.64)

[17] This is not a call to undo, dismantle, or dismiss a classification system based on valid scientific knowledge production. What is overdue is a critical study of its shortcomings. What needs to be developed is an understanding of how to reintegrate what has been omitted by its field-specific gaze. This can only be achieved through a transdisciplinary effort, probably even a trans-cosmological effort.

[18] A classification system that, like Sylvia Wynter in her thinking, moves between different discursive frameworks and modes of understanding. A system recognising the importance of embracing diversity and engaging with multiple forms of knowledge in order to achieve a more nuanced and inclusive understanding of ourselves and our world.

[19] Of course, institutional collections are archives that should serve scientific research and contribute to knowledge transfer in a wide variety of disciplines. It would be limiting to consider them exclusively in terms of historical and current inequalities—their usage must be multidirectional, addressing i.e. socio-political matters without silencing other forms of knowledge production inherent in these formats. Socio-political readings, however, are at present among the most underdeveloped perspective in this format of collections.

[20] The Lecture Intervention of Prof Dr Dr Helmut Maier, Interdisciplinary Centre for Science and Technology Studies, University of Wuppertal will address this knowledge production and circulation during German colonial times (Maier, 2023). Hosted at the Institute of Geosciences, Department of Geochemistry/Petrology, Lecture Hall, Poppelsdorfer Schloss, Bonn, Germany. Video recordings of all Lecture Interventions are available in the *Transpositional Geologies* online archive: www.transpositional-geologies.org (Basu, 2023b; Hearth and Robbins, 2023; Hill, 2023; Imalwa, Kaluwapa and Gramlich, 2023).

[21] Otavi Minen- und Eisenbahn-Gesellschaft.

[22] This approach will be deepened and extended in the forthcoming *Transpositional Geologies* exhibition in Tsumeb, Namibia—hosted by, and in collaboration with, the Tsumeb Municipality, Nov. 2023—and this *Transpositional Geologies* publication.

[23] Supported by *Institut für Auslandsbeziehungen (ifa)* programme 'Künstlerkontakte' and *Kunststiftung NR W*.

[24] An artist in residence will take place during July 2023 at the Mineralogical Collections of the Technical University of Berlin focusing on specimens from the Rössing uranium mine, Namib Desert, Arandis, Namibia, followed by a long-term residency from 2025 to 2027.

[25] Besides, Tsumeb as a mineral location has been a major research focus for Hugo Strunz.

[26] For detailed references see the earlier elaboration on classification systems, under (iii).

[27] Yusoff (2018) argues that traditional understandings of geology often overlook or obscure the ways in which racial and colonial violence have shaped the Earth's landscape. She proposes the concept of 'geologies of race' to highlight the entanglements of race, colonialism, and geology, emphasising how processes such as mining, resource extraction, and environmental degradation are deeply intertwined with racialised systems of power and oppression.

[28] O —> O–I —> I–O This transformative process allows for the mobilisation of localised, factually static archived object. As digital *Object-Image* (O–I) they can circulate, travel beyond transnational borders without bureaucratic constrains and re-materialise in context specific settings as Image-Object (I–O).

[29] 'Transposition' refers to a process of reworking and re-contextualising elements, particularly within art and artistic research. It involves shifting from one state or position to another, often resisting stable categorisation or representation. Transpositions operate within concrete situations, challenging traditional disciplinary boundaries and historical narratives. They highlight the dynamic relationship between art, knowledge, and institutional contexts, emphasising the unique and transformative nature of creative practice.

[30] This materialises for example as moving image as in the experimental film *transposition 001 (Happy Banality of Everything) (2022)*, or using a printed format as in *transposition 504 (I-O_ MMdUB-50215)* which generates a different kind of intimate

____2023. *transposition 633 (Prof. Dr. Selby Hearth (US) / Dr. Carrie Robbins (US): If You Don't Grow it, You Have to Mine It: Geologists, Knowledge Production, and Colonial Legacies)* [video online] Available at: <http://transpositional-geologies.org/archive/tg-t633> [Accessed 25 May 2024].

Hecht, G., 2012. *Being Nuclear: Africans and the Global Uranium Trade*. Cambridge, London: MIT Press.

Hill, C. R., 2023. *transposition 792 (Dr. Chris Hill (UK): Critical Earth: Uranium and the Namib in the Nuclear Imagination).* [video online] Available at: <http://transpositional-geologies.org/archive/tg-t792> [Accessed 25 May 2024].

Hill, C. R. and Ashipala, S. N., 2024. 'Follow the Yellowcake Road': Historical Geographies of Namibian Uranium from the Rössing Mine. *Historical Social Research*, 49(1), pp. 32–54. [online] Available at: <https://www.jstor.org/stable/27290653> [Accessed 23 May 2024].

Hoheisel, H., 1995. *Memorial for the murdered Jews of Europe (Berlin 1995).* [online] Available at:<http://www.knitz.net/index.php?option=com_content&task=view&id=27&Itemid=32&lang=en> [Accessed 16 March 2021]

Imalwa, E. S. U., Kaluwapa , P. S. and Gramlich, N., 2023. *transposition 495 (Elli Stella Udda Imalwa-Nangolo (NA), Philemon Sheya Kaluwapa (NA/DE), Noemie Gramlich (DE): German Colonialism, Environmental Injustice, and Ecological Reparation: The Case of Tsumeb).* [video online] Available at: <http://transpositional-geologies.org/archive/tg-t495> [Accessed 25 May 2024].

Kaluvi, J., Nangolo, H. and Jauch, H., 2025 'We Toiled for Zilch'. In: S. Mikloweit, ed., 2025 *Transpositional Geologies: Spectres Of Coloniality*. Berlin: Kerber Publishers.

Klare, M. T., 2012. *The Race for What's Left: The Global Scramble for the World's Last Resources*. New York: Metropolitan Books.

Kuhn, T. S., 1970. *The Structure of Scientific Revolutions: Second Edition*. Chicago: University of Chicago.

Last, A., 2015. Fruit of the Cyclone: Undoing Geopolitics Through Geopoetics. *Geoforum*, 64, pp.56–64.

Lloyd, D., 2019. *Under Representation: The Racial Regime of Aesthetics*. New York: Fordham University Press.

Maier, H., 2020. Circulations: Colonial Geology, the Prussian State Geological Institute and the South West African Copper (1850–1919). *Journal of Namibian Studies: History Politics Culture*, 28, pp.39–63.

____2023. *transposition 135 (Prof. Dr. Dr. Helmut Maier: Circulations: Colonial Geology, the Prussian Geological State Institution, and the South West African Copper (1850–1919)).* [video online] Available at: <http://transpositional-geologies.org/archive/tg-t135> [Accessed 25 May 2024].

Mandarino, J. A., 1998. The New IMA Nomenclature: A Summary for Mineralogists. *The Canadian Mineralogist*, 36(3), pp.905–908.

Mason, R. and Sayner, J., 2019. Bringing Museal Silence Into Focus: Eight Ways of Thinking About Silence in Museums. *International Journal of Heritage Studies*, 26(1), pp.5–20.

Metzger, G. 2016. *Gustav Metzger—Act or Perish!—A Retrospective*. Rome: Nero.

Mignolo, W., 2000. *Local Histories/Global Designs: Coloniality, Subaltern Knowledges, and Border Thinking*. New Jersey: Princeton University Press.

____2011. *The Darker Side of Western Modernity: Global Futures, Decolonial Options*. Durham and London: Duke University Press.

Mignolo, W. andWalsh, C. E., 2018. *On Decoloniality: Concepts, Analytics, and Praxis*. Durham and London: Duke University Press.

Mikloweit, S., 2010. *If You Are Frightened We'll Protect You: Second As..., Third As...* .[exhibition] Central Saint Martins College of Art & Design, London. 01–06 September 2010.

____2011. *If You Are Frightened We'll Protect You.* [exhibition] Associação Brasileira de Empresários de Diversões, Sao Paulo. 18 November–18 December 2011.

____2015a. *15yrs Cluster Mission Anniversary Artist-in-Residence European Space Agency (ESA).* [artist in residence] European Space Agency's (ESA), European Space Operations Centre (ESOC), Darmstadt. August–September 2015.

____2015b. *All Parts Of The Subsystem Are Present.* [exhibition] European Space Agency's (ESA), European Space Operations Centre, ESOC), Darmstadt. 22 September–6 October 2015.

____2017. *Some Parts Of The Subsystem Are Present*. In: Who will own Space?: Back to the moon. [exhibition] Copernicus Science Centre, Warsaw. 14–17 September 2017.

____2019. Research Artist-in-Residence 2019–2024. [artist in residence] *Mineralogical Museum University Bonn*, Germany.

____2023a. *We Build Our Languages Out Of Rocks—Transpositional Geologies.* [exhibition] Mineralogical Museum University Bonn, Germany. 20 April 2023–11 June 2023.

____2023b. *Before Thought Hardens Into Language—Transpositional Geologies.* [exhibition] Tsumeb Municipality, Namibia. 19 October 2023–30 November 2023.

____2024. Research Artist-in-Residence 2024–2027. [artist in residence] *Mineralogical Collections of the Technical University of Berlin*, Germany.

Mindat.org, n.d. Tsumeb Mine (Ongopolo Mine). [online] Available at: <https://www.mindat.org/loc-2428.html> [Accessed 09 Juli 2024].

____2020. *Most prolific type localities*. [online] Available at: <https://www.mindat.org/toptypelocs.php> [Accessed 15 March 2024].

Moako, S. Z., 1989. We Come and Die as Numbers. In: S. Z. Moako, ed., 1989. *Tell Them of Namibia: Poems from the National Liberation Struggle*. London: Karia Press.

Nickel, E. H., Grice, J.D., 1998. The IMA Commission on New Minerals and Mineral Names; Procedures and Guidelines on Mineral Nomenclature, 1998. *The Canadian Mineralogist*, 36(3), pp.913–926.

Nickel, E. H., Mandarino, J.A., 1987. Procedures Involving the IMA Commission on New Minerals and Mineral Names and Guidelines on Mineral Nomenclature. *American Mineralogist*, 72, pp.1031–1042.

O'Kane, P., 2025 A Non-Violent Explosion in the Taxonomic Schloss: Toward a Neobaroque. In: S. Mikloweit, ed., 2025. *Transpositional Geologies: Spectres Of Coloniality*. Berlin: Kerber Publishers.

Parikka, J. 2015. *A Geology of Media*. Minnesota: University of Minnesota Press.

Phillips, M., 1997. *The Dancing Face*. London: Harper Collins.

Quijano, A., 2007. Coloniality and Modernity/Rationality. *Cultural Studies,* 21(2–3), pp.168–178.

Quijano, A. and Ennis, M., 2000. Coloniality of Power, Eurocentrism, and Latin America. *Nepantla, Views from South*, 1(3), pp.533–580.

Rancière, J., 2009. *The Emancipated Spectator*. Verso Books.

____2011. *Aisthesis: Scenes from the Aesthetic Regime of Art*. Translated from French by Z. Paul, 2003. London: Verso.

Rogoff, I. and Scherman, D. J. eds., 1994. *Museum Culture: Histories, Discourses, Spectacles*. Mineapolis: University of Minnesota Press.

Rumsey, M., 2024. *Personal conversation with the Principal CIC, Minerals and Planetary Sciences, Natural History Museum London*. [phone and email] Interview by Sascha Mikloweit. 15 July–29 August 2024.

Schnorbus, H., 2000. *Die Gesellichte der Otavi Minen AG 1900–2000: Ein ereignisreiches Jahrhundert*. Eschborn: Otavi Minen AG.

Schuyler, G. S., 1931 (2012). *Black No More: Being an Account of the Strange and Wonderful Working of Science in the Land of the Free, A.D. 1933-1940*. Boston: Northeastern University Press.

Schwab, M., ed., 2018. *Transpositions: Aesthetico-Epistemic Operators in Artistic Research*. Leuven: Leuven University Press.

Scott, D., 2000 The Re-enchantment of Humanism: An Interview with Sylvia Wynter. *Small Axe*, 8, pp.119–207.

Shiner, L., 1994. 'Primitive Fakes,' 'Tourist Art,' and the Ideology of Authenticity. *The Journal of Aesthetics and Art Criticism*, 52(2), pp. 225–234.

____2001. *The Invention of Art: A Cultural History*. Chicago: The University of Chicago Press.

Sierra, S., 2005. *Position Exchange for two Distinct, 30 metre columns of earth.* [online] Available at: <https://www.santiago-sierra.com/200503_1024.php?key=20 [Accessed 14 March 2021].

Spivak, G. C., 1988. Can the Subaltern Speak? In: C. Nelson and L. Grossberg, eds., 1988. *Marxism and the Interpretation of Culture*. Baltimore: University of Illinois Press.

Strunz, K. H., 1941. *Mineralogische Tabellen zur Bestimmung der wichtigsten Mineralien*. Stuttgart: E. Schweizerbart'sche Verlagsbuchhandlung.

Strunz, K. H. and Klein, C., 1977. *Mineralogische Tabellen*. Stuttgart: E. Schweizerbart'sche Verlagsbuchhandlung.

Strunz, K. H. and Nickel, E. H., 2001. *Strunz Mineralogical Tables: Chemical-structural Mineral Classification System*. 9th ed. Stuttgart: E. Schweizerbart'sche.

Strunz, K. H. and Tennyson, C., 1982. *Mineralogische Tabellen*. 8. Auflage. Leipzig: Akademische Verlagsgesellschaft Geest & Portig KG.

transposition 001 (Happy Banality of Everything), 2022. [film] Directed by Sascha Mikloweit, sound by I. Kiziria, animation by H. Risse. Germany: Transpositional Geologies.

Veltmeyer, H. and Petras, J., 2014. *The New Extractivism: A Post-Neoliberal Development Model or Imperialism of the Twenty-First Century?* London: Zed Books.

Wallace, M., 2013. *A History of Namibia: From the Beginning to 1990.* New York: Oxford University Press.
Wildcat, D. R., 2009. *Red Alert!: Saving the Planet with Indigenous Knowledge*. Colorado: Fulcrum Publishing.
Williams, E., 1944 (2022). *Capitalism and Slavery*. London: Penguin Books.
Wilson, W. E., Pinch, W. W., Weber, D., Keller, P., Key, C. L. and White, J. S., 1977. Tsumeb! The World's Greatest Mineral Locality. *Mineralogical Record*, 8 (3), pp.1–111.
Wynter, S., 1989. Beyond the World of Man: Glissant and the New Discourse of the Antilles. *World Literature Today*, 63(4), pp.637–648.
____1995. 1492: A New World View. In: *Race, Discourse, and the Origin of the Americas.* Hyatt, V. L. and Netteford, R., eds., 1995. Washington: Smithsonian Institution Press.
____1997. Columbus, the Ocean Blue, and Fables That Stir the Mind: To Reinvent the Study of Letters. In: *Poetics of the Americas: Race, Founding, and Textuality*. Cowan, B. and Humphries, J., eds., 1997. Baton Rouge: Louisiana State University Press, pp.141–163.
____2003. Unsettling the Coloniality of Being/Power/Truth/Freedom: Towards the Human, After Man, Its Overrepresentation – An Argument. *The New Centennial Review*, 3(3), pp.257–337.
____2006 The Ceremony Must Be Found: After Humanism. *Textual Practice*, 20(3), pp.397–441.
Wynter, S. and McKittrick, K., 2007. Unparalleled Catastrophe for Our Species? Or, to Give Humanness a Different Future: Conversations. In: McKittrick, K., ed., 2015. *Sylvia Wynter: On Being Human as Praxis*. Durham: Duke University Press.
Yusoff, K., 2013. Geologic Life: Prehistory, Climate, Futures in the Anthropocene. *Environment and Planning D: Society and Space*, 31(5), pp.779–795.
____2015. Geologic Subjects: Nonhuman Origins, Geomorphic Aesthetics and the Art of Becoming Inhuman. *Cultural Geographies*, 22(3), pp.383–407.
____2018. *A Billion Black Anthropocenes or None*. Minnesota: University of Minnesota Press.
____2019. Geological Realism: On the Beach of Geologic Time. *Social Text*, 37(1(138)), pp.1–26.
____2020. *Geo-Logics: Natural Resources as Necropolitics*. [video online] Available at: <https://www.youtube.com/watch?v=QM8B-XZG8OQ> [Accessed 20 November 2020].
____2021a. The Inhumanities. *Annals of the American Association of Geographers*, 111(3), pp.663–676.
____2021b. Mine as Paradigm. *E-flux Journal*, 6(21). [online] Available at <https://www.e-flux.com/architecture/survivance/381867/mine-as-paradigm> [Accessed 17 January 2023].

proximity to the spectator, while simultaneously shifting the context it is placed in.

[31] A brief remark regarding the titling system, which is a play between scientific order and chaos (the latter in Elizabeth Grosz's understanding (2008)). Each transposition is assigned a randomised number between 1-999 in its title—i.e. *transposition 741* might be followed by *transposition 064*, followed by *transposition 001*, thus creating an emphasis on non universality and non linearity, on the fragmented nature of knowing and accumulating knowledge, on the gaps between obtained facts, on the never to be finished business of knowing or attempting to know. Our knowledge of things will always be ideologically influenced, fragmentary, and quintessentially subjective and should never be imposed as universal truth onto a diverse humanity, but rather be applied as tools to work towards the realisation of an inclusive humanism in Sylvia Wynter's spirit (i.e. Wynter, 2006).

[32] Regarding the concept of 'coloniality' see Quijano and Ennis (2000), and Quijano (2007).

[3] The Otavi area was insulated from early European colonisation because of its location: 600km inland, north east of the harbour town Swakopmund. To reach it, Europeans had to cross the Namib Desert, then the Namib Savanna, both of which require specialised environmental knowledge to cross. As a result, the first archived records of Otavi copper date to 1850, which for this analysis will start the era of 'European impact.' However, even before travellers arrived in Otavi, the Ovambo Kingdoms to the north were interacting with Portuguese traders and Finnish missionaries, and the Nama and Herero to the south were interacting with Boer, English, and German travellers, missionaries, traders, and hunters coming up from the Cape Colony.

[4] Saarelma-Maunumaa (2003) gives a good overview of terminologies surrounding 'Ovambo'. The term has variously been presented as Ambo, Aawambo, Ovampo, Owambo, among others. This essay uses 'Ovambo' to refer to the larger ethnic group, and names individual nations when they are known, e.g., the Ndonga. Similarly, 'Ondonga' is used to refer to the state, and 'Ndonga' to the people of that state.

[5] Early European writers referred to 'the Bushmen:' a group of several communities with similar languages and shared heritages, who today refer to themselves broadly as the San people (Gordon 1992, p.5; South African San Institute 2017). The Haiǁom are the San nation around Tsumeb today, and likely in 1850 as well (Gordon 1992, pp.15–24; Dieckmann 2007, p.35).

[6] The Herero claimed cultural relationships with the land and copper in Otavi (Schinz 1891, p.350; Gebhard 1999, p.69).

[7] Hearth (2021) summarises these early visitors.

Contested Land and the Extraction of Geologic Knowledge: Tsumeb and the Otavi Mountains as Reconstructed through White (pre-)colonial Reporting

Selby Hearth

Abstract: The Tsumeb Mine of northeastern Namibia is one of the most famous mineral localities on Earth, and has played important economic, political, and scientific roles throughout its nearly 100-year history. However, prior to the start of industrial mining in 1906, the land and the copper were extensively contested. This analysis reviews early written records about land, copper, and geologic knowledge production at Tsumeb and in the Otavi Mountains more broadly. European records from 1850 to 1895 reported that at least three communities had extensive relationships with land and copper in Otavi: the San community of Hai||om, the Ovambo nation of Ondonga, and the Herero. These communities also played critical roles in the first European geologic mapping of the Otavi Mountains, from 1892–1895. This mapping enterprise illustrates many themes in the relationships between European geologists and colonisation.

Keywords: Tsumeb, Otavi Mountains, mineralogy, copper, history of geology, history of mining, geologic colonialism, Namibia, South-West Africa, Deutsch-Südwestafrika (DSW)

In 1977, the *Mineralogical Record* declared Namibia's Tsumeb Mine 'the world's greatest mineral locality' (Wilson, et al., 1977). Located in the Otavi Mountains of Namibia's north, Oshikoto region, with today's Tsumeb Municipality to its south, the mine was worked from 1906 to 1996. With 72 type specimens[1] and 342 individual mineral species, Tsumeb is nearly the most mineralogically diverse site in the world (Mindat.org, 2020). The site's economic wealth brought significant geologic attention to the region, which has gained global importance in recent Snowball Earth debates[2] (Hoffman, et al., 2021).

However, the copper in and around Tsumeb has also been the source of centuries of conflict. Prior to the start of industrial-scale mining in 1906, copper formed the economic centre of a network of Otavi communities involved in mining, smelting, smithing, and trading. Copper lured the earliest colonists to the region, and instigated their removal (Hearth, 2021). It was the earliest extractable that Germany could find that might make the new colony pay for itself. Especially in the early colonial period, Otavi copper illustrates the interconnectedness of geologic knowledge extraction and conflict over land.

This essay reviews the early written records of geologic knowledge produced in Tsumeb and in the Otavi Mountains. In those years, when land was so actively contested, what were the relationships between geologic knowledge and the wider political and social processes in Otavi?

Based on European reporting from 1850 to 1895, this analysis suffers from the limitations of colonial archives: the perspectives of the Otavi communities involved are almost entirely absent. European writers wrote through the prejudiced lenses of their era, often encountering worldviews they did not understand, especially surrounding non-Western relationships with land and resources. Archives also preferentially preserved materials produced by Europeans, further skewing the imbalance of information now available. This imbalance in agency—*whose* perspectives are represented *by whom*—limits what can be accomplished in an analysis like this one. Ideally, this work would not *examine* these early Otavi communities, but find a way to *think with* them through actively deconstructing the European lens. However, the scant nature of the records available limits our ability to see the realities of this time period. The following therefore focuses on the records that are available, and tries to read them against the grain, looking for what is omitted.

Knowledge production prior to European impact: 1740–1850[3]

Geologic knowledge comprises narratives and explanations about the Earth and all its behaviours. Given the centrality of copper in the area, it is likely that pre-contact Otavi communities developed their own systems, explaining why the Otavi Mountains rose between the flat Etosha Pan and the Kalahari Desert. They likely developed explanations for why the copper outcropped where it did, how it had formed, and what it was materially. They likely noticed, named, classified, and had explanations for the different formations of limestone and the granites that underlay them. European records relate that at least three communities had relationships with Otavi land and copper during this period: the Ovambo Kingdom of Ondonga[4], the San community of Hai||om[5], and the Herero[6]. The surviving records from this period were all written by Europeans who only passed through the region.[7] These early records hint at broader systems of geologic knowledge within Otavi communities only through the lens of copper smelting

[1] A 'type specimen' is the reference specimen that defines a mineral. A type locality is the place where the specimen was found. Tsumeb yielding 72 'type specimens' means that, for 72 minerals, their first published descriptions were based on a specimen from Tsumeb.

[2] 'Snowball Earth' is a model of a climate event in Earth history. It suggests that Earth froze over completely between 717 and 660 million years ago, then again between 654 and 632 million years ago. The Otavi carbonates hold significant records from these time periods.

and trade: observable processes that an outsider could witness and write down without communicating deeply with the individuals performing a range of—partially highly specialised or laborious—tasks. The Portuguese trader Bernardino José Brochado travelled through the Ovambo Kingdoms (never to the Otavi Mountains) in 1850, writing:

> The Kwankhala [Haiǁom San, to the south] possess large copper mines and have copper in abundance. Only the Ndonga trade with them trading for copper for tobacco, beads, and pungo [cannabis]. However, the Ondonga do not precisely know where the mine is. (Gordon, 1992, p.25)

This baseline narrative—the Haiǁom mining the copper and trading it north to the Ndonga coppersmiths, who smelted and traded it further—is repeated throughout the early European writings on the Otavi Mountains. Seven years after Brochado, in 1857, linguist and Rhenish missionary Carl Hugo Hahn reported that he frequently met parties of Ovambo and Haiǁom transporting copper ore in the Otavi area:

> They carried the ore in nicely woven baskets made of palm leaves, evenly distributed on a palm tree trunk. Each load weighed 90 pounds and over, but they also carried their provisions, water and weapons, all together about 110 to 120 pounds, and they walked 6 to 8 hours or 15 to 20 English miles every day. The working of the copper mines seems to have been very lively, because [we] often met later Ovambo and Bushmen who carried the ore to Ondonga. There the Bushmen swap it for grain, tobacco and calabashes and the Ovambo sell it processed into large ankle rings to the neighbouring nations up to the Ovambuengo. (Hahn, 1859, p.301)

In a later record, Hahn estimates that the Haiǁom extracted and traded 50–60 tons of copper ore from the Otavi Mountains every year and that 'the Bushmen are so jealous of this trade that they have not allowed any stranger, not even the people of Ondonga, to see the place where they dig' (Hahn, 1867, p.286). This is probably an over-simplification. Geologist Matthew Rogers wrote in 1893 that frequent groups of Ovambo labourers were sent south to collect copper ore in baskets, and returned it north (Hearth, 2023). Similarly, Gerald McKiernan (1954, pp.53–54) reported finding a copper mine with evidence of smelting on site in 1875, and Francis Galton (1853, p.223) reported Haiǁom trading and smelting copper ore at Lake Otjikoto, just 20km west of Tsumeb. Contrary to what Hahn suggests, these accounts suggest that the Haiǁom did not guard the locations of the copper from all outsiders, nor that the Ndongans were the only copper smithers. The occasional reports of Ndongan coppersmiths and their smelting processes were only given in broad strokes. Andersson (1856, p.203) wrote of Ovambo smiths using 'fire-proof crucibles', bellows, hammers, and anvils made of stone, though it was unclear whether he was referring to Ndonga coppersmiths or ironsmiths in neighbouring Ovambo kingdoms. Hahn wrote much later of the Ndonga copper- and ironsmiths: 'Native smiths undergo a long period of training, and ultimately show a remarkable degree of skill with their craft, notwithstanding the very primitive design of the anvils, hammers, and tongs and other tools used by them' (Hahn, Fourie and Vedder, 1928, p.35).

Modern authors have summarised copper smelting techniques used in other areas, which may or may not have been used in Ondonga (e.g., Söhnge, 1967, p.12). Gebhard (1999, pp.15–16), for instance, describes a copper smelting process that involved using bellows to heat copper ore in a termite hill. Overall, the surviving records only *hint* at Otavi communities' geologic knowledge systems. The records treat communities as homogeneous; in reality, though, geologic knowledge production and use was likely specialised, as indicated by the existence of specially-trained smiths. Beyond the observations of copper use, the European writers failed to record the communities' wider geologic conceptual framings.

Constraining time

Constraining *when* copper smelting emerged in this area is difficult. Ovambo oral histories told that the Ovambo carried knowledge of *iron* metallurgy when they arrived in the Etocha area (Williams, 1991, p.36), an event that Vansina (2004, p.53) estimates at 300–400 BCE; however, this is considerably earlier than the first evidence of *copper* smelting in southern Africa.

Communities across southern Africa developed systems of knowledge around copper long before European contact. De Maret (1982, p.7) locates the earliest evidence of copper smelting in southern Africa at 345 +/– 75 CE. The Phalaborwa Mine in modern South Africa evidenced copper excavation going back to the 700s CE (van der Merwe and Scully, 1971, p.183). The Thakadu Mine in Botswana dates to between 1480 and 1680 (Huffman, et al., 1995, p.60), and Morton (2010, p.24) reported that the Bakgatla and the Makaba Tlokwa of Marothodi in what is now north-west South Africa were smithing copper since at least the 1600s. The Matchless Mine near modern Windhoek shows evidence of copper smelting dating to 1500–1600 (Kinahan and Vogel, 1982). There were also written accounts of Damara coppersmiths smelting in southern Namibia from the mid- to late-1700s (Kinahan, 1980).

There is evidence that smelted copper goods found at the Nqoma site in southern Namibia might have been sourced from Otavi copper, which would place an excavation date for the copper between 700 to 1180 (Vansina, 2004, p.71). When precisely local communities began copper extraction in the Otavi area is not found in this survey of existing records.

Knowledge production after European impact: 1850–1895

During the period of 1850 to 1895, the Otavi Mountains were increasingly visited by Christian missionaries (especially the German Rhenish Missionary Society), white hunters from the Cape Colony, and merchants. One—the hunter and trader William Worthington Jordan from the Cape Colony—negotiated a land deal with the King of Ondonga, Kambonde kaMpingana; a contract was signed on 21 April 1885. The deal allowed Jordan to establish the *Republic of Upingtonia*[8] near Grootfontein on 20 October 1885, and to bring Boer settlers to farm the area. The story is usually told that King Kambonde sold Jordan an area twice the size of modern Belgium for 25 muskets, an immunised ('salted') horse, and a barrel of brandy, with the Finnish Lutheran missionaries in Ondonga witnessing the deal.[9] However, Eirola (1991, pp.411–414) assessed that European concepts of land sales, ownership, and written contracts likely struck Kambonde as absurd, as land was variously used by different community members at different times. Kambonde was almost certainly not selling the land in an European understanding, but rather *leasing* it, as this was how land use was typically negotiated in Ondonga in that time period (Williams, 1991, p.147).

Jordan also claimed that King Kambonde had given him the rights to copper in the Otavi Mountains, though this was almost certainly a lie.[10] The copper from the Tsumeb region was one of Ondonga's oldest and most important trade goods (Hahn, 1867, p.286; Siiskonen, 1986, p.41), and Kambonde had already refused three other Europeans who had come seeking rights to the copper (Eirola, 1991, p.411). Jordan himself had admitted that Kambonde was reluctant to sell the area, and only agreed once Jordan promised he wanted only the land, not the copper. Records from the Finnish missionaries make no reference to the copper, nor is the copper mentioned in the deed of sale, which adds doubt to Jordan's claims (Eirola, 1991, pp.409–413).

Rights to Otavi copper were highly contested during the brief Upingtonia period (1885–1887). Gordon (1992, p.41) speculated that Jordan had brought 'a large number of whites' to the area specifically to support his claims to the copper, in the hopes that Western governments might intervene on his behalf. The Hai||om continued mining the copper (Schinz, 1891, p.293), and actively resisted Boer settlers on the land (Gordon, 1992, p.40–42). The Herero claimed the land and copper as historically their own (Schinz 1891, p.350; Gebhard 1999, p.69), and Chief Samuel Maharero had given the British trader Robert Lewis a 30-year concession on 'the Otavi Mine' (though it is unclear today which mine Maharero was referring to). King Kambonde grew increasingly irate with Jordan until, finally, Kambonde's brother Nehale had Jordan murdered.[11]

The year-long Upingtonia copper conflict benefited only the German colonial enterprise, both the government and, eventually, the South West Africa Company (SWAC). After Jordan's murder in 1886, the Boers accepted the German government's offer of protection, which, as far as the German federal and colonial governments were concerned, brought Otavi land and copper under the jurisdiction of Germany (Eirola, 1991, p.425). In 1890, the German Empire declared *Deutsch-Südwestafrika* a Crown Colony.

After considerable political intrigue, summarised in Dreyer (1987) and Hearth (2021), Germany awarded Otavi's mineral rights to SWAC, a joint British-German corporation. SWAC knew there was copper, but not how much, the quality, or even *where* exactly it was. So, in 1892, they sent geologist Matthew Rogers to map the Otavi Mountains.

Geologic knowledge production and Matthew Rogers: 1892–1895

From 1892 to 1895, Matthew Rogers carried out the first assessment of the geology of the Otavi region—enabled by a team of seven to 13 Europeans (miners, a mapmaker, a surveyor, and a mining foreman), and a large[12] team of workers from different communities between the Namibian coast and the Otavi Mountains. Rogers named almost no individual Africans in his letters, but he did reference guides, servants, cooks, assistants, wagon drivers, firewood carriers, water carriers, hunters, and miners hired from local communities. He rarely named which communities workers came from; however, given his path, most of these people were probably San or Herero. He also likely hired people from the coast and western interior during the early part of their trek inland. The pivotal role of the guides in knowledge production will be discussed below, but it should also be noted that the expedition would have failed immediately without the environmental knowledge of the team of Africans who kept Rogers alive across the inland trek. Geologic knowledge production cannot be separated from the environmental knowledge needed to keep geologists fed, watered, and sheltered.

Geologic mapping is the foundation of a geologist's work: walking the outcrops, identifying the rock types, classifying units, recognising larger structures. Rogers and his team produced the first topographic and geologic maps of the Otavi region between 1892 and 1895, representing an area spanning from the Waterberg in the south to Tsumeb in the west. For Rogers, this mapping allowed him to identify two parallel limestone ridges running east-west through the mountains. Today, this is recognised as the Otavi Syncline (see, for example, Germs, 1974; Lehmann, et al., 2015; Hoffman, 2021). The limestone that Rogers mapped is now classified as part of the Hüttenberg Formation within the Tsumeb Subgroup of the Otavi Group (Kamona and Günzel, 2007, p.399).

Rogers' mapping was foundational geologic knowledge production—but his *purpose* in making maps was to find

[8] The Republic of Upintonia and its relationship with Otavi copper is outside the scope of this analysis, as they did not contribute to geologic knowledge production in that area. However, the details of Upingtonia are examined in Hearth (2021).

[9] For a more detailed analysis see Eirola (1991) and Hearth (2021).

[10] Hearth (2021) reviews the controversy over the copper and land rights in greater detail.

[11] Hearth (2021) reviews the various speculations on why Nehale had Jordan murdered.

[12] Based on the archival documents, it is unclear how many local people were a part of this team.

the copper for SWAC. For this, his mapping was (as it turned out) not useful. Instead, he relied almost exclusively on local guides. Though Rogers' letters usually failed to identify the communities his guides came from, they were most likely Haillom San and Herero, as those were the two communities he had most contact with in that area. Evidence of his dependence of these guides is plentiful in his letters to the SWAC Board of Directors: On 20 December 1892, when Rogers first encountered Otavi copper, he wrote in his diary only: 'Mr. Eslick and I proceed to the mines on horseback, a native guide accompanying us'.[13] On 27 December 1892: 'The entrance [to the Otavi mines] on the south western side is by a very narrow passage, and it would be a difficult thing for any stranger to find it, unless shown by someone else'.[14] On 9 January 1893, with his guide being hungover, his lack of independence is evident: 'Having no interpreter, and not knowing the road, I have no alternative but to wait [...]'.[15] On 21 January 1893: 'Not being able to speak the language, I can only follow whithersoever I am led'.[16]

For Rogers, these Haillom and Herero guides, with their depth of knowledge of the region, were the key to locating and mapping the broad zone of copper mineralisation that runs through the Otavi Mountains. This mapping led to nearly 100 years of industrial exploitation and became foundational to future generations of geologists operating in the Otavi area. However, the contributions of the Haillom and Herero guides to this foundational work was unrecognised. This is not an exception, but rather the historical norm: indigenous guides were critical to geologic knowledge production throughout this time period, and were rarely named or recognised for their contributions.

In addition to the pivotal role of local guides, the Rogers expedition highlights several other themes in the relationship between geologic knowledge extraction and European colonialism. These are examined in detail in Hearth (2023), and will be summarised in the following.

The socio-political context: 1892–1895

Besides local knowledge for navigating the terrain and mapping, one of the most common themes in Rogers' letters was his dependence on local leaders for negotiating access to field sites. The German government had declared itself in charge of the region in 1890; however, they had almost no actual power beyond the coast. Rogers had to negotiate with Herero and Haillom leaders to gain access to the Otavi Mountains, particularly Herero Chiefs Manasse Tjiseseta and Samuel Maharero. Once there, other Herero and Haillom leaders continually challenged his right to stay—and Rogers' only bargaining chip to remain was the permission he had obtained from Manasse and Samuel Maharero. His status as a British citizen working for a British-German company in German-claimed territory bought him nothing; his relationships with Manasse and Samuel Maharero bought him occasional, guarded permission. Rogers' geologic work in Otavi was highly political—and the Herero and Haillom leaders were aware of that.

Ondongas' involvement in this region further complicated matters. With the Kingdom of Ondonga, which continued to assert their rights to both the land and the copper of Otavi, Rogers had only tangential interactions. The 1890s was a politically complex time period in Ondonga due to increased pressures from Portuguese traders (McKittrick, 2002) and this, combined with the 240km distance between Ondonga and Otavi, resulted in their appearing only occasionally in Rogers' letters. On 20 December 1893: 'Again, a report is prevalent, the Ovambos being displeased at our being here, are coming in force to make war on us'.[17] On 20 January 1894: '[The Ovambos] are incensed at our being here, and if they do not come in bodily force to drive us away, will steal from the [Herero], and otherwise harass them, that they will be obliged to recall us'.[18] Even though these threats never materialised, they clarify that the Ndonga still considered themselves actors in Otavi copper.

Intelligence gathering: 1892–1895

Rogers also understood the political nature of his geologic work. In his letters back to the SWAC Board of Directors, he returned his opinions on how to best proceed with colonisation of the country. He attempted gardens, and returned his observations on how well crops would grow in various areas (not well). He attended to water availability for farming and soil quality (both low). And, throughout his letters, he speculated on which communities would make the best labourers for SWAC's colonisation, and how the Germans could subdue those communities. For example, in a 17 April 1893 letter, Rogers wrote to the SWAC Board: 'I am enquiring after native labourers with the view of testing the possibility of bringing them to do mining work. Doubtless, with patience and perseverance, the Berg Damaras might be brought to do something in this way, but it will be some time before they will become efficient mine labourers.'[19] In another letter from 6 October 1894, Rogers speculated that the Germans would not be able to establish a colonial government in Deutsch-Südwestafrika because 'they have a stubborn and rebellious people to deal with, and that it is far from an easy matter to bring them into subjugation' (Rogers quoted in Hearth 2023, p.407). Someone at SWAC summarised in the marginalia next to Rogers' assessment: 'Native races not easily conquered' (Figure 20). Rogers and the SWAC Board of Directors both understood this geologic survey to be the vanguard of colonisation.

Expiration: Dec.1894 – Jan.1895

This political situation eventually drove Rogers out. In his 20 December 1894 letter to the Board, as he was camped near

[13] BArch R 1001/1480, p.112

[14] ibid., p.116

[15] ibid., p.155

[16] ibid., p.164

[17] BArch R 1001/1482, p.119

[18] BArch R 1001/1481, p.103

[19] BArch R 1001/1482, p.58

[20] BArch R 1001/1482, p.102

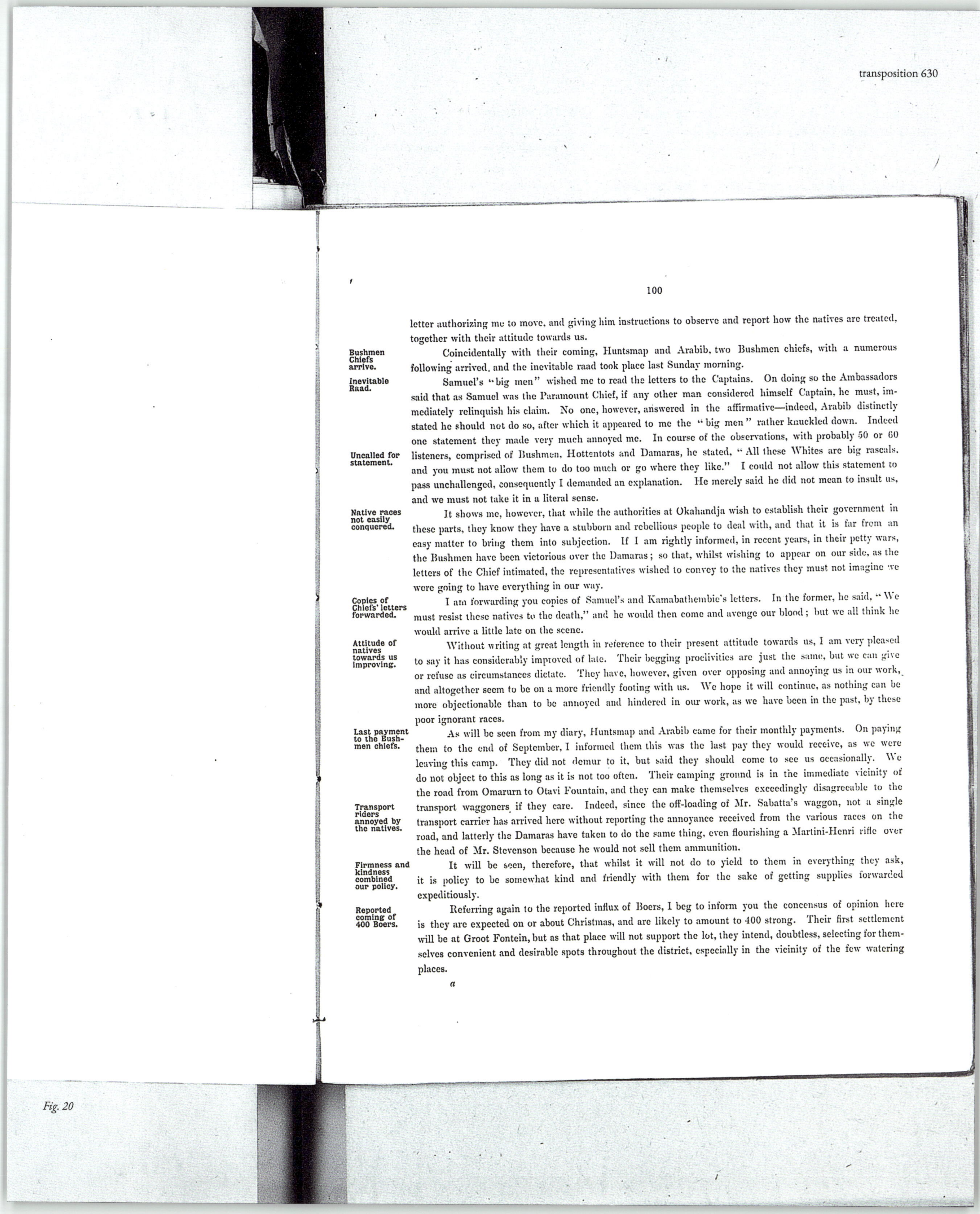

100

letter authorizing me to move, and giving him instructions to observe and report how the natives are treated, together with their attitude towards us.

Bushmen Chiefs arrive. Coincidentally with their coming, Huntsmap and Arabib, two Bushmen chiefs, with a numerous following arrived, and the inevitable raad took place last Sunday morning.

Inevitable Raad. Samuel's "big men" wished me to read the letters to the Captains. On doing so the Ambassadors said that as Samuel was the Paramount Chief, if any other man considered himself Captain, he must, immediately relinquish his claim. No one, however, answered in the affirmative—indeed, Arabib distinctly stated he should not do so, after which it appeared to me the "big men" rather knuckled down. Indeed one statement they made very much annoyed me. In course of the observations, with probably 50 or 60 listeners, comprised of Bushmen, Hottentots and Damaras, he stated, "All these Whites are big rascals, and you must not allow them to do too much or go where they like." I could not allow this statement to pass unchallenged, consequently I demanded an explanation. He merely said he did not mean to insult us, and we must not take it in a literal sense.

Uncalled for statement.

Native races not easily conquered. It shows me, however, that while the authorities at Okahandja wish to establish their government in these parts, they know they have a stubborn and rebellious people to deal with, and that it is far from an easy matter to bring them into subjection. If I am rightly informed, in recent years, in their petty wars, the Bushmen have been victorious over the Damaras; so that, whilst wishing to appear on our side, as the letters of the Chief intimated, the representatives wished to convey to the natives they must not imagine we were going to have everything in our way.

Copies of Chiefs' letters forwarded. I am forwarding you copies of Samuel's and Kamabathembie's letters. In the former, he said, "We must resist these natives to the death," and he would then come and avenge our blood; but we all think he would arrive a little late on the scene.

Attitude of natives towards us improving. Without writing at great length in reference to their present attitude towards us, I am very pleased to say it has considerably improved of late. Their begging proclivities are just the same, but we can give or refuse as circumstances dictate. They have, however, given over opposing and annoying us in our work, and altogether seem to be on a more friendly footing with us. We hope it will continue, as nothing can be more objectionable than to be annoyed and hindered in our work, as we have been in the past, by these poor ignorant races.

Last payment to the Bushmen chiefs. As will be seen from my diary, Huntsmap and Arabib came for their monthly payments. On paying them to the end of September, I informed them this was the last pay they would receive, as we were leaving this camp. They did not demur to it, but said they should come to see us occasionally. We do not object to this as long as it is not too often. Their camping ground is in the immediate vicinity of the road from Omarurn to Otavi Fountain, and they can make themselves exceedingly disagreeable to the transport waggoners if they care. Indeed, since the off-loading of Mr. Sabatta's waggon, not a single transport carrier has arrived here without reporting the annoyance received from the various races on the road, and latterly the Damaras have taken to do the same thing, even flourishing a Martini-Henri rifle over the head of Mr. Stevenson because he would not sell them ammunition.

Transport riders annoyed by the natives.

Firmness and kindness combined our policy. It will be seen, therefore, that whilst it will not do to yield to them in everything they ask, it is policy to be somewhat kind and friendly with them for the sake of getting supplies forwarded expeditiously.

Reported coming of 400 Boers. Referring again to the reported influx of Boers, I beg to inform you the concensus of opinion here is they are expected on or about Christmas, and are likely to amount to 400 strong. Their first settlement will be at Groot Fontein, but as that place will not support the lot, they intend, doubtless, selecting for themselves convenient and desirable spots throughout the district, especially in the vicinity of the few watering places.

a

Fig. 20

the Tsumeb outcrop, surrounded by workers angry at him for their lack of sufficient pay and food, he wrote:

> [...] another strike has occurred with our native employés, this time for more food [...] Our difficulties now appear to have taken a more critical form. Before, our enemies were from without, now they are from within the camp [...] Again, a report is prevalent, the Ovambos being displeased at our being here, are coming in force to make war on us [...] Really, gentlemen, this continual worry and anxiety is not good for me. I feel acutely the responsibility of my position here, with so many men under me, and so much property, that oftentimes I feel myself unequal to the occasion. I thought on coming to this district, and being comparatively free of the [Herero], we should prosecute our work in peace, but I find it is not the case, and now I really see no end to our trouble. Looking at all these matters I must ask you gentlemen to take into consideration the propriety of relieving me of my duties. In the present state of feelings here, it is impossible to carry on our work with any degree of vigour or satisfaction.

Rogers returned to England in January 1895. SWAC sent the geologist Christopher James to replace him in 1900, though he too lasted only a short time (until 1901).

Conclusions

In the era before industrial-scale production began at Tsumeb, copper, land, and the production of geologic knowledge were entwined with the larger-scale social forces operating around the Otavi region. The copper was a motivating force for early colonisers like William Worthington Jordan and the Upingtonian Boers, and for the German Empire. Its potential economic value enabled the geologic mapping that Rogers carried out—and the resulting conflict among Germans, British, SWAC, Herero, Ovambo, and Haiǁom.

The early records of copper in Otavi illustrate the commodity-oriented way that European hunters, merchants, and geologists engaged with indigenous peoples: their reports focused on how the Otavi communities used copper, never on indigenous geologic world views.

What the early records do show is that the knowledge of local communities was essential to foundational work of these early European geologists. Rogers' expedition would never have made it to Otavi without the dozens of local guides and hunters who assisted them, nor would they have found the copper once they got there. Indeed, it is likely they would have died en route to Otavi without the knowledge and active skills of the Haiǁom and Herero people.

Acknowledgements

This essay draws on conversations with Sascha Mikloweit, Carrie Robbins, Marianne Weldon, and the students in Bryn Mawr College's Fall 2022 360° program (Minerals, Museums, and Western Colonialism): Aha Anderson, Rosa Bieber-Stanley, April Chernila, Helen Christ, Hannah Cosgrove, Morgan Hanson-Rosenberg, Carly Hill, Maya Hofstetter, Emily Lazo, Izzie Ludlow, Samantha Lyster, Rachel Myers, Jasper Nash, Georgia Reed, and Julia Saint-Amour.

Archives

This essay cites research from the German Federal Archives, abbreviated in footnotes here with its code: BArch.

References

Andersson, C., 1856. *Lake Ngami, or Explorations and Discoveries in the Wilds of Southern Africa.* New York: Harper and Brothers.

de Maret, P., 1982. New survey of archaeological research and dates for West-Central and North-Central Africa. *The Journal of African History*, 23(1), pp.1–15.

Dieckmann, U., 2007. *Haiom in the Etosha Region: A History of Colonial Settlement, Ethnicity and Nature Conservation*. Basel: Basler Afrika Bibliographien.

Dreyer, R., 1987. Whitehall, Cape Town, Berlin and the economic partition of South-West Africa: The establishment of British economic control, 1885–94. *The Journal of Imperial and Commonwealth History*, 15(3), pp.264–288.

Eirola, M., 1991. 'Janomaan vaeltajat' ja Upingtonian buuritasavalta 1885–1887. *Acta Societatis Historicae Finlandiae Septentrionalis: Faravid*, 15, pp.409–431.

Galton, F., 1853. *Narrative of an Explorer in Tropical South Africa.* London: John Murray.

Gebhard, G., 1999. *Tsumeb, a Unique Mineral Locality*. GG Publishing, Grossenseifen, Germany.

Germs, G. J. B., 1974. The Nama Group in South West Africa and its relationship to the Pan-African Geosyncline. *The Journal of Geology,* 82, pp.301–317.

Gordon, R., 1992. *The Bushman Myth: The Making of a Namibian Underclass.* Conflict and Social Change Series. Boulder, Colorado: Westview.

Hahn, H., 1859. Reise der Herren Hugo Hahn und Rath im südwestlichen Afrika, Mai bis September 1857. *Mittheilungen aus Justus Perthes' Geographischer Anstalt über wichtige neue Erforschungen auf dem Gesammtgebiete der Geographie*, pp.295–303.

____1867. Hugo Hahn's Reise von Otjimbingue zum Cunene, 1866. *Mittheilungen aus Justus Perthes' Geographischer Anstalt über wichtige neue Erforschungen auf dem Gesammtgebiete der Geographie*, pp. 284–312.

Hahn, C. H., Fourie, L., and Vedder, V., 1928. *The Native Tribes of South West Africa.* Frank Cass and Co: Abingdon, U. K..

Hearth, S., 2021. The 'World's Greatest Mineral Locality': Haiǁom, Ndonga, Herero, and the Early Colonial Histories of Tsumeb, Namibia. *Earth Sciences History*, 40(2), pp.433–460.

____ 2023. Geologists as Colonial Scouts: The Rogers Expedition to Otavi and Tsumeb, Namibia, 1892–1895. *Earth Sciences History,* 42(2), pp.385–415.

Hoffman, P., et al., 2021. Snowballs in Africa: sectioning a long-lived Neoproterozoic carbonate platform and its bathyal foreslope (NW Namibia). *Earth-Science Reviews,* 219. DOI: 10.1016/j.earscirev.2021.103616.

Huffman, T., et al., 1995. Early copper mining at Thakadu, Botswana. *Journal of the South African Institute of Mining and Metallurgy,* 95(2), p.53.

Kamona, F. and Gunzel, A., 2007. Stratigraphy and base metal mineralization in the Otavi Mountain Land, northern Namibia – a review and regional interpretation. *Gondwana Research,* 11, pp.396–413.

Kinahan, J., 1980. Eighteenth-century coppersmiths in central Namibia: Comments on some sources and syntheses. *Namibiana*, 2, pp.17–22.

Kinahan, J. and Vogel, J.C., 1982. Recent copper-working sites in the Khuiseb drainage, Namibia. *The South African Archaeological Bulletin*, 37, pp. 44–45.

Lehmann, J., et al., 2015. Structural and geochronological constraints on the Pan-African tectonic evolution of the northern Damara Belt, Namibia. *Tectonics*, 34. DOI: 10.1002/2015TC003899.

McKiernan, G., 1954. *The Narrative and Journal of Gerald McKiernan, 1874–1879.* Cape Town: Van Riebeeck Society.

McKittrick, M., 2002. *To Dwell Secure: Generation, Christianity, and Colonialism in Ovamboland.* Portsmouth, NH: Heinemann.

Mindat.org, 2020. *Most prolific type localities*. [online] Available at: <https://www.mindat.org/toptypelocs.php> [Accessed 15 March 2024].

Morton, F., 2010. Fenders of space: Kgatla territorial expansion under Boer and British Rule, 1840–1920. In: P. Limb, N. Etherington, and P. Midgley, eds., 2010. *Grappling with the Beast: Indigenous Southern African Responses to Colonialism, 1840–1930.* Leiden: Brill.

Saarelma-Maunumaa, M., 2003. *Edhina Ekogidho–Names as Links: The Encounter between African and European Anthroponymic Systems among the Ambo People in Namibia*. Helsinki: Studia Fennica Linguistica.

Schinz, H., 1891. *Deutsch-Südwest-Afrika: Forschungsreisen durch die deutschen Schutzgebiete Gross-Nama- und Hereroland, nach dem Kunene, dem Ngami-See und der Kalaxari, 1884–1887.* Saarbrücken: Fines Mundi.

Siiskonen, H., 1986. *Studying the Northern Namibian Past: Research Seminar in Tvärminne, 2–4 December 1985.* Joensuu: Joensuun yliopisto, Humanistinen tiedekunta.

Söhnge, P. G., 1967. *Tsumeb: A Historical Sketch.* Windhoek, Namibia: Committee of the South West Africa Scientific Society.

South African San Institute, The, 2017. San Code of Research Ethics. In: D. Schroeder, et al., 2019. *Equitable Research Partnerships: A Global Code of Conduct to Counter Ethics Dumping*. Cham: Springer. [online] Available at: <https://link.springer.com/chapter/10.1007/978-3-030-15745-6_7#citeas> [Accessed 15 March 2024].

van der Merwe, N. and Scully, R., 1971. The Phalaborwa story: Archaeological and ethnographic investigation of a South African Iron Age group. *World Archaeology*, 3(2), pp.178–196.

Vansina, J., 2004. *How Societies Are Born: Governance in West Central Africa Before 1600.* Charlottesville: University of Virginia Press.

Williams, F. N., 1991. *Precolonial Communities of Southwestern Africa: A History of Owambo Kingdoms, 1600–1920.* Windhoek: National Archives of Namibia.

Wilson, W. E., Pinch, W. W., Weber, D., Keller, P., Key, C. L.,and White, J. S., 1977. Tsumeb! The World's Greatest Mineral Locality. Mineralogical Record, 8 (3), pp.1–111.

transposition 509

Colonial Geology, the Prussian State Geological Institute, and the South West African Copper (1850–1919)

Helmut Maier[1]

Abstract: The rich copper ore resources of the Otavi mountains in Namibia have been exploited for centuries. From the mid-1880s onwards, the tradition of colonial geology (Kolonialgeologie) was founded. The expectation of copper imports from SWA legitimised the German colonial expansion. The Prussian State Geological Institute (Preußische Geologische Landesanstalt) became the pivotal German centre for research on colonial geo-resources. Up to 1919, extensive studies of the geology of the region were carried out, resulting in a new theory on the genesis of ore deposits. The article examines the interdependence between geological knowledge production and the establishment of the flourishing copper-mining industry. It outlines how the colonial-geological knowledge system emerged through the commission of experts, with the establishment of departments by the government and by industrial interests. The article sheds light on the pivotal role of science and scientific institutions, and the transnational colonial-geological system in colonial exploitation.

Keywords: colonial history of Namibia, copper ore resources, colonial geology, Prussian State Geological Institute, colonial knowledge system, Hans Schneiderhöhn, Otavi Minen- und Eisenbahngesellschaft (OMEG), government geologists, Tsumeb 'Green Hill', Herero and Nama War

The copper ore resources of the Otavi mountains in Namibia have been exploited for centuries (Herbert, 1984; Lau 1987b, p.viii; Sandelowsky, 2004, p.73).[2] Missionaries reported on Herero and Ovambo trading ore for metallic copper. During 'the copper mining mania of the 1850s' (Smalberger, 1975, p.9; Levinson, 1976, p.14) the first commercial copper mine was established (Henrichsen, 2011, pp.2, 61, 70 f., 89–91). Expectations of abundant copper occurrences (Cairncross, 2004, p.79) in South West Africa (SWA) attracted adventurers, geologists and mining engineers. With the rise of the electrical industry from the 1880s onwards, copper became a strategic resource. Germany increased copper imports from 12,300 tons (1880) to 225,400 tons (1913) (Warnack 1914, p.81). The expectation of copper imports from SWA legitimised the German colonial expansion (Kaulich, 2003, p.383). From the mid-1880s onwards, a growing number of experts visited SWA, founding a tradition of colonial geology (*Kolonialgeologie*). The Prussian State Geological Institute (*Preußische Geologische Landesanstalt, PGLA*) was founded in Berlin in 1873. By 1919, it had become the pivotal German centre for research on colonial geo-resources (Udluft, 1968, p.20).

Due to the disappointing results of mining activities in SWA, however, German companies became reluctant to invest there. It was only after 1900, when the rich deposits of copper in the Otavi mountains (Tsumeb) promised profits, that German industrialists, along with South African and English enterprises, became heavily involved. From 1907 to 1914, the Otavi Mining and Railway Company (*Otavi Minen- und Eisenbahngesellschaft, OMEG*) paid exceptionally high dividends. Moreover, the mine was a scientific singularity: 'Tsumeb ranks as one of the highest grade base metal deposits in the world. It is geologically and mineralogically famous for 41 copper, lead, zinc, vanadium, and germanium ore minerals.' (Ratledge, Ong and Boyce, 1955)

In 1914, OMEG hired the young geologist Hans Schneiderhöhn (1887–1962) (Gehlen, 1988) to further explore the ore deposits of Tsumeb and the Otavi mountains. Up to 1919, Schneiderhöhn carried out an extensive study of the geology of the region, from which he developed a ground-breaking new microscopic method (Söhnge, 1976, p. 58). Eventually, he became famous for his theory on the genesis of ore deposits (Amstutz, 1988).

In the following, I examine the interdependence between geological knowledge production and the establishment of the flourishing copper-mining industry (Schneider and Seeger, 1993). I will outline how the colonial-geological knowledge system emerged through the commission of experts, with the establishment of departments by the government and by industrial interests. The article will shed light on the pivotal role of science and scientific institutions and the transnational colonial-geological system in colonial exploitation.

Research Frame and Master Narrative

According to Daniel R. Headrick, western science and technology were crucial for colonial expansion (Brahm, 2017, p.287). Thereby, the production of colonial knowledge even led to the formation of new scientific disciplines such as geography (Gräbel, 2015, pp.9 ff.). Despite the function of the PGLA in the geological exploration of SWA from the 1880s onwards, this chapter of its history has been largely neglected. Only in 1987 did H. J. Kamps stress the relevance of the PGLA in the context of German imperialistic aggression up to 1914 (Kamps, 1987, p.149).

The master narrative on the history of Namibia from 1884 to 1919 emphasises the failure of German colonial rule and

[1] *Transpositional Geologies* version of Helmut Maier's essay 'Circulations: Colonial Geology, the Prussian State Geological Institute and the South West African Copper (1850–1919)' (Maier, 2020).

[2] For the Drierivier site furnaces (mid- 17th century), Rehobot District, see Sandelowsky (2004), p.73.

genocide. The behaviour towards the African peoples was dominated by fraud, cruelty, racism, rape, murder, and land grabbing. The Herero and Nama uprisings from 1903 to 1907 resulted in a genocide that was carried out by the Germans. Of up to 100,000 Otjiherero speakers before the Namibian War, there were only around 20,000 survivors in 1911. According to Marion Wallace, 'using even the most conservative of these figures, more than half the population classified as Herero was wiped out.' Of around 20,000 Nama, roughly one third was killed. Other Nama-speaking groups were sent to concentration camps (Shark Island) and 'almost wiped out' (Wallace, 2011, pp.177 f.).

The Herero and Nama War against the German invaders was triggered, among other factors, by OMEG's decision to build a railway connection between Tsumeb and the Atlantic Ocean (Wallace, 2015, p.250). In fact, OMEG forced the Herero to hand over parts of their most fertile pastures, including their precious water rights (Drechsler, 1991, p. 49). The question is whether the geological expeditions that had brought to light the abundant copper deposits of Tsumeb were crucial in the chain of events that led to the genocide. With the inauguration of the railway in 1906 and the ensuing copper rush, science and the railway were conducive to the capitalistic exploitation of the colony that followed (Wallace, 2015, p.314).

Knowledge System, Experts, Expeditions

Expeditions were crucial for the generation of knowledge about colonial resources. The history of the numerous expeditions to SWA from the 18th century onwards can be reconstructed from a larger historiographical body of literature and published sources (Liesegang, 1940).[3] All pre-colonial commercial copper ventures, however, failed at an early stage (Lau 1987a, p.98).

The second phase of expeditions began in the early 1880s. The earliest expedition on ore deposits commissioned by the German government was conducted in 1882 (Fischer, 1972; s.n., 1901, p.35). Merchant and 'founder' of the colony, Adolf Lüderitz (1834–1886), who was mainly interested in gold and gemstones, initiated several expeditions. Lüderitz's second expedition was headed by a mining director, accompanied by six miners from Freiberg, Saxony (Lüert, 1971, p.178; Engelhardt, 2003). Lüderitz (Lüderitz, 1945, pp.10, 100 f., 109) sold his prospecting rights to the German Colonial Society for South West Africa (*Deutsche Kolonialgesellschaft für Südwestafrika, DKG*).

Several studies have revealed the extent to which colonial enterprises could become driving agents for geological knowledge production. These enterprises – e.g. the South West Africa Company (SWAC) (Voeltz, 1988) and OMEG (Schnorbus, 2000) – and their experts formed a collective that became responsible for the production of colonial knowledge. In 1888, mining engineer Philipp Braumüller undertook an expedition for the German-African Mining Company (*Deutsch-Afrikanische Minengesellschaft*) (Braumüller, 1875; Weidmann, 1894, p.18). In the same year, mining expert and chemist Eduard Fleck (1841–1917) (s.n., 2020) was sent out, and in 1889, he supposedly discovered 'valuable mines' in the Rehoboth region (s.n., 1893).

After the breakdown of the German rule in the Herero territory from 1891 onwards, the Germans started to establish a military and civil administrative structure, and a new governmental mining authority was created. There was, however, very little thorough geological research during the first two decades of German colonial rule (Drechsler, 1996). None of the copper mines – Kuyas, Hope, Matchless, Ebony, and Pot – had been successful (Stromer von Reichenbach, 1896, pp.141 f.). All deposits of gold, copper, lead, and iron ore, up until 1903, were accidental discoveries without any following mining activities – except in one case (Kaulich, 2003, pp.81, 87, 384–387, 620).

The outstanding exception to the rule was the expedition of Matthew Rogers (SWAC 1892–1895). The reconstruction of the events provide an impression of the sensation felt by Rogers when he 'discovered' the 'green hill' of Tsumeb:

> I [...] have visited various countries of the world, inspected mines, mineral outcrops, and prospecting for minerals; have been associated with the minerals gold, silver, tin, copper and lead; but in the whole of my experience I have never seen such a sight as was presented before my view at [Tsumeb], and I very much doubt that I shall ever see such another in any other locality. (Söhnge, 1976, p.18; Wilson et al., 1977, p.11; Schnorbus, 2000, p.59)

The hill was green because of its main mineral component, malachite, a copper mineral.

Judith Fait (2019, p.72) stresses the interplay of 'copper, colonialism, [and] capital', which led to the exploitation of the region. To transport the Tsumeb ores, a railway had to be built to connect the mine with the Atlantic. In 1900, OMEG was founded, with 60% of its shares held by SWAC. The ore deposit was estimated to be almost 300,000 tons, and a profit of over 300,000 pounds sterling per year was expected (Schmeißer, 1908, p.12); which corresponds to a 2023 value of around £45.85 billion. In the following years, the sensational mineral singularity of Tsumeb and the Otavi region was visited by at least seventeen other experts and their teams from different fields (Maier, 2020, p.50).

From the 1880s onwards, the PGLA intensified its colonial activities. Beginning in 1900, especially under its president

[3] For the pre-colonial expeditions, see Maier (2020).

Karl Schmeißer, the PGLA was dedicated to colonial geology and 'German colonisation' (Kamps, 1987, p.150). In 1902, Schmeißer was appointed a member of the colonial council of the German foreign office, and became a prominent protagonist of the exploitation of SWA (Gräbel, 2015, pp.64–68). In 1902, the PGLA founded its own 'colonial department' (Udluft, 1968, p.20) and began to build up a geological-mineralogical collection from the colonies (*geologisch-mineralogische Kolonial-Sammlung*) (s.n., 1928b, p.XXIV, XXIX). These new departments must be perceived as the cornerstones of the geological colonial knowledge system. The PGLA collected, examined, and prepared the geological material that flowed in from the colonies. These collections served in the professional preparation of the experts who were to be sent out to the colonies in the future (s.n., 1902, pp.106 f.).

Nudging Exploration and Mining: A New Mining Regulation (1905)

With the reluctance of the commercial colonial societies (Kaulich, 2003, p.307), and also of German banks, mining, metallurgical, and electrical industries to prospect and invest in SWA, the German government sought to stimulate activity. Thus, the legal regulation of mining in SWA was revised. This change is congruent with the conception of a transition of colonial rule from trade and theft to well-planned exploitation (Osterhammel and Jansen, 2012, p.73). With the introduction of the new mining regulation (*Kaiserliche Bergverordnung*) in 1905, the right of public prospection (*öffentliche Schürffreiheit*) was introduced (Kaulich, 2003, pp.386 f.). This expanded the opportunity for prospection beyond governmental agencies and colonial societies to the general public.

For the implementation of the new mining regulation, the SWA mining authority was reinforced by additional personnel:

> Although their progress was initially greatly impeded by the rebellions of the native population and the ensuing military campaigns, a considerable number of publications nevertheless originated from these geologists, especially from Range (s.n., 1953, p.XXVIII f.), who was one of the most prolific writers in his field. Mainly occupied with the search for groundwater in the southern districts, one of his duties was the technical management of the government's drilling operation. Range, however, found the opportunity to describe in detail the 'Namaland' and its geological formations, of which the Nama and Karoo successions received most attention. (Hugo and Schalk, 1982, p.5)

For the involvement of government geologists in the genocide it might be indicative that Paul Range's successor, Hans von Staff (1883–1915) (s.n., 1916), who volunteered for the war in 1914, bore the title war geologist (*Kriegsgeologe*). Several other individuals have been identified as belonging to this pool of experts (Krenkel, 1933, pp.144–146; Hugo and Schalk, 1982; Anhaeusser, 1997).[4]

Apart from the colonial knowledge and personal experience being fed back into circulation, universities, institutes of technology, mining academies, and commercial colleges increasingly offered lectures about the mineral deposits and geology of the colonies (s.n., 1928a, pp.VI). Furthermore, special attention was paid to the training of mining engineers for the colonial service. In this respect, the PGLA's collection of colonial minerals and ores had a crucial function (Schmeißer, 1908, p.37).

The discovery of the 'green hill' of Tsumeb, with its occurrences of high base metal content, triggered a mining rush. After mining commenced in 1900, further activities began. Shortly after the foundation of OMEG, samples of ores were shipped to Germany and England for test smelting (Schnorbus, 2000, p.71). The early procedures of ore processing were ineffective: 'Many companies examined Tsumeb but quickly lost enthusiasm because of the formidable metallurgical problems presented by the dump ore.' (Ratledge, Ong and Boyce,1955, p.374). However, new ore treatment technologies and the smelting process were developed by German and probably other companies abroad (Bürg, 1942, pp.106f.; Schnorbus, 2000, p.71).

Final Stages of the Colonial Knowledge System

With its collection of colonial minerals and ores, its educational function for those being dispatched to the colonies and the Imperial Colonial Office (*Reichskolonialamt*), and its laboratories and library, the PGLA had presumably become the pivotal German agency for consulting and research on colonial geo-resources. Due to the growing scope of its functions, the PGLA reorganised its colonial activities—a measure that can be seen as the final stage of building a smoothly-run and effective techno-scientific consulting agency. Shortly after 1908, the colonial archive of the PGLA (*Kolonial-Archiv*) was founded (s.n., 1936, p.18). Finally, in 1912, the Central Geological Department for the German Protectorates (*Geologische Zentralstelle für die deutschen Schutzgebiete*) of the PGLA was inaugurated. The new department built up and administered the colonial archive, library and map collection and the collection of colonial minerals and ores (s.n., 1928a, p.XXIX).

From 1907 onwards, OMEG was able to pay remarkable dividends. In 1913-14, ore mining in Tsumeb reached its peak (70,000 tons). Surprisingly, it was not until 1913 that a geologist for the thorough exploration of the mines' deposits was hired. Initially, Hans Schneiderhöhn from

[4] Further geologists were mentioned by Krenkel (1933), Hugo and Schalk (1982); for a comprehensive compilation of the South African geological community and its achievements, see Anhaeusser (1997).

the University of Berlin was only scheduled to spend three months in Tsumeb. Due to the outbreak of war, he did not return to Germany until 1919 (Schnorbus, 2000, p.64). In 1915, the Union Defence Force took over the region, but did not interrupt Schneiderhöhn's research. After the war, he published the results on Tsumeb and the Otavi region and conveyed an enormous mineral collection to Germany (Schneiderhöhn, 1958).

With the colonial knowledge system in full swing up to 1914, Tsumeb's eruptive body of rock even challenged the established geological paradigms. It became the nucleus for a new pre-eminent scientific method and theory, both created by Schneiderhöhn:

> By adapting and modifying a biological and a petrographical microscope he fashioned an experimental reflecting ore microscope and developed the technique of opaque mineral identification for which he soon after became world famous. (Schneiderhöhn, 1922; Söhnge, 1976, p.58)

Scientists later celebrated Schneiderhöhn for decoding the complex mechanisms of the genesis of eruptive ore bodies. At Tsumeb, he had developed a genetic system for the classification of mineral deposits on a physicochemical basis (Gehlen, 1988). Thus, Tsumeb not only became the source of vast profits for OMEG, but also the starting point for a new era in geological theory and economic geology (Radmann, 1928; Amstutz, 1988).[5]

Conclusion

The formation of the colonial knowledge system was a necessary condition for the establishment of colonial rule and the successful exploration and exploitation of the copper deposits of SWA. The extent of the involvement of government geologists and mining experts in the genocide is unclear. Taken together, the complex created tremendous techno-scientific and institutional effects in SWA, as well as in Germany and probably other colonial powers. Little by little, commercial colonial societies and metallurgical companies entered the system, but the PGLA took a leading role in knowledge generation, accumulation, circulation, collection, consulting and training, and colonial propaganda.

[5] For an international comparative discussion of the new theories up to 1928, see Radmann (1928).

References

Amstutz, C. G., 1988. Hans Schneiderhöhn – ein Versuch einer Standortbestimmung seines erzgenetischen Lebenswerkes. *Erzmetall*, 41, pp.375–378.
Anhaeusser, C. R. ed., 1997. *A Century of Geological Endeavour in Southern Africa*. Pretoria: Geological Society of South Africa.
Brahm, F., 2017. Imperialismus und Kolonialismus. In: M. Sommer, S. Müller-Wille and C. Reinhardt, eds., 2017. *Handbuch Wissenschaftsgeschichte*. Stuttgart: J.B. Metzler Verlag, p.287–294.
Braumüller, P., 1875. *Der geognostische Aufbau Syderöe's, ein Beweis, daß die nordischen Basaltgesteine neptunischen Ursprungs sind*. PhD. Universität Rostock.
Bürg, G., 1942. *Die nutzbaren Minerallagerstätten von Deutsch-Südwestafrika*. Mitteilungen der Gruppe Deutscher Kolonialwirtschaftlicher Unternehmungen, 7. Berlin: Walter de Gruyter & Co.
Cairncross, B., 2004. *Field Guide to Rocks & Minerals of Southern Africa*. Cape Town: Random House Struik.
Drechsler, H.,1991. Südwestafrika 1885–1907. In: H. Stoecker, ed. 1991. *Drang nach Afrika. Die deutsche koloniale Expansionspolitik und Herrschaft in Afrika von den Anfängen bis zum Verlust der Kolonien*, 2nd ed.. Berlin: Akademie Verlag, pp.36–58.
____1996. *Südwestafrika unter deutscher Kolonialherrschaft: Die großen Land- und Minengesellschaften (1885–1914)*. Stuttgart: Franz Steiner Verlag.
Engelhardt, D. v., ed., 2003. *Biographische Enzyklopädie deutschsprachiger Naturwissenschaftler*. München: Saur.
Fait, J., 2019. *Kupfer, Kolonialismus, Kapital. Das Bergwerk Tsumeb, Namibia*. Hamburg: Diplomica Verlag.
Fischer, W., 1972. Hoepfner, Carl. In: *Neue Deutsche Biographie, Volume 9*. Berlin: Dunker & Humblot, p.348.
Gehlen, K. v., 1988. Zum 100. Geburtstag von Hans Schneiderhöhn (1887–1962), dem bekannten Erzlagerstättenforscher und Mineralogen. *Erzmetall*, 41, pp.373 f.
Gewald, J.-B., 1999. *Herero Heroes. A Socio-Political History of the Herero of Namibia 1890–1923*. Oxford: Currey.
Gräbel, C., 2015. *Die Erforschung der Kolonien. Expeditionen und koloniale Wissenskultur deutscher Geographen, 1884–1919*. Bielefeld: Transcript.
Henrichsen, D., 2011. *Herrschaft und Alltag im vorkolonialen Zentralnamibia*. Basel, Windhoek: Basler Afrika Bibliographien, BAB; Namibia Wissenschaftliche Gesellschaft.
Herbert, E. W., 1984. *Red Gold of Africa. Copper in Precolonial History and Culture* Madison: University of Wisconsin.
Hugo, P. J. and Schalk, K. E. L., 1982. *1880–1980: A Century of Earth Science in South West Africa/Namibia*. s.l.: Department of Economic Affairs, Geological Survey of South Africa.
Kamps, H. J., 1987. Die Bildung und Entwicklung zentraler staatlicher geologischer Institutionen in Berlin in der Periode von der Entstehung des deutschen Imperialismus bis zum Inferno des faschistischen Krieges. *Zeitschrift für angewandte Geologie*, 33, pp.148–153.
Kaulich, U., 2003. *Die Geschichte der ehemaligen Kolonie Deutsch-Südwestafrika (1884–1914). Eine Gesamtdarstellung*. 2nd. ed.. Frankfurt: Lang.
Krenkel, E., 1933. Zur Geschichte der geologischen Erforschung Ost- und Südafrikas. *Zeitschrift der Deutschen Geologischen Gesellschaft*, 85, pp.134–146
Lau, B., 1987a. *Namibia in Jonker Afrikaner's Time*. Windhoek: National Archives, Department of National Education.
____ ed., 1987b. *The Matchless Copper Mine in 1857. Correspondence of manager C. J. Andersson*. Archeia. Windhoek Archives Publication Series, 7. Windhoek: National Archives, SWA/Namibia.
Levinson, O., 1976. *South West Africa*. Cape Town: Tafelberg.
Liesegang, C., 1940. Deutsche Bergmannsarbeit im alten Südafrika. *Glückauf*, 76, pp. 308–312.
Lüderitz, C. A., 1945. *Die Erschließung von Deutsch-Südwest-Afrika durch Adolf Lüderitz*. Oldenburg: Stalling Verlag.
Lüert, H., 1971. *Deutscher Bergbau im Ausland in der Vergangenheit, Gegenwart und Zukunft*. Köln, Berlin: Gote.
Maier, H., 2020. Circulations: Colonial Geology, the Prussian State Geological Institute and the South West African Copper (1850–1919). *Journal of Namibian Studies: History Politics Culture*, 28, pp.39–63.
Osterhammel, J. and Jansen, J. C., 2012. *Kolonialismus. Geschichte, Formen, Folgen*. München: C.H. Beck.
Radmann [Bergassessor], 1928. Neuere Theorien der Erzlagerstättenbildung und ihre praktische Anwendung. *Metall und Erz*, 25. p.543–549.
Ratledge, J. P., Ong, J. N. and Boyce, J. H., 1955. Development of Metallurgical Practice At Tsumeb. *Transactions of the American Institute of Mining and Metallurgical Engineers*, 202, pp.374–382.
Sandelowsky, B., 2004, *Archaeologically yours. A personal journey into the prehistory of Southern Africa, in particular Namibia*. Windhoek: Namibia Scientific Society.
Schmeißer, Karl [Kgl. Berghauptmann, Breslau], 1908. *Die nutzbaren Bodenschätze und die Entwickelung des Bergbaus in den deutschen Schutzgebieten*. Breslau: Korn.
Schneider, G. I. C., Seeger, K. G., 1993. Copper, In: Republic of Namibia. ed., 1993. *The Mineral Resources of Namibia*. Republic of Namibia: Ministry of Mines and Energy, Geological Survey, Ch. 2.3. pp.3–6.
Schneiderhöhn, H., 1922. *Anleitung zur mikroskopischen Bestimmung und Untersuchung von Erzen und Aufbereitungsprodukten besonders im auffallenden Licht*. Berlin: Gesellschaft deutscher Metallhütten- und Bergleute.
____1958: Zur Erforschungsgeschichte der Erze der Tsumeb-Mine und der geologischen Verhältnisse des Otaviberglands, Südwestafrika, *Neues Jahrbuch für Mineralogie*, pp.125–136.
Schnorbus, H., 2000. *Die Gesellichte der Otavi Minen AG 1900–2000: Ein ereignisreiches Jahrhundert*. Eschborn: Otavi Minen AG.
Smalberger, J. M., 1975. *Aspects of the History of Copper Mining in Namaqualand 1846–1931*. Cape Town: C. Struik.
s.n., 1893. Ludwig von Lilienthal. *Export*, 15, p.382–384.
____1901. Biographische Mitteilungen. *Leopoldina*, 37, p.32–40.
____1902. *Festschrift zum Deutschen Kolonialkongreß*. Berlin, am 10. und 11. Oktober 1902, Leipzig: Bibliographisches Institut.
____1916. Über die Tätigkeit des Kriegsgeologen in Deutsch-Südwest und die letzten Lebenstage von Hans v. Staff. *Centralblatt für Mineralogie, Geologie und Paläontologie*, pp.229–232.
____1928a, Curt Gagel †. Jahrbuch der der Preußisch Geologischen Lehranstalt, 48, p.I–XIX.
____1928b. Willi Koert †. Jahrbuch der der Preußisch Geologischen Lehranstalt, 48. pp.XX–XXXVII.
____1936, Franz Beyschlag †. Jahrbuch der der Preußisch Geologischen Lehranstalt, 56(2). pp.2–36.
____1953. Paul Range †. Geologisches Jahrbuch 67 (1953), p.XXIII–XXXVI.
____2020. Dr. Eduard Fleck, n.d. ZOBODAT [online] Available at: <www.zobodat.at/personen.php?id=7612&bio=full> [Accessed 9 July 2020].
Söhnge, G., 1976. *Tsumeb. A Historical Sketch*. Windhoek: Committee of the SWA Scientific Society.
Stromer von Reichenbach, E. Frhrr v., 1896. *Die Geologie der deutschen Schutzgebiete in Afrika*. PhD, München.
Udluft, H., 1968. Die Preußische Geologische Landesanstalt 1873–1939. *Beihefte zum Geologischen Jahr, 78*. Hannover: Bundesanstalt für Bodenforschung.
Voeltz, R. A., 1988. *German Colonialism and the South West Africa Company, 1884–1914*. Athens: Ohio University, Center for International Studies.
Wallace, M., 2011. *History of Namibia: From the Beginning to 1990*, Oxford: Oxford University Press.
____ 2015. *Geschichte Namibias. Von den Anfängen bis 1990*. Basel: Basler Afrika Bibliographien.
Warnack, M., 1914. *Unsere Kolonialwirtschaft in ihrer Bedeutung für Industrie, Handel und Landwirtschaft*. Berlin: Kolonial-Wirtschaftl. Komitee e.V..
Weidmann, C., 1894. *Deutsche Männer in Afrika*. Lübeck: Bernhard Nöhring.
Wilson, W. E., Pinch, W. W., Weber, D., Keller, P., Key, C. L., White, J. S., 1977. Tsumeb! The World's Greatest Mineral Locality. *Mineralogical Record*, 8 (3), pp.1–111.

transposition 261

'We Toiled for Zilch'

Contributing on behalf of the Tsumeb Municipality, Namibia:
Juuso Ndinolune David Kaluvi as recorded and introduced by Hidipo Nangolo
With a historical commentary in footnotes by Herbert Jauch

Abstract: This chapter outlines the experiences of a Namibian contract worker in the copper mine of Tsumeb who spent 40 years at the mine. The central component is his testimony, which is located within the specific colonial labour conditions to provide the reader with a glimpse of the de-humanising, highly discriminatory and exploitative practices that contract workers had to endure. The chapter presents a brief introduction to Tsumeb and the mining activities there historically and today. The background to the colonial contract labour system and its effects on workers is sketched and workers' experiences are recalled, such as the journeys in cattle trucks, life in the compounds, the dangers of underground mining, low pay, the lack of job security and the lack of consideration for workers' interests. The importance of collective resistance is shown with reference to the general strikes of 1971-72 and 1988.

Keywords: contract labour system, general strike, exploitation of workers , workers' hostels, Tsumeb copper mining, Tsumeb-economy, Tsumeb Municipality , Tsumeb Corporation Limited (TCL), South West Africa Native Labour Association (SWANLA)

Around 400 kilometres north of Namibia's capital Windhoek lies Tsumeb—a small mining town, strategically located along the main B1 road network that connects all parts of Namibia, in all directions. My name is Hidipo Nangolo (born 1968), a veteran journalist and activist. As a young college student at Ongwediva in the north during the 1980s, from time to time I visited relatives in Tsumeb. The facade of the once busy and prosperous mining town I glimpsed four decades ago had dissipated: exposing the underlying consequent calamities of business foreclosures and dwindling economic activities. Today, only a smelter remains in Tsumeb, as the famous underground mine closed in 1996, after 100 years of copper mining. The closed mine is now part of the central business district with its symbolic De Wet shaft seen in the main street of the town. The smelter, which was owned by Dundee Precious Metals Inc. (DPM), was recently sold to a subsidiary of the Chinese-owned Sinomine Resource Group for US$49 million. This deal, which has created hope and uncertainty at the same time about the future of the smelter, is expected to be finalised in the third quarter of 2024. Even though the smelter employs around 700 workers, Tsumeb Municipality and its economy is in decline. This is not unique to Tsumeb: Namibian mining outfits in general benefit from pre-independence neo-liberal mining policies, by which they are only eligible to pay royalties to the government, and are not obliged to contribute any finances to the town or the area they operate in. The only money that has flowed from DPM to the municipality of Tsumeb has been ridiculously small amounts of a philanthropic nature, which are in no relation to potential tax payments. In a current reality in which the globally operating DPM (to which the Tsumeb Smelter still belonged in 2023) just announced its 2023 revenue of US$520.1 million, and net earnings of US$182 million, the Namibian taxation logic leaves a Tsumeb Municipality financially alone and incapacitated. This becomes pertinent in a range of issues, such as Tsumeb's water sources reportedly being unfit for consumption. The Tsumeb Municipality is in a dilemma—it does not have the financial capacity to maintain or expand the town's crucial amenities, which were originally built for 15,000 people, but now have to serve around 40,000 residents.

The contract worker system is key to understanding the complexity of Tsumeb as a mining town. Until the late 1980s, Tsumeb was still the centre of contract labour brokerage, whereby workers had to work for the Tsumeb Corporation Limited's (TCL) copper mining and smelter for almost two years, without visiting their families and without job security after conscription. Contract workers were not allowed to freely reside south of the 'Red Line' without a permit or contract work. Their only permitted means of accommodation was in 'compound hostels'. After Namibia gained independence in 1990, these compound hostels, regarded as symbols of oppression and exploitation, were partially demolished, while others were sold to private businesses.

A compound hostel room in Tsumeb, in which the second president of Namibia (2005–2015) Hifikepunye Pohamba stayed as a contract worker at TCL from 1956–1960, has been designated as a heritage museum under the Tsumeb Municipality. The Tsumeb Municipality asked me to find and interview former TCL workers to collect their testimonies, which will become part of the Pohamba Room collections. To do so, in January 2019 I travelled 190 kilometres north of Tsumeb to find former TCL employee Juuso Kaluvi (1939–2023) at his village of Onalunike near Omuthiya in the Oshikoto Region, where we spoke about his job at TCL. Kaluvi was a 21-year-old graduate from Odibo St. Mary's Mission School when he was first employed by a contract labour brokerage company in the Time Office of the TCL in 1956. In his career spanning four decades as Time Keeper and Paymaster (1956–1996), he met almost every worker who entered and exited the underground mine. Kaluvi was employed in the Time Office because he was fluent in English, only one of a few blacks, usually graduates from Odibo, who worked in an office. In his position he signed job tickets, kept records of shifts and time schedules for the American company Newmont Mining Corporation, which, at the time, owned TCL. Newmont recruited workers through the South West Africa Native Labour Association (SWANLA). For his testimony, Juuso Kaluvi spoke in his native language of Oshiwambo: I recorded it on video, which can now be found in the Tsumeb Municipality archives.

To gain a comprehensive understanding of the contract workers' system, I invited the scholar, author, trade union expert and labour researcher Herbert Jauch to comment on Kaluvi's testimony. I met Jauch as a co-trade unionist

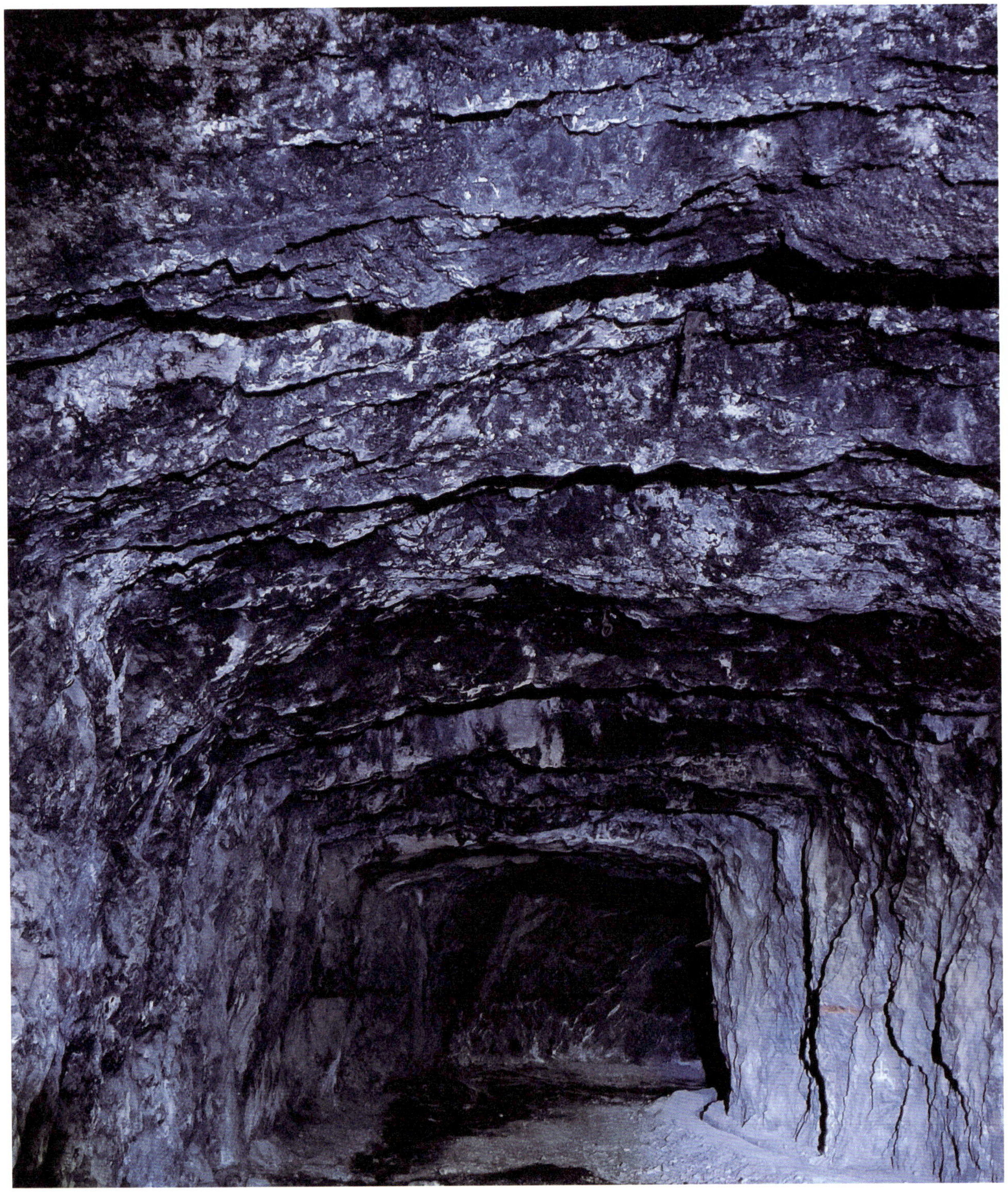

Fig. 21

in the mid-1990s, when I was the head of the shop steward council, which represented media workers of the national broadcaster—the Namibian Broadcasting Corporation (NBC)—on behalf of the Namibian Public Workers Union (NAPWU). Jauch was then the director of the Labour Resource and Research Institute (LaRRI), a research and education outfit helping our myriad unions affiliated to the union federation, the National Union of Namibian Workers (NUNW). During Jauch's reign, LaRRI assisted me, through the NUNW, to revive the union newspaper, *The Namibian Worker*, for which I was the Editor-in-Chief, but I had to resign two years later due to political reasons.

Testimony of Juuso Ndinolune David Kaluvi[1]

'When I first arrived in Tsumeb there was only a small residential area for whites in the east, where blacks were not allowed to roam, and in the west was a small native area with a few tiny two-room houses. One could not live wherever you wanted to in those days, and the only way to find employment south of the "Red Line"[2] was to be a contract labourer.

Up until the 1960s, steam-powered trains were still in use. Back then, we thought steam trains were the most senseless machines ever invented, as they were notorious for causing field fires and pumping out black smoke. Trains powered by diesel engines were only introduced in 1967. For new recruits, riding in the back of a cattle truck on the road from Ondangwa to Grootfontein was quite a new experience. First they would be driven to SWANLA's head office in Grootfontein, and from there they were sent to various workstations around the country, including Tsumeb.

From day one, we were told not to ask questions at work. The life of a contract labourer started in the compound, a place like no other for newcomers, particularly because many had never used electricity or seen neon lights before. Twelve miners shared steel bunk beds with no mats in a cramped room of around four square metres. At nine o'clock, the siren would sound, and all miners would be locked in the compound.[3]

In the 1960s, TCL employed around 800 manual workers. The compound was extended three times to accommodate around 3,000 to 4,000 workers at its peak, when the smelter was built from 1961 and started operations in 1964. At the time, pass laws[4] were still in force. Outside the compound, the municipality bar was the only place allowed to sell Tombo (fermented millet ale) to the natives and mine workers. Miners had to carry pass tickets to visit any other place around Tsumeb and anyone found without one would be arrested.

Underground mining was always hazardous and dangerous work. We were just numbers like the ones on our contract labour tags we were forced to wear on our arms, or around our necks. Many died underground from falling rocks due to negligence and poor communication. There was a saying among workers, "wealth at the backburner and eminent death in front", which basically meant you would never know each day whether you will return from work alive or not.[5]

Mine workers had to sign a disclaimer before entering the underground mine, effectively absolving the company from any responsibility in case of death. In case of injury, the mine hospital was always helpful to an extent. Although there were stipulated working hours, in reality, miners worked for 15 hours or longer a day, and some died of fatigue. We were working very hard but the salaries were inadequate, which today would be equivalent to six Namibian dollars a month, or one dollar fifty cents every week. Every Friday was payday.[6]

Labourers were contracted to work 15 months before they would be allowed to visit their families. There was no job security. When you went home after your contract had expired, you would have to wait for the person who replaced you to also complete his contract, otherwise you would be posted elsewhere. There was no communication regarding when you could return to work—you had to start all over again at Ondangwa. In most cases, the mining company would call labourers with TCL documents, and that's how you could return to your old job. Some never went back after their first time on contract, and those are the people who began to go into exile—for example, former President Hifikepunye Pohamba, and many others.

For us as contract workers, it didn't matter where you came from in the north. We protected each other fiercely. The ultimate goal was to save money, so much so, that friendship among mine workers often soured, especially with those that used their money to buy alcohol. Most had the ambition to work, return home, buy cattle and get married.

The wind of change started to blow in 1971 when TCL workers downed tools, bringing mining operations to a standstill, and only restarted in 1972.[7] Migrant workers demanded better working conditions, and thereafter some groups started to intensify their underground trade union activities. But the ultimate strike came years later in 1988,[8] which was the game changer, as we could now live wherever we wanted.

The miners of today are much better off than the miners of our generation. We salute those that started during the era of steam technology in Tsumeb. They were downtrodden, but didn't die in vain, as they contributed to the establishment of the mine in the first place. Without cheap labour, the mine would not have existed today. The mining company had the plan, but the contract workers implemented it.'

[1] The historical commentary in footnotes begins on p.80

Historical commentary in footnotes by Herbert Jauch

[1] The system of institutionalised labour recruiting developed from 1925 onwards under the Northern Labour Organisation (NLO) and the Southern Labour Organisation (SLO). The latter mostly supplied workers for De Beer's diamond industry in southern Namibia. In 1944, the NLO and SLO were merged into the South West Africa Native Labour Association (SWANLA). The NLO and SLO were owned by the major colonial capitalist enterprises, which made profits out of their operations. Therefore, they could run the recruitment agencies on a non-profit basis with the purpose of supplying a constant flow of cheap black labour. Black workers from Namibia's northern regions could not choose their place of work, but instead had to apply for work through these recruitment agencies, which would decide where workers were sent to. Workers were subjected to dehumanising medical examinations—they had to strip naked, their private parts examined, their cultural norms violated and their dignity disrespected (Ndeikwila 2019, Amupanda 2022). They were classified into categories according to which they were allocated work. Only those classified in the 'A' category could work in mines—the most sought after work due to the higher pay rate. In the 1940s, for example, the pay rate for workers in the 'A' category was 15 shillings [about €44 in 2024*] a month, compared to 10 shillings [about €30.60 in 2024] for category 'B' workers, and 8 shillings [about €23 in 2024] for category 'C' workers (Ndadi, 2009). Ndadi (2009) recalled how health and strength was all that mattered during the examination, and how workers were given a number and their classification, which had to be carried on a string around the neck. Those in the 'A' category were regarded as strong enough for jobs in the mining industry and the fishing industry. Those categorised 'B' were regarded as fit for lighter mine work, or factory work, as well as work in houses, hotels, restaurants and offices, while the 'C' category were labourers for farm work, such as milking cows, or herding flocks of sheep or goats (Kauluma, 1977).

* [Historical currency conversion processed with the 'Currency Converter: 1270–2017' of the National Archives, Richmond, UK and the 'Inflation Calculator' of the Bank of England based on 1940 shilling.]

[2] The 'Red Line' is Namibia's veterinary cordon fence, which was established as a stock disease control mechanism to protect Namibia's beef exporting industry from stock diseases, such as foot and mouth disease and lung sickness. Its name originated from the fact that it was simply a red line drawn on maps. Since its establishment in 1897, the 'Red Line' was controversial, because it did not only protect the cattle herds of German settlers from epidemics originating in the northern regions. Instead, it became a political boundary in colonial Namibia, with the area south of the 'Red Line' known as the 'Police Zone', which was under the direct control of the German colonial state. The area north of the 'Red Line' was ruled indirectly through appointed traditional authorities. Practically, the 'Red Line' separated Namibians in the north from white settlements and indigenous communities in central and southern Namibia.

Under South Africa's apartheid rule in Namibia, the 'Red Line' served not only as a veterinary control measure, but also as a barrier to restrict the movement of black Namibians south of the line. It still exists today and has become a living reminder of Namibia's colonial past. Its continued existence has been criticised, especially by livestock farmers in the northern regions, who are denied the opportunity to sell their meat south of the 'Red Line', and are also excluded from export opportunities. (Legal Assistance Centre, undated)

[3] Living conditions in the migrant workers' compound were horrific. Many workers slept in bunk beds in one room, and they had no choice regarding food—they had to accept whatever was available as they were not allowed to cook their own. The compounds had a company store that held a monopoly and gave workers little choice. They were not allowed to shop at the 'Whites only' store in town, and black domestic workers could only go there with a shopping list from their employer (Kauluma, 1977). The compounds were often so overcrowded, workers on different shift had to share the same room. A report from the International Labour Organisation (ILO) from 1977 described the conditions in Katutura, Windhoek:

> Five thousand Ovambos are being housed in circumstances that would disgrace a nineteenth century prison [...]. A visitor can only be appalled by the compound's unbelievable bleakness, the barbed wire fences; the concrete bunks in dark overcrowded rooms in which up to 20 men sleep; above all the overwhelming stench of urine which hangs over the compound. (Hishongwa, 1992, p.63)

Similar conditions existed in compounds elsewhere, including in Tsumeb. Workers had no privacy as their toilets and bathrooms were in large public buildings. Migrant workers' wives and children were not allowed to live with them.

[4] 'Pass laws' existed in apartheid South Africa and colonial Namibia. They required black people to carry a 'pass book' with them at all times, consisting of the person's name, portrait photo, place of origin and information on where the person was permitted to live. The movement of black people was tightly controlled, and they were arrested if they did not have the pass book on them, or if they were found in an area other than the one dictated by the pass book.

[5] The socially and economically destructive nature of contract labour on workers' lives was expressed clearly during an argument that Ndadi (2009, p.94) had with his father. He argued for the need to fight contract labour because:

> [...] if I'd been getting paid fairly, I'd be rich... we'd all be rich. [...] [My children will also suffer] if we don't get rich of [sic] contract labour. That's why we've got to fight. Not much difference if I suffer in prison or on contract.

[6] Contract workers experienced blatant discrimination and systemic levels of inequality at the workplace. Whites always had to have a job above blacks, even if they lacked the required skills and knowledge. Whites received exceptionally high wages and generous benefits, while black workers received wages below the household subsistence level. In cases of disputes, courts nearly always found the black workers guilty on the basis of the contracts, which gave them no right to negotiate, and no right to refuse employers' instructions. Workers' complaints about their conditions were ignored and often resulted in beatings. Workers had no recourse to fight such injustices, and experienced how the whole economic, social and political colonial system was stacked against them.

[7] The general strike of 1971–1972 was the most significant strike in colonial Namibia. The decades-old and unresolved grievances against the contract labour system led to the strike, which took place without workers being organised in any trade union. Strikes and other collective protest actions had occurred on many occasions before, and workers understood the scope and limitations of their bargaining power. They deliberately restricted their action to single demands. Systematic go-slows were chosen over open strike action, in order to protect workers from victimisation. The compounds became the main location for organising base shop-

floor campaigns (Moorsom, 1997; Ndadi, 2009). In the run-up to the general strike, the contract workers had forged alliances with student South West Africa People's Organisation (SWAPO) activists who played an essential role in the mass uprising. Some of these student activists had been expelled from high schools in August 1971 for demonstrating and some of them mobilised for a general strike against the contract labour system. Some found contract work in Walvis Bay and cooperated with worker activists and SWAPO branches. Mass meetings were held from November 1971 onwards and the strike was a calculated political campaign to which the state responded with severe repression. Posters on walls, mouth-to-mouth communication and the radio were used as means of communication. Various towns and workplaces confirmed their support of the strike which was a great success. Over 20,000 workers—the majority of migrant workers at the time—participated in the strike action in 23 workplace centres: 11 of them mines, including the TCL mine in Tsumeb. Activists set up a strike committee that set out the workers' demands, which were presented to a mass meeting of 5,500 strikers. They were approved unanimously and centred on the replacement of the South West Africa Native Labour Association by a free labour market, and an end to influx control (Moorsom, 1997). Employers and the colonial state responded with violence to the strike. By mid-January 1972, 13,500 striking contract workers had been transported to 'Ovamboland' by rail, and soon the struggle against contract labour broadened into a general resistance against colonial rule there. In the second half of January, peasant grievances were raised at mass meetings, government transport was stoned off the road, and over 100 kilometres of border fence with Angola was destroyed. These events signified the political nature of the strike, as the strikers' demands confronted the cornerstone of the colonial dispensation, namely cheap contract labour, pass laws and influx control (Moorsom, 1997).

[8] In 1988, colonial Namibia experienced political protest actions when students embarked on a class boycott, beginning in early March. Students under the umbrella of the Namibia National Students Organisation (NANSO) demanded the removal of South African military bases from school vicinities. On 17 March 1988, more than 700 students from the Ponhofi Secondary School in the North of Namibia embarked on a class boycott, after numerous meetings with parents and authorities where the removal of army bases from school vicinities was demanded. By March and April 1988, more schools in the North joined the class boycott. The Kassinga Commemoration on 4 May turned into one of the largest protest marches that the capital Windhoek had ever witnessed, when all schools in Katutura shut down and joined the procession through Katutura (the black township of Windhoek into which black people were forcefully removed in 1959). The protest march of 1988 was violently broken up by the police once the students tried to leave Katutura and proceed towards Windhoek's city centre. Several secondary schools declared a permanent school boycott until military bases were removed from the vicinity of schools, as well as military personnel withdrawn. Students also demanded an end to the militaristic 'kadet' system at schools. The state responded with increasing police brutality, arresting more than 45 students, alongside trade unionists and SWAPO branch leaders, in early June 1988. On 11 June 1988, more than 500 shop stewards of the newly formed trade unions, the Namibia Food and Allied Workers Union (NAFAU); the Namibia Public Workers Union (NAPWU); and the Metal and Allied Namibia Workers Union (MANWU) met at their trade union offices in the Katutura Community Centre to discuss the ongoing student uprising and possible support action by the workers. Country-wide consultations with workers were held, and their key demands were formulated: remove all military bases near schools, release the detained activists, and end the violence of the colonial police and Koevoet forces. Koevoet was the counterinsurgency branch of the South West African Police (SWAPOL), and was widely feared for its brutality meted out against those resisting apartheid and colonial rule.The colonial regime ignored these demands. As a result, trade unions and their members decided to embark on a national stay-away on 20 and 21 June. Despite heavy repression and ongoing detention of union leaders and political activists, the stay-away was successful and brought many workplaces to a standstill. About 80% of the workers in Windhoek, and almost all workers at Rossing Uranium (near the town of Arandis) and the Consolidated Diamond Mines (CDM) in Oranjemund, as well as the majority of workers in Tsumeb, followed the call from the country's oldest and largest trade union federation, the National Union of Namibian Workers (NUNW), to stay away from work. Most businesses and industries throughout Namibia were either forced to close their doors, or maintain production at a limited scale. The mining companies lost millions of US dollars as a result of the stay-away (LaRRI 2010).

References

Amupanda, J. S., 2022. *Diamond Warrior in Colonial Namibia: Diamond Smuggling, Migrant Workers and Development in Owamboland.* Basel: Basler Afrika Bibliographien.

Hishongwa, N., 1992. *The Contract Labour System and Its Effects On Family and Social Life in Namibia: A Historical Perspective.* Windhoek: Gamsberg Macmillan.

Kauluma, J. B., 1977. *The Migrant Contract Labour Situation and the Response to it. [manuscript] Archives of Anti-Colonial Resistance and the Liberation Struggle.* Windhoek: National Archives.

Labour Resource and Research Institute (LaRRI), 2010. *Strikes in Namibia: A Long History of Struggle.* Windhoek: Labour Resource and Research Institute.

Legal Assistance Centre, (n.d.). The Infamous "Red Line': Why Does Namibia Have a Veterinary Cordon Fence?. Windhoek: Legal Assistance Centre. [online] Available at: <https://www.lac.org.na/news/probono/ProBono_64-RED_LINE.pdf> [Accessed 30 March 2024].

Moorsom, R., 1997. Underdevelopment and Labour Migration: The Contract Labour System in Namibia. Working Paper No 10, 1997. Bergen, Norway: Chr. Michelsen Institute. [online] Available at: <https://open.cmi.no/cmi-xmlui/bitstream/handle/11250/2435915/WP%201997_10%20Richard%20 Moorsom-07112007_1.pdf?sequence=2&isAllowed=y> [Accessed 30 March 2024].

Ndadi, H. V., 2009. *Breaking Contract: The Story of Helao Vinnia Ndadi.* [manuscript] Archives of Anti-Colonial Resistance and the Liberation Struggle. Windhoek: National Archives.

Ndeikwila, S., 2019. *The Agony of the Truth: Autobiography of Samson Ndeikwila.* Windhoek: Kuiseb Publishers.

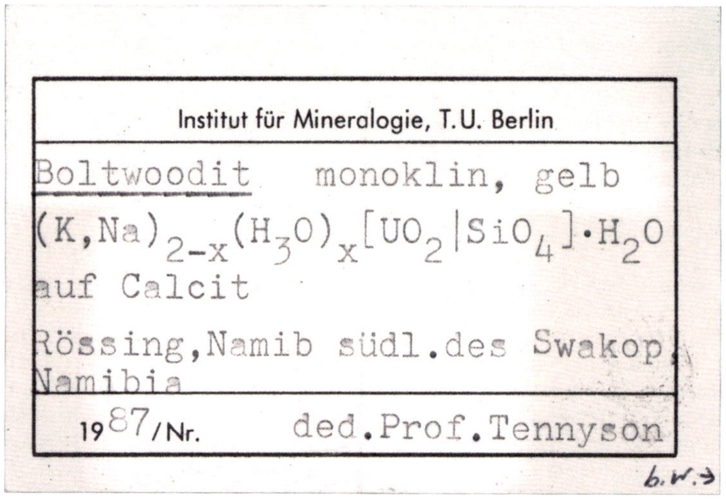

Institut für Mineralogie, T.U. Berlin

Boltwoodit monoklin, gelb

$(K,Na)_{2-x}(H_3O)_x[UO_2|SiO_4]\cdot H_2O$

auf Calcit

Rössing,Namib südl.des Swakop,
Namibia

1987/Nr. ded.Prof.Tennyson

b.w. →

Out of Space and Place: Tales of Namibian Uranium through Mineral Objects

Christopher R. Hill

Abstract:This chapter explores the significance of African uranium for visions of land usage and statehood in the history of Namibia. It suggests that the cosmic, radioactive origins of uranium imbued the metal with the promise of magical properties among the land's colonisers and traditional custodians. These inhabitants ranged from Herero, who associated uranium-rich lands with infertility, to German and Afrikaner settlers, who saw in uranium the power to transcend the hardship of the desert. The trajectory of these 'geo-cosmologies' was heavily shaped by German colonialism, British capital, and South African apartheid, leading to the rise of Namibia as a uranium-producing state. In this chapter, I unpack stories about radioactive earth through three mineral objects, including two uranium-bearing minerals stored in Berlin, and a uranium cube in Bonn, originally fabricated by the Nazis.

Keywords: Namibia, Germany, Britain, South Africa, mining, uranium, mineralogy, museums, Rio Tinto, Rössing Uranium Limited

The geology of Namibia was a source of wonder to both the land's colonisers and its traditional custodians. Their respective visions of the earth fed into the geopolitics of southwestern Africa: the region from which the Namibian state was formed. In this region, geological materialism went hand-in-hand with the sublime. African and colonial practices involving minerals, metals, and stones were linked to ideological, religious, and spiritual systems. For European settlers as much as African inhabitants, these earth relations were cosmological as much as geological: the subterranean helped explain one's place in the universe (Bashford, Kern and Bobbette, 2024). In this relationship between the geological and the sublime—between mundane matter and spectacular phenomena—the cosmic origins of certain metals, minerals, and stones took on particular importance. Just as Nama communities forged tools out of iron retrieved from meteorite showers, European prospectors also pondered whether the Namib Desert's diamond deposits could have originated as fragments from fallen stars (Cloos, 1954; Buchwald, 1975). In southwestern Africa, the underground and the transcendental shared a special relationship, where the earth proved fundamental to speculations about the environment, society, and progress.

In this essay, I suggest that uranium, the heavy metal needed to produce nuclear energy and weapons, was one of the 'out of space' elements around which African and European 'geo-cosmologies' materialised. The radioactive ores of central Namibia marked a key terrain around which inhabitants and settlers envisioned futures for themselves and the land. After describing the cosmic origins of uranium—and reflecting on the implications of its radioactivity for perceptions of place—I draw on three mineral objects to reveal the role of uranium in African and European worldviews of Namibian land. The first and second objects, uranium-bearing rocks kept by the Mineralogical Collections for the Technical University of Berlin, were collected in the 1980s from Rössing, which upon opening in 1976 was the world's largest uranium mine, located in the Namib Desert. The final object, displayed at the Mineralogical Museum of the University of Bonn, is one of fourteen surviving 'Heisenberg Cubes': 664 blocks of uranium fabricated for the Nazi nuclear weapons programme (Zacke, Friedeberg and Schumacher, 2018). While the cube's uranium originated from Jáchymov in the Ore Mountains of Czechoslovakia, it was during this period that strategic interests began to intensify around Namibian uranium: not only among British and South African imperialists, but also among German and Afrikaner settlers in Namibia. By focusing on mineral objects from German universities, I risk inflating the German influence over uranium mining in Namibia: an international project led initially by Britain, South Africa, and more recently, China (DeBoom, 2018). The Rössing mine only became operational in 1976, over 60 years after the end of German colonisation in southwestern Africa in 1915, when Union Defence Forces claimed the territory for South Africa, the occupying power until Namibian independence in March 1990. The company established to run the mine, Rössing Uranium Limited, was also controlled by South Africa and Rio Tinto Zinc (RTZ), a British mining house with a vast portfolio of mineral assets and subsidiary organisations around the world. West German and German settler support was nonetheless instrumental. In a letter explaining why West Germany had been given a preferential rate on Namibian uranium, Roy Wright, Deputy Chairman of RTZ, described how the state-supported Urangesellschaft Mbh & Co had negotiated a package deal 'covering exploration, feasibility studies, financing and supply of natural uranium'. He added that 'no other country or customer in the world would have contemplated it [...] bearing in mind all the circumstances of Rössing' (Wright, 1973). Shortly after this letter, the UN

Fig. 22

Council for Namibia passed Decree no. 1 for the Protection of Natural Resources, which sought to preserve the country's mineral wealth for Namibians. It ruled that 'no person or entity' can prospect for and extract natural resources within Namibian territory (Schrijver, 1995, p.136).The object of study here, however, is uranium mineralogy rather than the industrial mining of uranium. While two of this essay's mineral objects were procured with the support of the Rössing mine in the 1980s, the search for uranium minerals far precedes the history of the mine, stretching back even before the Afrikaner settler, Petrus (Peter) Schabort Louw, stumbled across an intriguing specimen of uraninite in 1928. If it is during exploration that 'future extractive projects take shape in the political imagination and modes of storytelling'—as Robyn D'Avignon (2022) has contended—then it is necessary to reach back before the Rössing mine to trace its founding myths and narratives. In the colonial search for pure or rare uranium minerals, it is possible to discern an ideological as well as scientific motivation: such minerals were prototypes not only of chemical classifications and compositions, but also of ideas about knowledge and power. That such minerals were sent to European laboratories and museums is revealing for the process by which colonisers came to comprehend and re-position the role of African earth within economic, ideological, and scientific systems. Such minerals were and are microcosmic, their distilled, exemplary structures serving as an index of abstracted knowledge about African earth and the riches it holds. How does a mineralogical perspective on Namibian uranium reshape its history? On the one hand, it enables uranium to be interpreted in a longer tradition of earth relations, where uranium serves as a successor, perhaps, to other Namibian resources that were seen as having powerful properties for African or European geopolitics. It is easier to grasp these relations and the futures they entailed before the emergence of Rössing: before economic and technological rationalisation laid down a template of what uranium could signify for Namibians. On the other hand, this perspective can marginalise the operational history of Rössing. While it is not possible to do justice to the immense and multifaceted role of RTZ, it could be argued that the foundations of the multinational's work had already been laid. Rössing Uranium Limited, for example, based its history on Peter Louw's uraninite discovery. The land on which the mine was constructed had been stolen from Herero (Drechsler, 1996). The mine was linked to a German-built rail line that was used to prosecute a genocidal war against these people (Bravenboer and Rusch, 1997). By the time Rössing entered operation, a robust system around land and minerals had been put into practice, its cultural and legal principles surviving well into Namibian independence.

Space and Place

Uranium is created in the aftermath of a supernova, a massive stellar explosion triggered by the collapse of a star's core. Before a supernova, the life of a star is shaped by a series of nuclear reactions, beginning with the fusion of hydrogen and then passing through a sequence in which heavier elements are fused, such as helium, carbon and oxygen. The sequence ends in iron, because the fusion of iron consumes more energy than it releases, placing it at the top of the nuclei binding energy curve. This explains why elements with an atomic number greater than iron are described as 'heavy': their formation requires conditions of heat and pressure beyond those found in the rising temperature of a star. These conditions are provided by a star's death, which occurs due to unsustainable pressure around its iron core. The ensuing supernova enables rapid nuclear reactions, allowing atoms to attain mass until they reach uranium, the heaviest known element. This element, launched into space by the supernova, carries the signature of its violent birth; its unstable structure shedding energy through radioactive decay. In space, it forms one of the primordial elements through which celestial bodies are made, including Earth 4.55 billion years ago (Comins, 2018, pp.462–489).

The cosmic origins of uranium provide a useful backdrop from which to think about human-land relations in uranium-rich environments. Across a range of social as well as scientific traditions, these environments have been regarded by inhabitants and settlers as pregnant with possibilities. Uranium heralded the future, its latent energy inspiring rival visions of destruction and abundance. Perhaps because of uranium's cosmic and radioactive secrets, it has also seemed out of place: an alien intruder uninitiated by geological mechanisms on Earth. From this perspective, it could be argued that uranium lands possess an 'uncanniness', a term theorised by Joseph Masco to refer to 'the material effects, psychic tension, and sensory confusion produced by [...] radioactive materials' (2006, p.21). This uncanniness can be discerned in what D'Avignon has called 'ritual geologies': 'practices, prohibitions, and cosmological engagements with the earth that are widely shared and cultivated across a regional geological formation' (2022, p.17). In Anishinaabe territory in Canada, for example, Lianne C. Leddy has highlighted how uranium was regarded by First Nation Elders with trepidation, its orebodies having been compared to the eggs of a dormant serpent, never to be awoken or disturbed (Leddy, 2021, p.3).

Similar rituals must have prevailed in the landscape featured in this essay: the uranium corridor that intersects with the middle section of the Namib Desert in Namibia. Jack Boulton has shown how uranium is often likened to a 'monster' among Oshiwambo speakers, many of whose livelihoods have become entangled with the uranium industry. In the process of its unearthing, this monster devours black bodies through labour and the ingestion of dust particles; it devours natural environments through the re-routing of natural resources and toxic modes of ore extraction (2021a; 2021b, p.49).[1] For the Afrikaners and Europeans who began to settle in Namibia

[1] Boulton refers to the following phrases: *eemina oda ndituli*, (Oshiwambo), *mughodhi urikutidya* (Shona), or *libulu aliyaka nzoto na ngai* (Lingala), all of which translate as *the mines are eating us* (Boulton, 2021b, p.49).

Fig. 23

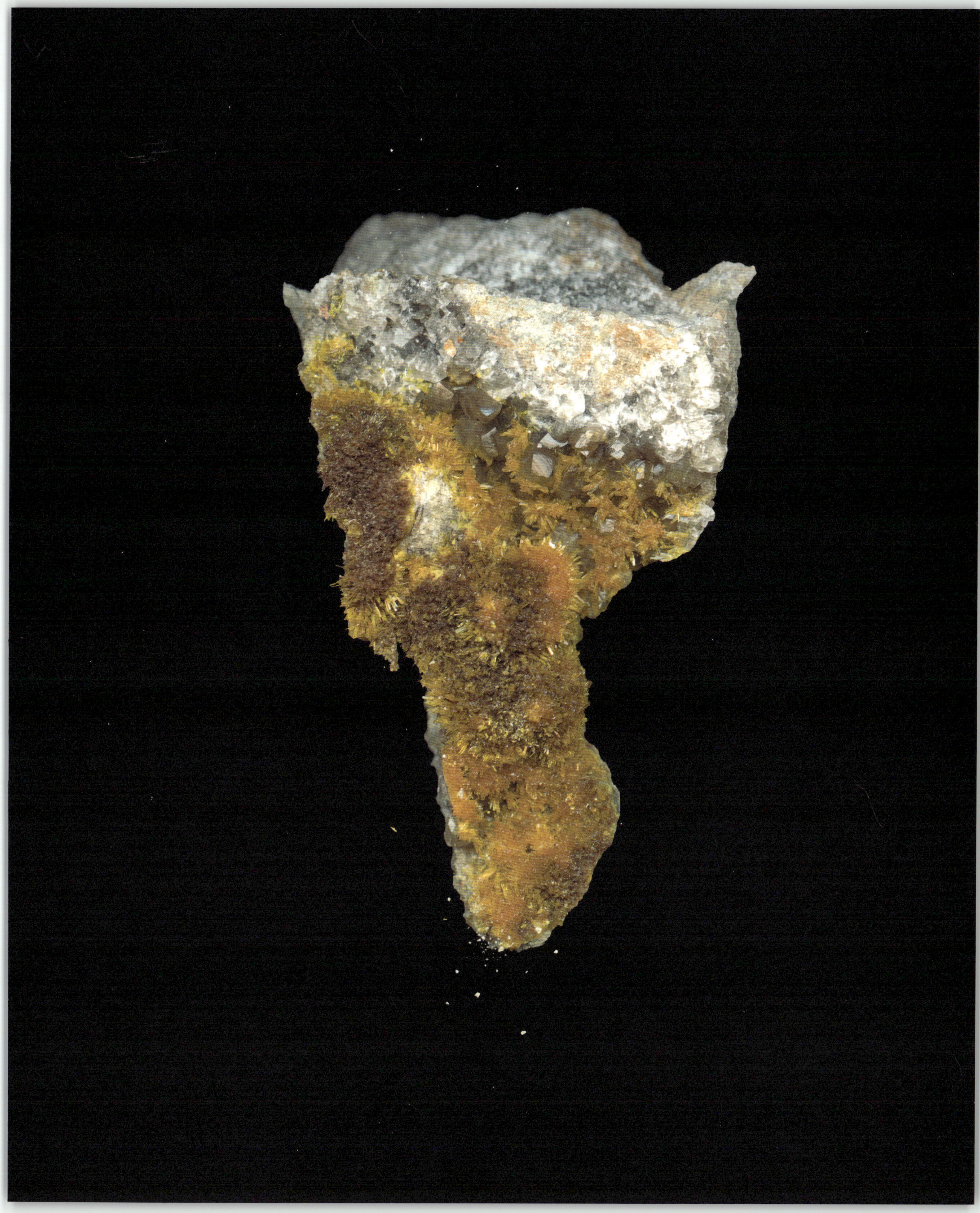

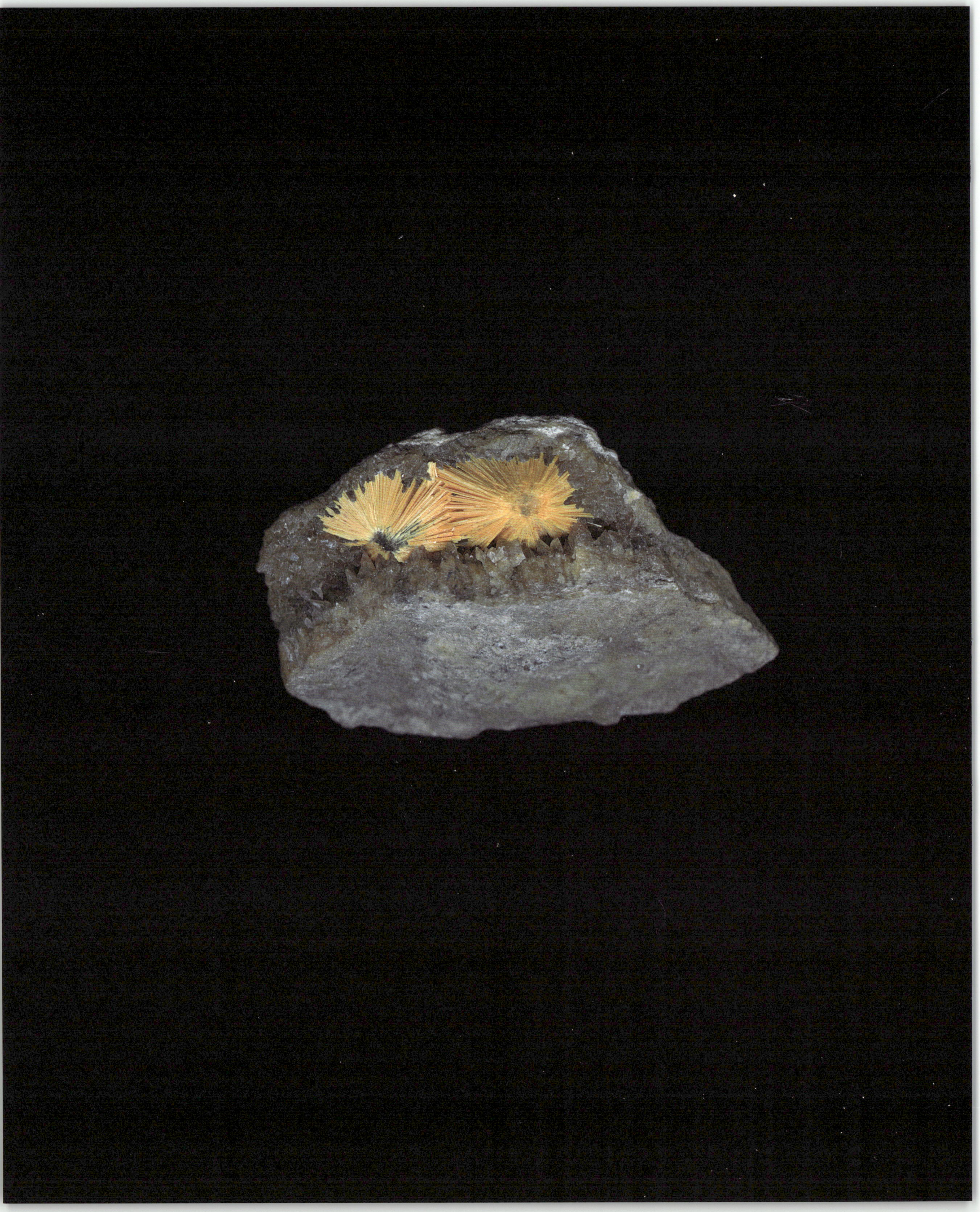

[2] Rio Tinto Zinc, which had its roots in the Rio Tinto Company founded in 1873, was renamed the Rio Tinto Group in 1995.

Fig. 24

from the mid-nineteenth century, however, this uranium came to promise a far brighter future: a portal to untold riches and technologies that could press the 'desert wilderness' into the service of western civilisation. (Louw, 2018).

The role of uranium in storytelling about future environments preceded a full understanding of its status as the only element in nature that can induce a nuclear reaction. The implications of this discovery, made by Otto Frisch and Lise Meitner in December 1938, were momentous, with uranium searches after 1945 being described as involving 'more man-hours' than have 'been spent in seeking all other metals in history' (Heinrich, 1958, p.vii). Yet interactions with radioactive orebodies were informed by tacit knowledge long before the emergence of sub-atomic physics in the 1890s (Parr, 2010; Shiga, 2019). This knowledge also preceded the identification of uranium by the German chemist, Martin Klaproth, who in 1789 isolated the element from pitchblende or uraninite. Known as 'Pechblende' in German, Klaproth had procured samples of this mineral from the world's oldest uranium mine: Jáchymov—formerly a silver mining operation known as St. Joachimsthal. The word 'Pech' came to be interpreted as 'bad luck', primarily because of the absence of silver, but also, perhaps, because of the unfortunate fate of the miners, who often succumbed to 'Joachimsthal miner's disease' (Veselovský et al, 1997; Wesch et al, 1999). The experience of living in and making sense of uranium-rich environments was as old as human history itself. Long ago, when natural uranium contained a higher proportion of its most unstable isotope, U-235, these environments could even become nuclear. The Oklo deposits of Gabon, for example, once functioned as natural nuclear reactors (Rigali, 2018).

Namibia's uranium, estimated at five percent of world supply, was also from what might be classed as an ancient, radioactive environment, even if radiation from its low-grade ores was extremely mild next to places such as Jáchymov and Shinkowoble, the Congolese mine that provided most of the uranium for the Manhattan Project (World Nuclear Association, 2023). The majority of Namibia's uranium stems from a transitional zone of the Namib Desert, extending north-west from the German-founded coastal town of Swakopmund and following the central belt of the Damara orogen: an inter-tectonic structure protruding inland towards the Congo. Johann W. von Backström, a geologist who surveyed the site of Namibia's first uranium mine, Rössing, described the zone's radioactivity in a 1970 report for Rio Tinto Zinc (RTZ), the British multinational that established and oversaw the Rössing mine with Apartheid South Africa.[2] The geologist noted how these ores were affected by the cold, offshore Benguela current, which pushes northwards from the Antarctic and collides with the heat of the desert. This results in a heavy dew that coats the desert's rocks and percolates downwards to uranium-bearing coarse-grained rocks, known as pegmatites, where groundwater containing natural solvents 'leached' the uranium from its host (Bäckstrom, and Jacob, 1970). In this way, the uranium and mild radiation it emits is part of the desert eco-system: a geophysical system stretching back into the mists of history, shaping situational knowledge and visions of Namibian uranium among both African and settler communities.

Mineral Tales

The two uranium-bearing minerals pictured in Figures 23 and 24, known as uranophane and boltwoodite, were taken from Rössing in the mid-1980s. To see the specimens in their display cases, labelled in the neutral terms of mineralogical science and with reference to their chemical composition, an onlooker would be forgiven for thinking that the minerals were of no more than specialist interest: the objects are rendered objective and stripped of their problematic histories. Yet, as with other specimens from Rössing in mineralogical collections around the world, these objects, locked up safely in their taxonomical trays, were wrenched from their spaces and places. In a similar manner to the colonial artefacts of western museums, the specimens were requisitioned under a foreign system of coercion and violence, as well as against a raft of international rulings. The legal status of the objects was dubious, since permission for their removal was granted by Rössing Uranium Limited, which itself was regarded as contravening Decree No. 1. The uranophane and boltwoodite are better referred to, in fact, as mineral subjects rather than objects: like the black labour that fed into and made possible their eventual extraction, the specimens were put in service to an outside regime of industry and science. Their role as objects of knowledge, ostensibly dissociated from German colonialism and apartheid occupation, was in fact rooted in this violent past, the promise of mineral secrets and wealth forming a wellspring of colonial desire.

The histories of the two uranium minerals are illustrated by their initial unearthing, their pretence to scientific isolation having been ca*ught up in a coloni*al web of geological knowledge and extractive power. Both the uranophane and the boltwoodite were collected either directly by or under the auspices of Karl Hugo Strunz (1910–2006), Professor at the Institute for Mineralogy at the Technical University of Berlin and mastermind of the Strunz system, a classification system developed with Professor Christel Tennyson (1925–2010), whose ingenuity and labour have gone uncredited (Mikloweit, 2025). Strunz's pre-eminence as a mineralogist was shaped by work he carried out at Tsumeb, the site of Germany's most important colonial-era mine and named 'the world's greatest mineral locality' by the Mineralogical Record (Wilson and Pinch, 1977; Hearth, 2021). Just as colonial operations at Tsumeb enabled Strunz to unearth new minerals such as söhngeite, so too did the onset of the Rössing mine enable progress in the field of uranium mineralogy. While the uranophane specimens at Berlin are unattributed—and may have been procured by Strunz himself—the boltwoodite is credited to Christel Tennyson who donated the lone specimen to the Strunz collection in 1987. By re-interpreting these minerals in relation to their geological and human histories, it is possible to counteract the neutralising effects of the Strunz classification system: a system that Tennyson devoted her life to maintaining.

The relationship between mineralogical knowledge and extractive power in Namibia was also highlighted by Johann W. von Backström, who pioneered theories of uranium mineralisation in the Namib as a freelancer for Rössing Uranium Limited (RUL): the company established by RTZ to manage the Rössing mine. Bäckstrom was able to explain minerals such as uranophane and boltwoodite within the geology of the Namib. As he relayed to RUL, both were 'secondary minerals' in that they stemmed from uranium-rich rocks. These richer rocks had supposedly been weathered in the deep-time of the desert, the Namib's ground waters 'reconcentrating' the uranium 'under suitable conditions in favourable receptors' (Bäckstrom, and Jacob, 1970). The result at Rössing was a deposit in which as much of forty percent of the uranium came from secondary minerals, with uranium making up only 0.04 percent of the ore (Bäckstrom and Jacob, 1979). As Rössing prepared to open and the forecasts for nuclear power seemed highly optimistic after the oil crisis of 1973, the mine's low-grade uranium was described by the International Atomic Energy Agency (IAEA) as representing a 'deposit of the future'. In the words of an IAEA geologist, it appeared that 'nuclear fuels will largely replace fossil fuels', causing 'a large demand for uranium'. Due to this demand, it seemed that deposits like Rössing would 'become [increasingly] economically exploitable' (Armstrong, 1974).

By the time that IAEA referred to Rössing as a 'deposit of the future', the Namib's uranium had already become harnessed to white settler visions of an energy and security revolution. In large part, these visions emanated from the pre-history of the Rössing mine, which was based on deposits surveyed and pegged by Peter Louw, an Afrikaner who settled in Swakopmund after serving for the Union Defence Force in South Africa's victory over Germany in July 1915. For Peter and his three sons, Graham, John and Alan, the unfolding possibilities of uranium likely coalesced with the opening up of Namibia, or South West Africa (SWA) as it was then known, as a frontier ripe for development and investment. Peter himself secured rights for a range of mineral prospects that have been regarded as somehow miraculous or transformative, from the guano deposits of seabirds on the Namib coast to the diamonds scattered across the desert. Since the Louws' accounts of Rössing were all written after the deposit was developed by RTZ, they tend to have a romantic and triumphalist tone, beginning with a sentimental note of Peter's in 1965 (Louw, 1965). This triumphalism, born out of the victory of a settler family over the hardship of the desert, still informs the dominant narrative of uranium history in Namibia, as reflected in the publications of RTZ and more recently the China National Uranium Corporation (Rössing Uranium Limited, 2022).

The significance of this settler story has obscured deeper, more complex roots of uranium development and mineral knowledge in the Namib: roots that pre-date Peter Louw's discovery of pitchblende near Rössing in 1928. For example, the role of German geologists in the identification of Namibian uranium has been largely sidelined. Ernst Reuning, who was employed by the German Colonial Company of South West Africa,3 was particularly influential in revealing uranium mineralisation around the granite intrusions that formed the Rössing deposit, carrying out field work on the site from 1910 (Reuning, 1911; Wagner, 1921). Particularly after Germany's defeat in 1915 and the subsequent publication of the British and South African 'Blue Book' about German colonial atrocities, it was perhaps more convenient for the uranium industry to downplay the German contribution, even though German capital, expertise and political networks were pivotal to the Rössing mine (Hill and Ashipala, 2024). Since uranium mining emerged as the largest contributor to Namibian GDP (Gross Domestic Product), the industry sought to develop a more consensual narrative around uranium, free from the taint of German colonialism and Nazism (Jones, 1986).

As western physicists began to realise the strategic importance of uranium from the late interwar period, however, the extent of uranium prospecting among German 'Southwesters', documented by the Inspector of Mining Works, seems substantial. In November 1945, the South Africans even intercepted a telegraph message about the power of uranium between a German husband and wife who had settled in SWA, known to the administration as 'ardent Nazis'. 'They talk so much about Uran [sic] having such a destructive effect', the husband, Otto Irle confided to his wife: 'We have enough of this stuff in South West, not only Uran, but also Kurnotit [carnotite]. Of the latter I shall bring you bags full when I have been in that district again'.4 In a passage where Irle refers to 'his carnotite' as the 'Gute Hoffnung' ('good hope'), he goes on to stress hopes that their 'children ... shall remain in touch with the ideals of their fathers ... at least ideologically' (National Archives of Namibia, 1945). In this regard, a fleeting German vision of the future, linked to the success of the Nazi cause, had also been invested in the development of Namibian uranium. [Figure 25].

The importance of the Louw family to the 'discovery' and projected vision of Namibian uranium has obscured the role of black African as well as German history. The only traces of that history prior to the Rössing mine can be found in fleeting references to African labour, beginning in articles about Reuning's field research from 1910 (Wagner, 1921). In fact, the only direct reference to the role of Africans in the discovery process comes from Henno Martin, the celebrated geologist who defied the Nazis in the Second World War by taking shelter in the Namib, where he and his compatriot,

[3] The Deutsche Kolonialgesellschaft für Südwest-Afrika (DKGSWA)

[4] Carnonite 'is a radioactive, bright-yellow, soft and earthy vanadium mineral that is an important source of uranium' (Britannica, 2006).

Fig. 25

Hermann Korn, persevered with their geological research (Martin, 1957). In his capacity as Director of the SWA Geological Survey from 1947, it was Martin who surveyed claims pegged by the Louw family as they sought to locate the site of their discovery in 1928. As Martin described, 'it was learned that [...] the claims [...] [were] the result of samples brought to [...] [Peter] by coloured folk who eke out a precarious living on the banks of the Swakop river in the neighbourhood of Salem', the temporary location of a Rhenish mission in the 1860s.[5] 'Only two of the five claims could be found, despite the fact that the coloured men, who made the original discovery, and Louw's son [...] searched for several days' (National Archives of Namibia, 1949).

From this snippet of text, the reader gains an insight into a historical reality that receives not a single mention in the published literature: the significance of African labour and knowledge for the discovery of Namibian uranium. Yet who were these 'coloured folk' of Salem, and how did they possess this knowledge of land and earth? Perhaps they were ǂAonin or 'Topnaar': 'people of the point' or 'furthest margin', as the name translates, who tended to travel across the Namib through the dry beds of the Khan, Kuiseb, and Swakop rivers, foraging and gaining an intimate knowledge of the environment as they did so (National Archives of Namibia, no date; Vigne, 2000). To reassemble their land relations and mineral cosmologies would require long-term archaeological and ethnographic work against a history of land alienation. Yet such cosmologies have been resurrected by research into the mining, metallurgical and trading practices among other African communities in Namibia. Hiltunen 1993; Hearth, 2021; Kinahan, 2022). It has been suggested that nineteenth-century Herero also associated the uranium corridor with infertility, which seems all the more striking given that Peter Louw's first wife, a former radiographer from London named Margery, was intrigued by the possibilities that Namib minerals might offer for western medicine (West, 1972; Louw, 2018). These uranium-inspired visions stemmed from unequal land and earth relations, where the repercussions of large-scale mining engendered conflicting experiences and outcomes for African and settler communities.

The modern development of Namibia can be interpreted in relation to the fantastical qualities that African and settler communities assigned to uranium in the period of the metal's scientific discovery. The astronomical, 'out of space' origins of uranium interacted with competing perceptions of its 'place' in environmental, geopolitical, scientific and museological systems. Where uranium tended to forebode futures of environmental ruin in indigenous traditions, the resource's 'stored potentialities' also presented a medium through which settlers visualised future progress and wealth (Simpson, 2019). Yet the right to divine the properties of this celestial metal—and by extension to shape the future of Namibia—became a matter of who controlled the land and earth: a matter of geopolitics. As in a well-worn tale of science fiction, it was the alien colonisers who were able to enforce this right to divine, the power of uranium being deemed far too valuable to be left locked up in indigenous lands (Vint, 2021). The journey of uranium thus mirrored that of other 'cosmic' objects from Namibia, such as the Gibeon meteorites, which despite having been used by Nama for tools and weapons, were sent back to London for testing by the British explorer, James E. Alexander, in 1836 (Buchwald, 1975). The cosmic secrets of Namibia's minerals were rendered a colonial preserve, their destiny to be extracted and transferred to western factories, laboratories and museums, often regardless of the repercussions for the environments where they were unearthed.

The speculative possibilities of Namibia's future were influenced and limited by human interactions with its rich orebodies: uranium orebodies in particular. In a similar manner to the looting of artefacts, the theft of Namibian minerals had a debilitating effect on African cultures and land relations. It severed links with the earth that were vital to environmental knowledge and a sense of the future. For the dispossessed, the mineral treasures of the Namib were reduced to a wasteland. When Namibia gained independence in March 1990, the new government, led by the South West Africa People's Organisation (SWAPO), stood as the inheritor of a settler future structured around uranium. The manifestation of that future in the state of Namibia's economy, infrastructure and nationalism has since set the parameters of political action. Yet a Namibian future beyond or without uranium—or a Namibia realised through other earth and mineral relations—remains alive, even against a century of land alienation. In the uranium corridor, around sixty kilometres from the Rössing mine, a community of artisanal miners still 'eke out a precarious living' by selling mineral specimens at the Ûiba Ôas Crystals Market. Their knowledge of the land and struggle to survive serves as a reminder that Namibia's future has never been settled, no matter how closed and shackled it may have seemed.

[5] The Rhenish Missionary Society (Der Rheinischen Missionsgesellschaft) began operating in SWA in 1829.

References

Armstrong, F. C., 1974. Uranium Resources of the Future—“Porphyry” Uranium Deposits. In: International Atomic Energy Agency (IAEA). *Formation of Uranium Ore Deposits*. Vienna: IAEA.

Backström, J. W. von and Jacob, R. E., 1970. *The Rössing Uranium Deposit Near Swakopmund, South West Africa*. A Preliminary Report, AB 48/1278, U3O8 Agreement between UKAEA and Rössing Uranium Limited of South Africa, The National Archives, UK.

____ 1979. Uranium in South Africa and South West Africa. *Phil. Trans. R. Soc. Lond.* A. 291, pp.307–319.

Bashford, A. Kern, E. M., and Bobbette, A., eds., 2024. *New Earth Histories: Geo-Cosmologies and the Making of the Modern World*. Chicago. University of Chicago Press.

Boulton, J., 2021a. *Have Your Yellowcake and Eat It: Men, Relatedness and Intimacy in Swakopmund*. Basel: Basler Afrika Bibliographien.

____2021b. Uranium as Monster in Swakopmund. In: Caterina Nirta and Andrea Pavoni, eds., 2021. *Monstrous Ontologies: Politics, Ethics, Materiality*. Wilmington: Vernon Press.

Bravenboer, B. and Rusch, W. K. E., 1997. *The First One Hundred Years of State Railways in Namibia*. Windhoek. TransNamib Museum.

Britannica. 2006. *Carnotite*. [online] Accessible at: <https://www.britannica.com/science/carnotite> [Accessed 13 December 2023].

Buchwald, V. F., 1975. *Handbook of Iron Meteorites: Their History, Distribution, Composition and Structure*. Berkeley: University of California Press.

Cloos, H., 1954. *Conversation with the Earth*. Translated by E. B. Garside. London: Routledge & Kegan Paul.

Comins, N. F., 2018. *Discovering the universe*, 11th ed. New York: W.H. Freeman.

D'Avignon, R., 2022. *A Ritual Geology: Gold and Subterranean Knowledge in Savanna West Africa*. Durham, North Carolina: Duke University Press.

DeBoom, M., 2018. Nuclear (Geo)Political Ecologies: A Hybrid Geography of Chinese Investment in Namibia's Uranium Sector. *Journal of Current Chinese Affairs*, 46(3), pp.53–83.

Drechsler, H., 1996. *Südwestafrika unter deutscher Kolonialherrschaft: Die großen Land- und Minengesellschaften (1885–1914)*. Stuttgart: Franz Steiner Verlag.

Hearth, S., 2021. The “World's Greatest Mineral Locality”: Hai‖om, Ndonga, Herero, and the Early Colonial Histories of Tsumeb, Namibia. *Earth Sciences History*, 40(2), pp.433–460.

Heinrich, E. W., 1958. *Mineralogy and Geology of Radioactive Raw Materials*. New York: McGraw-Hill Book Company, Inc.

Hill, C. R. and Ashipala, S. N., 2024. “Follow the Yellowcake Road”: Historical Geographies of Namibian Uranium from the Rössing Mine. *Historical Social Research*, 49(1), pp.32–54.

Hiltunen, M., 1993. *Good Magic in Ovambo*. Helsinki: The Finnish Anthropological Society.

Jones, 1986. *The Role of RUL in Income Generation in the South West African/Namibian Economy*. The National Archives, UK. FCO105/2254.

Kinahan, J., 2022. *Namib: The Archaeology of an African Desert*. Melton, Suffolk: James Currey.

Leddy, L. C., 2021. *Serpent River Resurgence: Confronting Uranium Mining at Elliot Lake*. Toronto: University of Toronto Press.

Louw, G., 2018. *A Tiger by the Tail: The Story of the Discovery of Rössing Uranium*. Windhoek, John Meinert Printing.

Louw, P. S., 1965. *Discovery of Uranium on the Namib Desert of South West Africa*. 4 July, 1965, Correspondence, reports and analyses on the discovery of ‘Pitchblende’ in the Namib in 1929/30. Sam Cohen Library (SCL), Scientific Society of Swakopmund, 2004.168.3.

Martin, H., 1957. *The Sheltering Desert*. Translated 1958. Edinburgh: Thomas Nelson and Sons.

Masco, J., 2006. *The Nuclear Borderlands: The Manhattan Project in Post-Cold War New Mexico*. Princeton, New Jersey: Princeton University Press.

Mikloweit, S., 2025 (2023). Thinking Through Minerals: We Build Our Languages Out Of Rocks. In: S. Mikloweit, ed., 2025. *Transpositional Geologies: Spectres Of Coloniality*. Berlin: Kerber Publishers.

National Archives of Namibia, No date. *Topnaars – ǂAonin: Tradisionele Geskiedenis*.

____1 December 1945. *Correspondence Suspected to Require Special Attention*. Union of South Africa Censorship, 1 December, 1945.

____1949. *Radioactive Ores in the Swakopmund Area*. Supplement to the Report made by Dr Martin.

Parr, J., 2010. *Sensing Changes: Technologies, Environments, and the Everyday, 1953–2003*. Vancouver: University of British Columbia Press.

Reuning, E., 6 October, 1911. *Frankfurter Zeitung*.

Rigali, M. J., 2018. Oklo Natural Nuclear Reactors. In: W. M. White and M. William eds., 2018. *Encyclopaedia of Earth Sciences Series: Encyclopaedia of Geochemistry*. Springer: Berlin.

Rössing Uranium Limited, 2022. *A Lived Legacy: Reflecting on 45 Years of Working for Namibia (1976–2021)*, Rössing Uranium Limited, Namibia.

Schrijver, N. J., 1995. *Sovereignty over Natural Resources: Balancing Rights and Duties in an Interdependent World*. PhD Diss., University of Groningen.

Shiga, J., 2019. The Nuclear Sensorium: Cold War Nuclear Imperialism and Sensory Violence. *Canadian Journal of Law and Society / La Revue Canadienne Droit Et Société*, 34(2), pp.281–306.

Simpson, M., 2019. Resource Desiring Machines: The Production of Settler Colonial Space, Violence, and the Making of the Athabasca Tar Sands. *Political Geography*, 74.

Veselovský, F., Ondruš, P. and Komínek, J., 1997. History of the Jáchymov (Joachimsthal) Ore District. *Journal of the Czech Geological Society*, 42(4), pp.127–132.

Vigne, R., 2000. The Hard Road to Colonization: the Topnaar (ǂAonin) of Namibia, 1670–1878. *Journal of Colonialism and Colonial History*, 1(2).

Vint, S., 2021. *Science Fiction*. Cambridge, MA: The MIT Press.

Wagner, P. A., 1921. On Some Mineral Occurrences in the Namib Desert. *Trans. Geol. Soc. South Africa*, XXIV, pp.71–97.

Wesch, H., Wiethege, T., Spiethoff, A., Wegener, K., Müller, K.-M. and Mehlhorn, J., 1999. German Uranium Miner Study: Historical Background and Available Histopathological Material. *Radiation Research*, 152(6).

West, R., June 1972. South Africa's Atom Factory, *New Statesman*.

Wilson, W. E., Pinch, W. W., 1977. [Tsumeb, Namibia] Minerals: A Descriptive List. *The Mineralogical Record*, 8(3), pp.17–37.

World Nuclear Association. 2023. *Uranium in Namibia*. [online] Accessible at: <https://world-nuclear.org/information-library/country-profiles/countries-g-n/namibia.aspx> [Accessed 13 December 2023].

Wright, R. W., 1973. *Letter to E.J.S. Clarke*. 19 January. AB 48/1278, the National Archives, UK.

Zacke, A., Friedeberg, F. and Schumache, R., 2018. *Mineralogical Museum of the University of Bonn: The Collection – Exhibition Catalogue*. Bonn: University of Bonn.

[4] Urangesellschaft mbH & Co., of the Federal Republic of Germany was established in 1967 through a partnership between Veba AG, Metallgesellschaft AG and Steinkohlen-Elektrizität AG. The West German government had a key stake in Veba AG. Urangesellschaft acquired a 10% shareholding in Rössing Uranium (Rogers, 1975, p. 3).

[5] Decree No. 1 for the Protection of the Natural Resources of Namibia was enacted on 27 September 1974 by the United Nations Council for Namibia.

[6] Ernst Reuning trained in geology at the Technical University, Darmstadt, and obtained a D.Phil in geology from the University of Giessen, in Germany. His work in the former German South West Africa began in 1909 when he was recruited by the Deutsche Diamanten Gesellschaft to prospect for diamonds on the southern coast of the colony. After two decades of work in southern Africa, he returned to Germany and took up a lecturing position at the University of Giessen (Plug, 2020).

[7] Daniel (1995, p.22), writes that the Louw Syndicate was established in December 1954 by Captain Peter Louw (who is credited for the discovery of the Rössing deposit), his son Graham Louw, and Peter's military friends: Edwin Beecroft and Major Archie Maclaren. The members of the syndicate each held an equal shareholding, and agreed to pull personal resources to aid in the exploration of their uranium discovery.

[8] Converted to Euro in 2023, this would be approximately €12.5 million (Bundesbank, 2024).

[9] In October 1970 it was expected that the German state will use approximately 30,000 tons yellowcake (BAK B/196/33620, 1970)—this would be a volume of 6.224m^3, or comparable to the volume of about 81 Lo ndon double-decker buses, and the yearly need for 1980 was estimated to be up to 8000 tons,the equivalent of 1.660m^3 or approximately 22 London double-decker buses.

'Rössing' Uranium Mine: The German Link to a British Enterprise

Saima Nakuti Ndahangwapo

Abstract: In 1968 Rio Tinto Zinc (RTZ) established Rössing Uranium Limited through a British-South African partnership, with capital input from Canadian, French, German, Iranian and Japanese entities. The German participation in Rössing was minimal, but the choice of the company name inadvertently gave the impression that the mine was managed by German interests. The name Rössing was historically associated with the rail network since the late 1890s, when the territory was known as German South West Africa (now Namibia). In naming its venture Rössing RTZ, not only invoked the links associated with eponyms, which the Aawambo refer to as 'edhina ekogidho', but also the colonial networks of power that linked the German and South African colonial periods, most visibly through the mineral extraction industry. This essay seeks to uncover these links in order to understand how a British mining enterprise came to be associated with the German name Rössing.

Keywords: eponyms, Germany, Namibia, Rössing, Rio Tinto Zinc (RTZ), uranium, networks

In 1970 the British multinational corporation Rio Tinto Zinc (RTZ) registered a company called Rössing Uranium Limited through which it would exploit the uranium deposits of Namibia, then under South African occupation (Berning, 1999, p.3).[1] Rössing Uranium came into production in 1976, at a time when 'the UN specifically mandated against the export of Namibian minerals' (Hecht, 2004, p.155). British investment in the production and export of Namibian uranium was thus especially condemned (Hill and Ashipala, 2024).[2] Rogers (1975, p.4), for instance, argued that the responsibility for the establishment of the Rössing Uranium mine 'rested primarily with the purchasers and notably the UK [United Kingdom]'.[3] The company's list of shareholders, however, reveals that there were others, aside from the UK, whose investments in Rössing Uranium were not beyond reproach. Urangesellschaft MbH of Frankfurt, was among these shareholders[4] and it inspired the question on whether Germany, and specifically the Federal Republic of Germany, could also be implicated in the exploitation of Namibia's uranium resources.[5]

Prospecting for mineral deposits in colonial settings required 'considerable resources and infrastructure' and resulted in the establishment of collaborative mining ventures, which produced what Hecht (2004, p.148) described as colonial networks of power, that is the transmission of social, institutional, political and technological power within colonial dimensions. It is these networks of power that make it difficult to hold a single entity or country, such as the UK, responsible for the exploitation of a colony's mineral resources. The German role in the discovery and exploitation of the Rössing deposit can be traced to the German colonial period in Namibia when the presence of radioactive minerals in the Swakopmund district was uncovered by the geologist Dr. E. Reuning (1881–1961) in 1910 (Berning et al., 1976, pp.351–368).[6] It was only, however, in the late 1960s when the German role in the discovery of Namibia's uranium deposits was revisited through RTZ's intensive investigations of the Louw Claims, which it had acquired in 1966 from the Louw Syndicate.[7] Between 1966 and 1968, RTZ carried out preliminary work on the Louw Claims, resulting in the conclusion that further prospecting work would be required in order to determine their economic viability. This second phase would be achieved through collaboration, which was in keeping with 'a long-standing colonial practice' which 'could maintain the economic viability of prospecting missions' (Hecht, 2004, p.53). It was at this stage, in mid-1969, that RTZ approached Urangesellschaft MbH & Co. of Frankfurt to collaborate in the prospecting of the Louw Claims. RTZ already had an exploration team in Southern Africa, so what was required from Urangesellschaft was essentially the capital to produce a feasibility study that would confirm the economic viability of the project (Ndahangwapo, 2025). Urangesellschaft, for instance, secured 6 million Deutschmarks[8] from the German government for use in the prospecting of the Rössing deposits (BAK B/196/33620, 1970). It was during this period that RTZ established the Rössing Uranium mine, an entity whose name created the impression that it was operated by German interests (Hill, 2025). For RTZ, collaborating with Urangesellschaft was ideal, not only because of that company's responsibility to secure uranium supplies for the German government's nuclear power programme[9], but also because of the history of Anglo-German collaboration and financial networks in colonial Namibia (Murray et al. 1974, p.91). These networks can be traced to the turn of the 20th century through the mining of the colony's copper and diamond resources, as well as through the construction of the infrastructure necessitated by these mining ventures. The collaboration between RTZ and Urangesellschaft was thus built upon the colonial networks of power, which

[1] Rössing Uranium Ltd was initially registered in South Africa. The company was jointly owned by Rio Tinto and the South African government. It had shareholders from Canada, France, Iran and West Germany. Japan and the United Kingdom were among the company's key customers.

[2] Condemnation against British investment in Rössing was based on Decree No. 1 '[...] to protect the natural resources of the people of Namibia and of ensuring that these natural resources are not exploited to the detriment of Namibia, its people or environmental assets [...]' (UN, 1974), which was enacted on 27 September 1974 by the United Nations Council for Namibia. Protests groups such as the Campaign Against Namibian Uranium Contracts (CANUC) based their condemnation of British investments in Rössing on Decree No. 1.

[3] The United Kingdom Atomic Energy Agency (UKAEA) was among Rössing Uranium's first customers. The UKAE signed a purchasing contract with RTZ in 1968 for the supply of over 7,000 tons of uranium from the Rössing mine.

exited from the German colonial period and continued into the South African colonial period and well into post-independence Namibia. With these colonial networks of power in mind, this essay sought to uncover the German link to a predominantly British enterprise, namely Rössing Uranium Limited, with a particular focus on how the name 'Rössing' came to be associated with the uranium mine.

Freiherr von Rössing: the Triple Namesake

If the name 'Rössing' sounds German, written and pronounced as it is with an 'ö' umlaut, it is because it is. It is not exactly clear how RTZ came to the decision to adopt the name 'Rössing'. What is clear, however, is that the name can be traced to the period 1884 to 1914, when imperial Germany colonised the then German South West Africa (now Namibia), and thus had very little to do with Urangesellschaft and its partnership with RTZ during the exploration stage (1968–1970). The name Rössing is linked to the construction of the state railway line in the colony. The imperial German government was initially not interested in constructing a rail network in South West Africa, but the rinderpest (cattle plague) of 1897 presented the colony with logistical challenges as the cattle and ox populations were wiped out, and with them the means of transporting people and goods in the colony. The construction of the first railway line 'connecting the port city of Swakopmund and the capital Windhoek between 1897 and 1902' (Decker, 2020, p.12) strengthened the power base of the colonial administration and enabled it to lay hold of the land, subdue indigenous populations like the OvaHerero, Nama-Damara and San communities who lived in central and southern Namibia, while facilitating the exploitation of the colony's mineral resources. The various economic regions and especially the points of mineral extraction and the points of export were also bound together by the railway line (ibid., p.77). The railway line from Tsumeb to Swakopmund for instance, linked the point of copper extraction and other base metals to the point of export and onward shipment to countries like Germany.

The construction, maintenance and operation of the state railway line was the responsibility of the German imperial government's Railway Brigade, which was established in 1890 (Kinahan, 1988). In 1895, eleven years after the colony of GSWA was established, the Railway Brigade was commanded by Lieutenant General Baron Arnold August Nonus Freiherr von Rössing (1841–1899). Born into Prussian nobility, von Rössing embarked upon a military career in the Prussian army, gradually rising through its ranks until his promotion to commander of the Railway Brigade (1895–1899). The Prussian railway troops who arrived on the GSWA shores to construct the state railway line in the 1890s were thus under von Rössing's command. Von Rössing personally never set foot on GSWA's soil, but in 1898 the railway troops petitioned the German imperial government to name a railway station after von Rössing. According to Kinahan (1988, p.5), in January 1898 drivers of goods trains in GSWA complained to the Railway Brigade 'that the designation KM39 was insufficient for the railway station at which the line from Swakopmund terminated'. Permission was then sought from the German Emperor, Wilhelm II, to rename the railway station from KM39 to Rössing (Kinahan, 1988, p.5). The renaming of the station occurred in mid-1898, and a year later in 1899, Freiherr von Rössing died at the age of 58. Von Rössing did not get to see the completion of the state railway line, three years later, in 1902, but in naming a station after their commander the Railway Brigade ensured that his name lived on.

Beyond the railway station, the name Rössing also continued to live on through a mountain range located close to the railway station. The Rössing Mountains, according to Jill Kinahan, were first sketched in 1786 by Lieutenant Home Riggs Popham (1762–1820) of the British Royal Navy, and it would undergo several name-changes before it became known as the Rössing Mountains in the early 1900s (Kinahan, 1988, p.1). In 1837, Captain James Alexander of the British Royal Navy, named the mountain range Mt. Colquhoun '[...] after [his] valued friend and connexion, Gideon Colquhoun, Esq., late Resident at Bussorah' (Kinahan, 1988, p.3). The change from Colquhoun to Rössing occurred in 1918, presumably because of the mountains' location 'to the immediate west of Rössing railway station' (Kinahan, 1988, p.3). By the early 1920s, the mountain range was referred to as the Rössing Mountains by the Office of the Surveyor General under the South African administration. It was thus the name Rössing that prevailed from the German into the South African colonial period, and von Rössing became the namesake and eponym of landmarks which he had never laid eyes on.

Decades later, in the late 1960s, the railway station and the Rössing Mountains became the inspiration for the name of the first uranium mine in South West Africa (SWA), Rössing Uranium Limited. The mountain range was especially pivotal to the naming of the mine because since the early 1900s, geologists had identified the presence of radioactive minerals in the vicinity of the Rössing Mountains.[10] The British-South African couple, Margery and Peter Louw,[11] who are credited by Rössing Uranium and its former parent company, RTZ for the 1928 discovery of what was initially known as the Louw Claims, later renamed the Rössing deposit, also 'found' what they identified as radioactive minerals in the vicinity of the Rössing Mountains. What these studies omit, however, was the role of indigenous communities in the discovery of these radioactive minerals. The significance of indigenous labour and knowledge, as Hill (2025) notes, is neither acknowledged in the geological studies, nor in the official history produced by Rio Tinto's Rössing Uranium mine.[12] The geologist Henno Martin's (1910–1998) reference to 'coloured folk' from Salem who brought samples of radioactive material to Peter Louw (Martin, 1950, p.3), as discussed in Hill (2025),

[10] Notably: Dr E. Reuning (1910), Dr P.A. Wagner (1921), Prof. T.W. Gevers (1932) C.M. Schwellnus and G.S.J. Kuschke (1940). See Rössing Uranium (n.d., p.43), The Extent of Current Mining Operations 3.1 Exploration History.

[11] Captain Petrus Schabort (Peter) Louw (1886–1978) was born in Beaufort West in the Cape Province of South Africa. Annie Margery Louw (née Burns) (1893–1953) was born in Sunderland, United Kingdom. Peter and Margery Louw moved to Swakopmund in 1925.

[12] This official history has been carried forward by the China National Uranium Corporation Limited (CNUC), Rössing Uranium's current majority shareholder and parent company.

confirms earlier studies carried out by the Geological Survey of South Africa[13], which found that radioactive material identified as 'a heavy black mineral' was 'scattered over a very large area extending from the Rössing Mountains in the west to Salem in the east' (Rössing Uranium, undated, p.43). It is thus not surprising that the indigenous communities of Salem, who are assumed to be a people of Nama descent known as the Topnaar, were able to identify this heavy black mineral and present samples to Peter Louw, whom Rössing Uranium credits for the *discovery* (see Hill, 2025). The omission of the indigenous role in mineral discoveries in colonial Namibia is not unique to the Rössing case. Press, for instance, writes that when diamonds were discovered in the Lüderitz Bay area, it was the German settlers and specifically August Stauch the German engineer who was 'hired to maintain the train track' from Lüderitz Bay, who were credited for the discovery and valorised as 'the colony's mineral pioneers' (Press, 2021, p.57). It was in fact an African migrant worker, Zacharias Lewala, who discovered the diamonds 'while shoveling sand around construction on the rail line from Lüderitz Bay' (Press, 2021, p.57). Throughout the German and South African colonial periods, however, the official histories on the discovery of diamonds in Namibia credit the engineer and not the migrant labourer whose role would only be acknowledged in post-colonial Namibia. This emphasis on the settler, and the focus on the professional, thus then explains why the discovery of the Rössing deposits is credited to Peter Louw and the consistent emphasis on his wife Margery Louw's experience as a former radiographer from London, which is repeated in every account produced by RTZ (and now CNUC) on the discovery of the Rössing deposits. The accounts on the railway station and the Rössing Mountains, on the other hand, illustrate that RTZ's decision to name its mining venture 'Rössing' was in keeping with the historical legacy of the namesake of the existent landmarks and particularly the identification of radioactive minerals in the vicinity of the Rössing Mountains.

In the early 1980s, Rössing Uranium 'took a step back into the past' and invited the descendants of Nonus Freiherr von Rössing to visit the mine. The timing of the invitation had to do with Rössing Uranium's newly adopted open-door policy, which saw a previously inaccessible mine host visits from prominent individuals, a move that was viewed to be in Rössing's interests, as the company prepared for the transition to Namibian independence (Ndahangwapo, 2025). The visit from the von Rössings was meant to shed information on the mine's namesake, information that could be shared with Rössing's prominent visitors as they took a tour of the mine, and learned about its history. Rössing Uranium's initiation of this German link interestingly coincided with the delivery of uranium supplies to its German shareholder and customer, Urangesellschaft. It was remarkably also around the same time of the visits from the von Rössing descendants that the mineral objects, the 'uranium-bearing rocks kept by the Mineralogical Collections for the Technical University of Berlin' discussed in Hill (2025), were collected from Rössing Uranium. The German role in the predominantly British enterprise, Rössing Uranium, had thus facilitated links which permeated through academic, familial, financial, infrastructural and mineral connections.

Colonial Networks of Power

Building on Tom Hughes 'Networks of Power' (1983),[14] Gabrielle Hecht argues that the nature of mining ventures in colonial settings 'produced a distinctive blend of [...] colonial networks' (Hecht, 2004, p.149). These colonial networks were especially visible through the capital investments of, for instance, British financial houses in Germany's overseas possessions (Henderson, 1945, p.57). In the case of Namibia, the best illustration of this was the South West Africa Company Ltd. (SWAC), which was founded in London in 1892 by 'Hamburg financiers headed by Dr Scharlach and Herr Wichmann who were unable to raise money for their venture in Germany' (Henderson, 1945, p.61).[15] Voeltz (1984, p.623) writes that the origins of SWAC 'were profoundly influenced by the colonial policies of the German, British, and Cape governments, as well as the schemes of British and German financiers' and that SWAC 'represented the continued expansion of British capital in southern Africa'. These British, Cape, and German economic and political interests were cemented when SWAC secured the exclusive right to exploit the minerals of the Otavi region, which included the Tsumeb copper mines from the German government. Cairncross (1997, p.111) adds that 'in 1900, a second company, Otavi Minen Und Eisenbahn Gesellschaft [sic] (OMEG), was established and an agreement was drawn up in which SWACO [sic] ceded mineral rights to OMEG [...] which included Tsumeb' on condition that a railway line is constructed from Tsumeb to Swakopmund to facilitate the transportation and onward shipment of copper from the interior. This Anglo-German colonial network was thus responsible for the construction of the Tsumeb-Swakopmund railway line, which commenced in 1903 and was completed in 1906 (Henderson 1945, p.61).

The continuity of the Anglo-German economic, colonial, and political networks that were exploited in the establishment and operations of SWAC and OMEG were therefore visible in the Rössing venture. RTZ's decision to name its mining venture 'Rössing' gave the impression that the mine was operated by a German company (Hill, 2025). Whether or not this was a deliberate strategy of protecting the mine from 'international efforts to halt the South African occupation of Namibia' and thereby distancing the mine from apartheid South Africa is not clear (Hecht 2004, p.155). The impression of German ownership, however, hinted at continuity, and particularly in terms of economic collaboration between British and German capital. Whereas SWAC required British financial investment in the 1890s, RTZ's Rössing Uranium benefited

13 The Geological Survey of South Africa was established in 1912 and it was based in Pretoria, South Africa.

14 Hughes' work makes a case for the importance of understanding systems and networks through which power is transmitted. This is not just in terms of electrical power, but also economic, political, technological and social power.

15 Drechsler (1966) explains that the Hamburg financiers actually started off as the Hamburger Consortium' (/Groll-Syndicate) which initially consisted of: Julius Scharlach (1842–1908), Carl Wichmann (1843–1920) (Director Nobel Trust, Hamburg), Adolph Woermann (1847–1911), Max Adolf Philipp (1836–1905), Nepomuk Johann Heidemann (1841–1913). By mid-1892 the Groll-Syndicate had failed. (Drechsler, 1966, p.80). By July 1892 only Scharlach and Wichmann were still members of the 'Hamburg Consortium' (ibid., p.83). SWAC was then officially registered on 18 August 1892, with finances raised in UK through individuals like Edmund Davis (1861–1939) and George Cawston (1851–1918) who were close associates of Cecil Rhodes (Ibid. pp.86–87).

from German financial resources through the acquisition of equity and the purchase of uranium from the Rössing mine. In 1970, Urangesellschaft acquired 10 per cent equity in Rössing Uranium for an estimated total of 11.5 million Deutschmarks[16] and signed a forward purchasing contract for the supply of over 6,000 tonnes of uranium, initially envisaged for a delivery period between 1979 and 1986 (RTZ, 1970).[17] Private and public German capital thus facilitated the German involvement in Rössing Uranium beyond the exploration phase and the continued exploitation of Namibia's uranium resources by Germany, as represented by Urangesellschaft. Unlike SWAC, Rössing Uranium displays colonial networks beyond the Anglo-German corporation, because it was jointly owned by RTZ and 'the South African government, with additional capital investment from French and Japanese sources' as well Iranian sources (Hecht, 2004, p.155). In both the SWAC and Rössing periods, however, the railway line remained the conduit through which these colonial networks of power flowed, allowing both entities to exploit their respective mineral deposits.

The historic railway line also continued to facilitate trade between these two historic regions of mineral exploitation and between the corporations that operated in them. In the 1980s, for instance, Rössing Uranium entered into a trade agreement with the Otjihase Mining Company, for the supply of pyrite from the Otjihase mine. Otjihase was co-owned by the Tsumeb Corporation Limited (TCL), which also operated the Tsumeb copper mines at the time, and the Johannesburg Consolidated Investment Company (JCI) (Weatherly Mining Namibia, 2013, p.6).[18] The pyrite was transported by rail from the Otjihase mine to the Rössing Uranium mine for the production of sulphuric acid used in the uranium extraction process at Rössing (Weatherly Mining Namibia, 2013, p.8). In the post-independence period when pyrite production ceased at the Otjihase mine, the Namibia Custom Smelter in Tsumeb became the main supplier of sulphuric acid to Rössing Uranium (Walker, 2015). The Tsumeb smelter was constructed in the early 1960s by the Tsumeb Corporation Limited (TCL), which was operated by the American multinational corporation, Newmont Mining. From the late 1980s onwards the Tsumeb smelter underwent several changes in ownership.[19] The sulphuric acid produced at the smelter was transported from Tsumeb to Rössing Uranium via the railway line, which was constructed under the SWAC-OMEG agreement (Figures 26 & 27).[20] These links thus speak to the colonial networks of power, which facilitated the exploitation of the territory's mineral resources, and which persist to date.

Conclusion

The Aawambo of northern Namibia, a people from which Rössing Uranium Limited would draw a great percentage of its workforce, have a common saying that goes 'edhina ekogidho'. Now, 'the noun *edhina* means *name* and *ekogidho* [means] *joining, connecting permanently together*' (Saarelma-Maunumaa, 2003, p.3). Loosely translated therefore, the saying means that names and especially shared names or eponyms serve as links. In the case of RTZ's Rössing Uranium, the joining of permanent connection was not only to the railway station and the mountain but also to the historic legacy of German colonial rule in Namibia. The Aawambo also believe that in transferring a name the recipient of the name will resemble the namesake and inherit the character traits of the namesake. Rössing Uranium would thus, in accordance to this belief, take on the traits of its eponym. In his professional and maybe even personal capacity, Arnold August Nonus Freiherr von Rössing represented colonial conquest and the subjugation of environmental terrains to technology and progress. Such conquest was essentially geared towards the facilitation of the exploitation of both the colony's natural resources and labour. It does not take long for one to recognise these traits in RTZ's Rössing Uranium. A once undisturbed terrain of roughly 1,200 square miles, located on the fringes of the Namib desert, was subdued with modern technology coupled with dynamite in order to develop a mining complex (Louw, 2018, p.35). This was particularly visible through the expansion of Rössing's star attraction, the open pit.[21] Rössing prided itself with having established a mining complex in a place which lacked infrastructure, such as electricity and a water supply, and the technical expertise and labour force to bring the mine into production. These are conditions which resonate with the earlier feats of the Railway Brigade under Freiherr von Rössing's command, who had to construct a colonial rail network through rugged terrain. That same railway network passes within miles of Rössing Uranium and connects the mine to the port of Walvis Bay, thus ensuring the continuity of the colonial exploitation of the territory's mineral resources. The link between past and present was revisited through Rossing Uranium's invitation to the descendents of Freiherr von Rossing to visit their vast mining complex in the Namib desert. These visits not only reaffirmed the mine's links to Germany, but they also reiterated the German legacy in the colonial networks of power, which enabled the exploitation of Namibia's mineral resources including its uranium. Rössing Uranium's initiation of this German link interestingly coincided with the delivery of uranium supplies to Urangesellschaft.[22] It was remarkably also around the same time of the visits from the von Rössing descendants that the mineral objects, the 'uranium-bearing rocks kept by the Mineralogical Collections for the Technical University of Berlin' discussed in Hill (2025), were collected from Rössing Uranium. The German link to the British enterprise had thus facilitated colonial networks of power, which permeated through academic, familial, financial, infrastructural and mineral connections.

[16] Converted to Euro in 2023 this would be approximately €23.8 million (Bundesbank, 2024).

[17] Rössing Uranium supplied uranium to its customers 'in a basic oxide form (U_3O_8) that could be further enriched for profit' (Hill and Ashipala, 2024, p.35).

[18] The development of the Otjihase Mine by the Otjihase Mining Company (OMC), a subsidiary of JCI, began in 1973 and came into production in late 1975. Due to the fall in the price of copper, mining operations were suspended in 1977. The mine was placed on care and maintenance until the end of 1980, when TCL entered into a joint venture agreement with OMC. The mine was reequipped, the mining method was changed and a new access tunnel was developed. Full production was achieved in 1982.

[19] Most recently, after 14 years of business, Dundee Precious Metals Inc. of Canada sold the Tsumeb smelter in March 2024 to a subsidiary of Sinomine Resource Group Co. Ltd. of China for US$ 40 million in cash (DPM, 2024).

[20] Sulphuric acid production in Namibia, its history, environmental impact, the implications for the health of workers and the communities of Tsumeb as well as the transportation of the sulphuric acid within and outside of Namibia is a complex tale worthy of an in-depth study. https://ejatlas.org/conflict/tsumeb-smelter provides a useful introduction to these matters.

[21] According to current estimates, the Open pit is approximately 3 kilometres by 1.2 kilometres and 390 metres deep.

[22] The uranium deliveries occurred in accordance to a uranium sales agreement signed between Urangesellschaft and Rössing Uranium in March 1979.

Fig. 26

Fig. 27

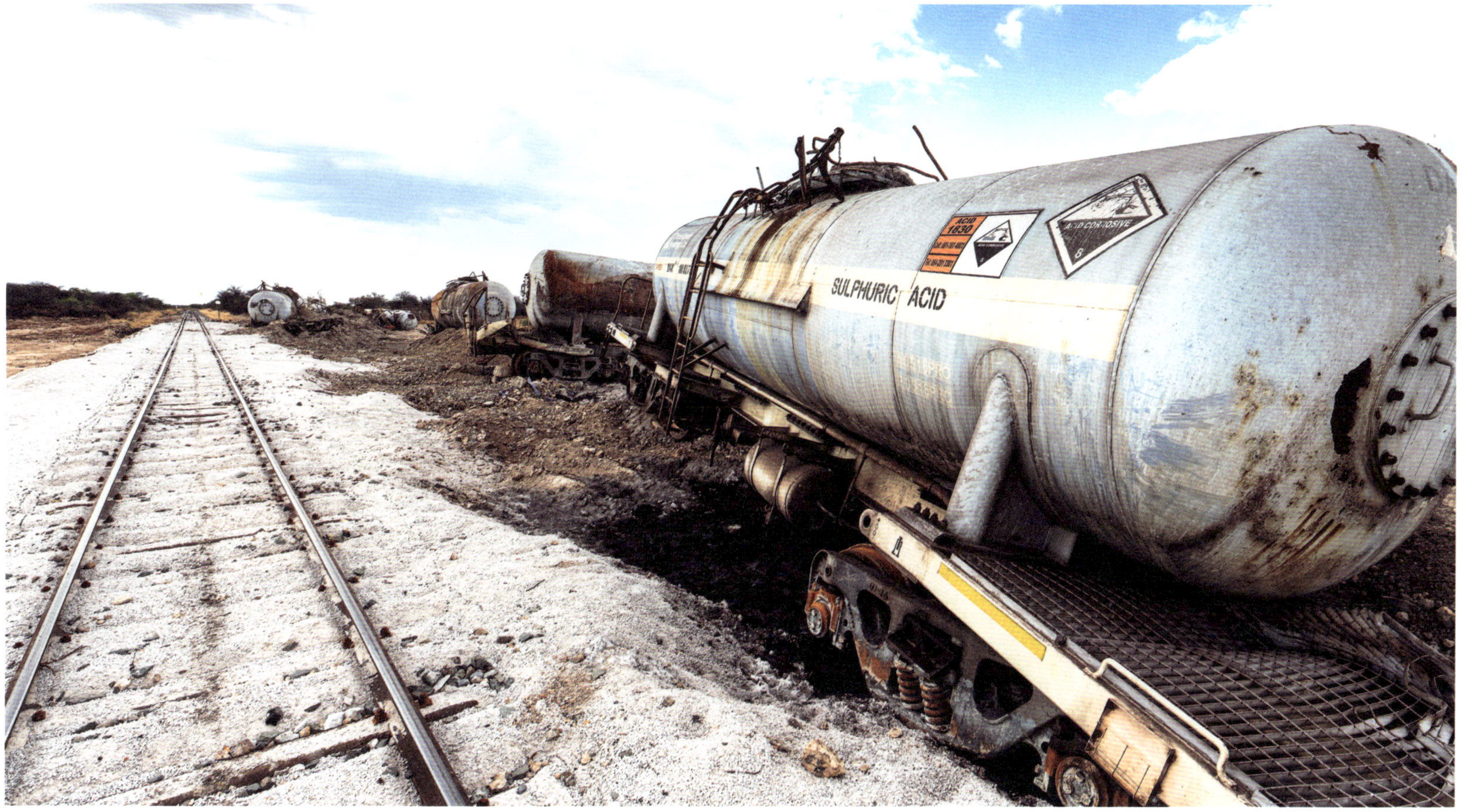
SULPHURIC ACID

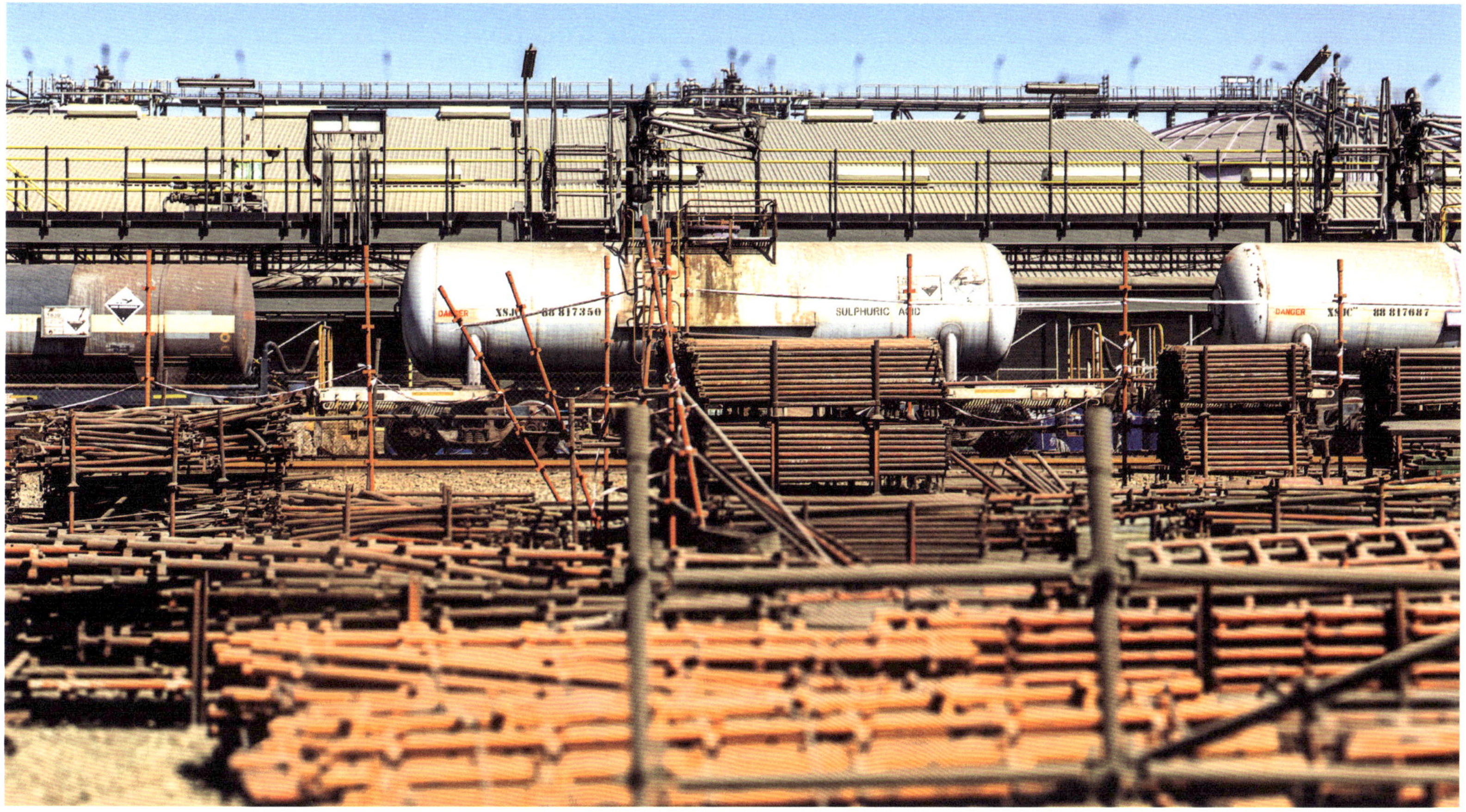

References

Author Unknown, 1970. *Langfristige Uranversorgung der Bundesrepublik Deutschland; hier: Bedeutung des Projektes Rössing, Bonn, den 23 October 1970.* [Document] BAK B/196/33620, Koblenz: Bundesarchiv.
Berning, J., 1999. The Road To Rössing Via Rio Tinto. Rössing News, December 1999. In: Rössing, 2006. *e-Rössing Bulletin: 30 Year Special Edition 1976–2006.* [online] Available at: <https://www.rossing.com/files/publications/1976-2006-E-Rossing-Bulletin-30yr-Special-Edition.pdf> [Accessed: 27 April 2024].
Berning, J., Cooke, R., Hiemstra, S. A. and Hoffman, U., 1976. The Rössing Uranium Deposit, South West Africa. *Economic Geology*, 71(1), pp.351–368. [online] Available at: <https://doi.org/10.2113/gsecongeo.71.1.351> [Accessed 27 April 2024].
Bundesbank, 2024. *Kaufkraftäquivalente historischer Beträge in deutschen Währungen. Frankfurt: Deutsche Bundesbank.* [online] Available at: <https://www.bundesbank.de/resource/blob/615162/94b87ff6d25eceb84c9cfb801162b334/mL/kaufkraftaequivalente-historischer-betraege-in-deutschen-waehrungen-data.pdf> [Accessed 23 May 2024].
Cairncross, B., 1997. The Otavi Mountain Land: Cu-Pb-Zn-V Deposits. *The Mineralogical Record*, 28(2), pp.109–130. [online] Available at: <https://www.researchgate.net/publication/303880412_The_Otavi_Mountain_Land_Cu-Pb-Zn-V_deposits_Namibia> [Accessed 13 May 2024].
Daniel, P., 1995. *Against All Odds: A History Of Namibia's Rössing*. Windhoek: Rössing Uranium.
Decker, J., 2020. Line In The Sand: Railways And The Archipelago Of Colonial Territorialization In German Southwest Africa, 1897–1914. *Journal of Historical Geography*, 70, pp.74–87. [online] Available at: <https://www.academia.edu/44476642/Lines_in_the_sand_railways_and_the_archipelago_of_colonial_territorialization_in_German_Southwest> [Accessed 27 April 2024].
Dierks, K., n.d. *The Development Of The Namibian Railway Network.* [online] Available at: <https://www.klausdierks.com/Namibia_Rail/2.htm> [Accessed 13 May 2024].
Drechsler, H., 1966. *Südwestafrika unter deutscher Kolonialherrschaft: Die großen Landgesellschaften und Minengesellschaften (1885–1914).* Stuttgart: Franz Steiner Verlag.
DPM (Dundee Precious Metals), 2024. *'Dundee Precious Metals Announces Sale of Tsumeb Smelter for US$49 Million'.* [press release] 7 March 2024. Available at: <https://dundeeprecious.com/news-media/news-releases/dundee-precious-metals-announces-sale-of-tsumeb-sm-8114> [Accessed 23 May 2024].
Hecht, G., 2004. Colonial Networks Of Power: The Far Reaches Of Systems. *Annales historiques de l'électricité*, 2(1), pp.147–157. [online] Available at: <https://www.cairn.info/revue-annales-historiques-de-l-electricite-2004-1-page-147.htm> [Accessed 08 April 2024].
Henderson, W. O., 1945. British Economic Activity In The German Colonies, 1884–1914. *The Economic History Review,* 15(1/2), pp. 56–66. [online] Available at: <https://www.jstor.org/stable/2590312> [Accessed 14 May 2024].
Hill, C. R., andAshipala, S. N., 2024. "Follow the Yellowcake Road": Historical Geographies of Namibian Uranium from the Rössing Mine. *Historical Social Research*, 49(1), pp. 32–54. [online] Available at: <https://www.jstor.org/stable/27290653> [Accessed 23 May 2024].
Hill, C. R., 2025. Out of Space And Place: Tales of Namibian Uranium Through Mineral Objects. In: S. Mikloweit, ed. 2025 *Transpositional Geologies: Spectres Of Coloniality.* Berlin: Kerber Publishers.
Hughes, T. P., 1983. *Networks of Power: Electrification in Western Society, 1880–1930.* Baltimore, Maryland: Johns Hopkins University Press.
Inspector of Mines, 1949. *Radioactive Ores in the Swakopmund Area: Supplement to the Report made by Dr Martin.* [Document] Inspector of Mines. IMW 15-1073-3. Windhoek: National Archives of Namibia.
Kinahan, J., 1988. The Story Of Rössing Mountain. *Rössing Magazine*, April, 1988. Windhoek: Rössing Uranium.
Louw, G., 2018. *A Tiger by the Tail: The Story of the Discovery of Rössing Uranium*. Windhoek, John Meinert Printing.
Martin, 1950. *Inspector of Mines, Radio-Active Ores in the Swakopmund Area. Supplement to the Report made by Dr. Martin, June 1950.* [report] NAN IMW 15-1073-3. Windhoek: National Archives of Namibia.
Murray, R., Morris, J., Dugard, J., and Neville, R., 1974. *The Role of Foreign Firms in Namibia*. London: Africa Publication Trust.
Ndahangwapo, S .N. (2025). *Defending The Investment: Rössing Uranium And The Business of Decolonisation In Namibia*. Basel: Basler Afrika Bibliographien.
Plug, C., 2020. *S2A3 Bibliographical Database of Southern African Sciences: Reuning, Dr Ernst.* [online] Available at: <https://www.s2a3.org.za/bio/Biograph_final.php?serial=2332> [Accessed 20 May 2024].
Press, S., 2021. *Blood And Diamond: Germany's Imperial Ambitions In Africa*. Cambridge, Massachusetts: Harvard University Press.
RTZ (Rio Tinto Zinc.), 1970. *Heads Of Agreement Between Rio Tinto Management Services South Africa, Rössing Mining Company And Urangesellschaft, Draft Part B – Sale Of Uranium B.3. Commencement Date And Quantity.* [Document] BAK, B/196/33619, Koblenz: Bundesarchiv Koblenz.
Rogers, B., 1975. *Namibia's Uranium: Implications For The South Africa Occupation Regime*. London: Indiana University.
Rössing News, 1983. Rössings Visit Rössing: Shed New Light On Our Namesake. Rössing News, 5(9), 22.7.1983. In: Rössing, 2006. *e-Rössing Bulletin: 30 Year Special Edition 1976–2006*, p.12. [online] Available at: <https://www.rossing.com/files/publications/1976-2006-E-Rossing-Bulletin-30yr-Special-Edition.pdf> [Accessed 27 April 2024].
____ 1984. *Baron And Baroness Von Rössing Visit The Mine*. Week 36, 07.09.84. Swakopmund: Rössing Uranium.
Rössing Uranium, n.d. *The Extent Of Current Mining Operations – Exploration History.* [online] Available at: <https://www.rossing.com/files/eis-rossing-mine/41-60.pdf> [Accessed 13 May 2024].
Saarelma-Maunumaa, M., 2003. Edhina Ekogidho – Names As Links: The Encounter Between African and European Anthroponymic Systems Among The Ambo People In Namibia. *Studia Fennica Linguistica,* 11. Helsinki: Finnish Literature Society. [online] Available at: <https://library.oapen.org/bitstream/id/4bd71c00-0fc3-4187-8b19-eb18cee3fcce/9789522228161_edhina_ekogidho.pdf> [Accessed 01 March 2024].
UN (United Nations General Assembly), 1974. *United Nations Council for Namibia: Decree On The Natural Resources of Namibia*. A/AC.131/33. New York: United Nations.
Voeltz, R. A., 1984. The European Economic And Political Penetration Of South West Africa, 1884–1892. *The International Journal of African Historical Studies*, 17(4), pp.623–639. [online], Available at: <https://www.jstor.org/stable/218904> [Accessed 20 May 2024].
Walker, S., 2015. African Acid: Turning The Market Around. *E&MJ Engineering and Mind Journal*. [online] March. Available at: <https://www.e-mj.com/features/african-acid-turning-the-market-around>[Accessed 13 May 2024].
Weatherly Mining Namibia, 2013. *Otjihase—Environmental Compliance Report: October 2019 – September 2022.* Windhoek: Ongopolo Mining Limited. [online] Available at <http://the-eis.com/elibrary/sites/default/files/downloads/literature/175_ECR_Otjihase%20Mine_Khomas%20Region.pdf> [Accessed 20 May 2024].

[4] The Massacre of Hornkranz is attributed to Curt von François—the then governor of German South West Africa—and his 200 members of the Schutztruppe, who attacked the Nama under Hendrik Witbooi in an ambush on 12 April 1893, killing old men, children and women (Olusoga and Erichsen, 2010).

[5] The war of 1904–1908 led to the first genocide of the 20th century, following an extermination order issued by Lotha von Trotha in October 1904. An estimated 64,000 to 84,000 Ovaherero were killed (Dedering, 1993, p. 82).

[6] The rinderpest epidemic is important background to the structural violence being experienced today through land inequality. After rinderpest wiped out a sizeable number of cattle, huge land dispossession through fraudulent agreements followed. Tizard (2020, p.10) maintains that 'rinderpest was introduced into sub-Saharan Africa in 1887 when infected cattle from India were imported into Ethiopia to feed Italian soldiers'. On his part on how the bovine virus came to afflict present-day Namibia, Miescher (2012) suggests that rinderpest entered the territory through Botswana. Similarly, Gewald (1999, p.111) observes that 'the disease was unwittingly introduced into the territory by infected trek oxen operating along the extensive trans-Kalahari trade routes, and by game, particularly eland and kudu [...]'.

[7] In this view, 'responding to international pressure that land reform must proceed within the law, Zimbabwean activists and intellectuals replied that private ownership was an artefact of colonial rule, imposed on the people of Zimbabwe through the territorial predations of the British South Africa Company in the late nineteenth and twentieth centuries' (Berry, 2002, p.663). Similarly, when tracing clashes in land ownership regimes, Peters (2013) points to a view that colonial officials held, viz. African landholding systems of communal ownership as backward and in need of change. Individual/private ownership of land was therefore alien, for the land belonged to the community, the village and the family.

Land Inequality as Structural Violence
Vignettes of Namibia's Enduring 'Land Question'

Ellison Tjirera

Abstract: The persistent land inequality in Namibia is a form of structural violence rooted in historical colonial dispossession. This analysis, drawing on Johan Galtung's concept of structural violence, illustrates how past and present land ownership patterns perpetuate social injustice. During colonial rule, violent expropriation and fraudulent land deals forced indigenous populations off their land, cementing economic disparities. Post-independence efforts, such as land reform policies and national land conferences, have largely failed to address these inequities, with legal structures such as the 'Willing-Seller, Willing-Buyer' principle and constitutional protections hindering significant change. Consequently, land ownership remains highly skewed, with the majority of land still owned by a minority of white farmers, exacerbating multidimensional poverty among indigenous communities. Addressing Namibia's land question requires dismantling these entrenched legal and structural barriers, implementing more effective land redistribution policies, and fostering genuine socio-economic transformation to promote equitable land distribution and alleviate persistent poverty.

Keywords: colonialism, land dispossession, land inequality, structural violence, legal system, land ownership

An analysis of the ongoing and unresolved 'land question'[1] in Namibia clearly shows that past and present distributions of landownership are forms of 'structural violence'.[2] Historically, physical violence shaped land rights through forced dispossession. Outside the context of physical violence that characterised the colonial epoch, thinking with Johan Galtung's conception of structural violence becomes a productive lens through which contemporary land ownership patterns can be understood. With regard to land inequality and the attendant structural violence that ensues, the antidote becomes land redistribution with emphasis on addressing material inequalities that are socio-spatially located (Francis and Webster, 2019; Forde, Kappler and Björkdahl, 2021; Gready, 2022). In other words, land redistribution efforts need to address landlessness at a broader societal level, rather than focusing on the political elites and their cronies. Namibia's land question insofar as it is a harbinger of prevailing social injustice is inextricably linked to more than a century's history of imperial interests. Literature on colonial history point to land as a central preoccupation of the period, so much so that it becomes apparent that the 106 years[3] of colonial rule in Namibia bear testimony to the observation that the fight against imperialism was, by and large, about land dispossession (Werner, 1993; Pankhurst, 1996; Botha, 2000). From the Massacre of Hornkranz in 1893[4] to the Genocidal War of 1904–1908,[5] colonisers forced indigenous people off their land—ultimately cementing subjugation in the central and southern parts of what was then German South West Africa.

The rinderpest epidemic[6] of 1896–1897 impoverished the Herero and Nama communities, forcing them into servitude. On account of land clearances and boundary enforcements preceding the spread of rinderpest, Herero cattle were forced into a smaller range, making them vulnerable to the disease (Gewald, 2000). 'Rinderpest and colonisation were processes that fed into and off one another', the economic devastation wrought by the epidemic ensured that society became indebted and this in turn led to further land sales and continued dispossession (Gewald, 2000, p.199). The impact of the rinderpest on the Ovaherero cattle herds led to the reorganisation of society and produced circumstances that placed them in a position of dependence, vis-à-vis the colonial state. New elites capable of profiting from the changed circumstances came to the fore and facilitated German colonisation of the territory as Herero, desperate for food, vaccines and medicines, were forced into a subservient position (Gewald, 1999). It is precisely for this reason that legitimate transactions of buying (by the colonisers) and selling (by the colonised) land becomes suspect. Understood within the context of imperialism and colonial occupation, the very idea of selling land becomes complex and in need of unpacking. Indigenous people's conception of land ownership at the time gives credence to shared and temporary usage defined by pastoralism as a mode of subsistence. By contrast, Eurocentric permanent private property regimes, colonial practices of buying, and land metrics were alien to indigenous people.[7] Every opportunity to expropriate land from the Ovaherero and Nama was seized upon; fraudulent agreements were signed with local chiefs alongside land confiscation through violent means (Bley, 1971; Gann and Duignan, 1977). Land agreements were signed under false pretences and coercive circumstances and can thus be regarded as fraudulent, such as in the cases of Samuel Maharero and Zacharias Zeraua—two prominent Herero chiefs of the 1800s—who sold tracts of land in areas which were beyond their control, in exchange for goods that allowed them to extend their patronage (Gewald, 1999). Chiefs were invariably paid in kind while the value of goods delivered was arbitrarily determined by settlers (Werner, 1998, pp.43–45). It was in this context that the German

[1] The term 'land question', refers to an unresolved issue, an ongoing structural violence that has provenance in colonial domination. Indeed, earlier colonial expansion is mirrored in today's gross inequalities regarding access to, and possession of land (Melber, 2005, p.135).

[2] I am using structural violence as defined by Galtung to mean social injustice. Structural violence refers to violence without a specific actor, built into societal structures, manifesting as unequal power and life chances. It differs from physical violence, being indirect and shown through exploitation, oppression, and unequal distribution of power. (Galtung, 1969, p.171).

[3] Present-day Namibia was first under 31 years of German rule (1884–1915), followed by 75 years of South African (1915–1990) rule.

colonial government consequently sold concessions to land companies for extensive tracts of land and encouraged settlers to move into the territory (Gewald, 1999). To compound matters, in 1903, the German Imperial Commissioner for settlement—Paul Rohrbach[8]—declared that '75% of the land owned by Africans had to be sold to Europeans and that the remaining 25% is proclaimed native reserves' (Mbuende, 1986, p.59). To be sure, land dispossession continued under South African rule, after the Germans surrendered in July 1915.[9] Indigenous communities had to vacate land which had been reclaimed after 1915 in favour of white settlers (GRN, 2020a). After 1915, Germany lost its effective power of decision making over its former colony, German South West Africa. The years 1915 to 1960 constitute the making of the Apartheid State, ushering in restrictions on mobility and ethnic zoning of urban and rural spaces (Du Pisani, 2010). In 1962, the final phase of entrenching racially structured land ownership was put in motion with the appointment by the Government of South Africa of the Commission of Enquiry into South West Africa Affairs, commonly known as the Odendaal Commission (GRN, 2020a).[10]

Since 1990—during the post-independence era—the situation has not changed much. This doesn't come as a surprise, if we take a closer look at the prevailing land distribution and its legal architecture. Patterns of land distribution have remained broadly intact. Figures published by the Namibian Statistics Agency (NSA) in 2018 (pp.12,30) reveal that pre-independence advantaged white farmers own 27.9 million hectares of freehold agricultural land, and previously disadvantaged indigenous farmers own 6.4 million hectares, resulting in a staggering imbalance of 70% to 16% landownership in favour of white farmers, who constitute less than 6% of the total population of Namibia.

Colonial Land Dispossession

Beyond the official figures on land ownership, Amoo (2020, p.15) posits that 'in a state where farm ownership is politically and racially charged, it is not easy to determine exactly who owns the land, because some ownership is concealed through various legal devices'. To compound matters, some of these legal arrangements are secret and private, making it difficult to ascertain how prevalent these devices are, let alone exactly what they are (Amoo, 2020).

Land dispossession in Namibia continued almost uninterrupted when South Africa took over the colony from Germany in 1915. The self-governing dominion of the British Empire found a smooth-running state apparatus and proceeded with forced expropriation of land from the indigenous communities. There were hardly any discontinuities between German rule and South African occupation and the transformation of the legal system is not to be underestimated here.

> The Republic of South Africa, just like their German predecessor, deployed similar administrative actions for the transplantation, adaptation and enforcement of foreign and colonial European legal discourses of property rights in land.
> (Kaapama, 2021, p.158)

As our closer look at the distribution of landownership has shown earlier, these continuities not only have outlived German and South African colonial rule, but their reverberations stretch into post-independent Namibia till now. Similar observations are possible regarding the legal property regime. A leap ahead in time from 1915 to 1982 will be helpful here, for the role and influence of the Western Contact Group (WCG)[11] in fundamentally shaping important elements of the *Constitution of Namibia,* with effect on matters of landownership, cannot be overemphasised. The Constitutional Principles of 1982 became the cornerstone of fundamental rights and freedoms contained in Chapter 3 of the *Constitution of Namibia* which was adopted in 1990.[12] On 16 March 1990, the United Nations Security Council (UNSC) published an addendum titled *Further Report of the Secretary-General Concerning the Implementation of Security Council Resolution 435 Concerning the Question of Namibia*. (UNSC, 1990). This addendum ascertained the extent to which the Constitution of Namibia complies with requirements of the Principles for a Constitution for an Independent Namibia, i.e. the 1982 Constitutional Principles.[13]

Post-Independence Land Reform Paths

To signpost the centrality of land reform for an independent and potential 'transformed' Namibia, the 1st Land Conference was held in Windhoek, a year after political independence in 1990 (Van Wyk, 1992; Kaapama, 2007). This conference was attended by no less then 500 delegates from a broad spectrum of stakeholders (government officials, traditional leaders, academics, farmers' unions etc.) and sought to provide a platform for fostering national consensus on the volatile issue of postcolonial land reform (Kaapama, 2007). Twenty-seven years passed before the 2nd National Land Conference was held in Windhoek in 2018 (SNLC, 2018; Melber, 2019). The two land conferences were interspersed by a number of policy initiatives aimed at redistribution. These include the National Resettlement Programme (NRP) and the Affirmative Action Loan Scheme (AALS). The NRP involves acquisition of freehold farms by the state and their transformation into small farming units for qualifying beneficiaries: Namibians with the necessary asset base—mainly black elites rather than ordinary citizens. The Affirmative Action Loan Scheme is catering subsidised loans to these elites to buy commercial farms. These two instruments of land reform have delivered results that leave much to be desired. Land distribution remains skewed and beneficiaries of land resettlement

[8] Paul Carl Albert Rohrbach (1869–1956), Imperial Commissioner for settlement in DSW 1903–1906 who after his return to Germany held the position of Professor of Colonial Economics from 1907–1919 at the Berlin School of Economics (1906–1945) (Anker, 2005).

[9] After the Battle of Otavifontein between the Schutztruppe and the Union Defence Force on 1 July 1915, Germany officially surrendered to South African troops at Khorab on 9 July 1915 (McGregor and Goldbeck, 2014).

[10] The Odendaal Commission of 1962–1963 went beyond racial segregation, for it created various homelands unto which each ethnic group was to be confined. In essentialising ethnicity and justifying segregation, the report of the Odendaal Commission stated that '[…] primarily the individual belongs to the ethnic group into which he is born. Within the setting of his own group he develops to maturity, and accepts as axiomatic the comprehensive whole or totality of the group thinking and group behaviour as built up, adhered to and upheld by his own people in their own characteristic way' (RSA, 1964, p.57).

[11] According to Karns (1987, p.93), in '1977, the United States and four other major Western governments embarked on a unique diplomatic exercise in the hope of negotiating an agreement for the independence of the territory of Namibia, or South West Africa. The 'Contact Group' as it became known (or Western Five), consisting of the United States, the United Kingdom, France, Canada, and the Federal Republic of Germany, functioned actively from April 1977 until mid-1982 as an ad hoc multilateral mediating and facilitating team in close proximity to but not directly linked with the United Nations'.

[12] Chapter 3 of the Constitution of Namibia encapsulates 23 substantive articles on rights and freedoms that include respect for human dignity; equality and freedom from discrimination; right to property; and apartheid and affirmative action, amongst others (Namibian Constitution, 1990). The rights and freedoms under Chapter 3 form the basis on which Namibia's independence was negotiated—and came about because of the 1982 Constitutional Principles conceived by the WCG in 1982 (Kaapama, 2007)

programme have not been capacitated to farm productively. Implementing various resolutions adopted at the two national land conferences, however, has proven difficult over the years. Most of the resolutions adopted at the 1991 1st Land Conference (to tackle issues such as absentee landlords, underutilised land, farm size and numbers) did not see the light of day. Whether or not the implementation of resolutions (e.g. resettlement policy criteria, pre- and post-resettlement support, accessibility to land by marginalised groups, removal of the veterinary cordon fence etc.) adopted at the 2nd National Land Conference will be any different remains to be seen. Of the 177 resolutions adopted at the 2nd Land Conference, only nine were implemented between 2018 and 2020 (GRN 2020b). Some of the resolutions that have been implemented include differentiating resettlement programmes for all Namibians and dispossessed communities; establishing a body to purchase livestock from Northern communal farmers (Red Line); and introducing pastoral land rights in communal land.

The 2nd National Land Conference resolved to abolish the Willing-Seller Willing-Buyer (WSWB)[14] doctrine and replace it with the phrase 'alternative acquisition methods' in the Progress Report on the Implementation of the Resolutions of the 2nd National Land Conference (GRN, 2020b: 1). That 'alternative acquisition methods' were left unspecified says a lot about the flippant approach pervading land reform policy directions. Two years after the 2nd National Land Conference, the government back-pedalled on the abolition of the WSWB principle, citing the entrenched nature of Article 16 (Property) of the Namibian Constitution (1990). This change of position warrants a quotation in full:

> Government consulted thoroughly on the implementation of this resolution and found that Willing-Seller Willing-Buyer is entrenched in Article 16 of the Constitution that bestows persons with the right to sell and buy property in Namibia. This means that the express deletion of the principle of Willing-Seller Willing-Buyer from the [Agricultural (Commercial) Land Reform Act (No. 6 of 1995)] would be a futile exercise as the right to buy and sell will remain entrenched in Article 16.
> (GRN 2020b: 2)

As per Article 24 (Derogation) of the Namibian Constitution, the entire Chapter 3 (Fundamental Human Rights and Freedoms)—which contains Article 16 (Property)—is entrenched, and therefore no derogation is permissible even under public emergency, state of national defence or martial law. In other words, no change is legally allowed as far as Chapter 3 is concerned, unless a state of emergency is declared to warrant the suspension of fundamental rights and freedoms—up to this 'state of emergency' the contemporary imbalance of landownership is protected by the Namibian Constitution. As Namibia's independence was a negotiated settlement, a lot of concessions were made by the Constituent Assembly.[15] These concessions are apparent in Chapter 3 and it becomes clear that the Western powers were acting in their best interests. Perhaps in their naivety or because of the circumstances of the time, members of Namibia's Constituent Assembly who negotiated for independence thought that they would be able to fix the legal architecture once colonialism was out of the way. Alas, concessions made that gave birth to Chapter 3 of the Namibian Constitution continue to haunt Namibian society. In this view, the legal system becomes constitutive of structural violence as it legitimises the status quo that has been passed on from the colonial period. A direct link between Article 16 and the WSWB doctrine is palpable and it shows how the legal architecture structures land acquisition process that favours those holding land ownership rights. What on a first view appears to be the protection of property as a 'Fundamental Human Right and Freedom' in Article 16 of the Namibian Constitution (and the WSWB), turns out to cement the staggering imbalance of 70% to 16% landownership in favour of white farmers, if looked at in the fullness of time. Analysing this legal architecture can help us to unearth the fundamental building-blocks of enduring multigenerational poverty and inequality in Namibia. Once we concede that land ownership is an important factor in generating wealth and reducing poverty, land inequality becomes a harbinger for various deprivations in human development. A driving factor in its reproduction are land laws that preclude radical change in land ownership structures. Structural changes require deliberate interventions that treat the causes and not the effect—addressing land inequality must therefore tackle the factors that contributed to skewed land ownership patterns head-on (such as multiple farms ownership by the economic and political elites and disparities in land sizes between freehold and leasehold tenure systems).

Multidimensional Poverty and Land

Findings from the Multidimensional Poverty Index (MPI) Report 2021[16] expose the reproduction of deprivations across social groups in Namibia. The San people, who were the first to suffer land dispossession, are at the very bottom when it comes to socio-economic indicators captured by the MPI. These indicators are grouped under three dimensions: 1) Education; 2) Living Standards; and 3) Health. In terms of language—a proxy used to identify ethnicity—the highest incidence of multidimensional poverty was reported amongst the population whose main language was Khoisan (93%), followed by Rukavango (68%) and Zambezi (54%). This contrasts sharply to populations whose main spoken languages were English and German (each with 3%) (NSA, 2021). Figures of multidimensional poverty and its distribution are not surprising: they rather suggest a trajectory of multigenerational poverty with historical antecedents. In disrupting structural factors through which

[13] Paragraph 3 of the Addendum's introduction reads: 'The full and definitive text of the Constitution as adopted in annexed to the present report (annex I), together with a note entitled "Comparison of the Constitution of the Republic of Namibia with the 1982 Constitutional Principles" (annex II).' (UNSC, 1990, p.1)

[14] The WSWB is a free market land acquisition method based on the notion that those owning land retain full discretion in deciding when and how to make their land available for sale, while prospective buyers can only contemplate buying commercial agricultural land when it is offered for sale (Moyo and Hall, 2004). '[...] Under the WSWB regime of agricultural land reform, the rights and interests of existing landowners are fully protected, as they are neither compelled to sell against their will nor at a price with which they are not fully satisfied' (Kaapama, 2007, pp.36–37). Within the parameters of the WSWB doctrine, the failure of land reform can therefore be attributed to reluctance in selling land from those who hold land ownership rights and exorbitant prices of land when/ if the land is availed by landowners. *Consequently*, it is apparent that the deck is stacked against the landless.

[15] The Constituent Assembly was the body tasked with writing the Constitution of Namibia in 1989 (LAC, n.d.)

[16] The MPI is a multidimensional measure of poverty developed by the Oxford Poverty and Human Development Initiative (OPHI) at the University of Oxford. The index assesses the simultaneous deprivations that are experienced by people in a society, based on several identified dimensions of poverty in a particular setting. The MPI reflects who is poor, the dimensions they are poor in and the breadth of their deprivations. (NSA, 2021, pp.9, 14.)

poverty reproduces, addressing the 'land question' could go a long way. A number of indicators under the dimensions of the MPI such as food security, child nutrition, and access to drinking water directly implicate the centrality of land in addressing various deprivations.

Conclusion

More than a century after colonialism, land ownership patterns that have been bequeathed unto contemporary Namibia create a gridlock for the majority of the population and remain difficult to dislodge. As we have seen, a radical change remains a mirage in the midst of a legal stranglehold. This is compounded by elite capture of opportunities aimed at redistributing land using taxpayers' money while at the same time violating laws that prohibit fencing off communal land. Or how else do we explain the tendency of the politically well-connected owning huge tracks of land while the majority is squeezed in crowded communal areas? Two national land conferences—in 1991 and 2018 respectively—have not been able to decisively resolve most of the pressing issues around the land question. Adopting resolutions without adequate implementation continue to hamstring the process of redressing skewed land distribution. And even though the two national land conferences have kept the land question in the public discourse, the enduring structural violence besetting Namibia requires revisiting the notion of privileging peace that sits on shaky ground. As Walter Rodney insightfully observes, 'colonial governments were repeatedly speaking about the maintenance of law and order, by which they meant the maintenance of conditions most favourable to the expansion of capitalism' (Rodney, 1972, p.164). Backtracking on abolishing the WSWB principle was certainly done with the express aim of maintaining the status quo. The social time bomb sustained by an enduring structural violence renders the maintenance of law and order shaky. For as long as the implicit goal is to maintain the class privilege of previously advantaged white Namibians, maintaining social cohesion will be elusive.

References

Amoo, S. K., 2020. Land Reform in Namibia: Beyond 2018. In: W. Odendaal and W. Weerner, eds., *Neither Here nor There. Indigeneity, Marginalisation and Land Rights in Post-Independence Namibia.* LAC: Windhoek, Ch.1, pp.13–34.

Anker, J., 2005. Rohrbach, Paul. In: Neue Deutsche Biographie 22. [online] Available at <https://www.deutsche-biographie.de/sfz107803.html?language=en> [Accessed 30 January 2024].

Berry, S., 2002. Debating the Land Question in Africa. *Comparative Studies in Society and History*, 44(4), pp.638–668.

Bley, H., 1971. *South-West Africa under German Rule, 1894–1914*. Heinemann: London.

Botha, C., 2000. The Politics of Land Settlement in Namibia, 1890–1960. *South African Historical Journal,* 42(1), pp.232–276.

Dedering, T., 1993. The German–Herero War of 1904: Revisionism of Genocide or Imaginary Historiography? *Journal of Southern African Studies*, 19(1), pp.80–88.

Du Pisani, A., 2010. State and Society under South African Rule. In: C. Keulder, ed. 2000. *State, Society, and Democracy. A Reader in Namibian Politics.* Gamsberg Macmillan.

Forde, S., Kappler, S. and Björkdahl, A., 2021. Peacebuilding, Structural Violence and Spatial Reparations in Post-Colonial South Africa. *Journal of Intervention and Statebuilding*, 15(3), pp.327–346.

Francis, D. and Webster, E., 2019. Poverty and Inequality in South Africa: Critical Reflections. *Development Southern Africa*, 36(6), pp.788–802.

Galtung, J., 1969. Violence, Peace, and Peace Research. *Journal of Peace Research*, 6(3), pp.167–191.

Gann, L. H. and Duignan, P., 1977. *The Rulers of German Africa, 1884–1914*. Stanford University Press.

Gewald, J-B., 1999. *Herero Heroes: A Socio-Political History of the Herero of Namibia, 1890–1923.* James Currey.

___2000. Colonization, Genocide and Resurgence: The Herero of Namibia 1890–1933. In: J-B. Gewald and M. Bollig eds., *People, Cattle and Land: Transformation of a Pastoral Society in Southwestern Africa*. Köppe: Köln.

Gready, S., 2022. The Case for Transformative Reparations: In Pursuit of Structural Sŏcio-Economic Reform in Post-Conflict Societies. *Journal of Intervention and Statebuilding*, 16(2), pp.182–201.

GRN (Government Republic of Namibia), 2020a. *Report of the Commission of Inquiry into Claims of Ancestral Land Rights and Restitution*. Windhoek: Government Republic of Namibia.

___2020b. *Progress Report (October 2018–November 2020) on the Implementation of the 2nd National Land Conference*. Windhoek: Government Republic of Namibia.

Kaapama, P., 2007. Commercial Land Reforms in Postcolonial Namibia: What Happened to Liberation Struggle Rhetoric? In: H. Melber, ed., 2007. *Transitions in Namibia. Which Changes for Whom?* Uppsala: Nordic Africa Institute, pp.29–49.

Karns M.P., 1987. Ad hoc multilateral diplomacy: the United States, the Contact Group, and Namibia. *International Organization,* 41(1), pp.93–123.

LAC (Legal Assistance Centre). n.d. *Overview of the Namibian Constitution*. Factsheet Series No. 1 of 8. Windhoek: LAC/HSF.

Mbuende, K., 1986. *Namibia, The Broken Shield: Anatomy of Imperialism and Revolution*. Malmö: Liber Förlag.

McGregor, G. and Goldbeck, M., 2014. *The First World War in Namibia. August 1941–July 1915.* Gondwana Publishers.

Melber, H., 2019. Colonialism, Land, Ethnicity, and Class: Namibia after the Second Land Conference. *Africa Spectrum*, 54(1), pp.73–86.

____2005. Land & Politics in Namibia. *Review of African Political Economy*, 32(103), pp.135–142.

Miescher, G., 2012. The Rinderpest Cordon of 1896–1897. In: *Namibia's Red Line*. Palgrave Series in African Borderlands Studies. New York: Palgrave Macmillan, New York.

Moyo, S. and Hall, R. 2004. *Conflict and Land Reform in Southern Africa*. Centre for Conflict Resolution: Cape Town.

Namibian Constitution, 1990. *The Constitution of the Republic of Namibia*. [online] Available at <https://www.lac.org.na/laws/annoSTAT/Namibian%20 Constitution.pdf> [Accessed 29 February 2024].

NSA (Namibia Statistics Agency), 2018. *Namibia Land Statistics Booklet*. Windhoek: Namibia Statistics Agency.

____2021. Namibia Multi-Dimensional Poverty Index (MPI) Report 2021. Windhoek: Namibia Statistics Agency. [online] Available at <https://www.unicef.org/esa/media/9041/file pdf/UNICEF-Namibia-Multidimensional-Poverty-Index-2.pdf> [Accessed 29 February 2024].

Olusoga, D. and Erichsen, C.W., 2010. *The Kaiser's Holocaust. Germany's Forgotten Genocide.* Faber and Faber.

Pankhurst, D., 1996. *A Resolvable Conflict? The Politics of Land in Namibia*. University of Bradford, Department of Peace Studies.

Peters, P. E., 2013. Conflicts over Land and Threats to Customary Land Tenure in Africa. *African Affairs*, 112(449), pp.543–562.

Rodney, W.,1972. *How Europe Underdeveloped Africa*. East African Publishers.

RSA (Republic of South Africa), 1964. *Report of the Commission of Enquiry into South West Africa Affairs (Odendaal Commission).* The Government Printer.

SNLC (Second National Land Conference 2018). Safari Hotel Windhoek, Namibia, 1–5 October 2018. [online] Available at <http://dna.nust.na/landconference/landconference.html> [Accessed 29 February 2024].

Tizard, I. R., 2020. *Vaccines for Veterinarians*. St. Louis: Elsevier.

Tjirera, E., 2022. Land Inequality in Namibia: White Indifference, Elite Capture and Policy Inadequacies. *Namibian Journal of Social Justice*, 2, pp.197–203.

____2023. Namibia's Intractable Land Question. *Current History*, 122(844), pp.167–171.

UNSC [S/20967/Add.2], 1990. *Further Report of the Secretary-General Concerning the Implementation of Security Council Resolution 435 Concerning the Question of Namibia. Addendum, 16 March 1990.* [online] Available at <https://documents.un.org/doc/undoc/gen/n90/053/43/pdf/n9005343.pdf?token=70gjdz9xSAIOWhqBVx&fe=true> [Accessed 29 February 2024].

Van Wyk, J. 1992. The Namibian Land Conference – A Step Towards Addressing a Burning Problem. *SA Publiekreg/Public Law*, 7(1), pp.30–39.

Werner, W., 1993. A Brief History of Land Dispossession in Namibia. *Journal of Southern African Studies*, 19(1), pp.135–146.

____1998. *'No one will become rich'. Economy and Society in the Herero Reserves in Namibia, 1915–1946.* Basel: Basler Afrika Bibliographien: Basel.

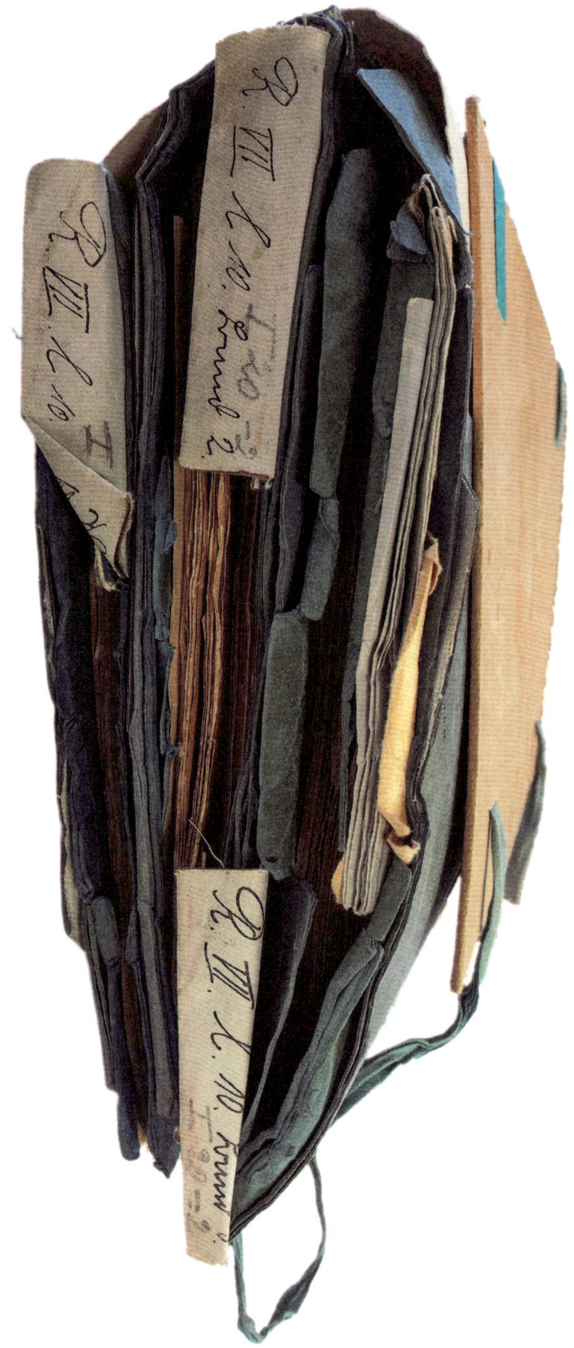

Extracts from the colonial archives

Bastian Herbst

Abstract: In parts, the history of mining in Tsumeb can be traced from the archives. In November 2022, the artist Sascha Mikloweit and the historian Bastian Herbst visited the National Archives of Namibia as part of Mikloweit's research project. Their aim was to get a more nuanced understanding of the history of mining in Tsumeb, and to add historical insights to what they have learned in Tsumeb itself, e.g., in conversations with people from the town. The consulted files shed light on the exploration of the Otavi region by white engineers and scientists, the building of the railways, the dangerous labour conditions in the mine, the difficulties of the OMEG to recruit workers (both Black and white), and the connections of the OMEG to the genocide against the Herero and Nama. The sources were carefully selected, condensed, and translated, in most cases, from the original German into English.

Keywords: colonial history of Namibia, history of mining in Tsumeb, building of railway, labour conditions, lead disease, Herero and Nama war, Otavi Mining and Railways Company (OMEG)

The following are extracts from historical sources at the National Archives of Namibia (NAN). For the purpose of this publication, they had to be considerably shortened. For a few days, the artist Sascha Mikloweit and I dipped into historical files to get a better understanding of the country's colonial history and its postcolonial setting. The extracts are therefore not an all-encompassing overview of the history of German colonialism in what was then named German South West Africa (GSWA). They serve as mere glimpses into the minds of those who colonised the lands of the people of Namibia—and, in a more limited way, the consequences of their decisions. This rather impressionistic approach was focused on the history of mining in Tsumeb, where for centuries Indigenous people of Namibia had extracted copper ores to use for weapons, tools, and jewellery. They did this at the exact same spot which Europeans supposedly 'discovered' in the 19th century to exploit the resources on an industrial scale.

The perspective given by the excerpts is naturally that of the colonisers (or rather, as the files show, the perpetrators of colonial violence). Stories that can be extracted from those files include: the first explorations of the region of Tsumeb by German and European scientists and engineers and their confrontation with, and treatment of, the first nations living there; the planning of mining and rail works in order to exploit, smelt, and ship copper and lead; the fight of the Indigenous inhabitants of the region, notably the Herero, against the exploitation of their land and its people(s)—and the genocide committed against them; the role of the Otavi Mining and Railways Company (OMEG) in the transportation of soldiers and military equipment in this war; the problems of the company in creating a profitable enterprise; the big profits they finally made; the harsh conditions under which European, and the even harder conditions under which Black miners had to work in the mines; the resulting problems to recruit workers (both European and Indigenous) to work in this hostile and very often deadly environment; and finally, the racist treatment of Black people and workers by the mining company, German administrators, and many of their white colleagues, in a system that would later be known as Apartheid. While the sources suppress many realities, they serve as a reminder that history is still being written in the region of Tsumeb: namely, into the hearts, minds, lungs, and bodies of its people through the ongoing devastation of the region, consequence of political and industrial decisions of the former colonisers and their successors.

Christopher James[1] (Tsumeb) to Otavi Mining and Railways Company [OMEG] (Berlin), 10 Dec 1900.[2]

Dear Sirs,

[...] Natives. Last month I paid some 250 boys wages, averaging M[arks] 15 per month each.[3] Owing to many Ovambos, and Hereros, having come to be useful with both hammer and drill, the wages [...] have increased [...] to M. 20 a month [...]. Some little trouble is caused, by boys suddenly leaving, and it is a little annoying, when you have spent time in teaching a boy, to find he has gone, first when he has become useful. I regret very much having no means of stopping this exit, and console myself with the thought, that no man has yet left through illtreatment[!]. More I cannot do, as the camp is not supplied with a surrounding wall and a locked gate!

Christopher James (Tsumeb) to v. Hansemann (OMEG, Berlin), 29 Dec 1900.[4]

Mining. Since my last letter, I have finished all work on the No.1 level and continued sinking the two new shafts. The East Shaft passed through a patch of loose earthy limestone, but is now in solid rock of no value, as usual. The West Shaft has encountered, latterly a soft clayey matrix, rich in carbonate of lead, with a little Copper, the percentages of both metals diminishing with almost every foot in depth. The present bottom is soft clay 14% Lead and 3% Copper. [...]

Foul air. One of my men going down too soon after the blower started, became insensible from carbonic acid gas and

[1] Christopher James was an American mining engineer who was contracted by OMEG in 1900 to explore the Otavi region in order to find exploitable copper deposits.

[2] NAN, ZBU 1547, R.VII, L.10, Vol. 1.

[3] 15 marks were worth about 0.73 pounds sterling in 1900 – an amount a skilled labourer in Britain would have earned in about 2-3 days.

[4] NAN, ZBU 1547, R.VII, L10, Vol. 1.

Fig. 28

was rescued with great difficulty by tying him in the bucket. [...]
Shipment of ore. Yesterday I consigned by bullock wagons 181 bags of ore, weighing about 9 tons and one box of samples, to the trading Company, to be forwarded by the first ship to Hamburg. [...]
With regard to the disposal of these samples of ore, I beg to advise that each of the 6 descriptions be crushed separately and passed through a half inch sieve and thoroughly mixed. Then I would suggest that the half of each be sent to Mssrs. Henry Bath & Sons,[5] Swansea and the other half retained in Hamburg. In this way the price from both markets can be tested. [...]

Christopher James (Guchab mine) to v. Hansemann (OMEG, Berlin), Sep 1901 (transl.).[6]
Dear Sir,
I have the pleasure to hand you herewith my report on the Tsumeb[,] Guachab[,] Asis[,] and Nagaib ore deposits[.]
The expedition, consisting of 33 persons, left Southampton on the 7 June 1900 and arrived at Swakopmund on the 28 June 1900 at 10/30 a.m.
Journey: After a very tedious journey the expedition reached Tsumeb on the 13th August 1900. [...]
Labour: There are five distinct races which I have had to deal with viz. [...]
Hereros: This is a fine race of men powerful and physically strong. They are willing and quick to learn, good natured and cheerful with their work and I look to these men in particular, if properly handled, to become in short time first class hammer and drill hands. [...]
Bushmen: [A]re light, active, intelligent, little men, quite unsuitable for heavy work, but most useful for surface hands, about machinery, and especially for cutting and handling logs.
Berg-Damaras and Hottentots: Several of my drivers belong to these races, and I have found them reliable men at their work.
Natives wages: [...] Twenty shillings per month[7] is the highest wage any man has received from me for working the mine: and I think that provided the rations of food are maintained on the same basis as I have instituted, this wage need not be increased. This matter of wages must be very carefully handled as the boys are quite capable of striking for more if they thought they could get it. [...]

OMEG. Annual report for the second financial year from 1 April 1901 to 31 March 1902 (transl.).[8]
The previous year's annual report of the management already mentioned all the expeditions which the company had undertaken both to investigate the copper deposits in the Otavi area and to determine the most suitable railway connection between this area and a port on the west coast. From August 1900 to September 1901, [...] Christopher James, assisted by several officials, 27 miners and several hundred indigenous labourers, carried out a thorough survey of all known copper mines in the Otavi area [...]. [...] The work at the Tsumeb mine under the leadership of Mr James was a great success. He calculates that 293, 330 tons of high grade ore [...] and 190, 519 tons of low grade ore [...] are in place and ready for extraction. With the mining equipment proposed by Mr James, it would take about 4 years and 8 months to extract the high-grade ore, and during the same period [...] a profit of £390,000 per annum[9] would be made on 7,761 tons of copper at a copper price of £60 per ton, and on 14,679 tons of lead at a lead price of £12 per ton. [...]

Meeting of the Board of Directors of the OMEG in the office building of Disconto-Gesellschaft, Behrenstr. 43/44 [Berlin], 19 Nov 1903 (transl.).[10]
[...] With regard to the draft report of the management, it was considered desirable, [...] [:] That the mention of the Herero chiefs be deleted [...] and [...] that the interest of the German Colonial Society for Southwest Africa in particular in meeting the demands of the Otavi Society be emphasised even more strongly; [...] The Chairman then made confidential remarks about a correspondence with the Colonial Department of the Foreign Office concerning the labour issue. He explained that in view of this and the unrest that had broken out in the protectorate, the possibility had to be reckoned with that the company would be forced to declare that it would not be able to meet the deadline set for the completion of the railway construction and the start of mining operations by the end of 1906.

OMEG. Annual General Meeting in Berlin, 17 Dec 1904 (transl.).[11]
[...] The reasons that prompted our company to connect the copper ore deposits at Otavi and Tsumeb with the sea by building a narrow-gauge ore railway with a track gauge of 0.6 m between Swakopmund and Tsumeb were explained in detail in the report for the third financial year. Under the contract concluded in July 1903, the Arthur Koppel company[12] took over the construction of the railway [...] and undertook to complete the entire line [...] within 2½ years [...]. [...] In August 1903, the general contractor sent the first engineers and officials to Swakopmund [...]. The preparatory work began in October; after the plans were approved by the supervising engineer [...] construction began in mid-December. [...] Thanks to the energetic commencement of the work, the final preparatory work had been completed up to km 130 and the earthworks and track laying had been completed up to km 9 when the Herero uprising broke out on 11 January 1904, which quickly affected most of the area touched by the Otavi Railway and naturally impeded the progress of our work in many ways. First of all, labour conditions were disturbed in the most sensitive way. Most of the indigenous labourers had to be removed for safety reasons; the few who remained were not sufficient to achieve a productive output. The general contractor therefore decided to continue construction with European labourers and made arrangements with a

5 Henry Bath & Sons is a company specialising in the global trade and warehousing of commodities. It was established in Swansea in 1794 as a (soon thriving) copper smelting company by the only 17-year-old Cornish Henry Bath (1776–1844). Bath would later become one of the co-founders of the London Metal Exchange. Cf. for an official history: https://www.henrybath.com/assets/downloads/Complete%20Henry%20Bath%20History.pdf

6 NAN, ZBU 1547, R.VII, L.10, Vol. 2.

7 According to the UK National Archives 'currency converter', 20 shillings would be worth approx. £75 in today's value, the equivalent of two days' wage for a skilled tradesman in 1900 Britain.

8 NAN, ZBU 1547, R.VII, L.10, Vol. 2.

9 Approx. £30 million in today's value.

10 NAN, ZBU 1547, R.VII, L.10, Vol. 2.

11 NAN, ZBU 1547, R.VII, L.10, Vol. 2.

12 Arthur Koppel & Co was a German railway equipment and construction company founded in 1885 by entrepreneur Arthur Koppel (1851-1908). Based in Berlin, it focused particularly on foreign and colonial markets. For its role in the construction of the Otavi line see Bravenboer and Rusch (1999), p. 60.

contractor who had a large number of Italian workers at his disposal.[13] This force arrived in the protected area in April; despite numerous difficulties and work stoppages, by the end of June the earthworks had reached kilometre 36 and the track up to kilometre 25.

The supply of building materials was then made considerably more difficult by the war, as the numerous military transports, which became necessary due to the deployment of considerable forces from the homeland to suppress the uprising, made heavy demands on the Woermann Line.[14] [...] We have endeavoured to promote the construction of the railway to the best of our ability despite these great difficulties. At the same time, we considered it our duty to provide the Imperial Government with our means of assistance to facilitate military transport in the protectorate. [...]

OMEG. Annual General Meeting in Berlin, 20 Dec 1905 (transl.).[15]

[...] In July 1904, the Imperial Government approached the company with the question of how the construction [...] could be accelerated in the military interest. The negotiations on this led to a contract concluded with the Imperial Government on 4 August 1904, in which the construction between km 80 and Omaruru (km 236) was promised in return for the granting of a kilometric surcharge and the assurance of sufficient military protection for the continuation of construction beyond Omaruru. At the same time, the company took over the construction of the 14 km branch line from Onguati to the main station Karibib of the government railway in return for the granting of public transport on the Swakopmund-Usakos-Onguati line and the establishment of a permanent military post at the terminus of the railway at the copper mines of Tsumeb. [...]

OMEG. Report of the management for the sixth financial year, 1 April 1905 to 31 March 1906, 17 Dec 1906 (transl.).[16]

[...] The continuation of the work beyond Omaruru was called into question after the departure of the Italian labourers, because the Ovambos came forward in increasingly sparse numbers and almost completely stayed away towards the end of 1905. The considerable difficulties which also hampered the progress of construction due to the long distances, the great difficulty in obtaining water for the work crews, etc., prompted the general contractor towards the end of 1905 to apply for a kilometric supplement to the construction work on the line from km 300 onwards, in return for which it undertook to have the entire line ready for operation by 31 December 1906 at the latest. [...] the terminus at Tsumeb was reached on 15 March 1906. [...] the labour issue gave cause for concern in that, following the departure of the vast majority of Italian workers, there was no guarantee that a replacement would be found. This replacement was eventually found in native labourers, who willingly went to work together with the foreigners who remained at work. The Imperial Government helped the company considerably by transferring Herero prisoners of war from the collection camps. [...] Experience gained during the construction period has shown that the originally planned joint use of the government railway station in Swakopmund by the Otavi Railway is not feasible, as the former is too heavily burdened by traffic for the Schutztruppe and also does not permit the necessary expansion for Otavi Railway traffic. [...] a separate railway station was therefore built for the Otavi Railway in the southern part of the city. [...] On 1 March 1906, a contract was concluded with the Imperial Government for the transport of large quantities of war and government goods to Karibib[...]. The transport volumes agreed by 28 February 1906 in the year under review on the basis of the contract of 4/5 August 1904 included: 1,098 passengers, 15,393 tonnes of goods, 2,515 horses and mules, 3,205 head of small livestock and 315 cbm of water for the cattle and horse collection points of the Schutztruppe. [...] Since the end of the year under review, the tireless efforts of the general contractor have succeeded in completing the earthworks by Tsumeb on 1 August and completing the track to the terminus by 24 August.

Appendix to the report on the meeting of the Board of Directors of the OMEG on 6 Dec 1907 (transl.).[17]

[...] The main aim of his [Mr Gaedertz's[18]] efforts in Tsumeb was directed towards the labour question. The white staff in Tsumeb, about 200 men, are of somewhat better character than the railway employees. Their constancy and integrity are not first class, but higher than among the railwaymen. [...] But the most difficult point is black labour. Many of his letters dwell on the appalling difficulties caused by the lack of numbers and the lack of permanence of black labour in Tsumeb and describe his tireless efforts to procure a larger number of black workers. In the end, he pinned his hopes on a meeting with Excellency von Lindequist[19] in Tsumeb [...]. I see from a memorandum [...] that Mr Gaedertz, after [...] pointing out that 1,000 black workers are necessary for the full working of our mine including the smelter, whereas in August only [...]388 men on average were employed in Tsumeb, seriously advocated the admission of Chinese workers to the governorate. [...]

25th meeting of the Board of Directors of the OMEG, 27 May 1908 (transl.).[20]

Privy Councillor Schöller [...] announces that [...] the smelter in Tsumeb has not delivered favourable results, whereas the mining and railway operations have delivered satisfactory results [...]. [...] Copper ore smelted in Tsumeb results in a loss which the management estimates at 480,000 marks; [...]. The management estimates the operating results of the railway for the current financial year [...]: [...] Profit approx. M 2,000,000.[21]

[13] On the history of foreign labourers in GSWA see Lyon (2024), esp. pp. 76-82.

[14] The 'Afrikanische Dampfschiffs-Actiengesellschaft (Woermann Linie)' was, from the late 19th century until the Second World War, one of the leading shipping companies in the world. Founded first as a trading house in 1837 by Adolph Woermann and based at 'Afrikahaus' in Hamburg, it specialised in the transportation of commodities and passengers between Germany and Africa. As such, it was closely entangled with German colonialism in West and South West Africa, shipping, for instance, German troops to Namibia during the genocide against the Herero and Nama. See Todzi (2023).

[15] NAN, ZBU 1547, R.VII, L.10, Vol. 3.

[16] NAN, ZBU 1547, R.VII, L.10, Vol. 2.

[17] NAN, BBW 11, Vol. 5.

[18] Baurat [German for 'government building surveyor'] Gaedertz was sent by the OMEG to inspect the railway, the mine and the landed property of the company. He arrived in Swakopmund on 2 June and departed from there on 21 September 1907. Having been injured in an accident in the 'protected area', he died shortly after the trip.

[19] Friedrich von Lindequist (1862–1945) was Governor in German South West Africa from November 1905 to May 1907, and as such the successor to the infamous Lothar von Trotha. As the first civil governor of GSWA (his predecessors had all been military men), he supervised the ending of the military campaign against the Herero and Nama, and, at the same time, interned many of them in concentration camps. After his return to Germany, he worked for the Colonial Office and co-founded the colonial revisionist 'Vaterländische Partei' (Patriotic Party). During the 'Third Reich' he served as superior advisor in the 'NSDAP Office of Colonial Policy'.

[20] NAN, ZBU 1547, R.VII, L.10, Vol. 2.

[21] Equivalent to approx. 100,000 pound sterling under the Gold Standard, or £7.8 million in today's value.

D.-No. 5870, The Imperial District Commissioner, Grootfontein, to the Governor, Windhoek, 3 Oct 1908 (Secret) (transl.).[22]
I take the liberty of bringing to Your Excellency's attention a number of abuses in the mining operations of the Otavi Mines and Railway Company. Recently there have been repeated cases of indigenous Ovambo and Herero labourers being seriously injured or killed while blasting rock. The investigations carried out have not revealed the guilt of any particular perpetrator[.] These accidents are more likely due to the fact that inferior explosives, especially poor-quality fuses, were used in the blasting operations. [...].

Furthermore: About 14 days ago, an Ovambo was killed when he tried to climb into a cage that was in the process of being lifted. If the cage had been operated properly by trained personnel, the incident should have been avoided.

Furthermore: About 4 weeks ago, 6 wagons loaded with ore rolled off a track within the minefield, derailed and were completely smashed. No people were injured, but this could easily have happened. This accident was only made possible by a complete lack of proper supervision. Furthermore: in the mine itself, as a worker told me, mining is supposed to be so careless and haphazard that falling rock poses a permanent and serious danger to the black and white workers. [...] the foreman in charge is said to have explained that it didn't matter, the only thing that mattered was to extract as much ore as possible. [...]

Otto Schmatolla (Pharmacist and Chemist, Berlin) to the Governor of German South West Africa, 5 Nov 1908 (transl.).[23]
Your Honour,
I kindly ask you to take note of the enclosed information. They concern the mining engineer Karl Schmatolla who was injured in an accident in Omaruru. I humbly request that the Otavi Mines and Railway Company be obliged to support the accident victim, as it is trying to evade this obligation in a dishonest manner [...]. [...]
Otto Schmatolla

Appendix: Report on the mismanagement of the OMEG in its mining operations in Tsumeb [...], by Otto Schmatolla, 30 Oct 1908 (transl.).[24]
[...] The abuses both in the village of Tsumeb itself and in the mine, caused by the management of the Otavi Mining and Railway Company, appear to be of such a blatant nature that they have already claimed the lives of numerous immigrant Germans, Italians and other members of white nationalities as well as natives. The directors [sic] Gattmann, mining engineer Haschke, mining assessor Hassinger and, above all, the local directors in Behrenstrasse are primarily responsible for this. As introductory evidence, a private letter addressed by the miner Risse in Tsumeb on 23 June 1908 to his colleague who had been dismissed from the Tsumeb mine should be singled out:

> 'Tsumeb, 23 July 1908.
> Dear friend Kintscher! [...]A lot has changed here, a new foreman arrived a fortnight ago. [...] August Ascheid [...] is leaving tomorrow, he was in hospital for two months with lead poisoning [...][.] Weiss and Schuecki are also in hospital and Klein and Schmidt were there too, the latter is very poorly. We have also had two serious accidents, one a hole in the first floor, two K[***] drilled dead on old dynamite, and the fat Italian Beligonda in the open-cast mine with powder which ignited when he was loading it, as he had smashed the cartridges on Hasc[h]ke's orders in order to be able to load them. The doctor stitched up a cornea in his right eye, everything else was burnt, arms, face, but he got off all right. I'm just curious when the law will finally intervene. [...]'

The mine management showed an egregious neglect of safety precautions in its mining operations. An Italian named Komisso fell into the shaft and died for lack of the most primitive safety precautions. The warning signals were neglected before blasting began, and a worker named Josua had his skull smashed in by a flying piece of explosive. Equally serious is the negligence of the mine management against the mine gases erupting in the mine. [...] Due to their volcanic nature, the air in the shafts is heavily poisoned by carbonic acid, so that work has to be interrupted at times and, according to various witnesses, the lights go out. As the only precautionary measure [...] the mine management has set up the blower of a field forge, to which the wheel of an ox cart is attached for propulsion. Director Gathmann has the audacity to refer to this blower as a mine ventilator, but the most outrageous thing is that if the miners [...] actually try to keep this blower in motion for a longer period of time by a number of natives, the superior engineers prevent this and drive the workers into the contaminated pits. [...]

'In the matter of Schmatolla against Otavi Mines', [court of?] Lüneburg, [trial?] transcript, 6 Oct 1908 (transl.).[25]
Appeared when called [...] witness Gathmann (Operations Director of Tsumeb Mine):

> [...] there have been a number of occasions when carbonic acid has erupted from the third level of the Tsumeb mine [...]. Work in the affected areas [...] had to be stopped [...] until fresh air was provided by natural or artificial means. [...] The fan used for irrigation was only small and was operated by hand. However, in combination with natural ventilation, it was sufficient to make the air suitable for breathing.

[22] NAN, ZBU 1547, R.VII, L.10, Vol. 3.

[23] NAN, ZBU 1547, R.VII, L.10, Vol. 3.

[24] ibd.

[25] ibd.

> A better device [...] was not available. [...] To my knowledge, the mining authorities in Windhoek never criticised the ventilation at the Zumeb [sic] mine. [...]

Let's hear what the injured mining engineer reported:

> 'Carbonic acid gases were stored in the deepest working places [...], so that the operation of the mine there had to be stopped [...]. Worth mentioning are the gas channels that the miners Schimmek and Fischer had to open. After this, operations had to stop for 14 days as no light would burn there. Later, oil lamps burned sparingly and candlelight went out at the slightest movement, which is why the latter was abolished by the management. During this time there were also cases of illness among the white workers, which in one case was [fatal] [...]. Since on average all the white labourers working in the mine fell ill, the management felt compelled to operate the mine underground only with indigenous Hereros and Ovambo N[***] and to place only one white foreman [...] in front of each of them. The foremen Piohl and Harwey [...] who only worked for a few weeks, complained of headaches and vomiting. Likewise, the natives No 1–60 in my shift book. A native Herero named Nauke, for example, also suffered from great emaciation, weakness and had a swaying gait. In this condition he was discharged from the company hospital there as cured. A young Herero Mika [...] fell ill as a result of inhaling the gases and died in the military hospital. The reason given was consumption [tuberculosis] (due to emaciation) although consumption does not occur at all in German South-West Africa. I myself lost 33 pounds in body weight [...]. This was followed by headaches and irritability [...]. And when the management ordered me to stand in for other sick white foremen [...] I began to vomit violently, which was followed by a fever that knocked me out. Complaints [...] were ignored by the management [...]. The white foremen had to give up their jobs within a few weeks in order to preserve their health. The natives also fled and looked for work elsewhere [...].'

Draft for a letter, District Commissioner (Grootfontein) to Dr Heimann (Otavi Mine, Tsumeb), 25 Nov 1909 (transl.).[26]

1. [I]n view of the great labour shortage from which the Otavi mine is currently suffering, I would like to support you. However, it will not be possible to exert any pressure on the Ovambos as [...] they prefer to go to L[üderitz]bucht because of the higher wages and because they do not have to work underground there. Of course, I have nothing against you sending a native to the Ovambos to recruit labourers. Perhaps you could also use some families of the captured Hottentots in Grootfontein for light labour. However, the people should not be employed underground – except with their express consent – as they are too weak and would soon die. The wages would have to be paid ¾ to the people themselves and ¼ to the camp treasury[?]. I would ask you to liaise with the camp [commandant] in Grootfontein, who may initially give you 3 to 4 of the Hottentot prisoners of war and their families. However, they must be closely watched to ensure that they do not run away. [...]

Link (Imperial District Office, Grootfontein) to Imperial Government, Windhoek, 30 Sep 1910 (transl.).[27]

As reported to the mining authorities by telegraph, a landslide occurred in the Tsumeb mine on Saturday morning [...]. In the third level, part of the rock broke away [...] and buried the natives below. 3 Owambo were dead, 3 injured. As a result of this break, the backfill on the second level sank. [...] This accident caused great excitement among the entire workforce of the mine, which was exacerbated by the fact that [...] a fissure was discovered which had opened up in the last few days. When I spoke to the director Heimann early on Tuesday the 27th, about 20 whites were celebrating, all the Kapboys and about 250 Owambo had left Tsumeb that morning. I had the Owambo brought back, which the police succeeded in doing after some effort; after I had promised them to go to the mine myself and see if everything was in order, they went to the work-yard and resumed work the next day. [...]

Heinrich Vogel[28] (Berlin, Hotel zum Reichstag) to [unknown], 14 Dec 1910 (transl.).[29]

Confidential!

Your Excellency

I have the honour to inform you that a number of miners from the Siegerland[30] had taken work with the Otavi Mining Company [...]. I am informed that they resigned from this work before the end of the agreed period [...] because it was concealed from them [...] that employment in the Tsumeb mine entailed the danger of lead disease [...] and that the people were more or less severely ill with this disease. Until now, my author considered the complaints about this to be exaggerated. Recently, however, they have been confirmed to him by a person who returned from South West Africa. According to the latter, lead sickness is indeed rampant in Tsumeb; the Otavi Mining Company is also said to have treated the Siegerland miners less than favourably. Some are said to have left work before the end of their contract due to illness. Others simply stayed away from work and were then sued by the company. Third parties are said to have demanded a pay rise from 12 M to 14 M because their health was at risk. This wage increase was allegedly rejected by the management in Berlin, but satisfied by the managing director out of his own pocket. There is also said to be great bitterness because the company has reserved the right to terminate the contract at any time in the inspections, but the miners are bound to the entire duration of the contract. Such unequal treatment no longer corresponds to the conditions under which miners

[26] NAN, ZBU 2065, W.VI, H.6, Vol. 1.

[27] NAN, ZBU 1547, R.VII, L.10, Vol. 5.

[28] Heinrich Vogel (1856-1934) was a mining assessor and mining director in different parts of Germany, as well as member of the Reichstag for the National Liberal Party (NLP) from 1909 to 1912.

[29] NAN, ZBU 1547, R.VII, L.10, Vol. 5.

[30] Region in the west of Germany in what is now the federal state of Rhineland-Palatinate. At the beginning of the 20th century, the Siegerland was one of the most important centres of iron ore mining and processing in Europe.

in Germany work and to which they are accustomed. [...]
Signed: Vogel (Member of the Reichstag)

Telegram by Dr Heimann (Tsumeb) to Imperial Governorate Windhoek, 4 Jul 1911 (transl.).[31]
[...] not a single Ovambo has arrived here for a month. [...] immediate remedial action required as factory interests are highly jeopardised.

Minute by Imperial District Office, Grootfontein, Subject: 'Ovambo labourers for the Tsumeb mine', 23 Apr 1912 (transl.).[32]
[...] It is a fact that the Ovambo do not like going to Tsumeb. The main reason for this is the work underground, which they do not like. It is also possible that the fact that they do not get blankets plays a part. [...]

Notes on the ore deposits near Tsumeb, no date (ca. 1913) (transl.).[33]
[...] To prevent lead sickness as far as possible, the mine management has forbidden eating and smoking in the mine. Furthermore, the company provides the white workers with free pit clothes, trousers and skirts, which are washed at the mine's expense. After the shift, there is an opportunity to take a bath in the smelter's large cooling water reservoirs [...]. Bathing is organised in such a way that the upper fresh water basin is reserved for the whites, while the lower cooling basins are available to the coloureds.

[31] NAN, ZBU 2065, W.VI, H.6, Vol. 1.

[32] Ibd.

[33] NAN, ZBU 1547, R.VII, L.10, Vol. 6.

Sources

National Archives of Namibia, Windhoek (NAN)
NAN, BBW 11, Vol. 5.
NAN, ZBU 1547, R.VII, L.10, Vol. 1.
NAN, ZBU 1547, R.VII, L.10, Vol. 2.
NAN, ZBU 1547, R.VII, L.10, Vol. 3.
NAN, ZBU 1547, R.VII, L.10, Vol. 5.
NAN, ZBU 1547, R.VII, L.10, Vol. 6.
NAN, ZBU 2065, W.VI, H.6, Vol. 1.

References

Bravenboer, B., Rusch, W., 1999. *The first 100 years of state railways in Namibia*. Windhoek: TransNamib Museum.
Lyon, W. B., 2024. *Forged in Genocide. Migrant workers shaping colonial capitalism in Namibia, 1890–1925.* Berlin & Boston: De Gruyter Oldenbourg.
Todzi, K. S., 2023. *Unternehmen Weltaneignung. Der Woermann-Konzern und der deutsche Kolonialismus 1837–1916.* Göttingen: Wallstein.

Conversions from Marks to British Pounds were made with:
https://www.historicalstatistics.org/Currencyconverter.html

The value of wages was calculated with the UK National Archive's 'Currency Converter':
https://www.nationalarchives.gov.uk/currency-converter

Fig. 29
Fig. 30 [p.114]
Fig. 31 [p.117]
Fig. 32 [p.118]
Fig. 33 [p.121]
Fig. 34 [p.122]
Fig. 35 [p.125]
Fig. 36 [p.126]

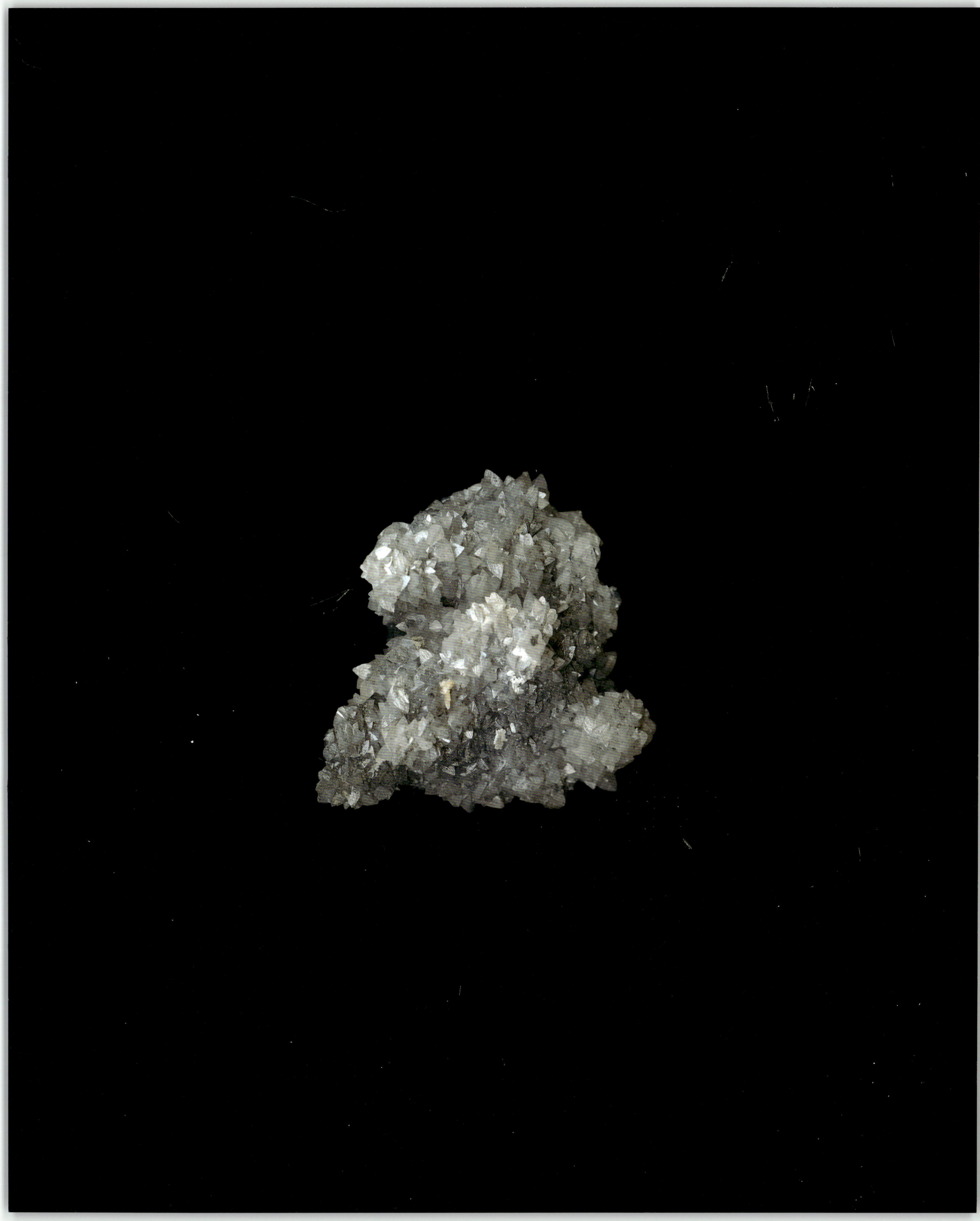

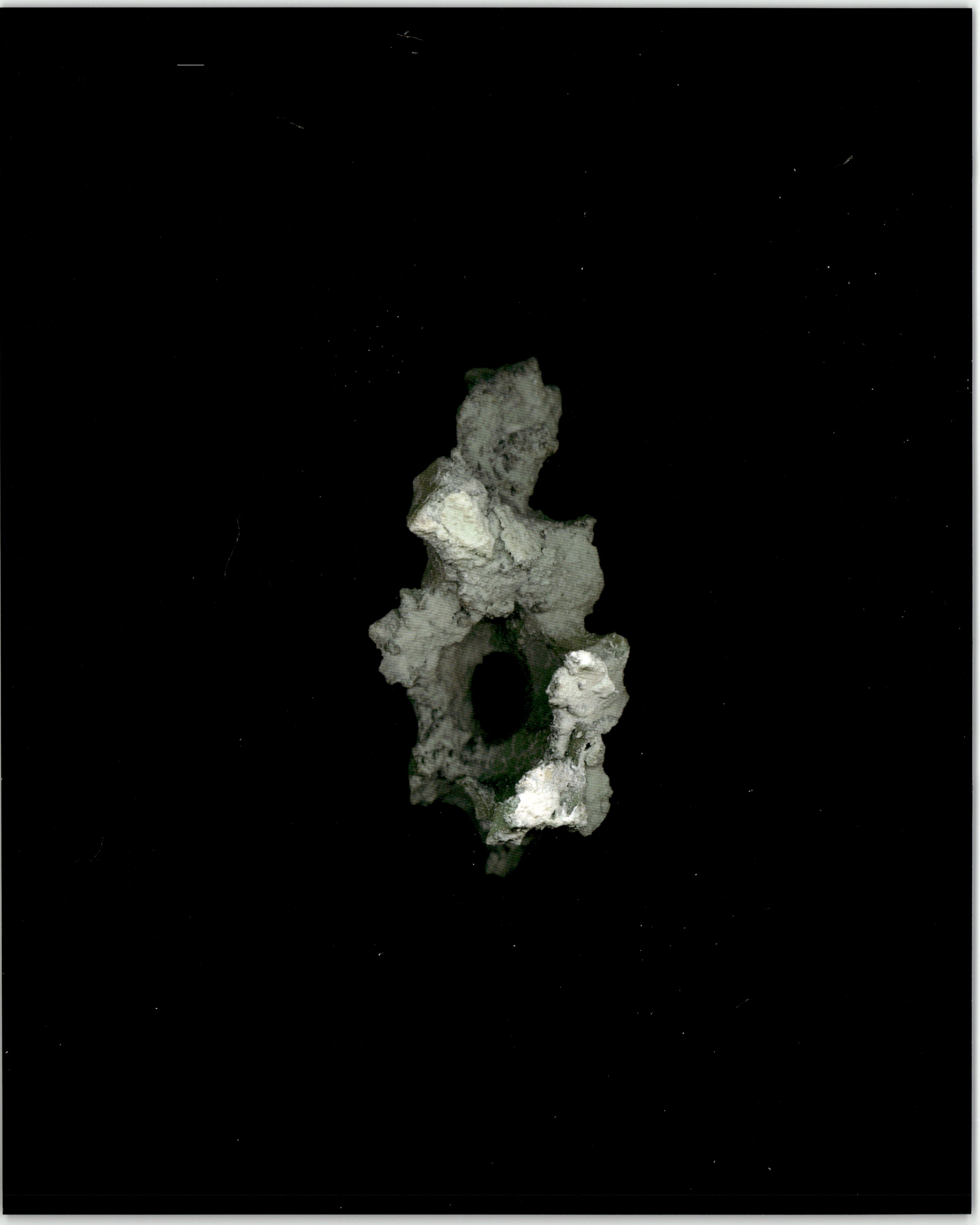

The Colour of Ripe Mangos

Veripuami Nandeekua Kangumine

My father wears a thick copper band
On the wrist of his right hand
That has faded from the fuchsia
Of hibiscus juice

To the warm salmon colour
Of a faded bruise
Like something he never removes
Never takes off

He says it is for his arthritis
For his back pains
For his knee that hurts when it rains
It is also to return lost love

His hands are wrinkled
In the way that is evident
Of someone who works around water
And rough from the hard labour of someone who digs

And unburies things for a living
Who spends months at a time
In the deepest part of the Earth's epidermis
Searching for God in metal form

But they still look nice to hold
And when he puts his hands
Around his himself
He looks like a young man

The very man my mother fell in love with
The father who taught us

To only pick mangoes
The colour of rubies
But weeks can go by without him talking to us
He thinks he is still imprisoned underground
Buried alive somewhere
Where the dark is both night and day

And the light of a torch comes nowhere
Close to the purplish red of a fire's eye
Or the bright golden hue
Of a summer's sunset

It is as if he is an excreted waste
The earth has spit out
His eyes sunken
His skin ash

He takes months to re-adjust back into
The rhythm of our home
Hiding away in the garden
Busying his hands
Always digging behind the house
Under the guava tree
Searching for the coral red
The inside colour of a bruise

The afternoon after we find him
Staring at his own blood
until he faints
He tells the doctor
It reminded him of the colour of rust

After years away from the mines
He still searches for anything
that resembles the deep magenta
that glows only in the darkness of a cave.

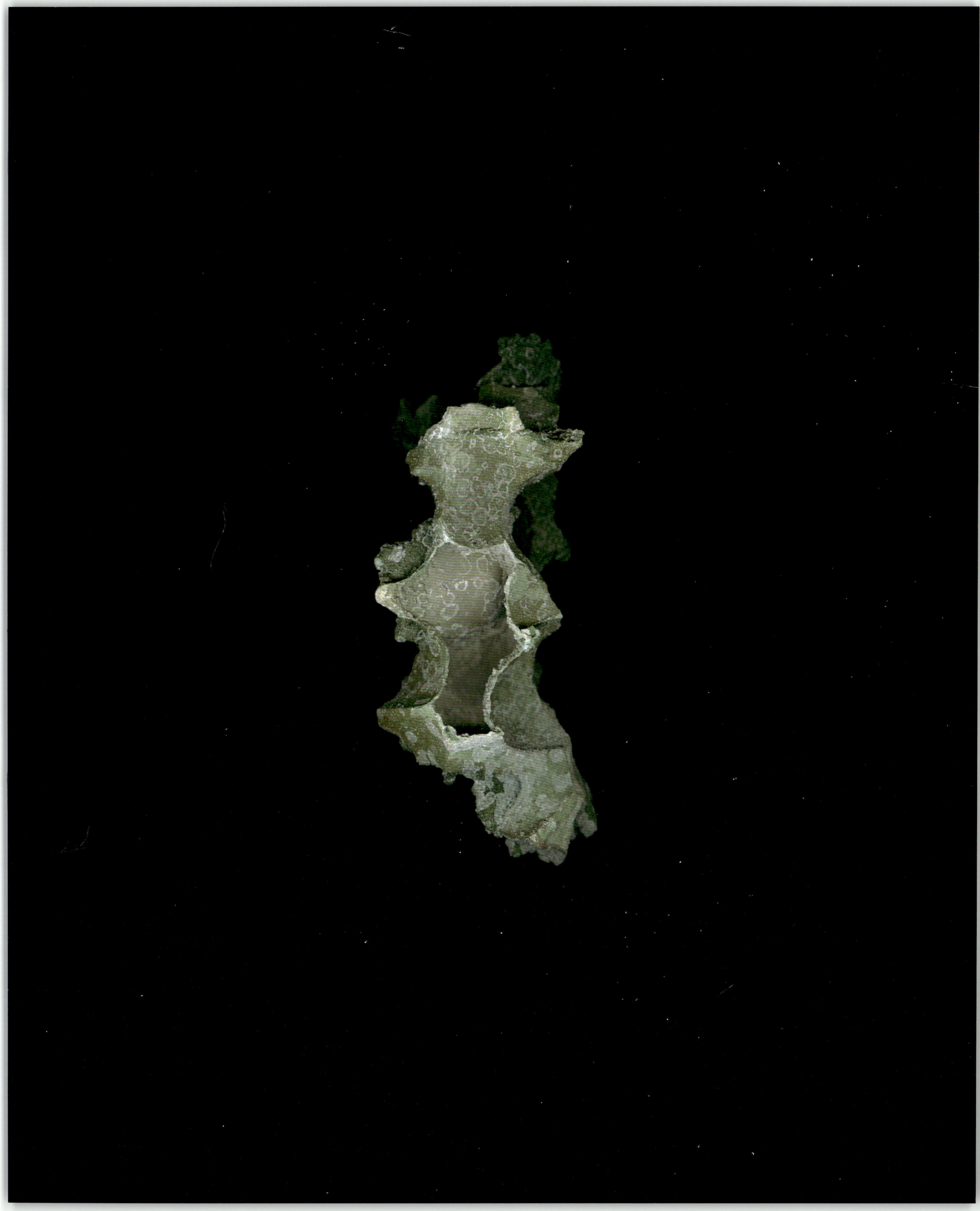

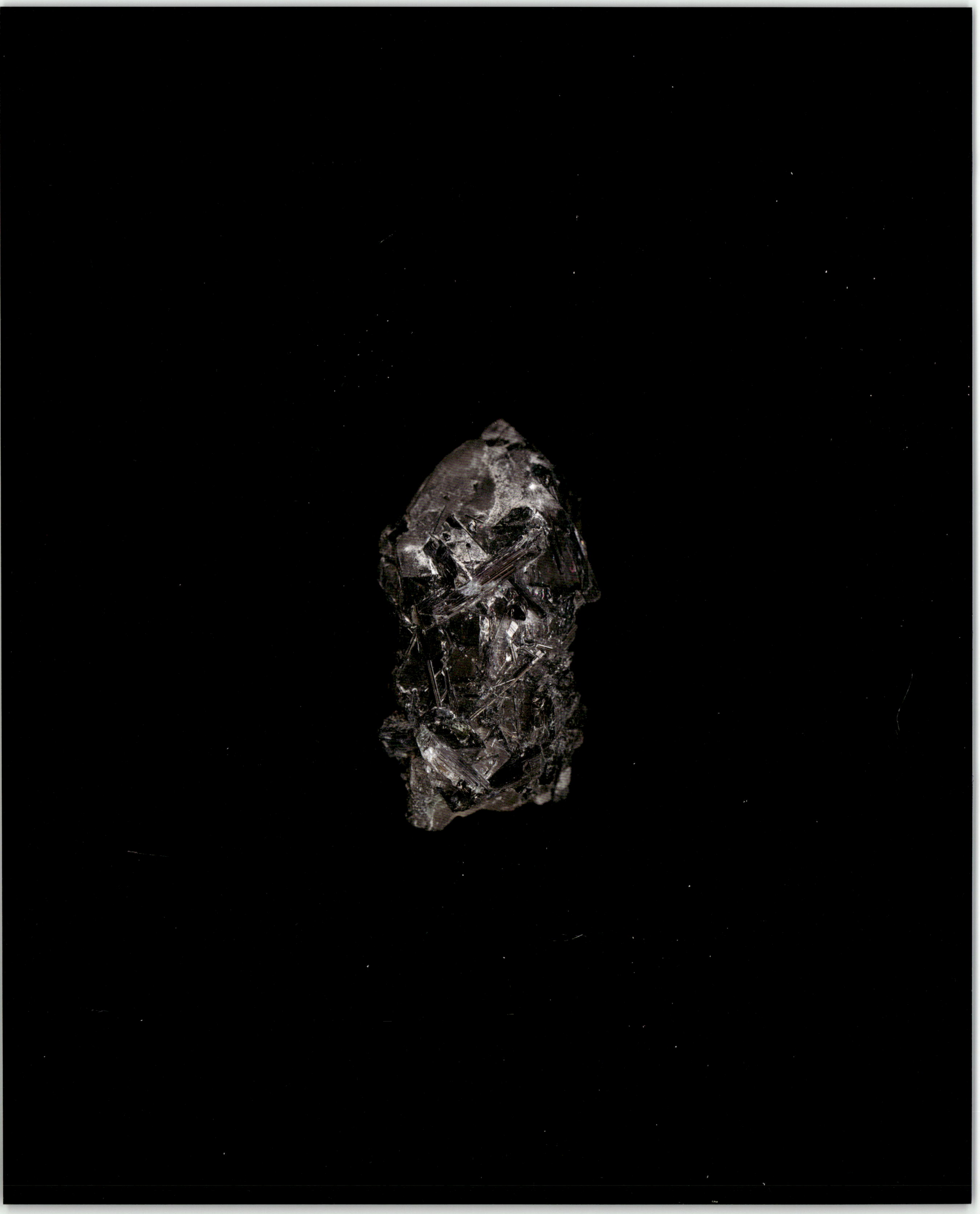

Pink Salt

Maria-Oo Haihambo

Born in the land of folded sediments
pushed and squeezed out of the ocean.
A glamorously exposed desert landscape
Today, the inhabitants, like the rocks, erode.
In the land younger than the Kalahari sands
to some people, water is a precious find,
containing ores, a land so wealthy
Nonetheless, a society impoverished.
Visitors are placed first,
in a liberation government
Locals are always third if they count at all.
Laws serve only as formalities.
Even race betrays you in your wage
at the cost of smoking all that dust
While your health limps, radiation walks.
Labour hard to afford them high-end cars
after all, they are businessmen
who work hard by knowing someone
Who knows someone
who knows someone else
In this nation
It is called "having connections", not "bribery".
Their greed ever so cancerous
They will hand you nothing but the smell of death
even long after you are gone
you still reek of the site.
She will offer you salt from the coast
to cleanse yourself
The land continues to provide
even after it has been exploited.

Desert Mother

Charmaine //Gamxamûs

I lay myself down at your feet, Great Mother
Though many have taken from you
You remain generous with the fruits of your Being,
Thank you.

I lay myself down at your feet, Great Mother
Flatten myself on your solid earth,
As if to 'lay hands upon' you
To join in universal prayer
For your healing

Though we venture to leave you, Great Mother
You remain constant,
Evergreen,
Ever present.

I lay myself down at your feet, Great Mother
Humbled,
Grounded by your presence,
Abiding in you.

You believed in me, Great Mother
My hope, prayer, and belief
Remains grounded,
Your healing is inevitable,
Incumbent
Immanent

The sons and daughters of your soul gather, Great Mother
Lay themselves at your feet,
Lay their hands upon your bare bones
With intention.

Blessed be, Great Mother
We can only thank you,
We can only give you praise,
Thank you for your unfailing grace.

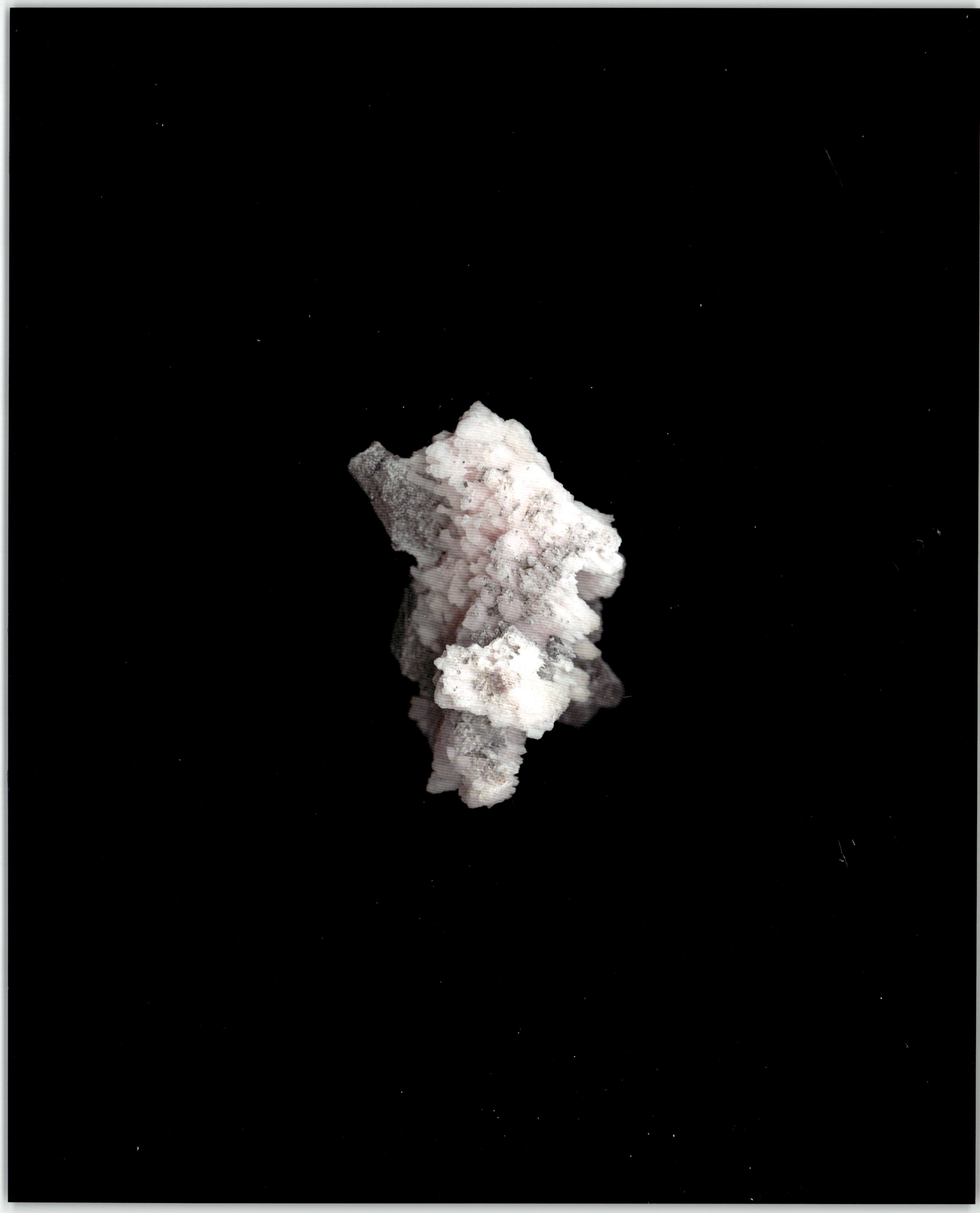

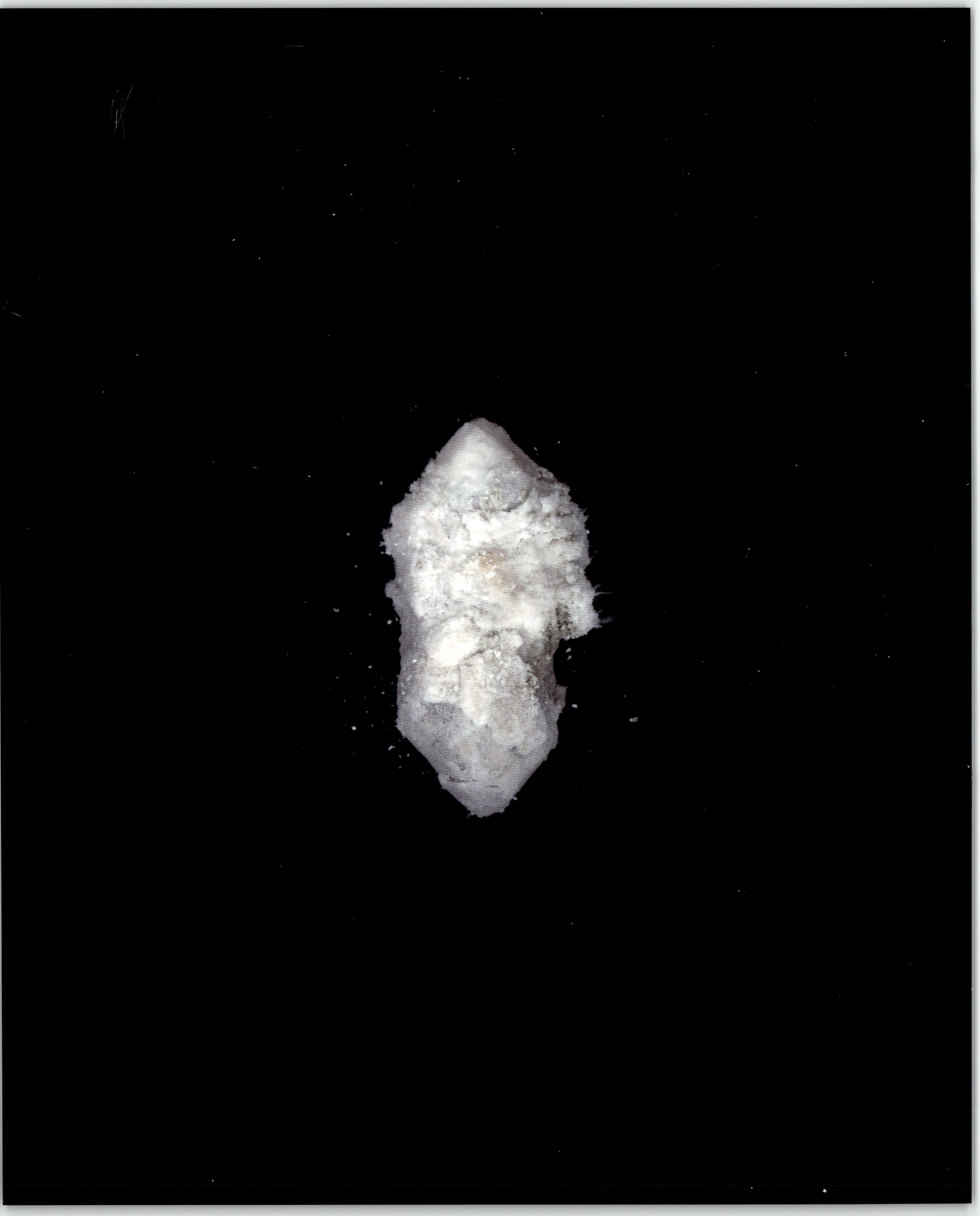

Wounded Earth

Kuhepa Tjondu

Cursed is the hand that wounded the earth
For minor material gains we forgot the primary laws
That everything is everything and we are all connected
The All in all and the One in the many

Incestuous deep wounds in the belly of the earth
The virgin forest ruined by the hand of man
As we continue to chop, dig and pound
Defiling all that was once sacred

Polluted waters, our source of life
Polluted air, from the plants to our contaminated lungs
Polluted earth, giving us our daily provisions
Exploited humans, the monsters we have become

For the sake of gold we have forsaken our Gods
For the sake of silver we sold our souls
For manganese we have mutilated and humiliated man
Remember that a diamond is but a rock in the empty desert

Could it be the deep hole in our collective consciousness?
That vacuum that will never be filled
Which drives you and I deep into the ground
Only to find hollow emptiness

Long after we are gone
This wounded earth will heal herself
The water, the air and the soil will be cleansed
Mother earth will be here.

TSUMEB – A Poem

Prince Kamaazengi Marenga I

To Lake **Otjikoto**
At the depth no hand can grab
A deep wound rapture
And the tears of mother earth flow
Creating a soothing sound-scape
Whose chorus is the beak of bird
The rattle of leaf running
Water ripples going and coming
The glittering stone and wet bank
Screaming under the searing sun

To the green hill
They sent an ant
To dig, grain after grain
In canal and drainage ditch
With feet and fingers
No secrets can reveal
Scabbing the flesh of foliage and soil
To reveal the great wealth below

The well of wisdom couldn't help but vomit out
Stones green and blue beneath the chromatic kaleidoscope
Unsung, the bushman stands in the middle, fixed like stars
Mouthing "Tsombtsou", mocking the earth's floor
Earliest branches of our human family – **ovakuruveha**
Bathed by the evening sun in to a golden glow

Motion and emotion in hieroglyphic paintings
Incredible whistling in cryptic click sounds
Naming the tomb, calling the womb
Earth and soil all fossil beneath their foot
Serving time past in silent tone and stone

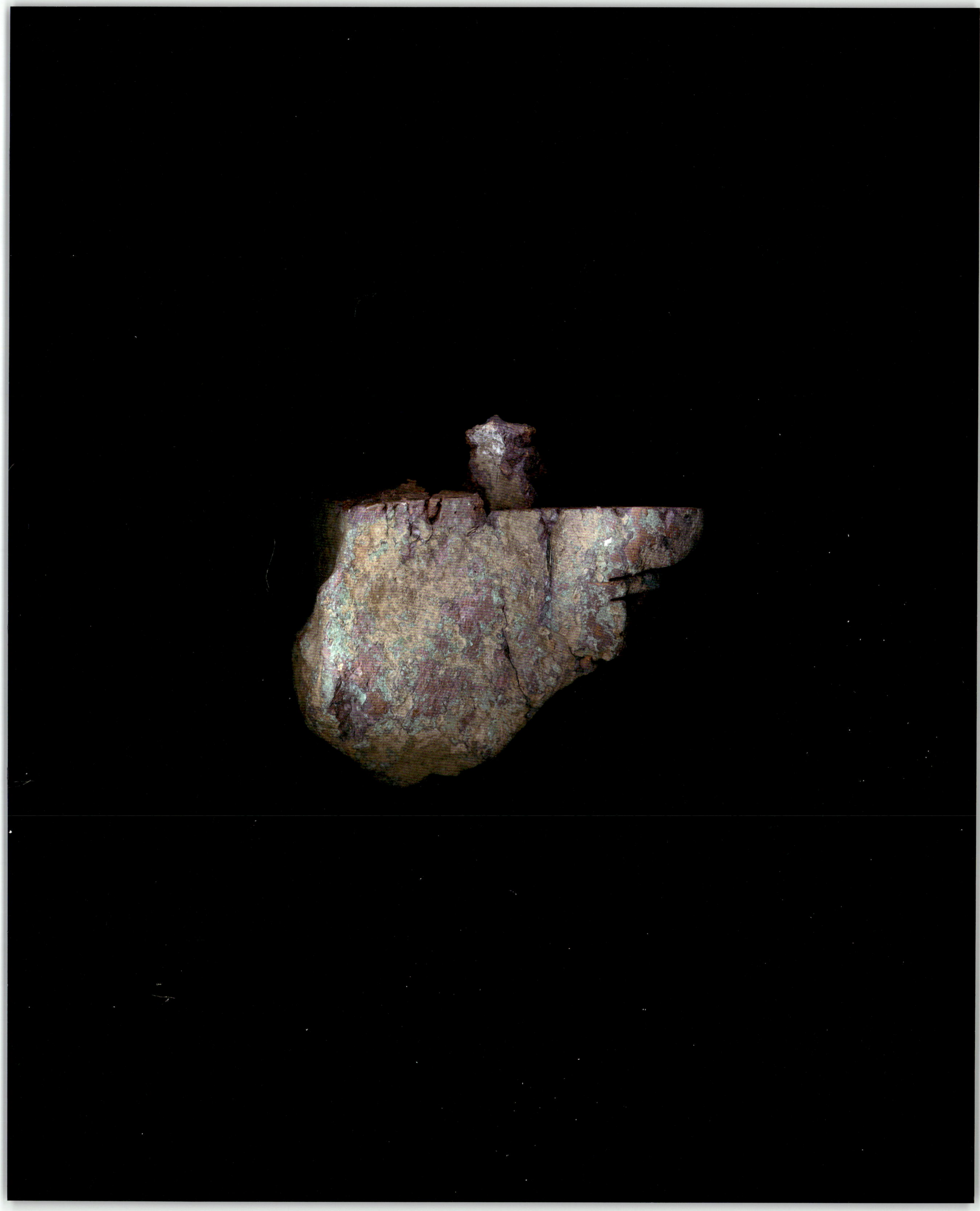

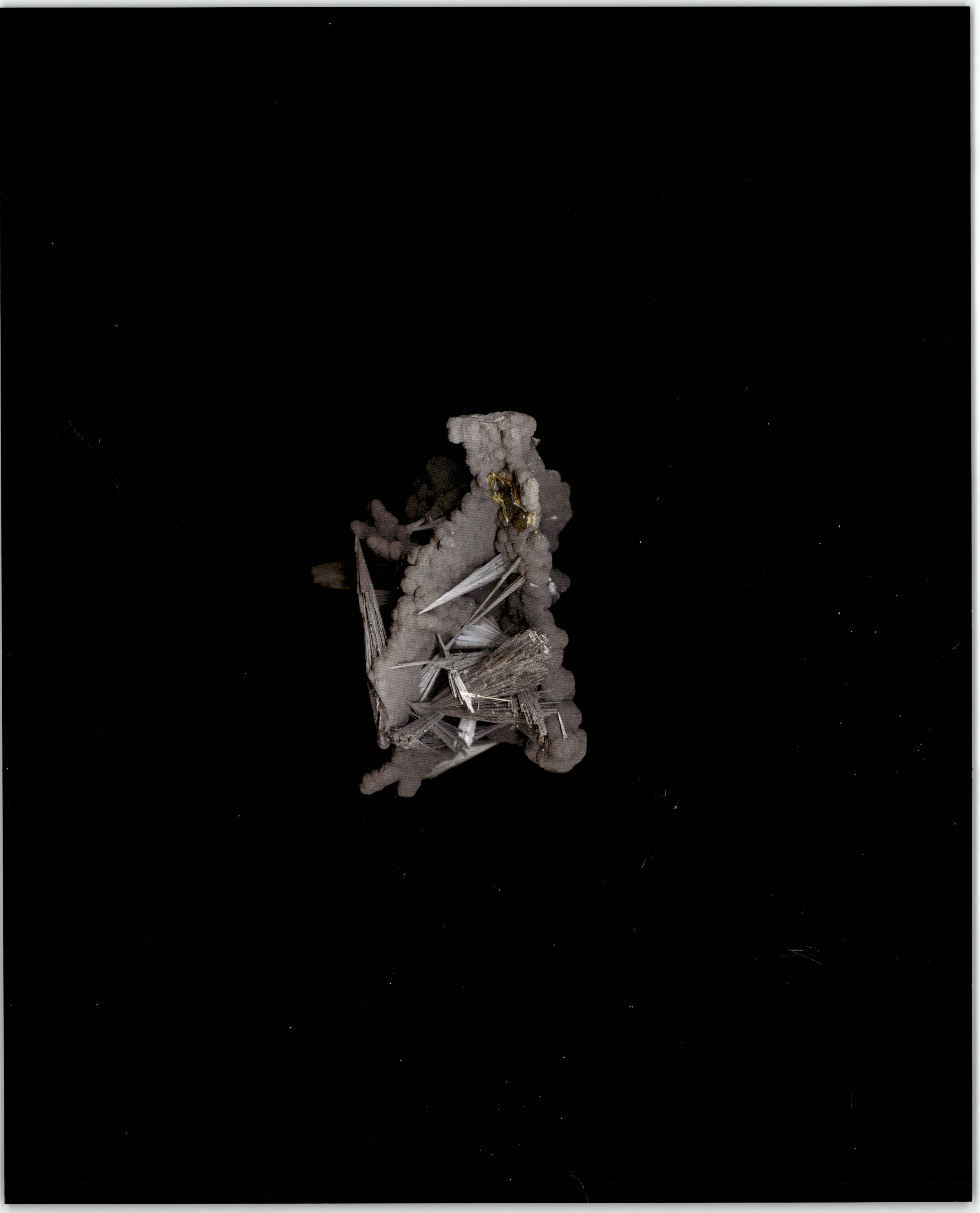

History! My Human Geography.

Jermaine Solunga

Please look for me, and be sure to find me easily,
In these difficult times.
I was last seen dressed in ancestry,
Some say I was seen running with the others into Bechuanaland.
Others say I was among the Mau-Mau.
But please do look for me!

Include every ridge of my inheritance,
And every rhythm of my dance.
Look for me, and look carefully,
Just as I do, trying but hopeful.
And just as I am, tired but grateful,
Look.

Ask for my name from the tongue that has come before,
Ask for me at every moment of the village,
But not when the elders are speaking, as it is sacred... to listen.

To seek for me, is to find not the treasures, but the value of my existence,
Overcoming the longest journey of my journey.
Through corners and circles, and a tired destiny.

I am the land, and the people, and the harvest
I am today and tomorrow and forever,
I am this history, please find me without fail.

transposition 123

On the Impurity of Epistemic Objects

Samo Tomšič

Abstract: The paper begins by recalling the grand narrative of the birth of modern science, which, despite criticism, continues to play an important role in the European 'epistemological exceptionalism'. The author then discusses the role of epistemic objects in the modern (accumulative) regime of knowledge, returning specifically to Descartes' deduction of *cogito*. The latter does not only stand for the first thoroughly modern philosophical conception of human subjectivity, but also and moreover for a key epistemic object. This philosophical scenario is contrasted with other critical examples of the transformation of a subject into an (epistemic) object, notably in the case of racialisation.

Keywords: subject, object, modernity, science, knowledge, accumulation, race, epistemology, critique, capitalism

For well over a century, various strands of European philosophy, epistemology, and history of science told the world the grand narrative of the scientific revolution that inaugurated modernity. This bifurcation in the field of knowledge essentially broke with everything that preceded it and introduced an entirely new way of making science. According to Alexandre Koyré (1957), who played a central, albeit not under-complex role in enforcing the narrative, an important aspect of this historical development consisted in challenging the Aristotelian epistemological tradition, which saw the main task of science in 'saving the phenomena' (i.e., taking the way reality appears to the conscious human observer as normative). Instead, science now becomes increasingly 'counter-intuitive' and its reality foreign to human experience. The 'experimental' is precisely *not* to be conflated with the 'empirical': the former finds its foundation in the technological setting and formal language, whereas the latter still presupposes a (conscious, sensuous) human subject. The human experience is thus no longer crucial for legitimising the authority of scientific knowledge, which henceforth relies on quantitative procedures, modelling, and experimentation. The most subversive achievement of the modern regime of knowledge thus seems to consist of the 'emancipation of science' from the human subject.

Today, the tale of the birth of modern science is one of those 'grand narratives' that the French philosopher Jean-François Lyotard (1984) once declared to be over. The scientific revolution in early modernity turned out to be an 'impure event' amounting to fabrication of speculative, even fictitious theories or elements of theory, signalling that scientific knowledge remains stained by traces of human subjectivity. Not to mention pseudo-scientific conceptions, such as race, and the normalisation of racism, through normative disciplines such as ethnology or biology, as well as more dubious ones such as phrenology and physiognomy, which played key roles in the 'mismeasure of Man' (Gould, 1996). In addition to this, the very purifying tendency of 'hard' or 'positive' sciences, as they are often labelled, produced a specific kind of epistemic fetishism: the tendency to remove all traces of human subjectivity from the register of knowledge ends up treating objects of science or 'epistemic objects' as containers of knowledge, which henceforth assume the curious position of 'inhuman' or 'post-human' subjects. To borrow a comment Marx once made describing the way humans treat or relate to commodities, and more generally to the economic abstraction that is 'value': knowledge is no longer seen as a 'social relation', but as something that pertains—*qua* positive quality—to the order of things, understood as the agencies of knowledge. These agencies are accumulated, collected, carefully curated, and finally put on display, where they gaze back on us from behind the glass, safeguarded, as it were, from the corrupting touch of human hands. Whether they are human-made objects, animals, plants, or minerals, they all share one feature: that of being vessels of knowledge in the modern epistemic machine.

In the grand narrative, the notion of 'epistemological break' played a key feature. Introduced by the historian of science Gaston Bachelard (1938), the notion became popularised with the rise of structuralism in mid-twentieth century France and beyond. Epistemological breaks have been observed everywhere, first in early modern physics and astronomy (16–18th centuries), where the disruption in question was associated with the gradual replacement of the geocentric cosmological model with heliocentrism, and later reframed in terms of the paradigm shift from the 'closed world to the infinite universe' (Koyré, 1957). Other readings applied the notion to major theoretical and experimental breakthroughs in the 19th and 20th century's European sciences and humanities, covering diverse disciplines such as biology, psychoanalysis, structural linguistics, even historical materialism. The common denominator of these various uses of the epistemological break is that they expose a decentralisation in the field of knowledge. The latter, as already mentioned, no longer needs to be anchored in the conscious human observer, even less, it requires the help of a divine being, which would guarantee the truthfulness of scientific knowledge. Rather, knowledge is sustained, verified, or rejected by the scientific discourse itself. Of course, there is still such a thing as scientific community, which performs the shared research work, but this community, and ultimately the very figure of the scientist, becomes a mere tool among other tools in generating knowledge. Meanwhile, the geographical centralisation or localisation of the scientific revolution remains consistently unquestioned, and this, too, signals the perseverance of a specific subjectivity in the field of knowledge: the subject of Enlightenment, taken as the measure of 'civilisational development'.

The question remains whether the epistemological grand narrative truly ended with the advent of postmodernity? Its ending would imply a less passionate, and more critical take on the modern understanding of 'scientificity', not only in the broader perception of sciences, but also and more importantly still, within the self-understanding of European or Western science. Characteristic of every grand narrative, is that it is a tale of origin, striving to provide insight in how something came to be. Different epistemologies have different takes on this issue. What nevertheless unites them is the assumption that scientific knowledge is a register in which human beings obtain objective access to reality—unburdened by subjective and social worldviews, convictions, fantasies, or prejudices. Not only the old dichotomies of truth and fiction, knowledge and fantasy or science and superstition are crucial for the self-legitimation of scientific rationality. Today's contrasting of scientific epistemology and indigenous epistemologies is no less prone to reproducing the European epistemological exceptionalism.

There is indeed one feature of the modern regime of knowledge that significantly differs from the premodern, and that is accumulation and/or growth.[1] This distinction is best addressed in the German term *Wissenschaft,* which, unlike the English 'science', covers natural sciences and humanities alike. The term moreover directly acknowledges that the 'essence of science' consists in creating (*schaffen*) knowledge: that it is a mode of production of epistemic value. Epistemic objects are thus defined as containers of a specific abstraction that is knowledge. The thesis about modern science as an epistemic mode of production has been, among others, proposed by Lacan (1964–1965, unpublished), who introduced it in the following manner:

> I would call the state of knowledge before Descartes pre-accumulative. With Descartes knowledge, scientific knowledge is constituted on the mode of production of knowledge. Just as an essential stage of our structure that one calls social but is in fact metaphysical, and which is called capitalism, is accumulation of capital, the relation of the Cartesian subject to this being, which is affirmed in it, is founded on the accumulation of knowledge. After Descartes, knowledge is what serves to make knowledge grow.

The once trendy notion of 'cognitive capital' certainly tackles the entanglement of epistemic accumulation with the accumulation of capital in its own manner. But Lacan's point seems to be more fundamental, addressing a structural homology between science and capitalism, knowledge and capital, epistemic value and economic value. Just like the economic sphere, the modern, accumulative regime of knowledge contains an ongoing tendency toward growth, and is thus marked by constant expansion and appropriation of reality. Its proximity with the accumulation of capital can be pursued further still, since modern science and capitalist economy share the same ideal: the 'mastery and possession of nature' (Descartes, 2006, p.51)—the destructive consequences of which can be observed at both margins of modernity. On the one hand, colonial expansion (and its ideological justification in terms of the progress of civilisation) and on the other, the climate emergency (where the problem of the 'closed world', or planetary finitude, returns like a boomerang). Nature, then, is postulated as the *Other* of science and economy; its domination, or domestication, is required for the accumulation of knowledge and capital to endure. Needless to say, the aggressive character of this relation becomes most evident in extractivism, which stands not only for the extraction of raw materials (minerals, rare earths, oil, gas, wood etc.) from natural environments, but also extraction of value, be it epistemic or economic, from bodies, materials, and environmental systems. Evidently, next to being a social-material practice, extraction also implies a specific relation to 'being' and can therefore, indeed, be called a metaphysical activity. Whereby the prefix 'meta' concerns precisely the surplus (value and knowledge) extracted from natural bodies, geographies, ecosystems, etc.

Descartes's place in historically transforming the scientific take on the world is exceptional insofar as he linked the growth of knowledge with the question of the subject's being. To recall the familiar story from *Meditations*, Descartes' methodological doubt began by putting the entirety of reality into question, dissolving it to the very point of the pure act of thinking. Nothing is certain, everything is a potential deception, coming either from my senses, my mind, or an imaginary 'malicious genie'. It is in this process of progressive questioning of reality, both external and internal, that Descartes is believed to have encountered a moment of pure certainty, from which the world could then be recomposed in a scientific manner, and hence populated by a countless amount of epistemic objects. This moment is epitomised in Descartes' famous formula: 'I think therefore I

[1] It is this dynamic that has significantly contributed to the idea of progress, to the assumption of exceptional status—or meta-status—of the European epistemological tradition, and consequently to the proclamation of all those epistemological traditions that are not 'growth oriented' as underdeveloped, ignorant, and finally, 'primitive'.

am', in which the certainty of being is inferred from the fact that an intellectual activity takes place and recognises itself as such activity. The content of thinking is here less significant (it can moreover be true or false). What is important is the form: the fact that I doubt, hence perform a cognitive action, which cannot unfold without something being there, a 'thing that thinks' *(res cogitans)* and knows to do so. This formulation exposes an important feature of the Cartesian, and more broadly scientific, procedure. On the one hand, the 'thinking subject' delivers itself to knowledge: it turns itself into an epistemic object. On the other, this reification grounds the aforementioned epistemic fetishism, according to which knowledge, or epistemic value, takes the appearance of a positive quality of things themselves. 'Epistemic object' thus comes to designate an 'object supposed to know'.[2]

One could say that, although it is not the first epistemic object to be produced by modern science, *cogito* is certainly an exceptional object of this kind. For it is the activity of *cogito* that performs the type of labour, whose surplus-product is knowledge. This knowledge is then detached from the labouring subject, and integrated in the broader framework of epistemic and economic accumulation. In this Cartesian setting, we can thus observe something comparable to social labour: extraction of surplus from a process, in which the labouring subject was made into a thing with double status: it is both an object among objects and a value-producing object (hence, an exploited subject). In Marx's language, labour-power is a commodity among commodities and a value-producing commodity. And just like labour-power reflects in all other commodities, the exceptional epistemic object *cogito*, too, reflects in every other object, area, or register of reality, appropriated by the scientific apparatus. Because science purifies its objects from every trace of subjectivity, it also mystifies (to use another Marxian term) the social status of epistemic objects, and more specifically, the link between production of surplus-knowledge and exploitation of labour. The accumulative regime proceeds by purifying the terrain that is to be appropriated, erasing above all the historical circumstances in which an object has been extracted from its geological, geographical, or social context.

Although *cogito* marks the point of objectification of subjectivity and its scientific appropriation, it contains further complications and inequalities. The point of *cogito* is not simply in equating the activity of thinking with the fact of existence (again 'I think therefore I am'), but moreover in stressing that the more I know the more I am certain to exist.[3] This means not only that my existence henceforth depends on science, but also that those who do not 'embark' the discourse of science are condemned to non-existence. To repeat, the key 'epistemic object' produced by modern science was the subject itself. This is the true value of the Cartesian procedure, which demonstrates the first act of extraction of knowledge from a subjective drama (questioning one's own status in the world, and the status of the world itself). But the procedure contains an important complication, since it distinguishes the subject, who delivers itself to the process of knowledge-making, from other subjectivities, who are considered unknowing: Descartes' notorious example is the madman or the dreamer. Looking at the broader historical circumstances, in which the modern way of producing knowledge has been implemented, we can immediately notice that the division between knowledge and non-knowledge obtained its privileged expression in the colonial distinction between the civilised and the primitive (Mills, 2007).

This distinction is coded in the very genesis of the modern epistemic object, obtained through an act of extraction from the 'to-be-dominated' nature, and on display in the ethnological, mineralogical, archeological, and other collections across the so-called global North. Objects of mineralogy are a case in point. On the one hand, they strike one as being entirely immersed in the 'ideal' epistemic space—the museum or the archive—which quasi-represses their historical and social context: in this space, the collection functions as a system of objects-agencies. On the other hand, every mineralogical object can be understood as a mute sign, or 'social hieroglyph' (to borrow Marx's description of commodities), whose history must be reconstructed or retold, if one wishes to break the perseverance of European epistemological exceptionalism in the present. To phrase it in the language of psychoanalysis, the objects in question are both markers of repression and the return of the repressed.

Considering all these aspects, what then is an epistemic object? The general answer would be: every object that is appropriated by science and integrated in the regime of knowledge. From this point of view, the epistemic object intersects with the empirical object studied by science. Of course, this is not the whole story, for an epistemic object is also something that is specifically produced or constructed by science. In this respect, there is strictly speaking no difference between the way science before Galileo and Descartes used the geometry of spheres to describe heavenly movements, and modern epistemic objects such as force, wave, or energy.[4] This does not mean to suggest that an epistemic object is a fiction with no ground in material reality. Rather, the critical point of an epistemic object is that it challenges the simple distinction between reality and fiction, objectivity and imagination, etc. It is a fusion of materiality and symbolism, whereby the modern use of epistemic objects is underpinned by the relation of mastery and extraction. By means of science, the 'Moderns' foster the illusion to be the 'masters and possessors of nature', to repeat the words of Descartes (2006, p.51). This relation of domination and possession itself divides and segregates humanity on the dominators and the dominated. Far from being exempted from the process of economic and epistemic appropriation of reality, the dominated were instead transformed into epistemic objects of economy: some into 'free' labour-power—as workers—others into forced labour-power—as slaves. The modern 'reinvention' of slavery

[2] To paraphrase Lacan's concept of the 'subject supposed to know'. Homologically, in the economic sphere, commodities are considered containers of value, according to Marx's well-known critique of commodity fetishism.

[3] A similar claim can be made for the economic sphere, where existential certainty results from the expansion of property relations ('the more I possess, the more I am certain to exist', the 'I' here being Adam Smith's *homo oeconomicus*, rather than Descartes' *cogito*). Furthermore, just like the growth of knowledge, accumulating value and expanding the scope of private property—hence, of expropriation, the necessary flipside of privatisation—,also becomes the expression of presumable cultural superiority: 'Having an abundance of "objects of comfort" is the litmus test that distinguishes "civilized and thriving nations" from"savage" ones, "so miserably poor" they are reduced to "mere want"' (Buck-Morss, 1995, pp. 124–125; the quotes within the quote are taken from Adam Smith's *The Wealth of Nations*).

[4] For which Lacan (2001, p.522) once remarked that it was not a substance, but a numeric constant, which serves physicists in order to be able to calculate, hence quantify physical reality and make their predictions.

may suggest that we are dealing with a return of premodern social relations at the core of humanist modernity. There is, however, one significant difference between the premodern and the modern slave: while the former was understood as a living tool, most controversially in Aristotle's 'pragmatic' attempt to naturalise slavery, the latter is the product of racialisation, an operation still unknown to the Aristotelian *episteme*.[5]

With the solidification and expansion of the modern regime of knowledge, extraction and accumulation become the key structural dynamics of scientific discourse and co-determine the normative way of making science. In contrast to this novel epistemic framework, practices in Greek and Roman antiquity, and the Christian and Muslim world in the Middle Ages, appear as semi-knowledge. Henceforth labelled 'premodern science', these epistemic practices and traditions still stand in direct historical, geographical and philosophical relation to modern science. This multi-layered relation allowed philosophers and epistemologists to construct the narrative of a continuous evolution and progress in cognition. The assumption of rupture or discontinuity merely confirms the hypothesis of historical progress.[6] Contrary to this framing, non-European forms of knowledge are placed outside of this linear historical movement and denounced as non-knowledge or ignorance. As the European assumption goes, they do not possess the capacity of forming epistemic objects.

More generally, the accumulation and expansion of the modern regime of knowledge is accompanied by purification and declaration of 'uncharted territories' that are to be integrated into the scientific and economic map of the world. The discovery of the so-called 'new world' remains the paradigm of such a purifying event, whereby the very expression implies the encounter of the unknown. *Terra incognita*, the unknown lands, were also considered lands inhabited by peoples in a presumably natural state of ignorance, the perfect negation of *cogito*, the objectified subject of European science (Schick, 1999, p.48–49). The colonisation of these lands meant their integration into the regime of knowledge as a multitude of epistemic objects; made such through racial classification. Indeed, a major function of this problematic epistemic object (since this is what race was for centuries) consisted in substantialising the difference between those who know, and therefore are, and those who do not know, and therefore are refused not only the status of personhood and self-ownership, but of existence. *Cogito* and the racialised subject are two sides of the same conversion of subjects into epistemic objects, albeit with opposing social consequences. While *cogito* is endowed with being, the subjects of race are targeted by forms of exploitation and violence that condemn them to non-being. Epistemic objects are memorials or traces of this subjective vicissitude.

It is, of course, anything but unproblematic to speak about modern science in singular. Such discourse is part of the grand narration of the modern scientific revolution, and reflects the attempt to impose the accumulative regime of knowledge as the ultimate measure of what counts as scientific, and what does not. In their epistemological exceptionalism, European sciences did not only claim for themselves a neutral position in relation to other epistemic traditions in the world: they understood themselves as the only true actualisations of science, whereas 'other' epistemologies, although acknowledged, are made redundant as still being anchored in an excessively subjective relation to the world.[7]

Scientific production thus strives to foreclose subjectivity from knowledge and from the realm of epistemic objects. Scientific collections exemplify this rejection, on which the ongoing accumulation of knowledge is grounded. But the geographical origin of the collected objects still offers an implicit trace of historical context, in which they have been extracted from their environment and transferred to the 'epistemic space' (museum, archive, university, institute). It is this historical context that allows reintroducing subjectivity into the overall picture of knowledge accumulation, thus pinpointing that epistemic objects remain 'social hieroglyphs' (to put it again with Marx). As such, epistemic objects are both expressions and mystifications of the social relation, in which they operate as 'objects supposed to know'. Contemporary debates around restitution of collections dating back to the colonial era and the sometimes more, sometimes less sophisticated argumentations to circumvent restitution, expose the perseverance of epistemic fetishism, which stubbornly detaches objects from social context and attributes to them, if not exclusively, then at least predominantly an epistemic value. Consequently, this resistance, to call it by its name, in advance excludes restitution as a possible scientific act (challenging epistemic fetishism and indicating an alternative understanding of 'scientificity').

In times of increased rejection of scientific knowledge through populist ressentiment, conspiracy thinking and other forms of denial of scientific knowledge, criticising science may create a wrong impression. Is 'our' science not already in enough of a weak position? And is it not equally threatened by capital, which strives to turn every scientific practice into a component of profit-making? A thoroughly critical epistemology certainly must insist in its positions, since the voices, which see science under threat from capital and populism, all too often assume science to be a neutral space of gaining knowledge for the sake of knowledge, and eventually for the sake of improving the human condition. Such understanding of science perpetuates the epistemological grand narrative. While Martin Heidegger (1954) recognised in the modern experimentation, mathematisation and technologisation the accomplishment of European metaphysics, the universalisation of the possessive relation to 'being', seeing in science a figure of uncorrupted and authentic rationality is no less problematic. Both expressions of epistemic fetishism blur the fact that sciences form a

[5] The fact that modern slavery historically precedes the conceptualisation of race in anthropology, ethnology, and biology merely confirms that the practice of racism does not depend on the category of race. Put differently and taking the developments in the late 20th and early 21st century into account, the phenomenon of racism reaches well beyond the 'scientific credentials' of race. What Étienne Balibar (1991, p.21) once called 'racism without races' stands both at the origin, and in the aftermath of the scientific concept of race.

[6] Kant famously made the same point in relation to the French Revolution. For Kant, the political event that abolished the *ancien régime* unambiguously demonstrated that humanity progresses toward the greater good. For a critical contextualisation of Kant's 'hypothesis', see Comay 2011, p. 26–50.

[7] Modern scientific subjectivity is constituted on the background of a differentiation between abstract rationality and what Immanuel Kant called the 'pathological'. The term covers the register of sensuality, as well as psychological or social motives, which would determine someone's relation to knowledge. As David Lloyd (2018) insisted in his critique of Kant, both sides of the Kantian divide—transcendental vs. pathological subjectivity—reproduce the distinction between the civilised and the primitive (or between whiteness and blackness), which was ultimately, to repeat, an epistemological distinction.

central terrain of social antagonisms, rather than being the realm of some imagined neutrality and freedom.

Scientists must be reminded of the troubled history from which their disciplines emerged and that the unifying success story they present to themselves is ultimately a fiction. As Lacan occasionally wrote, 'the fact is that science, if one looks at it closely, has no memory. Once constituted, it forgets the circuitous path by which it came into being' (Lacan, 2006, p.738). A significant part of this epistemic amnesia concerns the errors that a science assisted introducing in the social context, whereby these errors were always epitomised in epistemic objects.[8] The more a science understands itself as the most accurate description of reality, the more its errors turn into troublesome human destinies. Race is the paradigmatic example of a debunked epistemic object, which still determines our relation to geographical, cultural, and indeed corporeal difference. This problematic social category continues to be on display in scientific collections across the world, even when these do not comprise human-made objects. And all the voices, who react with ressentiment to the calls for a thorough decolonisation of knowledge, demonstrate that their presumable 'desire to know' is underpinned by a 'passion for ignorance'.

[8] Outsourcing critique of scientific rationality and European epistemic hegemony out of the scientific field into artistic research and humanities (whose scientific status is moreover continuously questioned) is an essential expression of this amnesia.

References

Bachelard, G., 1938. *La formation de l'esprit scientifique.* Paris: Vrin.
Balibar, É., 1991. Is There a 'Neo-Racism'?. In: É. Balibar and I. Wallerstein, *Race, Nation, Class: Ambiguous Identities.* London: Verso, pp.17–28.
Buck-Morss, S., 1995. Envisioning Capital: Political Economy at Display. In: L. Cooke and P. Wollen, eds. 1995. *Visual Display. Culture Beyond Appearances.* Seattle, WA: Bay Press, pp.111–141.
Comay, R., 2011. *Mourning Sickness. Hegel and the French Revolution.* Stanford: Stanford University Press.
Descartes, R., 2006. *Discourse on Method.* Oxford: Oxford University Press.
Dussel, E., 2000. Europe, Modernity, and Eurocentrism. *Nepantla: Views from South*, 1(3), pp.465–478.
Gould, S. J., 1996. *The Mismeasure of Man.* New York: W.W. Norton & Co.
Heidegger, M., 1954. *Vorträge und Aufsätze*, Pfullingen: Neske.
Koyré, A., 1957. *From the Closed World to the Infinite Universe.* Baltimore: Johns Hopkins.
Lacan, J., 1964–1965. *Le Séminaire, livre XII, Problèmes cruciaux de la psychanalyse.* (Unpublished).
____2001. *Autres écrits.* Paris: Éditions du Seuil.
____2006. *Écrits.* New York: Norton.
Lloyd, D., 2018. *Under Representation: The Racial Regime of Aesthetics.* New York: Fordham University Press.
Lyotard, J.-F., 1984. *The Postmodern Condition: A Report on Knowledge.* University of Minnesota Press.
Mills, C. W., 2007. White Ignorance. In: S. Sullivan and N. Tuana, eds. 2007. *Race and Epistemologies of Ignorance.* New York: SUNY Press, pp.13–38.
Schick, I. C., 1999. *The Erotic Margin: Sexuality and Spatiality in Alteritist Discourse.* London: Verso.
Wynter, S., 2003. Unsettling the Coloniality of Being/Power/Truth/Freedom: Towards the Human, After Man, Its Overrepresentation—An Argument. *The New Centennial Review*, 3(3), Fall 2003, pp.257–337.

Art and the Earth: Creativity, Geology and Extractive Economies

Kathryn Yusoff and Elizabeth Grosz

Abstract: This is an interview/ discussion between a philosopher and a geographer that addresses the social, cultural, political and economic impacts of mining, especially in impoverished nations in Africa and Asia. Among the questions addressed are those linked most directly to the bodies of miners, and their place in the extraction of wealth from the depths of the earth—questions about race, gender, class, age, and health. Central to this discussion is the question of the fascination, and desire, that gems and jewels exert, in spite of the human and inhuman cost of their extraction.

Keywords: extractivism and slavery, art, bodies (race, gender) and the earth, aesthetic gem objects, human and inhuman, geotrauma, environmental reparations, Deleuze, Bergson, Marx, Darwin

In a discussion between a philosopher and a geographer, what can we find in common? How can we discuss things that do not speak for themselves, objects and processes that pass between people (and perhaps even animals)?

EG: Let us start at the beginning. Can you talk briefly about the creation of minerals and especially those minerals we highly value—gems, stones, oil, objects taken from or found within the earth, whether through mining or exploration—that come to have some value, whether financial or personal, after their extraction? How are they formed, how are they extracted, and what are the costs of their discovery and refinement, and for whom? Is anyone untouched by mining and mineral extraction?

KY: This is a big question! Maybe we can bring in some corporeal scale and then open things up to a wider scope. We are used to thinking of bodies as biological, involved in historic and contemporary processes of life. And we have developed a somewhat biocentric view of life over the last two hundred years—a division of life and non-life that has served a particular form of colonial extraction, and structured its racialised politics. We are less used to thinking of ourselves as earth systems, or geologic subjects that participate and partake with the universe (Grosz, 2008; see also Deleuze and Guattari, 1996). Understanding all bodies as involved in the earth, in its geochemical attractions, and conversely, its forms of pollution or sickness, is one way to think about this queer geochemistry. Historically, bodies contain a history of their encounters with earth materials, both in terms of evolution and epigenetics. Bones literally hold the histories of their environmental encounters, exposure, and exploitation. Racialised labour practices mark the epigenetics of mineral deficiencies—just as, for example, indigenous people who were exposed to nuclear testing in the Marshall Islands by the US navy have traces of that nuclear radiation.

In the past, bones of animals were ground up to make paint, which was used to draw the first human depictions of human identity on cave walls. So, value is a relative term. These paintings have survived 40,000 years underground. And yet 'we' happily unearth coal and oil from deep within the strata, which has been in the earth for up to 300 million years, to make things that last seconds and end in a landfill. Anthropogenic climate change has made clear the cost of that 'unearthing'. More broadly, unearthing is a complicated transition zone in how value is made and stabilised. The organisation that perhaps connects the earth and particular categories of (often racialised and gendered) bodies, is that of 'permissible' extraction. Or what costs to people's lives, land and relation, can be externalised in that unearthing to generate value elsewhere. All mining is destructive, but the 'who' and 'where' of mining is governed by an extractive geologic that makes decisions around value (which organises hierarchies of value of whose lives matter and whose are rendered injurable). The question of value homogenises inhuman materialities and inhumane practices of extraction.

Minerals and metals are subject to selective narrativising and are implicit in the development of new forms of production. Diamonds are one example of the stabilisation of value. They are the hallmark of heterosexual imaginaries of reproduction, yet still require an underground of young black men to go deep into the earth under intense pressure that is caused by depth. Diamonds are some of the earliest formed minerals of the earth, being between one and 3.5 billion years old, which lends to the symbolism of stability that is narrativised around them—they 'solidify' marriage as an enduring institution.

Diamonds are perhaps the most over-determined rocks in terms of gender and reproduction, and they are also very hard and good at cutting things. So minerals and metals are subject to selective narrativising and are implicit in the development of new forms of production.

EG: What are the conditions, means, and effects of techniques of extraction: how many types of mine are there; how do they work and what do they require? And perhaps most significantly in the present moment, what are their effects as mining becomes more intensified? Here, I would like to discuss not only the intensification of our *need* for fossil fuels for electricity, cars, machines and automation more generally, but also of our *desire* for beautiful and rare objects, for gold, silver, gems of all kinds, ochres and tints that condition certain of the arts. Do commerce and art engender different mining techniques? Or do they require the same means of extraction, used for different ends? If machines and automation are directed to commercial production and its intensification, is the search for aesthetic objects—clays for pottery, ochres for colour, diamonds, sapphires and rubies for their intoxicating glistening and light refraction—directed to beauty and aesthetic pleasure? Are there two directions, two kinds of interests here, or two sides of one process?

KY: This is a difficult question because I think there are two kinds of aesthetics at work here. Firstly, in the case of aesthetic mineral, metal and gem objects, which is more immediate and organises the imagination and its exploration of temporality (through the connection to deep time). And secondly, there is another aesthetic that is organised through technology, which seeks to make precise interventions in those geological materialities, in which the intervention becomes the aesthetic—such as skyscrapers as an architecture of modernity and futurity. In both, there are aesthetic interests. I think there are different origin stories at work in the perception of these different orders of minerals and metals. One is about the origin of deep time, the other about the origin of technology. I'm not suggesting a Heideggerian/Arendtian earth and world here, brute (glitzy) matter versus technology, rather that the different articulations of temporal aesthetics *space* a different set of associative qualities, in relation to geologic time, that narrativise difference. The narrativisation of difference is perhaps more symbolic than the actual differences in material structures of extraction. If you go to a coal mine in India or a gem mine in Southern Africa, you will find children and the racialised poor operating in both. Racial poverty makes value in the material structures, stock exchanges and jewellery shops of urban centres.

I think the question of intensification and desire is more complex, in terms of how the incorporeality of geologic temporalities is part of the economy of value. If geology is about transforming time (how fossil fuels make speed, for example) as it is about transforming space (whether as architecture or planetary mining), then mining is a temporal equation (a way to ingest different feelings of duration, from marriage to new Anthropocene epochs). Again, this comes back to value and an account of value that can accommodate the dimensionality of these choices, as choices about attachments to time, as much as choices about the transformation of matter.

EG: In your wonderful and intense book, *A Billion Black Anthropocenes or None* (Yusoff, 2018), you make a convincing claim that the impulse to mine and the organisation of mining practices are not neutral to questions of race—indeed, there is an integral and historically enforced connection between blackness and the earth, which has justified and enabled the enforcement and proliferation of slavery, and the capture of goods and objects of value from the earth, to be regarded as two entwined elements: the construction of Blackness is what comes to mediate between white bodies and the earth itself, with its inhuman and at times anti-human forces. Blackness—a blackness represented in, and for, white social and economic life—you claim, is 'historically constituted and intentionally enacted deformation in the formation of subjectivity, a deformation that presses an inhuman categorisation and the inhuman earth into intimacy' (ibid., p.xii). I would love you to unpack and explain this interconnection of racism and extraction a little more. How do 'Black Anthropocenes subtend White Geology' (ibid., p.xii)? Does the inhumanity of minerals require the inhumanity of categories of human? Or is this an historical aberration, something correctable? Must there be another 'Billion Black Anthropocenes' for future economic growth? Is economy historically inherent in race and racism? How might 'political geology' look otherwise' (ibid., p.12)?

KY: I think that the natal moment of colonialism and the invention of race secured and stabilised a historical valuation of whiteness and devaluation of Blackness. Which is also to say that Blackness as a degraded subject position is what organised the generation of the value of whiteness. This dialectic of race was used to create a subterranean subject position that organises spaces of extraction, and the export of value (often to colonial centres). The plantation, too, can be understood as a strip mine, which follows the establishment of the mine as a paradigmatic form of racialised relations.

White geology is a way to understand how a set of colonial practices of surveying, knowledge creation and extraction came to transform the earth on a planetary scale, which required as its prerequisite the theft of indigenous land, and of enslaved energy and personhood to transform that earth through extraction. To regulate that theft, the terms of indigenous and Black subject-hood were invented as a racial deficit to unearth value for whiteness. Thus, white supremacy is also a material condition that relies on a specific racial deficit through extraction. While the terms of extraction have changed, they haven't changed that much. Extraction is a multi-ethnic activity (although some of the biggest mining companies are still based in colonial and settler colonial nation-states), but it still relies on undergrounds of racialised subjects to surface value. Historically and contemporaneously, there is an on-going intimacy between race and the inhuman (as a category of non-personhood), through the relation between inhumanity and the materialisation of the inhuman world. That intimacy is why environmental justice issues are always racial justice issues. And there is no environmental justice without first attending to questions of racial inequity. As a geographer, I understand now that this separation between environment and race needs to change spatially, as well as socially, for there to be any corrective to how geo-social worlds emerge. Which is not to say that other things did not also happen in that space of intimacy. Historically, acts of freedom relied on, and rely on, environmental knowledges and tacit understandings of enacting with the earth in ways that the overseer, enslaver or state could not. I think of these as 'tactics of the earthbound', which were developed alongside, and in resistance to, the organisation of colonial natures, and the segregation of culture and earth (human and inhuman). Often it is indigenous community leaders and queer collectives that are showing us what another political geology might look like, and these 'earth defenders' (as they are called) pay a heavy price for their activism. They are refusing the false values of colonial economies, and insisting that the earth and water is life (i.e., life is not exterior to the inhuman, but interior).

EG: To address a different kind of question, where do you think our excitement about, and desire for, gems, glittering stones or shiny metals, comes from? Are humans alone directed to such bedazzling objects? Or is there something more primordial about what glitters and shines? Is the desire for the extraction and creation of shiny objects, whether from the earth or above it, something we share with animal life, or something that distinguishes us from animals? Why do you think some of us willingly work in, say, gold or silver mines, in opal mines, with perhaps little hope of finding something? What is the 'hold' of the glittering? Is it linked to art as well as to economics? And is art thus a part of the deadly economy of extraction, or an exception? Or both? Here, I understand 'art', not simply as fine art or the high arts, or what has monetary value for others, but a more modest conception of art, as the transformation of raw materials into objects of reflection and exploration, more than use or practicality—a conception of art that does not distinguish art from craft, seeing a continuum in creative making, and its processes, techniques, materials and objects.

KY: I think this question comes back to some of your earlier work on the nature of difference. How difference incites, rather than the more normative accounts of difference as something to be overcome as an alienation (Grosz, 1994). Maybe there is an attachment to an appearance of a different temporality and permeance, or maybe something about how the first artworks involved touching cave walls and spitting the self into being with ground-up minerals. Also, as you have written about, birds and other non-humans are attracted to shiny things, and they have aesthetic preferences and engage with that thing we might call *desire* around different materialities (Grosz, 2008). Whole social worlds have been shaped in relation to materialities, from gunpowder to sky-metal, to the more obvious 'bling' of bars of gold, but I do not subscribe to an environmental determinism around these attachments (as some geographers do). The work of attachment is more complex, I think, and less determined in its outcome. Some attachments clearly have a utility, but as you have discussed, many do not. We can speculate on curiosity, imagination, and experimentation in relation to metals and minerals, but I don't see this as exclusive to human animals or living things. The earth is one big geochemical attractor, whose experiments produces the earth as an emergent axis of different recombinations of minerals and materials. I do think there is an ancestral memory of materiality though, which is often collective and socialised in particular ways. The dream of time and of other worlds in minerals and materials certainly initiates a compelling imaginative pull.
One of the questions that opened up for me when I was thinking about geology, that resulted in *A Billion Black Anthropocenes or None*, was how the inhuman category of gold and the inhumane category of slave can trade places without any disruption to the language of exchange. Or, rather, what is this particular kind of violent grammar and structure of thought that allows this to happen? And consequently, how do these grammars of geology persist as operative languages that continue to facilitate inhumane transactions? One of the things that becomes immediately apparent is that the human is an exclusive category, rather than a biological category, and that the division between human and inhuman is what allows value to be generated as value. The desire for value is clearly a brutal intoxication.

EG: In thinking about art more generally, and especially about the art of slaves, those who might otherwise be working in mines, extracting the materials needed for white artists, is there not also an art of liberation, the possibility and actuality of an art of the oppressed—the making of huge pottery vessels, the creation of paintings, sculptures, weaving, and above all, song, an art without necessary external objects, various kinds of art that express the specificity of black lives and black experiences, even if in disguise or indirectly?

KY: This is a complicated question, as perhaps there is a distinction between the romance of geology—elemental thought without extraction—as a narrative form in art and elsewhere, and the question of what art *is* under conditions of oppression. Art is a necessity of the earthbound because it is an imaginative space in which to rehearse and enact other possibilities of value in the context of devaluation of racialised life. The recent exhibition Hear Me Now (2022) at the Met (New York) of the Poetic jars by David Drake, an enslaved man from South Carolina that I think you are referring to, is but one example of the ways in which oppressed peoples made work as a means to live otherwise—to claim a poetic life that was violently refused by white supremacy. There were many blacksmiths too, who were practicing the craft of metallurgy in creative and distinct ways in the Southern states. Not to mention the quilters of escape routes, the writers and practitioners of the underground railroad and North Star. And, as you say, songs as both maps and routes of historical memory. The geographer Clyde Woods (2000) called this 'blues epistemologies', to describe the forms of knowledge production that the blues carried across and about geography. We also see this in the poetry of politically oppressed people, where the spaces for political dissent must find other languages and forms of communication that have a subterranean passage. Not all art can survive in the current political conditions of the day, so it must find ways to travel underneath existing structures and states. This is why aesthetic languages or geo-poetics are so crucial to the experience of subjectivity and the possibility for it to be otherwise.

EG: Can there be any kind of restitution from these extractive economies and the companies and individuals that have profited from them? If so, what might this look like? Can there be a restoration? Or a reconstitution of what has been mined?

KY: As the earth is emergent, I would say that there can be no reconstitution, but I think restoration as a mode of redress has to be part of the equation in addressing extraction. There is no collective planetary future without modes of redress that have restoration, understood as a form of responding to that which has been damaged, as a procedural ethic of engagement. That response needs to be led by those who have been impacted by extractive economies, or it simply repeats the structures of power that create these uneven environments. Without redress, there is no collective, there are just rich tech bros building bunkers or spaceships to Mars, selling their version of the future. I think about histories of colonialism as 'geo-trauma', and trauma is never just erasure, it is always about the future too, and what grows around that which is taken or has suffered violence. If we put geo-trauma at the centre of understanding environments, this also pushes forward a perspective that takes account of the way worlds are ecologically co-joined and how subjectivity is held together across more than human relationships. Geo-trauma also calls attention to the specificity of place-based relation, rather than trafficking in modes of equivalence, as the new economies of off-setting and carbon trading do. For example, you cannot off-set a mine, if that off-setting creates more extraction of indigenous land elsewhere. Environmental reparations are needed to off-set colonial reverberations and ongoing infrastructures of harm that continue to target indigenous and racialised populations.

As time has been a constant refrain through your writing and analysis, and more specifically the marks or nicks of time (as you call them) (Grosz, 2004), do you have an account of the economy of time that is able to stretch across the compression and densities of temporal exchanges? I have been thinking about the ways in which Marx's account of value falls short in its temporal range, as if he hasn't quite been able to properly account for inhuman materials, despite his proximity to mining events and magnates. I would like to know how you navigate Marxist materialisms in your work and how your more synthetic account of time challenges that.

EG: I do not have what you call an 'economy of time.' And in fact, I am not sure that time gives itself over very readily to a notion of 'economy'. That is, to techniques of calculation, regulation and ordering, although of course time is readily countable through various techniques of calendar, days, seasons, as well as hours and minutes, through any arbitrary means. I remain more Bergsonian (and thus Deleuzian) than I am Marxist on this. For Marx, it seems, time, or at least the events that occur in and through time, have a trajectory or an end, a *telos*, the end of class divisions and the eventual order of self-rule. I think his account of value falls short for many reasons, and clearly, he foresaw an end that still remains completely out of sight or even comprehension. For me, following Bergson, it is perhaps because of the nature of temporality itself that, although it follows causal trajectories, it also accumulates everything and concentrates all of history in each passing present. In my work, Marx was where I began. The notion of a dialectical passage through history, though, is where I can no longer go! There is no clearly discernible direction in history, only the movement of events and their collective, incalculable forces in the present and beyond. As your work indicates, there is no overcoming of the past, only a continually deferred reckoning with it, as its forces press on our present and future, without our control or recognition.

KY: Yes, Marx is inadequate, as Black Marxist and Black feminists amongst others have argued. Although, it is hard when you are talking about value around minerals to not get folded into a normative understanding of materiality that is intensified by racialised regimes of value (known as racial capitalism). You identify in your work moments of instability and thus possibility in the inhuman, so I'd like to ask how you have sought to imagine forms that capitalise on that instability to produce new insights into destabilising colonial forms of categorisation and dispossession (in gender, sex, race)? And how might we valorise other relations of value and use, which do not articulate a trans-historical set of relations and potentials (and thus fail to notice the contingent and resistance disruptions of located struggles)?

EG: These are very big questions, with no clear answer. I would think there are countless moments of instability, of halting and uncertainty that are economic, climatological and even, as we see currently, related to global and local health. All systems of power are, as Foucault (1977, 2008) reminds us, in perhaps a kind of supplement to Marx, partial, unable to cover all acts and all subjects in the same laws, forms of codification or normalisation. All societies have moments of eruption, transformation, upheaval, some of which we may understand as political resistance, and others are moments of internal disruption, such as technological changes or shifts in economic operations. We are, for example, currently aiming at a movement away from fossil fuels with the rise of electric cars and modes of transportation, with the use of hydrogen as fuel and so on—clearly developments that aim to transform capitalism and enable further production by other means. This is both progress and regress: it is a technological progress, an advancement of technologies, but in no way is it, in itself, a mode of political progress. At a more local level, at the level of self-formed groups—political, artistic, intellectual, technological—there are other modes of value and practice, even as they must operate within a larger background of capitalist and extractive labour. This, though, is how resistance has always occurred—in the interstices of other practices, in the labours and practices of minoritarian groups and rare singular individuals. There is no power without resistance; but also, no resistance without power in its ever-changing forms. This peculiar dance of forces continues, and the movements of resistance must invent places for their actions in the interstices of power, as a resistance to it, and a force that requires the terms of power itself to continually shift.

KY: One of the challenges that any geo-philosophy seems to face is the practice of philosophy itself, and its modes of abstraction, which often become difficult to reconcile to more exacting geographies and geologies. And, in this sense, philosophy often seems to parallel geology in the creation of values that can circulate as ideation—such as the price of gold—independent of their modes and places of extraction. I wanted to ask you about what you thought the challenges were to inventing an earthbound geophilosophy (if indeed this is desirable!)? And whether geoethics was an appropriate term for thinking about an orientation within geophilosophy? Or do we need new languages for the inhuman and its forces all together, which help us imagine an earth that is not predicated on extraction?

EG: I agree that philosophy and geography are themselves different kinds of conceptual and practical forces, forces that can sometimes align, as I think they do or might in the present (that is, in this ongoing 'moment' of climate catastrophe, which poses to philosophy its broadest and most pressing social, political and technological questions; as well as forces a new reckoning for geographers and those whose work is used by—or fights against—mining and other commercial interests which engage and finance them) but often go their separate ways without affecting each other. But equally, there can be no philosophy without a tacit geography and even geology: we implicitly think of philosophy as being located somewhere, as coming from somewhere—Greek philosophy, Medieval philosophy, French philosophy, Anglo-American philosophy—often the birthplace or location of a particular philosopher, which does indeed make its mark on the questions and methods of particular schools of philosophy. Given the sense of imminent climate catastrophe that many of us feel as the most pressing problem of the present, it is desirable, indeed urgent, for us to develop a geophilosophy, a geoethics, a way of addressing better and living with, not only other living beings, and their forms of oppression, resistance and force, but also of the non-living forces that surround us, enable and support us, and also threaten to drastically transform us and all other forms of life. We need to consider what is usually understood as 'the unspoken background' of our activities—the earth, the air, the oceans and waterways, our particular geographical locations, our movements on and over the earth and the effects they have on other forms of life, a new ethos, a geoethics and geopolitics to match the growing-closer of geography and philosophy: that is, a new geophilosophy, a way of living with and understanding as best we can our place on the earth and in the cosmos.

KY: This might seem like a strange question to non-philosophers, but in philosophy the earth always seems to appear as an abstract, as you say, a backgrounded concept rather than a foregrounded materiality, as in your work. I wanted to ask you about why you have been interested in the earth, and how a continuous engagement with the earth has situated your work through a number of shifts in understanding feminism, bodies and sexuality (alongside evolution, excess, and biologisms)?

EG: You are right: if the earth is even discussed in the work of philosophers, it is usually either an inert background or an abstraction. There is in fact very little that is focused on the earth as either a specific location—a particular terrain with its own qualities, or a planet with its own activities and forces. I think that I have been interested in the earth in its specific qualities for a long time, probably without directly realising it (maybe even from childhood) with explorations, long walks, adventures around cliffs, beaches and the outback. But more directly, intellectually and explicitly, it has become clear to me since my work on Darwin (Grosz, 2004). It was Darwin, and his fellow evolutionists, (along with colonial explorers and traders) who explored varying terrains, islands, 'exotic' and far-away lands, and the animals and plants that occupy them, in order to make visible the modes of co-existence of various forms of life and their particular habitats. The earth is not a series of abstract locations, points on a map, but innumerable dynamic interfaces, a series of massive and often unperceived (at least by us) forces, energies, substances and encounters, movements that, if not alive, are just as dynamic as life that is contained within a skin. All art is of the earth; and thus all thought, all creation, all that lives, is of the earth in a way that is generally unrecognised. Our task, as scholars, is to find ways of recognising and acting on these creative forces without, somehow, solidifying them into norms or commodities. We have always undertaken practices and created objects and relations that fall below the level of commodities or ascertainable values: discovering or developing practices of making, including self-making, may help us to create pockets of resistance, small sites of invention and performance. What do you think? Any final thoughts?

KY: I think the inventive geo-pulls of materiality is what inspires me about the earth and your work. I see the earth as a medium of struggle that is explicit in the formation of freedom dreams and more chaotic dreams, which disrupt economies of value and devaluation. It is these creative practices with the earth that block the trajectory of those that would weaponise geology for their own ends. In these creative temporal realignments of artful practice, what is given is the imagination and material existence of the possibility of change, which brings down the geo-logics of value, and contributes to the non-renewal of extraction.

References

Deleuze, G. and Guattari, F., 1996. *What is Philosophy?.* Translated from French by H. Tomlinson and G. Burchill. New York: Columbia University Press.

Foucault, M., 1977. *Discipline and Punish: The Birth of the Prison.* Translated from French by A. Sheridan. London: Allen Lane.

____2008. *The Birth of Biopolitics: Lectures at the Collège de France 1978–1979.* Translated from French by G. Burchill. London. Palgrave MacMillan.

Grosz, E., 1994. *Volatile Bodies. Toward A Corporeal Feminism.* Bloomington, Indiana: Indiana University Press.

____2004. *The Nick of Time: Politics, Evolution and the Untimely.* Durham, London: Duke University Press.

____2008 (2020). *Chaos, Territory, Art: Deleuze and the Framing of the Earth.* New York: Columbia University Press.

Hear Me Now: The Black Potters of Old Edgefield, South Carolina, 2022. [exhibition] The Met Fifth Avenue, Gallery 955, New York. 9 September 2022 – 5 February 2023.

Woods, C., 2000. *Development Arrested: The Blues and Plantation Power in the Mississippi Delta.* London, Verso.

Yusoff, K., 2018. *A Billion Black Anthropocenes or None.* Minnesota: University of Minnesota Press.

transposition 191

Geoscientific Collections as Spaces of Sedimented Histories

Malina Lauterbach, R. Johannes Giebel, and Sascha Mikloweit

Abstract: Mineralogical collections are a thorough embodiment of colonial history and its transhistorical geopolitical implications. As such, they provide access to study continuities of colonial power dynamics that grew out of imperial ideologies. While minerals once helped colonial powers to glorify their achievements, today these specimens bear witness to the violent extraction and unjust appropriation inherent in imperial sciences. By this nature, mineralogical collections are mattering coloniality. This renders popular concepts of restitution and decolonisation inefficient. Herein, however, resides their potential. Focusing on the Mineralogical Collections of the Technical University of Berlin, we emphasise the urgent need to critically reassess these collections in light of their sedimented colonial histories and legacies. Our re-evaluation questions the notion of geoscientific collections being repositories of purely 'natural' objects. Instead we shift our emphasis to the specimen's significance as 'objects of culture', and advocate for inter- and transdisciplinary collaborative approaches to address historical injustices, and redefine the role of geoscientific collections in today's society.

Keywords: Colonial histories, geoscientific collections, transdisciplinary research, critical museum studies, Mineralogical Collections of the Technical University of Berlin, Royal Mining Academy, imperial geosciences, German mineralogists, transhistorical empowerments, mineral extraction

> [...] geology is not just an exploration of deep time. It is an analysis of the past as an analogy of the present. (Chakrabarti, 2020, p.14)

Debates focusing on institutional natural history collections and their relation to colonial histories are at an early stage within German discourses. A comprehensive study on this topic was provided by the joint research project *Dinosaurs in Berlin! Brachiosaurus brancai as an Icon of Politics, Science, and Popular Culture* from 2015 to 2018 between the Humboldt University Berlin, the Berlin Museum of Natural History and the Technical University of Berlin (TUB, 2016). The researchers examined the colonial history of the Brachiosaurus brancai located in the Berlin Museum of Natural History collection. The skeleton of the dinosaur was excavated in 1909 in German East Africa (present-day Tanzania) under the leadership of the museum's palaeontologists, among them Werner Janensch (1878–1969) and Edwin Hennig (1882–1977). Reflecting on this study, Ina Heumann and Mareike Vennen argue that the common belief of natural history museums holding 'purely natural objects' is counterproductive to a critical examination of collection histories (Heumann and Vennen, 2018, pp.8, 15). In the *Guidelines for German Museums: Care of Collections from Colonial Contexts* published in 2021 by the German Museums Association, this issue manifests. Placing specimens outside of the realm of cultural heritage, natural history collections are assigned a status of exception here:

> Natural history objects and collections [...] were not of particularly significant importance for indigenous communities of origin until very recently; and they have so far rarely been considered to be part of the indigenous heritage. Moreover, natural history objects, in contrast to art items and archaeological finds, for example, are not artefacts and objects of art in the literal sense, but exemplary representations of nature. These objects and collections therefore stand in stark contrast [...]. (Glaubrecht, 2021, p.109)

This attested (in)significance of 'natural history objects'—or what he categorises as such—to their communities of origin, is situated in the context of a 'recent past' debate: that there had been no interest prior to this 'recent past' is his assumption. How precisely Glaubrecht arrives at such problematic ontological categories remains unclear, further, his understanding is simply contrary to contemporary German and European law.[1]

The European dualistic separation of natural history museums (including their geological and mineralogical collections)[2], and museums of art coincided with the colonial expansion of Europe and the emergence of 'modern' natural sciences in the 18th century (Te Heesen, 2012; Brunner, 2018). Natural history museums like the Muséum National d'Histoire Naturelle in Paris (founded 1793), or the Natural History Museum in Berlin (founded 1810) were conceived as modern scientific institutions with a specific focus on research and education. As 'national museums', however, they also embodied the political and ideological interests of the nation-state by promoting patriotic knowledge through the exhibition of their collections (Te Heesen, 2012, pp.51–52). In the feminist discourse of the 1980s, thinkers like Carolyn Merchant (1983/US), or Maria Mies (1986/DE) wrote about the intertwinement of modern natural sciences with colonialism:

> In the same measure as European conquerors and invaders 'penetrated' those 'virgin lands', these lands and their inhabitants were 'naturalised', declared as wild, savage nature, waiting to be exploited and tamed by the [predominantly]

[1] Even though the German Museums Association does not regard natural history objects (e.g. minerals) as cultural heritage, this position is held differently in German law, which is supplemented by the European regulations concerning natural history (e.g. mineralogical) collections. Reading the German Cultural Property Protection Act (FMoJ, 2016), in relation to Annex I A.13(a) of the Council Regulation (EC) No 116/2009 '[c]ollections and individual specimens from zoological, botanical, mineralogical or anatomical collections;' (CEU, 2008, p.4) are regarded as cultural objects. In this sense, the 'Guidelines for German Museums' and German—in dialogue with European—legislation apply two different standards. Due to their legal status as cultural objects in EU legislation, natural history objects such as minerals are subject to export regulations. For this very reason complex export authorisation procedures are mandatory if such objects become part of international institutional loans or exchanges. The German Museums Association's ontological exclusion of natural history objects from the category of 'cultural property' obstructs the necessity to debate the significance of such objects in relation to indigenous communities and against the backdrop of colonial extraction. What both applied standards have in common is 1. that they (directly or indirectly) enable the respective authorities to suppress a potential thematisation of the consequences of extraction and 2. that they prevent a potential 'migration' of these objects from Germany back to their countries of origin.

[2] As parts of natural history collections, geological and mineralogical collections are closely intertwined and often cannot be clearly separated within the category of geoscientific collections. In the literature, the term 'geological collections' is hence often mis-applied, while these in fact should be denominated as 'geoscientific collections', which does include both categories. In the following discussion, mineralogical and geological collections will be used interchangeably and referred to as geoscientific collections whenever possible. In specific cases, one term may be used based on the collection's name within a particular institution or for citation purposes when authors mention a specific collection.

> male civilisers. Similarly, the relationship between human beings and external nature or the earth was radically changed [...]. The taboos against mining, digging holes in the womb of Mother Earth, were broken by force, because the new patriarchs wanted to get at the precious metals and other 'raw-materials' hidden in the 'womb of the earth'. The rise of modern science, a mechanistic and physical world-view, was based on the killing of nature as a living organism and its transformation into a huge reservoir of 'natural resources' or 'matter', which could be analysed and synthesised by Man into his new machines by which he could make himself independent of Mother Nature. (Mies, 1986, p.75)

After Maria Mies, it took another 40 years for criticism to emerge in German research institutions and museums—and its pace is still slow. Innovative thoughts instead proliferate today from British discourses: in her 2018 publication *A Billion Black Anthropocenes or None*, Kathryn Yusoff—a transdisciplinary geographer focusing on inhuman geographies—exposes the entanglements of geological knowledge production and racist-colonial ideologies from a contemporary perspective. Her thoughts provide insight into the specific nature of geological collections, where the heritage of what she terms 'White geology' (Western colonial knowledge and extraction practices) is being preserved (Yusoff, 2018, pp.26–27). As US-based geologist Selby Hearth and art historian Carrie Robbins have pointed out: 'The same mines that motivated, funded, and fueled Western colonialism, imperialism, and industrialisation, also yielded the museum-quality mineral specimens now displayed throughout the West' (2022, p.13).

The entanglements between geological collections and colonialism still need to be addressed.[3] And yet, if we consider geological collections as the materialisation of colonial geological knowledge (Rogers, et al., 2021), we have to think of them in connection to the violent exploitation of mineral resources by European colonial powers, as well as the hegemonic order of the modern museum and the natural sciences (Balzar, 2022, p.55). If these collections, similar to ethnological collections, are 'stained with blood' (Savoy, 2019), they perpetuate the violent histories of European colonialism by failing to address these contexts. This raises questions: what is the specific significance of geoscientific collections in their relation to past and ongoing destructions and exploitations of land, and exploitations of labour and knowledge in mineral-rich regions? How can collection managers and curators today research, unpack and mediate these histories and continuities? How can such collections be re-contextualised to voice their silenced pasts, without negating their valuable achievements?

In the context of this essay, we cannot produce definite answers to a topic of such vast complexity. We will, nevertheless, as a starting point of an ongoing investigation, share our initial reflections regarding the potentials and limitations of a critical reassessment, taking the Mineralogical Collections at the TUB as an example.

The TUB's Mineralogical Collections[4] house approximately 200,000 specimens, with minerals and rocks from Namibia being a significant epistemic focus (about two per cent). The collections are closely linked to the founding of the Königliche Bergakademie (Royal Mining Academy) in the late 18th century and colonial research expeditions.[5] Since

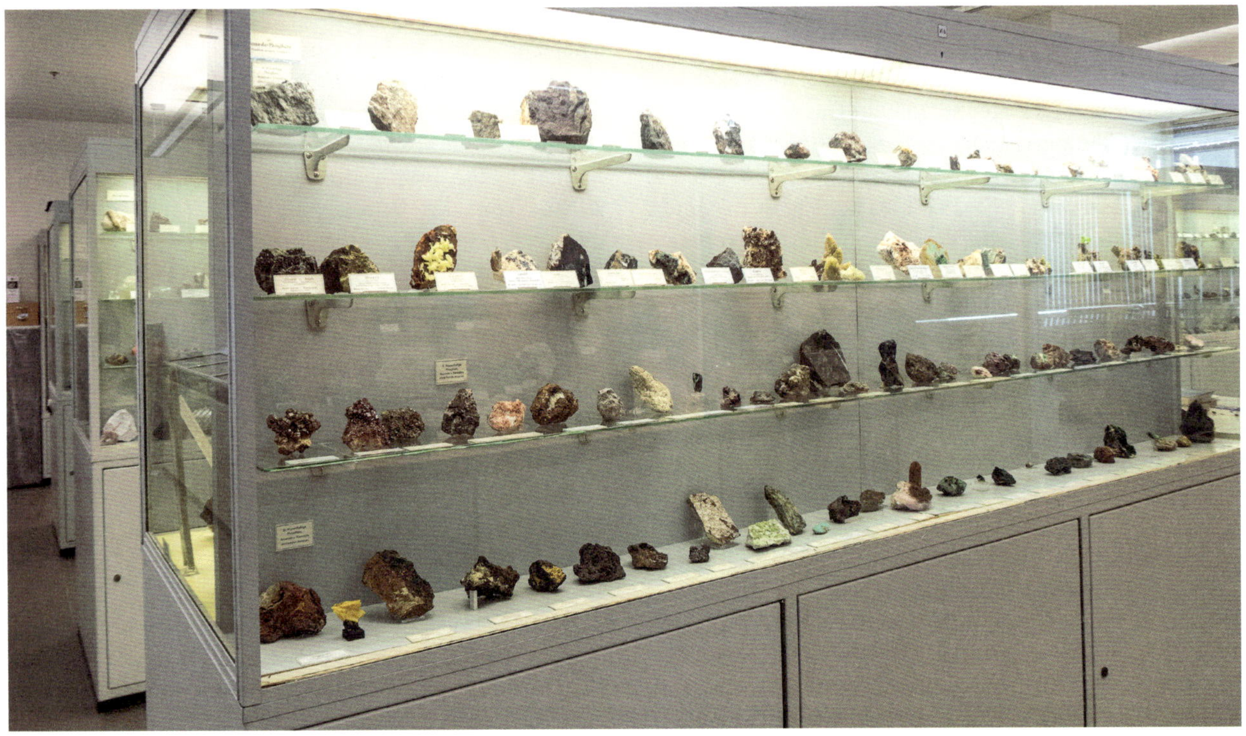

Fig. 37

[3] A discussion on this matter took place in 2021 during the public workshop Colonial Collections in Berlin Universities at TUB, funded by the Berlin University Alliance. In session 3, Johannes Giebel, Kristine Okoth and Yann LeGall gave their input on Colonial Extraction and Berlin Mineral Collections (Giebel, Okoth, and LeGall, 2001).

[4] The Mineralogical Collections of the TUB were constituted as Königliches Mineralienkabinett (Royal Mineral Cabinet) in 1781 by Carl Abraham Gerhard (1738–1821), who had founded the Königliche Bergakademie (Royal Mining Academy) back in 1770 by decree of Frederick the Great, King of Prussia. In 1916, the Königliche Bergakademie merged with the Königliche Technische Hochschule (Royal Technical Academy; a fusion of the Königliche Bauakademie (Royal Building Academy) founded in 1799 and the Königliche Gewerbeakademie (Royal Academy of Commerce) founded in 1821) to form the Technische Hochschule (Technical Academy). Both institutions had independent mineralogical collections, which have been united. After the Second World War, the Technische Hochschule was re-established as today's Technische Universität Berlin (Technical University of Berlin; Strunz, 1970).

[5] In 1908 Prof. Robert Scheibe (1859–1923) (Bergakademie Berlin) undertook an expedition on behalf of the 'Gibeon Mining and Trading Company' to investigate diamond deposits in German South West Africa in the areas of Gibeon and Berseba (1913, p.18–19).

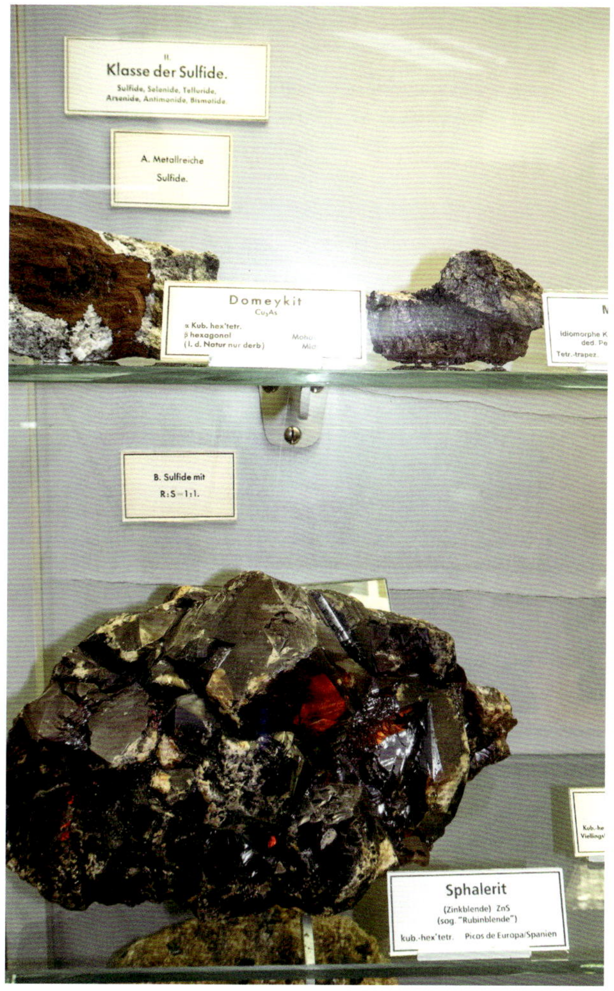

the early 20th century, specimens from Namibia have been incorporated into the collections, some of them related to the former German colony of German South West Africa (1884–1915), others to South West Africa (1915–1990 under South Africa rule), and present-day independent Namibia (1990–ongoing).[6] The TUB's Mineralogical Collections are no exception to what Hearth and Robbins attest for most museums in Europe, where the display arrangements seek to inspire curiosity in the viewer, showing an impressive number of specimens of various colours and forms (Hearth and Robbins, 2022, pp.11–12). In the same manner, the collections' most representative minerals are arranged in glass vitrines mainly based on 'natural' characteristics; their chemical and structural composition (Figures 37 and 38).[7] The most widely used mineral systematic[8] was developed in 1941–1978 by German mineralogists and former TUB professors Karl Hugo Strunz and Christel Tennyson[9] (Strunz and Nickel, 2001). The collections' archive is much less representative. The inventory books, which document the acquisition activities before 1945, were destroyed during World War II and then re-documented in 1953, leaving significant archival gaps. Only in some cases, the small boxes, in which the minerals are being stored, contain paper notes, which identify collectors and places of origin, referring to former colonial territories, such as 'German South West Africa' (Figures 39 and 40). Within the sterile confines of displays, this information is rendered invisible as it holds no value for their mineralogical classification. Instead, the 'natural' display arrangement constitutes an epistemic framework, which disregards the minerals' multidirectional histories and socio-political contexts. To relink these contexts, three concepts might proof useful: Stoler's reading 'along the archival grain' Ahmed's 'sedimented history', and Yusoff's 'geologies of race'.

In her analysis of colonial documents from the nineteenth-century Netherlands Indies (a territory that largely comprises the present-day state of Indonesia), anthropologist Ann Laura Stoler (2008), engages with archives as epistemological spaces of imperial power, where racial ontologies are being (re)produced. '[...] archival production is itself both a process and a powerful technology of rule [...]. We need to read for its regularities, for its logic of recall, for its densities and distributions, for its consistencies of misinformation, omission, and mistake—along the archival grain' (ibid., p.100). With this analytic approach, Stoler suggests to carefully engage with colonial archives not as sites of a 'grant narrative of history,' but of contested knowledge—of sentiment and speculation turned into fact. The TUB's Mineralogical Collections have been shaped by imperial sciences, war destruction and the divergent collection management of different generations of German mineralogists, pointing to the institution's very 'disordered' history. Reading the inventory books and labels 'along their archival grain', and thereby paying particular attention to the consistencies of gaps and confusions, confronts the universalism of scientific 'order'.

Sara Ahmed coined the term 'sedimented history' to describe the lasting effects of colonialism on contemporary society. The concept captures the tangible weight of histories of exclusion, racism, and oppression that continues to impact individuals today. 'To think of race as a sedimented history is to think of how race matters as matter. Something becomes sedimented, when it has settled, often near a barrier, as that which stops a flow. And race is precisely this: a congealing, a solidifying: a history that becomes concrete, a physical

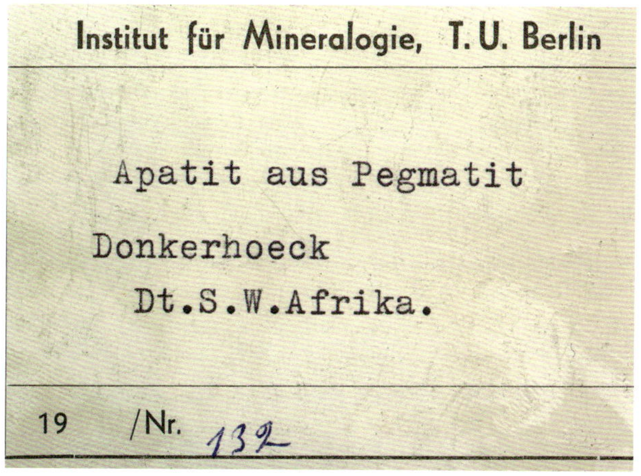
Institut für Mineralogie, T. U. Berlin

Apatit aus Pegmatit

Donkerhoeck

Dt.S.W.Afrika.

19 /Nr. 132

[6] These are, of course, the denominations for three different political phases of one and the same territory since 1884, which also mark the eventful history of Namibia. However, as 'mineral name' and 'place of origin' are the minimum requirements for the documentation of a specimen, the use of the different territorial designations at a particular time bear witness to the relative time in which the specimens were removed from their place of origin.

[7] The exhibition at TUB is mainly organised after the Strunz mineral systematics (~ 78%). However, a smaller part of the collection is displayed according to different mineral-specific characteristics, such as their affinity to certain types of deposits (~ 18%), mineral characteristics (e.g., twinning and epitaxy; ~ 2%), or industrial and commercial use (~ 2%). These types of presentation are autonomous and independent of the organisation according to the mineral systematics.

[8] Officially used by the International Mineralogical Association (IMA), which is authorised to accept or reject new mineral discoveries and designations.

[9] The 1st edition of the Strunz mineral classification system (Strunz Mineralogical Tables) was published in 1941 by K.H. Strunz himself. This classification was significantly revised and expanded in cooperation with C. Tennyson. C. Tennyson played a fundamental role in the 4th to 8th editions in particular. The 9th and last edition was published by Ernest Nickel (Canada, Australia).

Fig. 38

Fig. 39

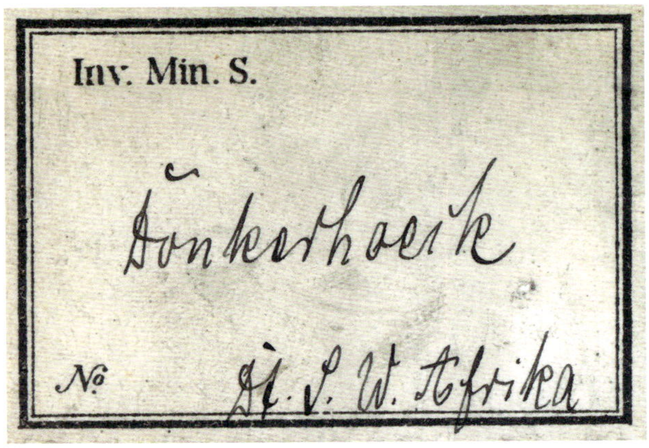

barrier in the present: stop. Or not: go' (Ahmed, 2015, p.95). Mineralogical collections are physical manifestations of 'sedimented histories' as they inhabit what Yusoff phrases 'geologies of race' (2021, p.664).

> The historic and geographical differences across which the questions of geology and race travel are specific and varied, and by no means exclusive, but across the plateaus of colonialism, race provided the political redemption for the white supremacy of matter. The discourse of "improvement" brought raced subjects and indigenous land theft into the economic terra-forming project of natural resources. This is to say that knowledge of the earth is also interned in a perspectivism and praxis of the supremacy of whiteness, which marks the capricious geography of the human. (ibid. p.665)

> [A]s geologic classification of earth minerals made matter as value, it also captured enslaved subjects in the brutal calculative enclosure of the inhuman [sic] as chattel' (ibid, p.666).

Thus, the history of colonial mining, and its representation in geoscientific collections, is a history of racial subjugation and dehumanisation. Bringing Stoler's, Ahmed's and Yusoff's thinking in dialogue forces a reckoning of how minerals matter as racialised matter, which have been sedimented (and preserved) in European collections. This can be determined not only through a critical examination of the collection's objects (in a material sense) and reading the archives 'along their grain', but also through an analysis of the entangled histories of imperialism and geoscientific knowledge production.

The latter appears in the book *The Mineral Kingdom* (1903),[10] which serves as a compelling example of how minerals become manifest as scientific illustrations of imperial ideologies. Published by German mineralogist Reinhard Brauns (founding member of the German Mineralogical Society, in 1908), the book became well-known in European mineralogy. Its colour illustrations emphasise the aesthetic qualities of the mineral specimens, which he sourced from museum collections in Bonn, Göttingen, Marburg, Stuttgart, Karlsruhe, Frankfurt and Hamburg (Brauns, 1912, p.III). His focus on their shape, lustre, and colour, generates an aura of the mineral objects as desirable, valuable commodities. Each mineral casts a shadow and hovers above the ground, symbolising its literal ascent from the 'raw' interior of the Earth (Figure 41). Following Brauns' introduction, the mineral world is a 'universal kingdom':

> In a sense the Mineral Kingdom is universal, for it embraces not only the whole earth, but also meteorites, those celestial messengers which come to us out of the depths of the stellar universe. Some members of this immense kingdom have, as it were, taken up their abode in other domains, and like the colonies to the mother-country retain only a loose connection with the mineral kingdom. Such are the fossils, those petrified remains of animals and plants which are found embedded in hard rock. While they show any trace of organised structure fossils are excluded from the mineral kingdom, for in that domain is included nothing which lives or has once lived. When, however, such changes in the fossil have taken place that it loses the shape and form originally impressed on it by the vital activity of the creature of which it formed a part, it ceases to be a fossil and is once more included in the mineral kingdom. The emigrant returns to the mother-country. (Ibid., p.5–6)

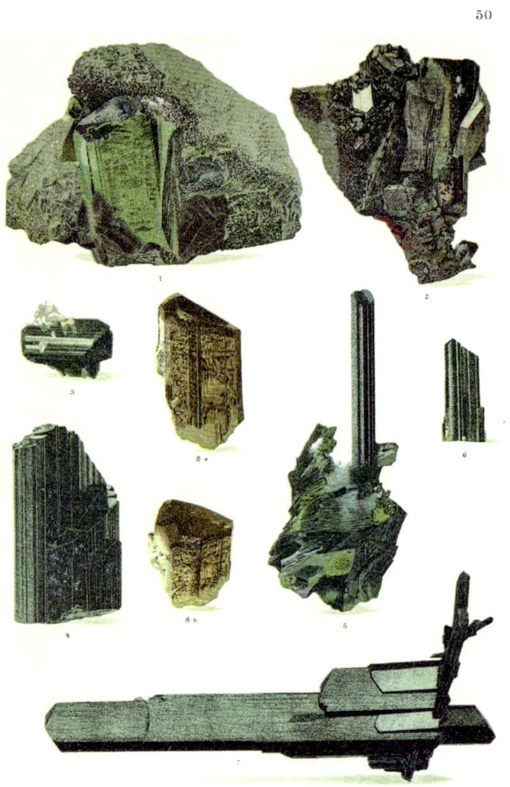

[10] Originally published in 1903 under the German title *Das Mineralreich* and translated into English in 1912 to reach a wider (European and selected international) audience.

Fig. 40

Fig. 41

Brauns continues to acknowledge the miner—the 'master of the Earth'—for his achievements: 'fruit of his labour furnishes the world with the material for the construction of tools and machinery, railways and telegraphs, and with gold, the precious metal which serves as a standard of currency' (ibid., p.5). The author neglects to mention the violence surrounding mineral extraction: from the exploitation of Black and Indigenous people's labour force and lives in the colonies at that time, to the destruction of land and ecosystems. By oscillating between the natural world of minerals, colonial rule (colonies vs. mother-countries), and the economic benefit of mining in the imperial centres, he legitimises a master-narrative of 'white geology' with its infrastructure of wealth.[11] Analysing these intertwined processes of a scientific geological discourse, its narrations with its naturalising effects and its socio-political dimensions, lead to a collection's violent epistemic framework.

Researching the histories of minerals, their acquisition contexts and epistemological transformation in European museums, as well as granting unrestricted access to their archives, can provide an opportunity to reestablish connections between colonial archives in Europe and Africa. In case of the TUB's Mineralogical Collections, the activities of Robert Scheibe (1859–1923) in colonial German South West Africa can provide such a point of entry for analysis that is in need of more scrutiny. From 1895 to 1916 he served as a professor of Geology at the Bergakademie Berlin, and later held the position of professor at Technische Hochschule Berlin from 1916–1923; both of which were predecessor institutions of the TUB (Strunz, 1970). Scheibe's scientific activities were closely linked to the exploitation of diamond deposits by the German Empire. In 1908–1909, Scheibe, on behalf of the Bergakademie, together with August Stauch (1878–1947), chief railway foreman and hobby mineralogist, later diamond prospector (Levinson, 1983), sought diamond deposits in German South West Africa (Von den Förderern, 1914, pp.18–19). During his expedition, he also collected a diverse range of mineralogical specimens, including fragments of the Gibeon meteorite, which he brought back to Berlin.[12] In his travel report, he writes on his diamond explorations with an adventure's voice:

> On New Year's Day in 1909, my friend Stauch and I, after an exciting two-and-a-half-day journey through sandstorms and other adverse events in the desert, reached the main valley of the Pomona area, which lay untouched. After initially unsuccessful searching, we suddenly found rich spots [of diamonds] beyond all imagination, and felt like we were in a fairyland. (Scheibe, 1913)

In German historiography it is often ignored that there was a veritable diamond rush in German South West Africa from 1908 after the German Reich established a restricted diamond area (Press, 2021b). Following Press, this 'was a compelling new reason for Germans to continue, and legally endorse, violent acts that left Germans as owners of nearly all Namibian property' (ibid., 2021a).[13] By 1914, the German Reich had extracted almost five million carats of diamonds worth 150 million Reichsmarks (Hollenbach, 2020; Schmitt, 2020).[14] Studying the activities of Robert Scheibe and August Stauch would produce a more comprehensive understanding of the economic dimensions of German colonialism in Namibia and its ties to geoscientific institutions.

Furthermore, a restitution case from Finland illustrates that the extraction of what historically has been classified and preserved in European museums as 'geological specimens,' can have serious consequences for a region's social stability. In 2015, a delegation from the Museums Association of Namibia visited the National Museum of Finland in Helsinki and identified a piece of the Ondonga Power Stone on one of the museum's archival photographs. Eight years later, the Finnish Museum returned two fragments of the Ondonga Power Stone to the National Museum of Namibia, owing to museum ethics, the provenance of the fragments and their special significance to the descendant community (Ondonga Traditional Authority, 2023). During their travels in the Kingdom of Ondonga in 1886, Finnish missionary Martti Rautanen (1845–1926) and Swiss geologist Hans Schinz (1858–1941) unlawfully removed fragments of a stone they believed to be a meteorite—which it was not. This stone held sacred significance to the Ovambo people and its removal occurred against the laws of the Kingdom. According to Erikka Nehunga (Councillor in the Ombalantu Traditional Authority) and Martha Mutambo (Headwoman of Omufitu weelo village), the stone historically served to pacify the nation, and its absence was believed to disrupt the nation's calmness (Silvester and Shiweda, 2020).

However, individual restitution cases such as the one mentioned above seem to be rare in the context of geoscientific collections. Rather we need to recognise that, as Hearth and Robbins state, 'the entire modern Western world is built on minerals mined from colonised land', and the voices represented in mineralogical collections predominantly belong to white Western male collectors, who were involved in mining activities (2022, p.13–14). Therefore, aiming at a critical examination of the collections' violent histories, the question arises as to whether 'provenance research' (Savoy, 2018) can be helpful in these hegemonic structures at all, or if it simply leads to novel strategies of deferral where essential agency still remains with the successor systems of former colonial powers. Referring to the restitution of looted artefacts from the Kingdom of Benin (Nigeria), philosopher Kwame Opoku (2021) critically frames provenance research as a 'German invention', which occurred as an institutional reaction to the demands for restitution from African countries and the influential restitution report by economist Felwine Sarr and art historian Bénédicte Savoy (2018). Opoku (2021, p.10) states: 'Provenance research and

[11] Until today, this contentious conception is actively commemorated as the mineral 'Reinhardbraunsite' has been named in tribute to the author in 1983 (46 years after his death; Kirfel et al., 1983). Furthermore, Braun's illustrations are displayed in mineralogical collections (i.e. Senckenberg Museum Frankfurt) for their aesthetic value and without being given any further context.

[12] Robert Scheibe brought four meteorite fragments of the Gibeon meteorite from German Southwest Africa back to Berlin. The smallest and the largest of the meteorite fragments (ranging from 160kg to over 500kg, respectively) are still on display at TUB's Mineralogical Collections. One of the meteorite fragments (~400kg) was probably given (donated in 1952) to the Deutsches Museum für Bergbau (German Museum for Mining; on display and listed in their inventory). The last meteorite was cut into slices, polished and etched in various attempts to make the inner structures of the specimen visible. The largest of these slices, together with the aforementioned meteorite fragments, is on display at TUB's Mineralogical Collections; another one is on display in Kiel. The whereabouts of the remaining slices and fragments are unknown. Initial reports assume that two meteorite fragments went to North America. However, a connection to those from TUB is not traceable (Scheibe, 1990).

[13] The acquisition of Lüderitzbucht and Lüderitzland (later known as the restricted diamond area) has become known in history as the 'mileage fraud'. In 1883 Adolf Lüderitz, represented by Heinrich Vogelsang, bought the bay of Angra Pequeña (later named Lüderitz Bay) and the area within a radius of 5 geographical miles for 100 pounds in gold (today about 12,251 pounds*/14,283 euro) and 200 rifles. In the same year, Lüderitz also bought the area from the Orange River northwards to latitude 26° and 20 geographical miles inland for 500 pounds (today about 61,255 pounds*/71,423 euro) and 60 rifles. Lüderitz led the Nama chief Joseph Fredericks to believe that he was referring to English miles (approx. 1.6km), but Lüderitz actually demanded Prussian miles (approx. 7.4km), which was almost 4.6 times the distance. Subsequent complaints and protests by the Nama were not accepted under German jurisdiction (Vogelsang, 1883a; 1883b). *calculated according to the conversion rates of Purchasing Power Parities published by the Bank of England, 2024.

[14] This corresponds to a 2023 value between 705 million and 945 million euro according to the Purchasing Power Parities conversion rates published by the Federal Bank of Germany, 2024 (Bundesbank, 2024).

digitalisation of artefacts are useful *per se* but are being used in this context to delay restitution in so far as the financing, researchers, the tempo, and objects are all determined by the Western governments and institutions that are not in any hurry to return the looted artefacts they have kept for so long'. As Jeremy Silvester and Napandulwe Shiweda (2020) examine in their reflections on the return of the Sacred Stones of the Ovambo Kingdoms: '*Provenance research* often concentrates on museum archives with the aim of reconstructing paper trails to provide legitimacy to *ownership*. A converse approach should start by evaluating the cultural and historical significance of artefacts today to a *descendant community*'. Ellen Namhila, Pro-Vice Chancellor of the University of Namibia, emphasises the significance of scientific exchange and discussions on colonial dimensions and injustices in collaboration between European and formerly colonised countries.

> We do not say, 'Empty your storehouses and museums of Namibian objects and send them all to us, case closed'. This case cannot be closed, and European museums should not be emptied of all colonial objects, to create a tabula rasa as if nothing had happened. (Namhila, 2019)

If we take these critiques into account, conducting provenance research that aims simply at 'returning' geoscientific objects will not address the real issues: that of epistemic restitution, and that of a redistribution of power and agency. And worse: it runs the risk of becoming an instrument of 'whitewashing'. Such mono dimensional procedures would further divert attention from the significant potential that is dormant in institutional geoscientific collections—that of working towards an epistemic restitution that produces deep understandings and far-reaching consequences, particularly in regards to matters of transhistorical 'coloniality' (Quijano, 2000).

In terms of redistributing agency, projects—ideally of inter- and transdisciplinary nature, in cooperation with partners from affected countries, should be endeavoured in order to hear voices that have been rather silenced thus far, and to create spaces in which new relations of a non-toxic, non-extractive nature can grow that may enable cooperations rooted in mutual respect and exchange.

An experimental approach thinking from such a position is provided by the *Transpositional Geologies* project at large, and was applied in the two linked exhibitions 'We Build Our Languages Out Of Rocks—Transpositional Geologies' (April to June 2023, Bonn, Germany) and 'Before Thought Hardens into Language—Transpositional Geologies' (October to November 2023, Tsumeb, Namibia) by the researcher and artist Sascha Mikloweit. Mikloweit collaborated with descendant communities in Namibia, Namibian artists and several international research institutes to create a network of relations, performances, and temporary

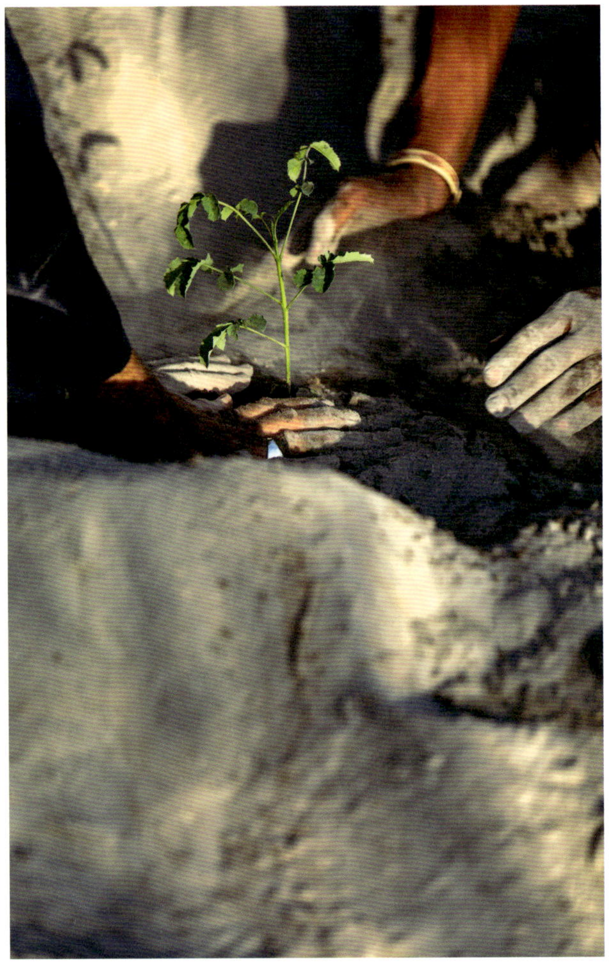

and permanent interventions in public space (Figure 42). In centring around minerals as epistemic objects, he shifts the focus of institutionalised mineral objects to their omitted trans-historical, socio-political content: he probes the format of institutional mineralogical collections, and reframes them as 'laboratories of thought where the individual mineral specimen becomes a thinking device to unearth power structures that facilitate historical and current inequalities' (Mikloweit, 2025). His work reflects the questions of recognising the losses of the Namibian people in the forgetful European mindset. In contrast to many other voices, his work poses the question: what is fair restitution on an appropriate scale, and which forms might it take?

While it was colonialism that boosted Europe to its current prominence, and the plundered countries suffered oppressions in their developments, is a simple restitution of objects, land and finances even enough? Which new forms of restitution are required? Mikloweit does not claim to find a solution in his work, but places the unpleasant questions as prominent elements in public and institutional spaces.

To re-think and re-link geological collections today means to pay attention to a paradigmatic shift towards unpacking the 'colonial DNA' of geosciences and its ideological

Fig. 42

Fig. 43

[15] Even after Germany lost its colonies at the end of WWI and under the subsequent administration of South Africa as part of the British Commonwealth, Tsumeb continued to be mined by the Otavi Mining and Railway Company (OMEG) under German management until the outbreak of WWII (Schnorbus, 2000).

sedimentation in museum and research institutions. This not only has the prospect to reappraise the epistemological potential of the collections and to communicate complex and often contradictory contexts to international students and visitors, but, in a best-case scenario, it may become an agent in redistributing power and agency more justly.

As art historian Sria Chatterje has argued, a connection to the present is an integral part of the true educational mission of colonial collections (Chatterje, 2021). Colonial extractivism is trans-historical as it developed production cycles and economic power relations around mining operations. Therefore, still ongoing in the present, everyday life is permeated with the material consequences produced by extractive processes, mattering in applications such as skyscrapers, computers, and mobile digital technology. This, however, reaches far beyond civil society—military and space operation technology, as well as various 'dual use' technologies, would be unthinkable without minerality.

Mineral specimens link geo-politico-economica activities of public and private, national and international operators through time, space, and place. This becomes tangible, if we take a closer look at the mineral germanite (a copper-iron-germanium sulphide; $Cu_{13}Fe_2Ge_2S_{16}$), which is widely represented at TUB's mineralogical collections, (Figure

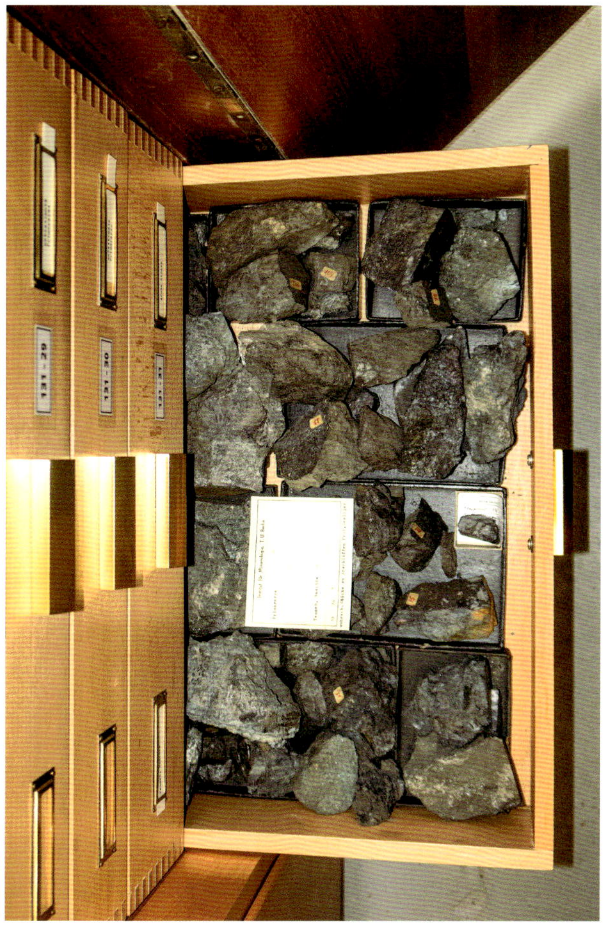

43). Germanite bears a violent history as it was excavated in Tsumeb (Namibia) during German colonial occupation.[15] The mining town was renowned for its abundant ore deposits, particularly zinc and copper ores containing the mineral germanite (Lombaard et al., 1986). Today, the significance of germanite has grown due to digitalisation and an increased energy usage: it is utilised in the highly competitive electronics industry as a crucial raw material for semiconductor (e.g. glass fibre) and solar cell production. However, as resources with the highest potential for germanium extraction, germanite (and other germanium mineral)-rich deposits are scarce globally.[16] As they are primarily located in remote (Indigenous) or historically exploited regions (e.g. the Black Angel deposit, Greenland; Kipushi, Democratic Republic of the Congo; and Tsumeb, Namibia), the extractivist cycle continues. Among many other raw materials (such as lithium and cobalt), germanium is a classic example of today's 'green' extractivism and climate colonialism (Dorn, 2023, pp.17, 21).[17]
While the methods may change, the imperial structures of political and economic power-imbalances and the instigation of conflicts by industrialised nations persist. It is within these continuities that minerals serve as a stark reminder of the ongoing exploitation of land, people and life. This predominantly takes place in 'Lower-middle-income countries', to phrase it in terms of the World Bank. Such transhistorical entanglements makes it very unlikely to ever encounter a truly post-colonial history of geoscientific collections that are holding specimens connected to global extractivism. The enduring injustices in resource-rich regions financed by the wealth of 'Upper-middle- and high-income countries'[18] and their power relations, argue against it.

A first step into a productive direction, however, acknowledges that unlike any other type of collections, geoscientific collections form access points to trace and render the coloniality of power in contemporary societies visible.

Then there is one final transdisciplinary question left: what other forms of reparation, what other forms of a systematic transformation rooted in dialogue between diverse modes of existence are necessary to stop the violence of modern scientific 'universalism' and its consequences?

[16] In addition to discrete germanium minerals (which potentially also occur as processed slags; highest production potential), germanium is mined as a by-product from sulphides (low to medium production potential), and coal and lignite (high production potential). This production is dominated by China and Russia (Melcher and Buchholz, 2012).

[17] During the colonial period, germanite was primarily mined for copper recovery and not used in significant quantities, because its applications were negligible. Hence, during historical industrial copper and zinc production in Tsumeb, germanium was handled as a waste product of the process, deposited on the dumps as a component of the furnace slags, mainly as germanium-bearing silicate glasses (Ettler, et al., 2021). In contrast, germanium today is an important raw material used in critical technologies and consistently in high demand. These materials, however, are currently not economically processable (ibid.) and therefore the 'germanium dumps' in Tsumeb are not yet considered a potentially lucrative resource. Nevertheless, when the Chinese company Sinomine Resource Group Corporation Ltd. took over the Tsumeb copper smelter in 2024 (DPM, 2024), all associated areas including the large germanium-rich dumps came along as part of the purchase. So far this is only a future resource, but once a way is developed to process such material and produce germanium concentrates, the Chinese company will expand the already existing monopoly position of China in the germanium mining industry (currently at about 60% of the world market). In principle, the possibility of processing is certainly only a matter of time. Supply chains for critical raw materials like germanium are highly vulnerable, due to monopolisation by certain countries. Export controls from these countries can disrupt markets in the short term. And such monopolisation allows these countries to exert political pressure and influence or control geopolitical developments.

[18] There is a paradox worth pointing out in relation to Namibia. The World Bank's categorisation of 'Upper-middle-income economies' includes both Namibia and China, despite China being one of the most aggressive investors in Namibia at present. A similar issue arises with the categorisation of 'Global North / Global South', where both Namibia and China fall into the latter category. Both classifications fail to adequately address the exploiter/exploited relationships, such as that between China and Namibia.

References

Ahmed, S., 2015. Race as Sedimented History. *Postmedieval*, 6(1), pp.94–97.
Balzar, C., 2022. *Das kolonisierte Heiligtum. Diskriminierungskritische Perspektiven auf das Verfahren der Musealisierung.* Bielefeld: Transcript.
Brauns, R., 1903. *Das Mineralreich*. Stuttgart: Fritz Lehmann Verlag.
____1912. *The Mineral Kingdom*, Vol.1. London: J. F. Schreiber.
Brunner, C., 2018. *Epistemische Gewalt. Wissen und Herrschaft in der kolonialen Moderne.* Bielefeld: Transcript.
Bundesbank, 2024. *Kaufkraftäquivalente historischer Beträge in deutschen Währungen*. Frankfurt: Deutsche Bundesbank. [online] Available at: <https://www.bundesbank.de/resource/blob/615162/94b87ff6d25eceb84c9cfb801162b334/mL/kaufkraftaequivalente-historischer-betraege-in-deutschen-waehrungen-data.pdf> [Accessed 23 May 2024].
Chakrabarti, P., 2020. *Inscriptions of Nature: Geology and the Naturalization of Antiquity.* Baltimore: Johns Hopkins.
Chatterje, S., 2021. The Long Shadow of Colonial Science. In: *Noema Magazine* [online] Available at: <https://www.noemamag.com/the-long-shadow-of-colonial-science> [Accessed 15 May 2024].
CEU, 2008. *Council Regulation (EC) No 116/2009 of 18 December 2008 on the export of cultural goods.* Document 32009R0116, Annex I, 13(a). [pdf] Brussels: Council of the European Union. Available at: <http://data.europa.eu/eli/reg/2009/116/oj> [Accessed 3 July 2024].
Dorn, F. M., 2023. Die Kolonialität der Energiewende und Grüner Etraktivismus in Lateinamerika. In: M. Schmidt, S. L. Middendorf, S. Purwins, and C. Walter, eds. 2023. *The Plurality of Political Ecology.* Augsburg: University of Augsburg, pp.17–26.
DPM (Dundee Precious Metals Inc.), 2024. *Dundee Precious Metals Announces Sale of Tsumeb Smelter for US$49 Million*. [press release] 7 March 2024. Available at: <https://dundeeprecious.com/site/assets/files/16376/030724_dundee_precious_metals_announces_sale_of_tsumeb_smelter.pdf> [Accessed 15 March 2024].
Ettler, V., Mihaljevič, M., Strnad, L., Jedlicka, R., Křı́bek, B., Bowell, R. J., Kamona, F. and Mapani, B., 2021. Mineralogy of Ga- and Ge-bearing metallurgical slags from Tsumeb, Namibia. *Mineralogical Magazine*, 85(6), pp.879-889.
FMoJ, 2016. *Act on the Protection of Cultural Property*. [pdf] Berlin: German Federal Ministry of Justice. Available at: <https://www.gesetze-im-internet.de/englisch_kgsg/englisch_kgsg.pdf> [Accessed 3 July 2024].
Giebel, R. J., Okoth, K. and LeGall, Y., 2001. Colonial Extraction and Berlin Mineral Collections. In: *Colonial Collections in Berlin Universities*. [online workshop] 27 April 2001. Berlin: Berlin Center for Global Engagement, Berlin University Alliance. Available at: <https://www.static.tu.berlin/fileadmin/www/10002011/Forschungsprojekte/Abgeschlossene_Forschungsprojekte/Forschungsnetzwerk__Koloniale_Sammlungen_in_Berliner_Universitaeten/BUABCGE_workshop_programme.pdf> [Accessed 15 March 2024].
Glaubrecht, M., 2021. Collection History: The different Types of Museums and Their "(Post-) Colonial Heritage"—Natural History Collections. In: *Guidelines for German Museums. Care of Collections from Colonial Contexts.* Berlin: German Museums Association, pp.108–110.
Hearth, S. and Robbins, C., 2022. Mineral Displays as Embodiments of Geologic Thought and Colonial Invisibility. *Journal of Natural Science Collections*, 10, pp.3–17.
Heumann, I. and Vennen, M., 2018: Fragmentieren. Dinosaurier und Geschichte. In: I. Heumann, H. Stoecker, M.Tamborini and M.Vennen eds. 2018. *Dinosaurierfragmente. Zur Geschichte der Tendaguru-Expedition und ihrer Objekte 1906–2018.* Göttingen: Wallstein, pp.7-21.
Hollenbach, M., 2020. *Deutschlands fast vergessene Kolonialgeschichte.* [radio] Deutschlandfunk Kultur. [online] Available at: <https://www.deutschlandfunkkultur.de/deutsch-suedwestafrika-deutschlands-fast-vergessene-100.html> [Accessed 23 May 2024].
Kirfel, A., Hamm, H.-M., Will, G., 1983. The crystal structure of reinhardbraunsite, $Ca_5(SiO_4)_2(OH,F)_2$, a new mineral of the calcio-chondrodite type Tschermaks. *Mineralogische und Petrographische Mitteilungen*, 31, pp.137–150.
Levinson, O., 1983. *Diamonds in the desert. The story of August Stauch and His Times.* Tafelberg: Cape Town.
Lombaard, A.F., Günzel, A., Innes, J. and Krüger, T. L., 1986. The Tsumeb lead–copper–zinc–silver deposit, South West Africa/Namibia. In: C.R. Anhaeusser and S. Maske, 1986. *Mineral Deposits of Southern Africa*, vol. II. Johannesburg: Geological Society of South Africa, pp.1761–1787.
Melcher, F. and Buchholz, P., 2012. Current and future Germanium availability from primary sources. In: *Minor Metals Conference*. Cologne, Germany, 24 April 2012. Berlin: Deutsche Rohstoffagentur.
Merchant, C., 1983. *The Death of Nature. Women, Ecology, and the Scientific Revolution.* San Francisco: Harper & Row.
Mies, M., 1986. *Patriarchy and Accumulation on a World Scale. Women in the International Division of Labour.* London: Zed Books.
Mikloweit, S., 2025 (2023). Thinking Through Minerals: We Build Our Languages Out Of Rocks. In: S. Mikloweit, ed. 2025. *Transpositional Geologies: Spectres Of Coloniality.* Berlin: Kerber Publishers.
Namhila, E. N., 2019. The Hendrik Witbooi Diaries: From Colonial War Booty to Memory of the World Treasure. *Historical Judgement,* 1, pp. 18–20.
Ondonga Traditional Authority, 2023. *Two Fragments of The Kingdom of Ondonga's Power Stone return to Namibia.* [press information guide] [online] Available at: <https://www.museums.com.na/entry/2023/04/two-fragments-of-the-kingdom-of-ondongas-power-stone-return-to-namibia> [Accessed 23 May 2024].
Opoku, K., 2021. *Will Provenance Research Delay Restitution of Looted African Artefacts?*. [online] Boasblogs. Available at: <https://boasblogs.org/de/dcntr/will-provenance-research-delay-restitution> [Accessed 23 May 2024].
Press, S., 2021a. As Germany Acknowledges its Colonial-Era Genocide in Namibia, the Brutal Legacy of Diamond Mining still Needs a Reckoning. [online] *Time Magazine*. Available at: <https://time.com/6072145/namibia-germany-apology-diamonds> [Accessed 23 May 2024].
____2021b. *Blood and Diamonds: Germany's Imperial Ambitions in Africa.* Cambridge: Harvard University Press.
Quijano, A., 2000. Coloniality of Power, Eurocentrism, and Latin America. *Nepantla. Views from South*, 1(3), pp.533–580.
Rogers, S. L., Lau, L., Dowey, N., Sheikh, H. and Williams, R., 2021. *Geology Uprooted! Decolonising the Curriculum for Geologists*. [online] Geoscience Communication. Available at: <https://gc.copernicus.org/articles/5/189/2022/gc-5-189-2022.html> [Accessed 23 May 2024].
Sarr, F. and Savoy, B., 2018. *Restituer le Patrimoine Africain.* Paris: Seuil / Philippe Rey.
Savoy, B., 2018. *Die Provenienz der Kultur*, Berlin: Matthes & Seitz.
____2019. *'An manchen Objekten klebt Blut'. Bénédicte Savoy im Gespräch mit René Aguigah.* [radio] Deutschlandfunk Kultur. [online] Available at: <https://www.deutschlandfunkkultur.de/kunsthistorikerin-benedicte-savoy-an-manchen-objekten-klebt-100.html> [Accessed 23 May 2024].
Scheibe, E. A., 1990. *Zwei verschiedene Eisenmeteoriten aus Südwestafrika – Namibia*. Der Aufschluss (41), p. 287-289.
Scheibe, R., 1913. *Vortrag über Diamanten*. [manuscript] Estate Ernst Albrecht Scheibe. GA 28/026 – 030. Freiburg: University Library Freiburg, p.21–22.
Schmitt, P. P., 2020. Sand in Sicht. [online] *Frankfurter Allgemeine Zeitung*. Available at: <https://www.faz.net/aktuell/gesellschaft/sand-in-sicht-16620702.html> [Accessed 23 May 2024].
Schnorbus, H., 2000. *Die Geschichte der Otavi Minen AG: Ein ereignisreiches Jahrhundert (1900–2000)*, Eschborn: Otavi Minen AG.
Silvester, J. and Shiweda, N., 2020. The Return of the Sacred Stones of the Ovambo Kingdoms: Restitution and the Revision of the Past. *Museum and Society*, 18(1), pp.30–39.
Stoler, A. L., 2008. *Along the Archival Grain. Epistemic Anxieties and Colonial Common Sense.* Princeton: PUP.
Strunz, K. H., 1970. *Von der Bergakademie zur Technischen Universität Berlin, 1770 bis 1970.* Berlin: Technische Universität Berlin.
Strunz, K. H. and Nickel, E., 2001. *Mineralogical Tables. Chemical-Structural Mineral Classification System* (9th Edition), Stuttgart: Nägele & Obermiller.
Te Heesen, A., 2012. *Theorien des Museums. Zur Einführung*, Hamburg: Junius.

TUB (Technical University of Berlin), 2016. *Dinosaurs in Berlin! The Brachiosaurus Brancai—a Political, Scientific, and Popular Icon*. [online] Available at: <https://www.tu.berlin/en/kuk/forschung/projekte/abgeschlossene-forschungsprojekte/dinosaurier-in-berlin-brachiosaurus-brancai-eine-politische-wissenschaftliche-und-populaere-ikone> [Accessed 05 Juli 2024].

Vogelsang, H., 1883a. *Letters to Lüderitz from Heinrich Vogelsang/Vogelsangs report on Angra Pequena and journeys to Bethanien, 1883–1885*. Accessions and Collections A. 0132. Microfilm, Findaid 2/087. Windhoek: National Archive of Namibia.

____1883b. *Verkoop Contract tusschen Kapt. Joseph Frederics van Bethanien Groot-Namaqualand en F.A.E. Lüderitz te Bremen in Duitschland....Bethanien d. 25 Augustus 1883.* [original handwritten copy] German Colonial Society - Colonial Image Collection. 060-0951-01; 060-0951-02. Frankfurt/Main: Library of the Goethe University.

Von den Förderern eds., 1914. *Die deutschen Diamanten und ihre Gewinnung. Eine Erinnerungsschrift zur Landesausstellung Windhuk*. Berlin: Reimer.

Yusoff, K., 2018. A Billion Black Anthropocenes or None. Minneapolis: University of Minnesota Press.

____ 2021. The Inhumanities. *Annals of the American Association of Geographers*, 111(3), pp.663–676.

TG-T931
transposition 931
(Geoscientific Specimen Refusing its Further Objectification)

Artist: Sascha Mikloweit
Year: 2019 / 2021
Type: HDR archival ink print on Hahnemühle Photo Rag Ultra Smooth 305g
Edition: 5 + 2AP
Dimension: 50 x 40 cm
Material: O—I_MMdUB-041-sm; MMdUB-041-sm (no archive No.), Beryl/Morganite, Afghanistan, Hindu Kush, Dara-i-Pech

Image showing a Beryl/Morganite object from the Mineralogical Museum of the University of Bonn, originating from Dara-i-Pech in the Hindu Kush region of Afghanistan. No additional information is available. The object lacks an official archive number, though its museum label records it as part of the 'Seifert' collection, with the identifier 'v.77.2'. The image was affected by a device error during the imaging process.

Fig. 44

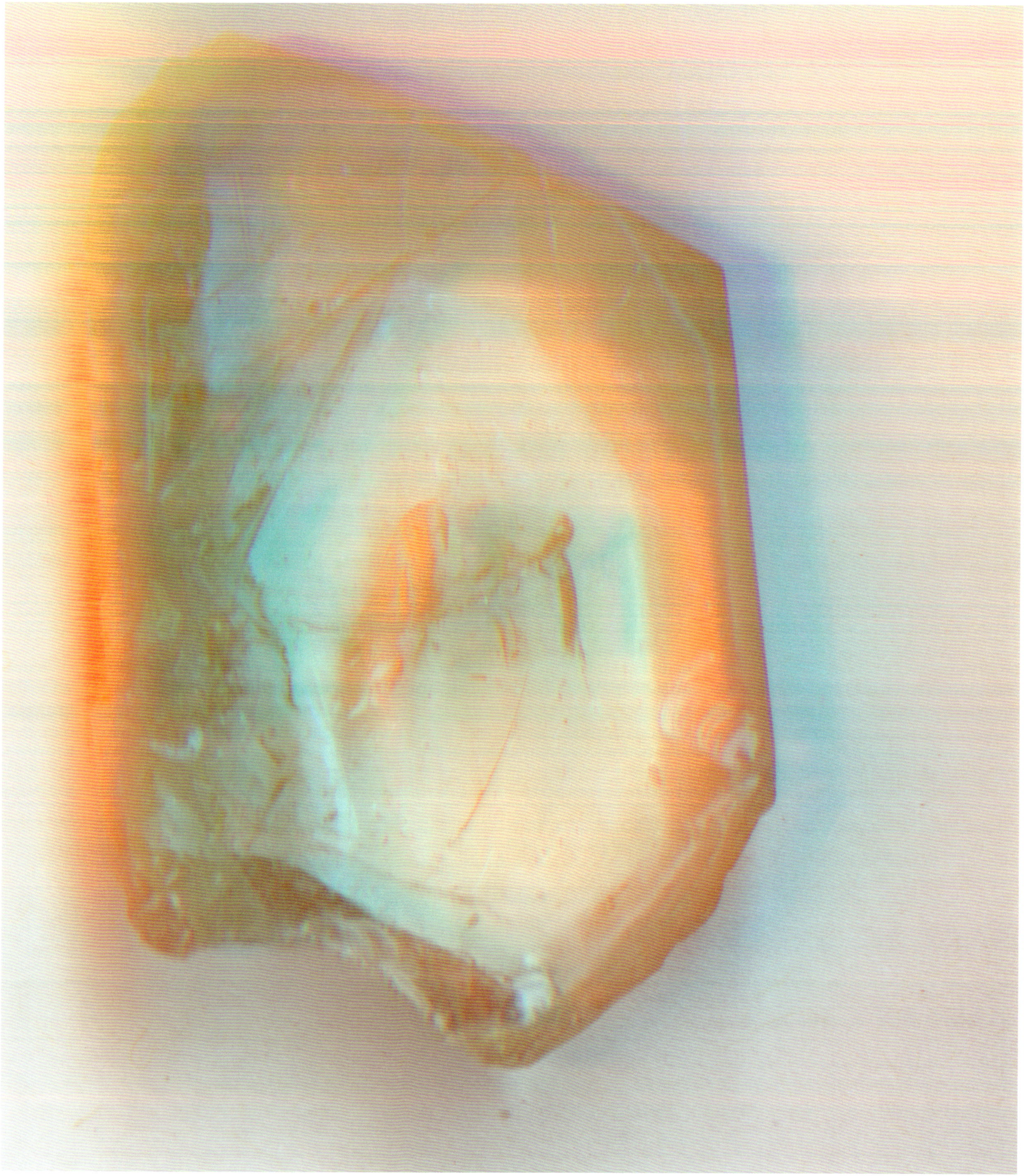

TG-T884

transposition 884
(22°40’56.1”S 14°32’15.1”E •<—>• KM334)

Artist: Sascha Mikloweit
Year: 2022
Location: Otavi Mining and Railways Company (OMEG) railway line Swakopmund-Tsumeb Namibia, kilometre 334
Type: Sculptural Intervention, Image—Object (I—O), Photography
Edition: 5 + 2AP
Dimensions: Variable, 40 x 22 cm
Material: Nameless grave, _DSC1550.tiff; reactive dye print on silk; OMEG Railways km334; HDR archival ink print

In the afternoon of 21 October 2023, as we returned from a guided tour of the B2Gold mine that I had organised, Stella (Imalwa, see this volume, p. 291) suggested we visit the Khorab Memorial. Located just outside Otavi, it required only a brief detour of a few kilometres. The memorial, established in the 1930s and renewed in 1973, commemorates the surrender of the Imperial German Forces (*Schutztruppe*) on 9 July 1915. After driving back and forth along the railway north of Otavi, we were still unable to locate the memorial and eventually gave up—the heat was simply too intense. Before leaving, however, I pushed through a few thorny bushes and found myself not in front of the memorial, but rather at kilometre marker 334 of the former Otavi Mining and Railways Company (OMEG) line, which connects Swakopmund and Tsumeb (it appears the kilometre count begins at the Karibib junction, as Swakopmund lies roughly 500km away).

I was carrying the silk printed with the nameless grave that was *transposition 064* (see this volume, pp. 156–159) during the Bonn exhibition. I draped it over the stone column marking kilometre 334 to establish a dialogue between the OMEG railway line, the Nama & Ovaherero Genocide Monument in Swakopmund, *transposition 064*, and the Khorab Memorial—the latter likely just metres away, though it eluded us.

Fig. 45

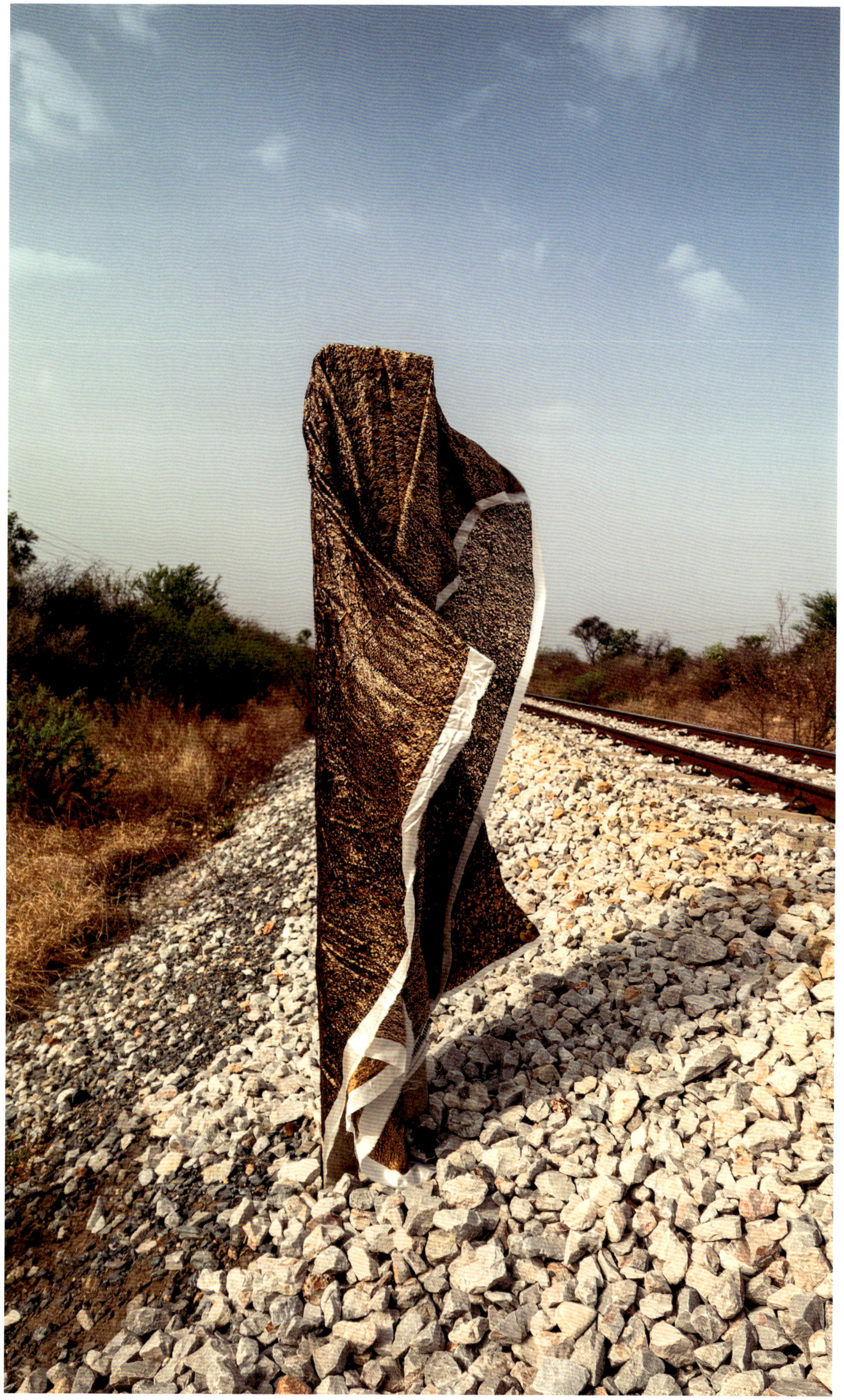

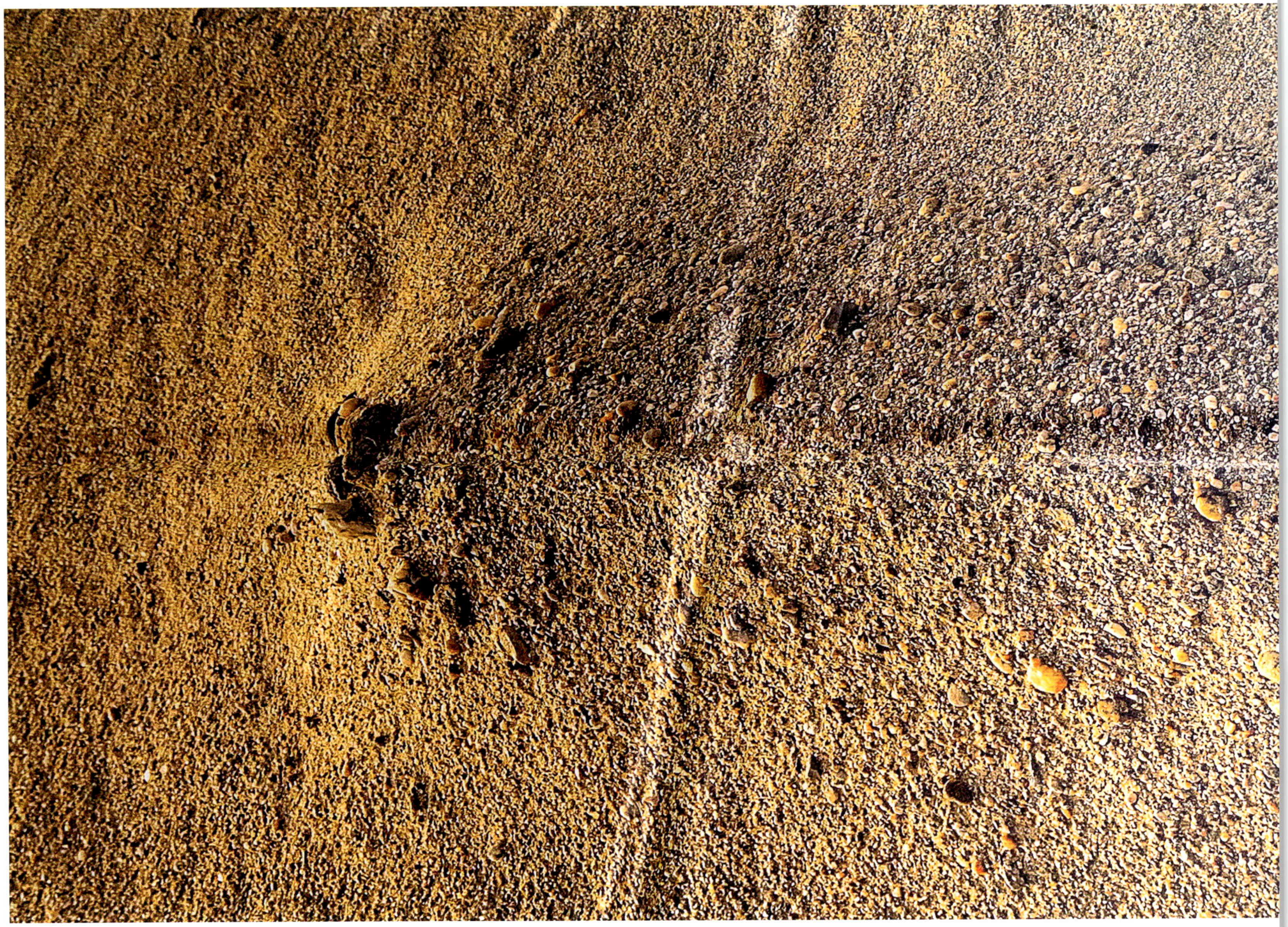

Fig. 46

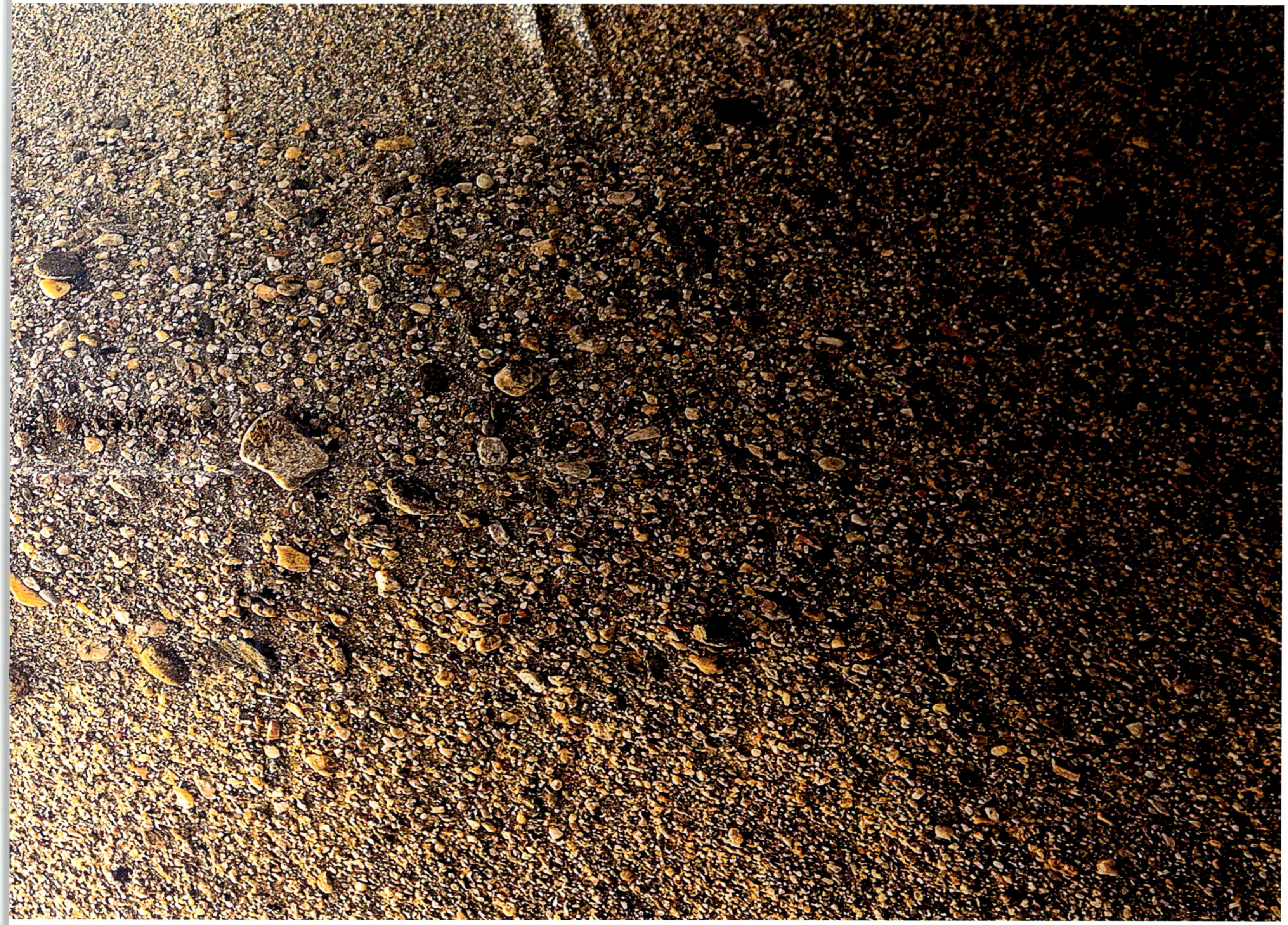

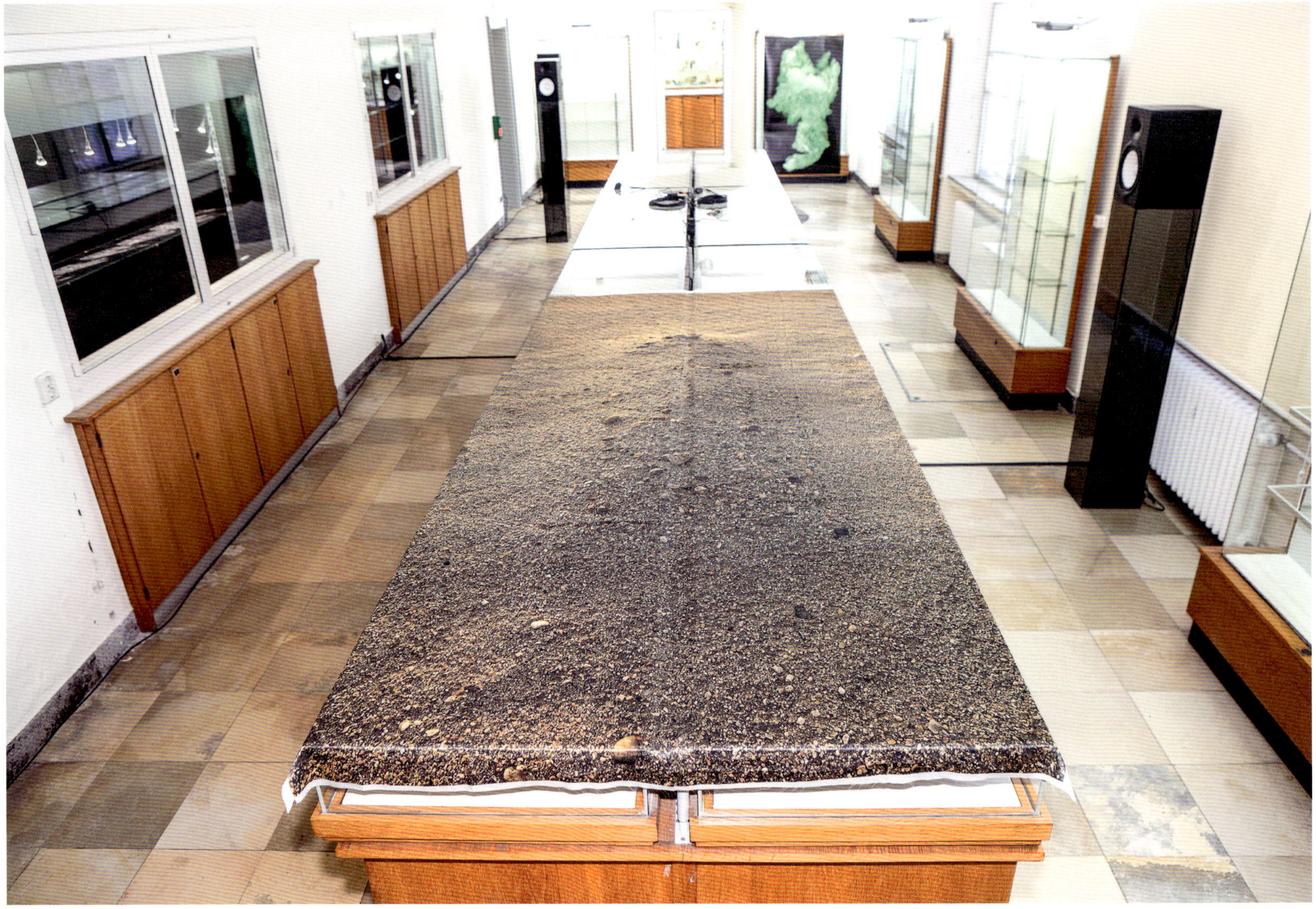

TG-T064
transposition 064
(22°40’56.1”S 14°32’15.1”E •<-->• 50°43’29.9”N 7°05’31.2”E)

Artist: Sascha Mikloweit
Year: 2023
Location: Mineralogical Museum of the University of Bonn, DE, Exhibition Hall 2, Special Exhibitions
Type: Sculptural Intervention, Image—Object (I—O)
Edition: 5 + 2AP
Dimensions: 115 x 140 x 750 cm, 147.5 x 279.5 cm (silk)
Material: Nameless grave, _DSC1550.tiff; reactive dye print on silk; museum vitrine

Fig. 47

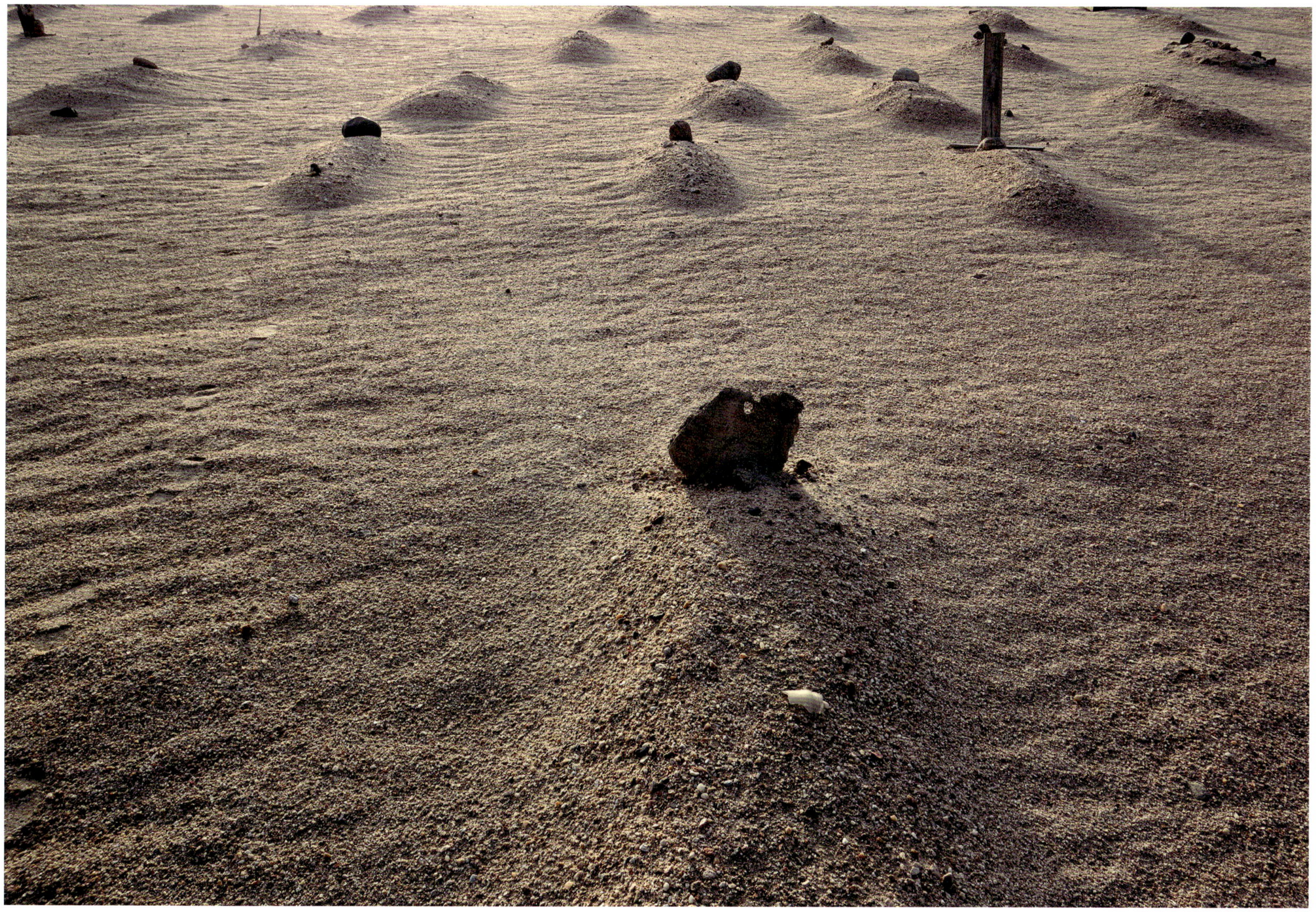

Fig. 48

Photography of a nameless grave resting at the site of the *Nama & Ovaherero Genocide Monument 1904-1908* in Swakopmund, Namibia, printed on pure silk, and layered over the emptied central vitrine of Exhibition Hall 2 at the Mineralogical Museum of the University of Bonn. The photograph was taken by the artist during his Namibia research trip in November 2022.

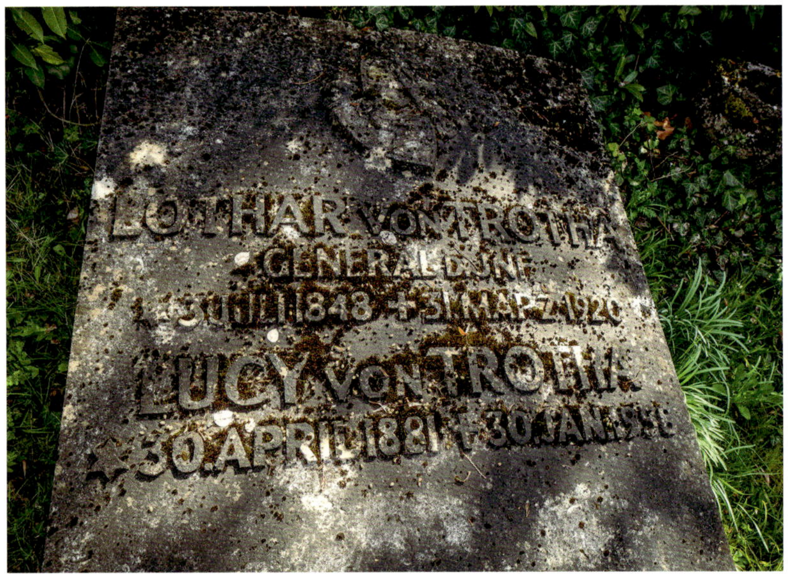
LOTHAR VON TROTHA
GENERAL D.INF.
LUCY VON TROTHA

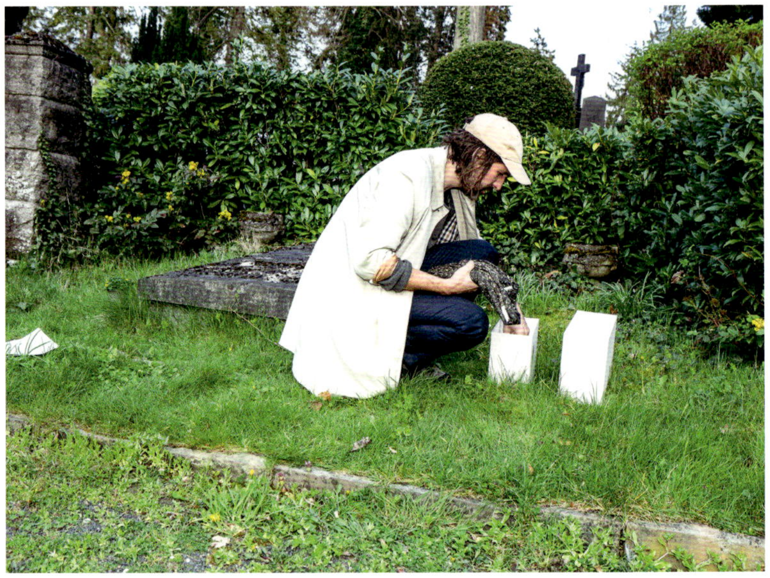

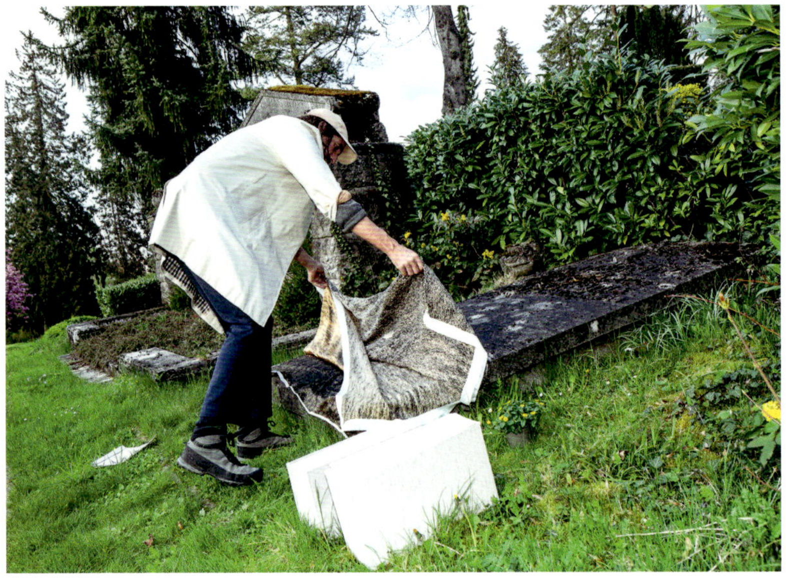

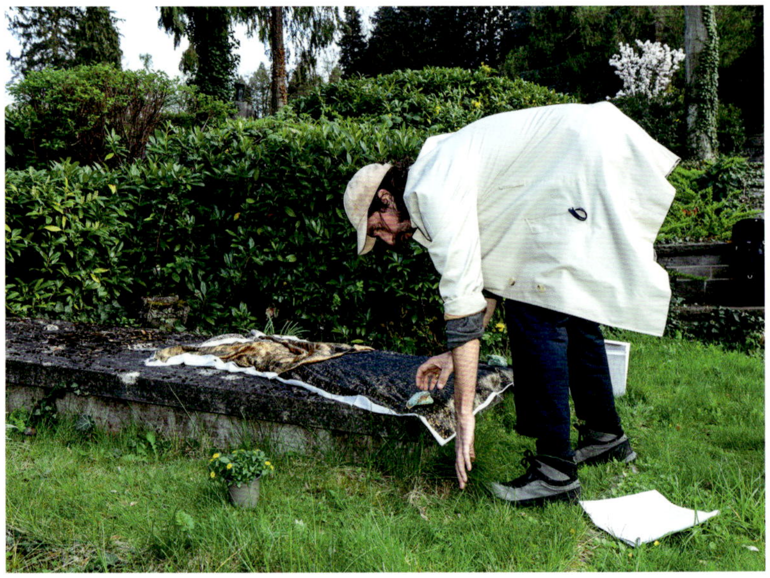

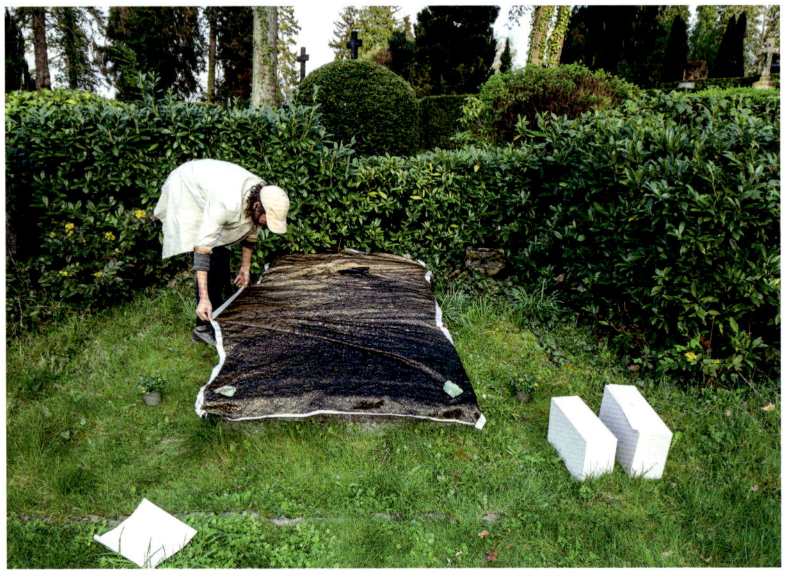

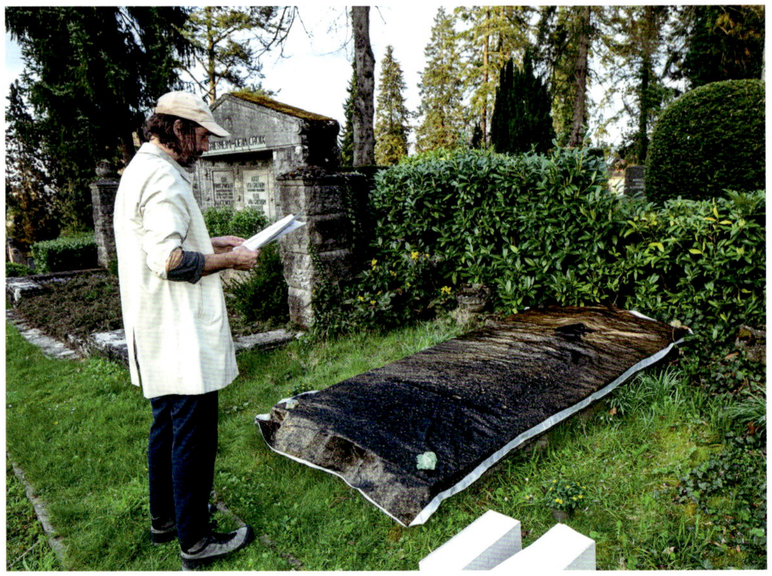

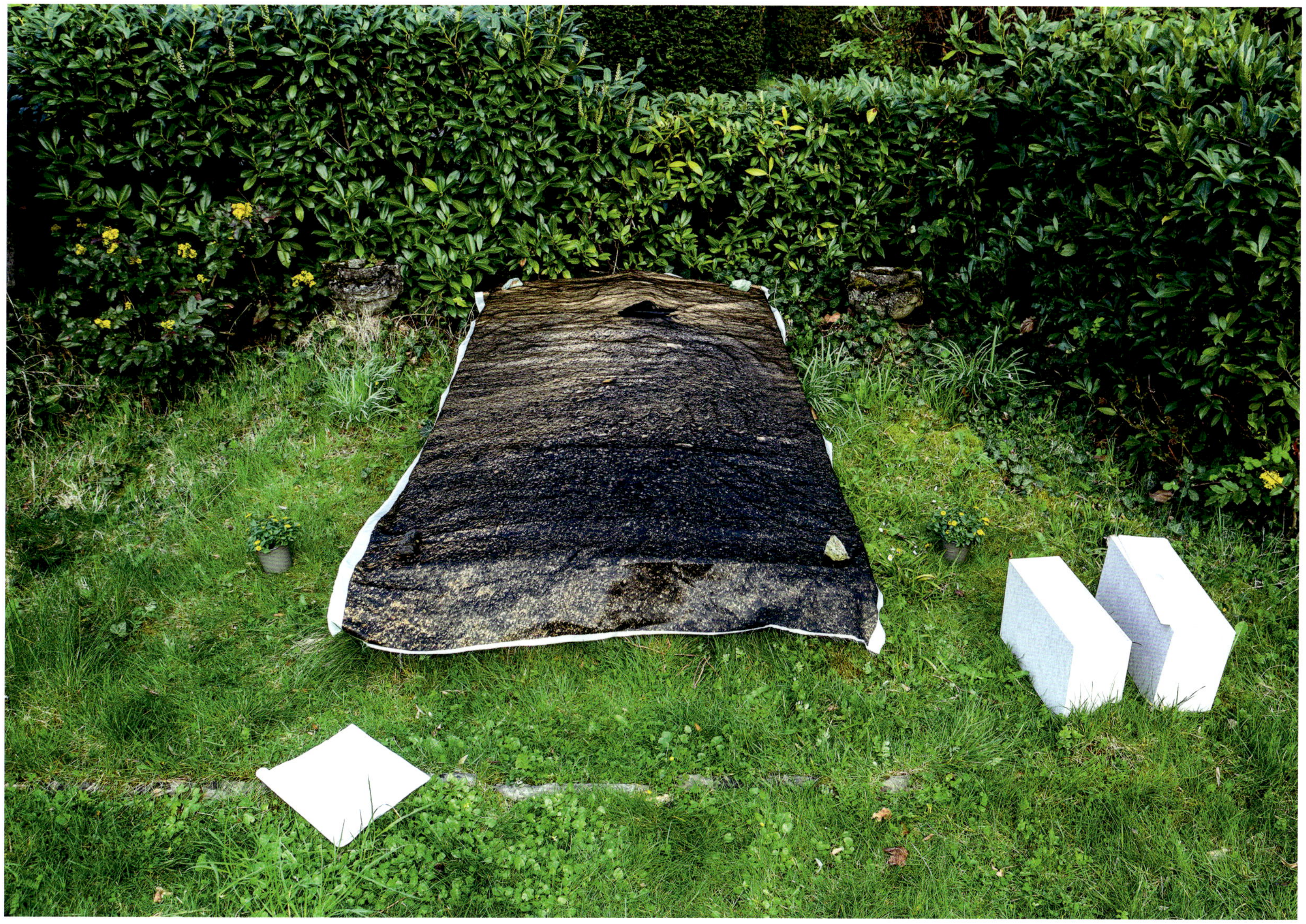

TG-T904
transposition 904
(22°40'56.1"S 14°32'15.1"E •<-->• 50°43'05.1"N 7°04'57.6"E)

Artist: Sascha Mikloweit
Year: 2023
Location: Mineralogical Museum of the University of Bonn, DE, Exhibition Hall 1, Systemic Collection: Cemetery Poppelsdorf, Bonn
Type: Sculptural Intervention, Performance
Dimensions: 134 x 25 x 81 cm (Module Museum), 20 x 147.5 x 279.5 cm (Module Cemetery)
Duration: Variable
Material: Mineralogical Museum of the University of Bonn's Systemic Collection, Tsumeb Minerals, Nameless grave, _DSC1545.tiff; reactive dye print on silk; custom box; script; glass; Lothar von Trotha grave; visitor

Fig. 49–54

Fig. 55

Presented as a sculptural intervention within the Systematic Collection in Hall 1 of the Mineralogical Museum of the University of Bonn, the work could be activated as a performance enacted by visitors. The performance script led participants to the grave of General Lothar von Trotha (1848–1920), located a fifteen-minute walk from the museum at Poppelsdorf Cemetery. General von Trotha commanded the genocide of the Herero and Nama people (1904–1908) in German South West Africa (now Namibia), an atrocity fuelled by imperial German economic ambitions and racist ideologies. Mikloweit's piece establishes connections between this first genocide of the 20th century, the contemporary Nama & Ovaherero Genocide Monument in Swakopmund, the colonial mining operations of the Otavi Mining and Railway Company (OMEG) in Tsumeb, the minerals extracted there—some of which are on display in the Mineralogical Museum's Systematic Collection—and the grave of von Trotha. This work was inhibited by the City of Bonn's administration (see this volume pp.176–191).

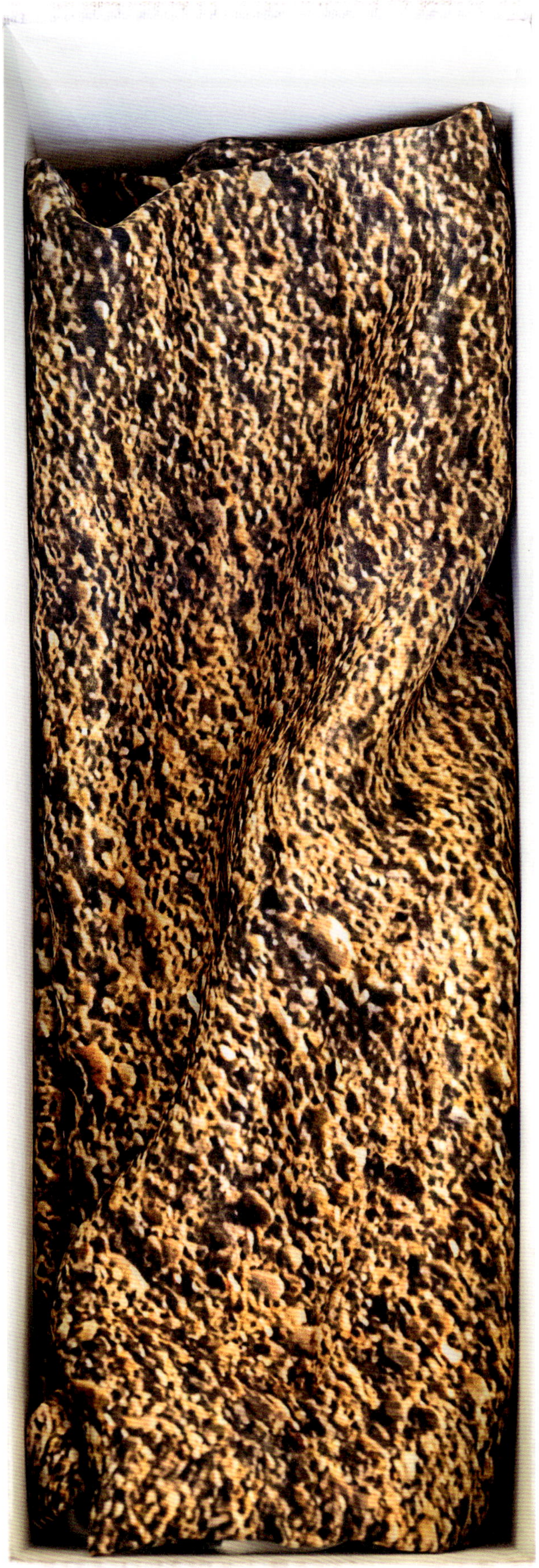

Fig. 56

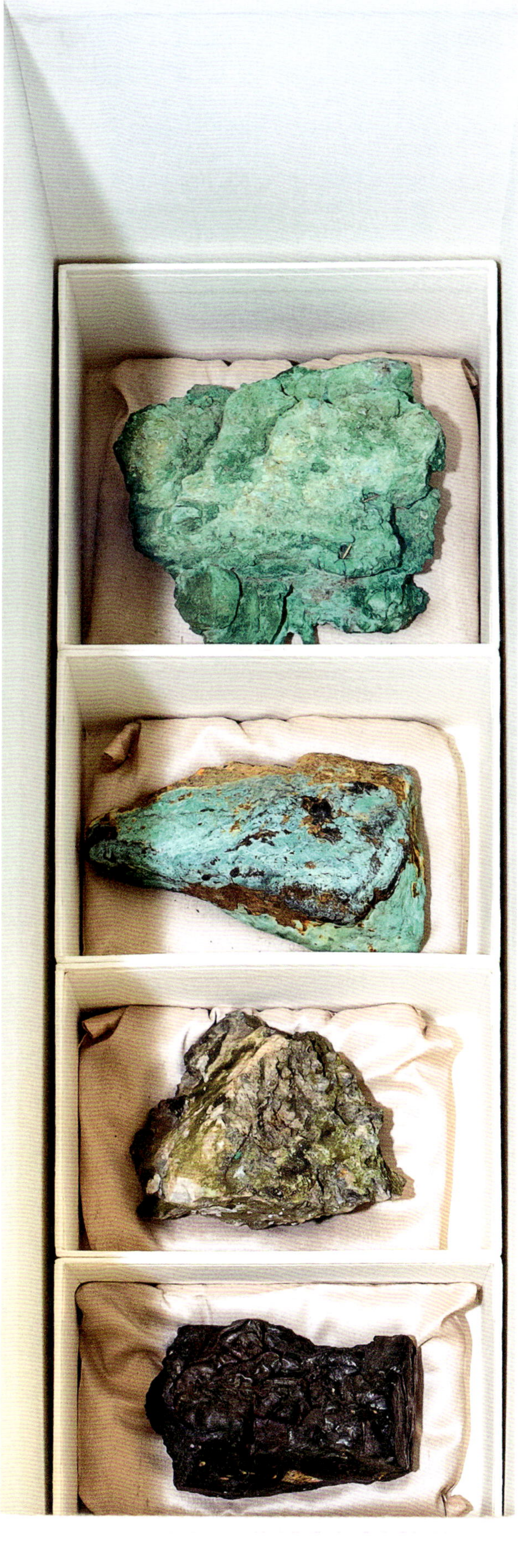

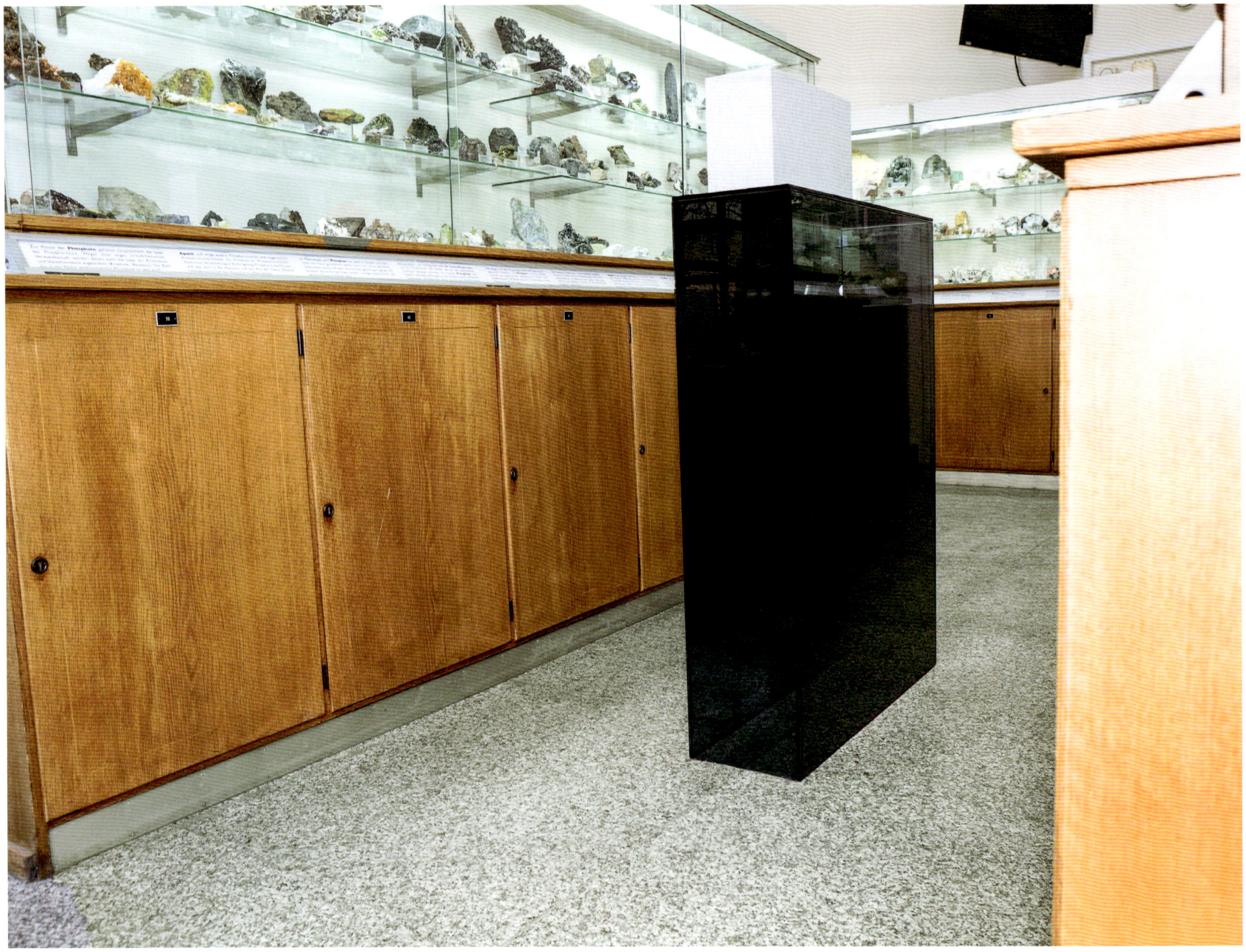

Fig. 57

Fig. 58

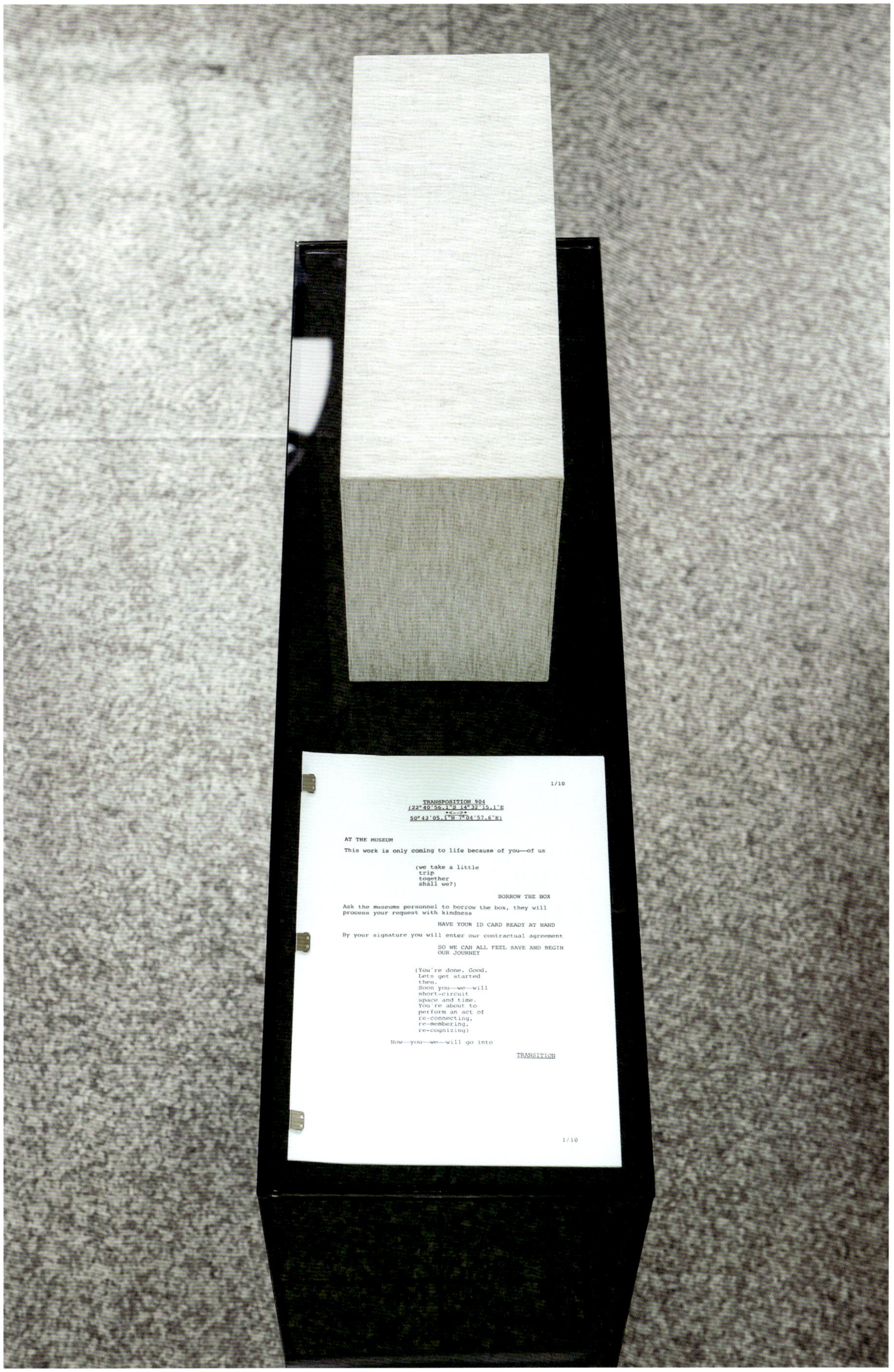
1/10
TRANSPOSITION 904
(22°40'56.1"S 14°32'15.1"E
•<-->•
50°43'05.1"N 7°04'57.6"E)
AT THE MUSEUM
This work is only coming to life because of you—of us
(we take a little
trip
together
shall we?)
BORROW THE BOX
Ask the museums personnel to borrow the box, they will process your request with kindness
HAVE YOUR ID CARD READY AT HAND
By your signature you will enter our contractual agreement
SO WE CAN ALL FEEL SAVE AND BEGIN OUR JOURNEY
(You're done. Good.
Lets get started
then.
Soon you—we—will
short-circuit
space and time.
You're about to
perform an act of
re-connecting,
re-membering,
re-cognizing)
Now—you—we—will go into
TRANSITION
1/10

1/10

TRANSPOSITION 904
(22°40'56.1"S 14°32'15.1"E
•<-->•
50°43'05.1"N 7°04'57.6"E)

AT THE MUSEUM

This work is only coming to life because of you—of us

(we take a little
trip
together,
shall we?)

BORROW THE BOX

Ask the museum's personnel to borrow the box, they will
process your request with kindness

HAVE YOUR ID CARD READY AT HAND

With your signature you will enter our contractual agreement

SO WE CAN ALL FEEL SAFE AND BEGIN
OUR JOURNEY

(You're done. Good.
Let's get started
then.
Soon you—we—will
short-circuit
space and time.
You're about to
perform an act of
re-connecting,
re-membering,
re-cognizing)

Now—you—we—will go into

TRANSITION

1/10

Fig. 59 [pp.166–157]

2/10

Take the box with you

BE CAREFUL
ITS CONTENTS ARE PRECIOUS
YOU——WE
ARE TRAVELLING WITH THE DEAD NOW
THEY MEAN YOU NO HARM

Let us leave the museum
Pass through its portal

HOLD ON
JUST FOR A SEC

(Let me take a guess;
you're a bit of a
high tech person,
aren't you;
aren't we all——a bit
at least——that's ok;
you're not alone in
this;
let them,
the minerals do the
work for you,
for us;
don't be shy; get
out your smartphone;
let them minerals do
the work for you,
for us
let THEM crush
the numbers;
hack in the location
data)

39M+63F BONN, GERMANY

(...or would you
prefer it the
old-school way; a
bit more private, I
suppose; I dig that;
we can do it
together; it will be
fun)

Let's turn left then and follow the rotunda

2/10

(straight ahead would
be the collection
with all those
stuffed animals; but
that's a whole
different story——or
is it?)

LET'S LEAVE IT
AT THAT THOUGHT

Turn left again and walk out of the Schloss through its main gate

(Isn't that a lovely
view ahead
such a gorgeous
sample of landscape
sculpting this
Poppelsdorfer Allee
Such a superior
achievement
What a culture!
That, however, is
not our path today
Don't get distracted
Let's get back on
track, shall we)

We are still in front of the Schloss now
Do you see the metal gates to your left
Pass through them onto

MECKENHEIMER ALLEE

Turn left here and follow this road for about 650m, till the 7th road appears to your right.

(Just so you know
after some time
MECKENHEIMER ALLEE
will be renamed:
CLEMENS-AUGUST-
STRAßE)

then turn right on to

WALLFAHRTSWEG

After about 140m take the path in the middle

STATIONSWEG

4/10

Walk about 270m up the hill, there will be a house to your right behind the fence, just after comes a gate to enter the cemetery same side

POPPELSDORFER CEMETERY

Walk about 104 steps

(You should be in
front of a lawn now)

Turn left and walk for about 18 steps

(Can you see it
already
It is the grave in
the corner on the
left
the horizontal one
with the big stone
plate
finally—you—we
arrived)

AT THE GRAVEYARD
IN FRONT OF THE GRAVE OF GENERAL ADRIAN DIETRICH LOTHAR VON TROTHA (1848-1920)

Now——you will go into

<u>TRANSFORMATION</u>

You are standing in front of Trotha's grave
facing it

Now open the box

HOLD ON FOR A
MOMENT

4/10

5/10

(In your hands you—we
are going to hold
another grave soon
It is printed on
silk
I thought
Some luxury
at last
It is a nameless
grave
It travelled
10,763km
Its home is the
graveyard of the
nameless
the
NAMA & OVERHERERO
GENOCIDE MONUMENT
in Swakopmund
Namibia
Here rest all
who fell prey to
Trotha's
extermination order
His? Our? a German?
genocide
the first one
of the 20th century
by German hands
that's for sure
It started in 1904
Of course—the
processes leading to
it began way earlier
Look into it
Inform yourself
Try to understand
the
non-understandable)

Take out the fabric

HOLD ON FOR A
MOMENT

5/10

6/10

(once unfolded
it will be the size
of the grave's stone
plate
have it overlapping
a bit
let it embrace the
stone)

Unfold the fabric on top of the grave

Take out the Tsumeb minerals

HERE ARE FOUR OF 'THOSE' MINERALS
WITH THEIR SMELL OF SWEAT AND DEATH

Place them at the four corners of the grave

THEY WILL HOLD DOWN THE DEAD
PUT THEM DOWN IN THEIR ASSIGNED
PLACES

Now—stand in front of the grave
facing it
seeing it
(can you smell it)

Take a moment for yourself and

read this poem

MAYBE YOU CAN SMELL IT THEN

When we enter the mines,
We receive our numbers;
Marked on a plastic "bracelet";
Welded onto our wrists.
For one or two years,
We are baptized as numbers.

We serve as numbers;
We clock in as numbers;
Our white bosses call us by numbers;
We clock out by numbers –
Our only identity on the mines.

We get our wages as numbers;
As numbers we have accidents;
(MORE)

6/10

7/10

We go to hospital as numbers;
As numbers we die;
And we enter our coffins as numbers.

We enter the compounds as numbers;
As numbers we leave the compounds;
We get our meals as numbers;
As numbers we sleep and wake up;
And we move around as numbers.

When we are sucked dry
By the murderous speed of work;
We are sent to the reservations as numbers,
When we are sick of poisonous gasses;
We are denied compensation –
For it is not worth numbers.
When the fine dust eats away our lungs;
We are sent to our graves -
as numbers.

So we come and die as numbers.

YOU SMELL IT NOW

7/10

8/10

(this poem was
written by Namibian
poet Simon Zhu Mbako
his life was short
he died young
1994
in his 44th year
on earth
unfortunately
I know very little
about him
it might have been
my ignorance
or limited time
or I missed the
resources to mine
this information
Nor do I know when
exactly these lines
where written
which kind of life
nourished
these thoughts
but
maybe this doesn't
matter
maybe these lines
are enough
maybe there is a
chance—you'll find
out more)

Maybe, however, you'd like to read his poem again
Maybe read it out loud

TO BREAK THE SILENCE

We Come and Die as Numbers
By Simon Zhu Mbako

When we enter the mines,
We receive our numbers;
Marked on a plastic "bracelet";
Welded onto our wrists.
For one or two years,
We are baptised as numbers.

We serve as numbers;
(MORE)

8/10

9/10

We clock in as numbers;
Our white bosses call us by numbers;
We clock out by numbers –
Our only identity on the mines.

We get our wages as numbers;
As numbers we have accidents;
We go to hospital as numbers;
As numbers we die;
And we enter our coffins as numbers.

We enter the compounds as numbers;
As numbers we leave the compounds;
We get our meals as numbers;
As numbers we sleep and wake up;
And we move around as numbers.

When we are sucked dry
By the murderous speed of work;
We are sent to the reservations as numbers,
When we are sick of poisonous gasses;
We are denied compensation –
For it is not worth numbers.
When the fine dust eats away our lungs;
We are sent to our graves -
as numbers.

So we come and die as numbers.

(If you've finished
your thoughts
and you feel you are
ready)

Place them, the minerals back into the box

CONCEAL THEIR SMELL OF SWEAT AND
DEATH

Fold the fabric back to fit
Place the dead back to cover the minerals of Tsumeb, Namibia

CONCEAL THEIR SMELL OF SWEAT AND
DEATH

9/10

Close the box

Now——you——we——will go into

TRANSITION

Pick up the box
We make our way back

REMEMBER

DON'T GET LOST
ON OUR WAY

Return the box to the museum's personnel

THEY WILL BE GENTLE
THEY KNOW WHAT YOU'VE JUST BEEN THROUGH
YOU'VE TRAVERSED TIME AND SPACE
YOU'VE RECONNECTED TO OUR SHARED HISTORIES

Collect your ID card

YOU STILL HAVE YOUR COPY OF OUR CONTRACT
DON'T FORGET
HOLD ON TO IT
SO WE CAN ALL FEEL SAFE
AND BEGIN
OUR JOURNEY

Now——you will go into

OUR FUTURE

transposition 846

Fig. 60

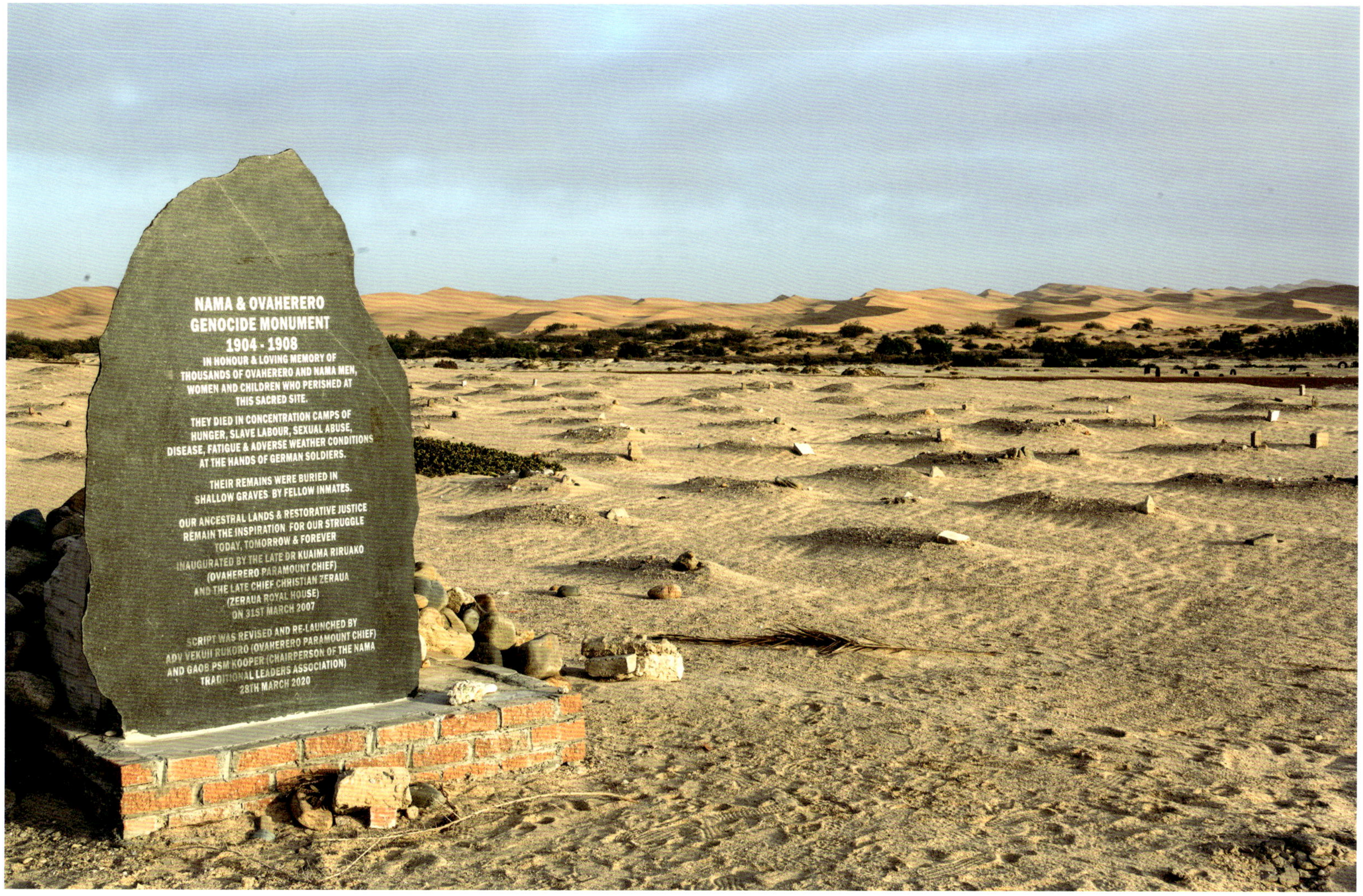
NAMA & OVAHERERO
GENOCIDE MONUMENT
1904 - 1908
IN HONOUR & LOVING MEMORY OF
THOUSANDS OF OVAHERERO AND NAMA MEN,
WOMEN AND CHILDREN WHO PERISHED AT
THIS SACRED SITE.
THEY DIED IN CONCENTRATION CAMPS OF
HUNGER, SLAVE LABOUR, SEXUAL ABUSE,
DISEASE, FATIGUE & ADVERSE WEATHER CONDITIONS
AT THE HANDS OF GERMAN SOLDIERS.
THEIR REMAINS WERE BURIED IN
SHALLOW GRAVES BY FELLOW INMATES.
OUR ANCESTRAL LANDS & RESTORATIVE JUSTICE
REMAIN THE INSPIRATION FOR OUR STRUGGLE
TODAY, TOMORROW & FOREVER
INAUGURATED BY THE LATE DR KUAIMA RIRUAKO
(OVAHERERO PARAMOUNT CHIEF)
AND THE LATE CHIEF CHRISTIAN ZERAUA
(ZERAUA ROYAL HOUSE)
ON 31ST MARCH 2007
SCRIPT WAS REVISED AND RE-LAUNCHED BY
ADV VEKUII RUKORO (OVAHERERO PARAMOUNT CHIEF)
AND GAOB PSM KOOPER (CHAIRPERSON OF THE NAMA
TRADITIONAL LEADERS ASSOCIATION)
28TH MARCH 2020

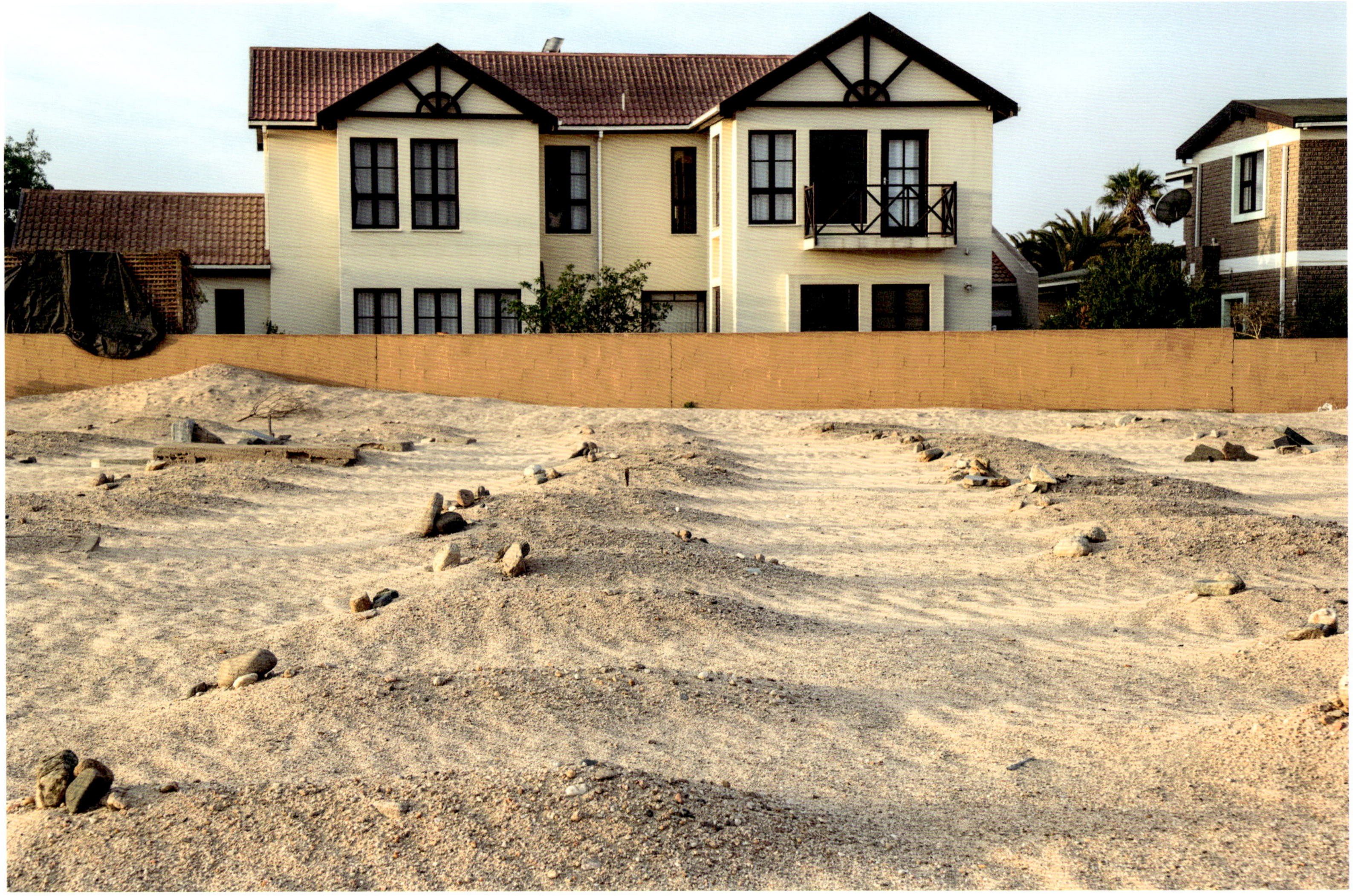

Fig. 61

Down the Cemetery Path

The Promise of Active Remembrance, a Reluctance of Colonial Memory, and the Federal City of Bonn[1]

Sascha Mikloweit

Abstract: As part of *Transpositional Geologies*, the exhibition project *transposition 904* in Bonn, Germany, navigated complex bureaucratic hurdles and public debates surrounding the memorialisation of Lothar von Trotha, a controversial figure associated with the first genocide of the 20th century. Following initial, in-kind support from the Federal City of Bonn's official 'Active Culture of Remembrance' framework, the project eventually encountered resistance from the city's Office for Environment and Urban Green Spaces. This stalled official approval for a non-invasive performance at von Trotha's grave, ultimately limiting a carefully planned artistic activity to an act of civil disobedience. In revealing a tension between historical accountability and bureaucratic inertia within administrative structures, this case study highlights the broader societal challenges in confronting colonial legacies and fostering a critical culture of remembrance.

Keywords: City of Bonn, urban planning, administrative processes, culture of remembrance, municipal memory practices, colonial legacy, genocide, Lothar von Trotha, performance art, mineral specimens

Swakopmund, Namibia

On 31 October 2022, once we reached the southern edge of the city, right where its fringe meets the Namib Desert, the temperatures already began to fall, and the sun started casting long shadows on the ground. Our little research group—the British-Mexican photographer Amanda Holmes, the German historian Bastian Herbst, and I—entered the Swakopmund Cemetery. It is located in the neighbourhood of 'Kramersdorf', a former 'whites-only' suburb of the Namibian costal town Swakopmund—a town that, by the way, resembles classic German seaside resorts such as the famous Imperial Spa Heringsdorf[2] by the Baltic Sea in eerie detail. As we stepped in, I remembered a local resident had provided directions to me earlier that day: "If you keep left and pass through the main cemetery it will lead you towards the grave field."

Once in, we started slowly moving along the east edge of the cemetery, looking at the graves left and right of us. Given the place's proximity to the Namib, I remember our surprise that the plant life was considerably well cultivated, and most graves were well tended. About twenty metres in, a large, neat arrangement appeared to our right: meticulously constructed from different types of stone, about the size of twenty to thirty standard graves, at its most erect point two or three heads larger than an average-sized human. This memorial too, was well maintained. Particularly its keystone was in a shiny polished condition, at its top-end crowns a large German *Iron Cross.*[3] The centrepiece's golden lettering, carved in old German font, commemorates 'Wilhelm Eduard Richard Heldt, Captain of the Imperial Schutztruppe for South West Africa and District Chief in Swakopmund, born on February 11, 1867, died on November 11, 1899. In loyal remembrance, the comrades and the residents of Swakopmund'[4]

The wide, wall-like substructure on which the centrepiece rests extents about five times the headstone's width. Written across this base, the large capitalised letters announce: 'They gave their lives for you'.[5] One and a half metres upfront of this main ensemble, left and right of it, two large erect stone plates present sixty military ranks, names, and death dates of soldiers, all members of the Imperial German colonial armed forces, the 'Schutztruppe' ('Protection Force'), their deaths ranging from 1895 to 1907.[6] Alongside this colonial memorial no further historical context was provided to the visitors in any form. The well-kept state of German memorials of this period in Swakopmund is not an exception, as the same accounts for the rest of the country:

> Even though several genocide memorials exist today, the German legacy remains more visible than the history of the suffering and oppression of all other Namibian people (Blum, 2024).[7]

We continued to proceed along the eastern edge toward the end of the official cemetery area. Here, the reach of formal care and maintenance ends, although a majority of grave sites extend beyond this boundary: the estimated 3,500 to 4,000 nameless graves (ibid., 2024) of the *Nama & Ovaherero Genocide Monument 1904–1908* stretch almost endlessly—forming a wavelike structure which starts unfolding about 150 metres away from where the comparably lush green of Swakopmund Cemetery ends for now. From there, nevertheless, the official cemetery slowly but steadily grows across the expanse, grave by grave, toward the Genocide Monument. The graves there are small heaps of desert sand, the majority of which bear no further signification aside from occasionally placed small, unhewn rocks. They stretch between the Riverside Avenue's affluent housing to the north—with their views upon the grave field—and the reddish silky dunes of the Namib Desert

[1] I borrowed this from the English idiom 'to lead somebody down the garden path'. It is used metaphorically to describe leading someone into a situation where they are deceived or misled, often by presenting information or ideas in a way that initially seems plausible or attractive but ultimately turns out to be incorrect or deceptive.

[2] A German coastal resort town 230km north of Berlin that was known for hosting Wilhelm II (1859–1941), Emperor of Germany and King of Prussia.

[3] The Iron Cross (German: *Eisernes Kreuz*) was a military decoration in the Kingdom of Prussia, and later in the German Empire (1871–1918) and Nazi Germany (1933–1945).

[4] Translation of the author, German original: '*Wilhelm Eduard Richard Heldt, Hauptmann der Kaiserl. Schutztruppe für Südwest-Afrika und Bezirkshauptmann in Swakopmund, geb. am 11. Februar 1867, gest. am 11. November 1899. In treuem Gedenken die Kameraden, sowie die Einwohner von Swakopmund*'.

[5] Translation of the author, German original: '*Sie gaben Ihr Leben für Dich*'.

[6] The recent use of this site for political purposes warrants a brief explanation, as it introduces a further contemporary dimension to the transhistorical context explored in this text, specifically in the German state of North-Rhine Westphalia, which is home to the City of Bonn. In July 2024, nearly two years after our visit to the Nama and Ovaherero Genocide Monument, an eleven-strong delegation from a cross-party delegation of the German North Rhine-Westphalian state parliament travelled to Namibia to study 'the consequences of colonialism and the genocide committed by the German occupying forces' and to honour its victims (Eberle and Weiland, 2024). One parliamentarian of the delegation placed a wreath at the Imperial German memorial, which commemorates the perpetrators, situated just 250 metres from the grave site of the victims, and documented this action on social media, accompanied by a citation of a controversial German war song. This gesture was widely seen as historically revisionist and sparked a diplomatic incident, which remains ongoing at the time of writing (Eberle and Weiland, 2024; Magoley, 2024; Oesterle, n.d.).

to the south: a sea of graves expanding towards the horizon. Not long ago '[...] tourists on quad bikes and horses, [...], would run past or over the unmarked graves, unaware of or disregarding their presence. A protective wall has recently been built to put a stop to this.' (ibid., 2024; see also Van Wyk, 2023). A German-Namibian in town, perhaps in his mid-sixties, told me that before the wall came up, the local youth, too, were quite fond of this area as a motocross range (Mikloweit, 2022).

As we advanced towards the monument, the site began overwhelming all three of us slowly but steadily. For a brief moment we paused in front of the memorial stone of the Genocide Monument and read its inscription:[8]

> 'Nama & Ovaherero Genocide Monument 1904–1908. In honour & loving memory of thousands of Ovaherero and Nama men, women and children who perished at this sacred site. They died in concentration camps of hunger, slave labour, sexual abuse, disease, fatigue & adverse weather conditions at the hands of German soldiers. Their remains were buried in shallow graves by fellow inmates. Our Ancestral lands & restorative justice remain the inspiration for our struggle today, tomorrow & forever. Inaugurated by the late Dr Kuaima Riruako (Ovaherero Paramount Chief) and the late chief Christian Zeraua (Zeraua Royal House) on 31st March 2007. Script was revised and re-launched by ADV Vekuii Rukoro (Ovaherero Paramount Chief) and Gaob PSM Kooper (Chairperson of the Nama Traditional Leaders Association) 28th March 2020' (see in this volume p.177).

After reading, we walked quietly and dispersed between the thousands of nameless graves. Amanda took photographs for the project archive, and so did I, increasingly filled with a feeling of sadness and despair.[9] In the presence of these silent witnesses, any utterance seemed unworthy and superfluous, so we did not exchange a word until after we reconvened at our car.

The Nama & Ovaherero Genocide Monument 1904–1908 is a place where omnipotent absence becomes present, and where transhistorical entanglements become matter. On this grave field, contrary to the site commemorating the white community, including the perpetrators, there is almost no regular maintenance, aside from that conducted by the Namibian Genocide Association (NGA). They had started organising clean-ups, assembling a diverse group of voluntary helpers who gather annually since 2018, to carefully restore the sand heaps, clean litter, and on occasions, build a grave for bones found in unexpected places during the process (Van Wyk, 2023). Swakopmund Municipality, however, only maintains the main cemetery:

> [...] NGA were motivated to start with the cleanings when they saw how graves belonging to the genocide victims were being neglected by the workers of the Swakopmund Municipality. 'These graves have never been given attention,' [...] 'The municipal workers there only take care of the graves of the whites, while forgetting about the graves of our ancestors. [...]' (ibid., 2023)

Bonn, Germany

The setting of the grave of the victims' executioner in Germany is a very different affair: a beautiful 20-minute walk from the Mineralogical Museum of the University of Bonn, the cemetery of Poppelsdorf is a green and well-kept place. The grave of General Lothar von Throta, who ordered the genocide in 1904 as commander of the German Imperial troops in German South West Africa, is positioned prominently. A solid stone plate of 145cm by 273cm, selected by his widow, marks the grave, which is partially covered in soft moss and fallen leaves, depending on the season and the cycle of routine maintenance.

Experiencing the stark contrast between these two grave sites, separated by 11,000 kilometres, within the context of my research, prompted my decision to make evident the connection between the Namibian Nama & Ovaherero Genocide Monument, von Trotha's grave, the Mineralogical Museum of the University of Bonn's collection, and the Tsumeb Mine in Namibia by incorporating them into my exhibition at the University of Bonn in 2023 (Mikloweit, 2023c). To date, these four sites and their (trans-)historical entanglements had not been put in relation—neither historically, nor culturally. My exhibition in Bonn presented a unique opportunity to re-link these histories—once separated by time, space, politics, and knowledge production—to the present in a novel way.

The Federal City of Bonn itself was West Germany's former capital until 1990. It is now the second seat of the German government with federal ministries and authorities, home to one of the top German universities, a major hub for United Nations organisations, significant national scientific establishments, major public broadcasters, offices of global corporations, and around 150 non-governmental organisations (City of Bonn, 2024a, 2024b).

The following account takes us from German colonial history to the present administration of the City of Bonn, eventually unearthing a latent reluctance of the City of Bonn's administration to address part of its colonial history.

(Trans-)Historical Context

To situate the reflection on this case, we must revisit the then German Empire's context of colonialism in German South West Africa. The current state of professional academic

[7] While reading *The Namibian*, Namibia's leading daily newspaper, in Swakopmund's German bakery that very morning, a headline on the first page had reported: 'Great-grandson says Von François "did a lot for this country"'. Curt von François (1852–1931) was a commissioner of the imperial colonial army of the German Empire in 'German South West' where he was responsible for the foundation of the city of Windhoek in 1890, and the harbour of Swakopmund in 1892. Only a few days before, the city council of Windhoek had decided to remove François' statue and put it on display in a museum. This happened after debates about the 'founding father's' and governor of Windhoek's role in the colonizing of Namibia. Even though von François' descendants publicly fumed at the democratic decision, the statue was finally removed in November, 2022. (Petersen, 2022; Chothia, 2022)

[8] This is a transcription of the monument's original inscription.

[9] Amanda would later describe her experience as: 'Although we were in the desert, the grave field felt cold. An eerie and sad feeling grew inside me as I walked around and took pictures. I wondered who was laying below, but of course, I had no way of knowing. These grave mounds seemed to fuse perfectly with their stark surroundings and the distant dunes. The contrast I felt between the marble tombstones [on the official part of the Swakopmund Cemetery] and the grave field mounds only emphasise to me the disparity this country [Namibia] had lived.' (Holmes, 2024)

[10] For German speakers, one of the many available sources is a backgrounder by the Federal Agency for Civic Education (Bundeszentrale für politische Bildung, 2021). It highlights the context of 'military violence' by the German Empire against the local population at the time, which 'culminated in the war of extermination against the Herero and Nama [...] [when] Lieutenant General Lothar von Trotha ordered the complete destruction of the Herero in October 1904 and of the Nama in April 1905. Approximately up to 100,000 people were murdered by the German troops, died of thirst in the Omaheke Desert, or perished in concentration camps. The genocide is considered the first of the 20th century' (ibid., 2021)

[11] Regarding the intertwinement of private business and public institutions during the German colonial era see Helmut Maier's contribution to this volume (p.70ff).

discourse about the German colonising of Namibia in the late 19th and early 20th centuries, and the genocide against the Herereo and Nama people between 1904 and 1907, has been extensively documented by e.g. Bley and Ridley (1971), Drechsler (1966; 1980; 1984), Dedering (1993; 1999), and Wallace (2013). For non-historians—whether citizens, researchers or, of course, officials from public administrations—the established (non-contested) consensus on historical facts can be accessed in general reference works available online by authors from both sides of the hemisphere, for instance in the South African History Online (SAHO) database and the Britannica online edition:[10]

> The scorched earth campaign conducted by [imperial] Germany against the Herero people in South West Africa (now known as Namibia) is considered among the first genocides in the twentieth century. […] Lieutenant-General Lothar von Trotha was deployed to Namibia to command the German forces there. […] In 1894, he was assigned to Tanzania, then known as German East Africa, where he served as commander of the Schutztruppe (colonial protection force) [sic]. […] he also […] was instrumental in the suppression of the revolt of the Wahehe people in 1896. On attaining the rank of major general [sic], von Trotha was sent to China to command the 1st East Asiatic Infantry Brigade. It was in this position that he helped suppress the Boxer rebellion of 1900. On 11 June 1904, von Trotha arrived in Swakopmund [Namibia] to take charge of the German forces there. He then proceeded to exterminate and subjugate the Herero tribes with brutal efficiency. This included poisoning water holes, shooting survivors on sight and confining the Herero that had managed to elude the German forces to desert regions of Namibia, where many succumbed to starvation and disease. […]. (SAHO, 2024)
>
> The first phase of the war was fought between the Germans and the Herero (with a single Ovambo battle at Fort Namutoni near the Etosha Pan). It reached a climax when General Lothar von Trotha defeated the main Herero army at the Battle of Waterburg and, taking no prisoners, drove them into the Kalahari, where most died. By 1910 the loss of life by hanging, battle, or starvation and thirst—plus the escape of a few to the Bechuanaland protectorate—had reduced the Herero people by about 90 percent (80–85 percent dead, 5–10 percent in exile). The Nama resistance war came late because a key letter from Maherero's son and successor, Samuel Maherero, to the Oorlam chief Hendrik Witbooi that proposed joint action had been intercepted. The resistance was finally crushed in 1907, and Nama survivors were herded into concentration camps. War, starvation, and conditions in the camps claimed the lives of two-thirds of the Nama. The Germans allocated about half of the usable—and apparently all of the best—ranchland [sic] (except that of the Rehoboth Basters) to settlers and restricted Africans to reserves. The Tsumeb copper and zinc mines opened in 1906, and diamond mining (more accurately, sand sifting) began near Lüderitz in 1908 and at the main fields at the mouth of the Orange River (Oranjemund) a few years later. Railways linked Lüderitz, Keetmanshoop, and Windhoek as well as Swakopmund, Windhoek, and Tsumeb. (Green and Louw, 1999–2024)

In this history, we encounter several business operations, institutions, events, and persons that became entwined with each other and with the subject of my enquiry at the Mineralogical Museum:[11] two colonial undertakings of an industrial key actor in 'German South West Africa', the Otavi Mining and Railways Company (OMEG), are relevant here. First, OMEG built a railway line from Swakopmund to Tsumeb (1903–1905) (Wallace, 2013, p.150) cutting through land used by various local communities, and second, OMEG started Tsumeb's first industrial scale mining operation (from 1907) (Mindat.org, n.d.), using the railway line to ship out the extracted copper from Tsumeb to Germany and other places.[12] The mine disrupted a long tradition of local, small-scale mining operations and their complex ecologies (Hearth, 2021). The disruptions caused by these two ventures to local customs and economies were partly responsible for the Herero and Nama genocide, with Lothar von Trotha as its commander-in-chief. The significant amount of Tsumeb minerals that are part of the Mineralogical Museum of the University of Bonn's collection—some of them stemming from German expeditions of the colonial era—therefore, connects the institution to these events, their racialised regimes, and to Lothar von Trotha and his grave in Bonn.[13]

After his 'annihilation' of an estimated 64,000–84,000 Herero, and 50 per cent of the Nama population (Dedering, 1993, p.82; 1999, p.252 inc. FN52)[14] general Adrian Dietrich Lothar von Trotha returned to imperial Germany and retired to Bonn. His military career had been marked by unscrupulous violence, justified by, and embedded in, the racist ideologies of his time (Dedering, 1993, p.84).[15] Decorated with the highest merits, he earned the praise of Chancellor Bernhard von Bülow (1849–1929), whose statement at the German Reichstag on 9 December 1905 provides an impression of the esteem that von Trotha enjoyed at the time:

> [Chancellor von Bülow] stated in the Reichstag that von Trotha had 'displayed personal bravery, energy, foresight, all the qualities of a capable and outstanding leader in a protracted and difficult campaign. […] Therefore, he rightly wears the highest military decoration awarded by Prussia, the order of merit *Pour le Mérite*.' (German Bundestag Research Service, 2021, p.11)[16]

The villas of the upscale area of Bonn Bad Godesberg became von Trotha's homes from 1906–1909 and 1913–1918 before

[12] This initiated German-Namibian economic relations of which we can find transhistorical continuations in Tsumeb till today: the Hamburg based *Aurubis AG* (formerly Norddeutsche Affinerie AG (1866–2009) for which the Tsumeb Mine was a key source of raw copper ore) is today the exclusive customer for the blister copper produced in the *Tsumeb Specialty Smelter, Namibia*. This information was provided to us in person by the Dundee Precious Metals Inc. staff during an introduction lecture that was opening a 'Smelter Tour' in Tsumeb—once to our research team during our second *Transpositional Geologies* field trip in November 2023, and again to students and researchers of the TU Berlin, Institute for Applied Geosciences, during their field trip in March 2024. On the *Tsumeb Specialty Smelter* and in regard to its extreme toxicity see: NIOH (2013), Kondarev (2015), Popov, Kohrs, and Bacheva-MacGrath (2016), Hasheela (2018), GAoEJ (2022), DPM (2024).

[13] These relations, however, are not communicated nor accessible otherwise to researchers or to the general visitors within the Mineralogical Museum of the University of Bonn's archives and public displays. The ontological emphasis is here on geoscientific objects as 'purely natural objects' rather than to read them as 'objects of culture' which could enable the visitor to read them within their transhistorical, socio-political contexts. (For further reading on the subject see in this volume (p.143ff) Malina Lauterbach, R. Johannes Giebel, and Sascha Mikloweit).

[14] These numbers should be read in recognition of the fact that '[…] all historians dealing with the 1904-1907 war […] have to work with the estimates given by the missionary and colonial sources.' (Dedering, 1993, p.82). Marion Wallace points out that the key sources are the 'native population counts' and the 1911 census. (2013, pp.177–178).

[15] Dedering emphasises that '[i]t was for political, not military reasons that von Trotha wanted to decimate the Herero population […]. This idea was justified in racist terms and came easily to the mind of a battle-hardened warrior, who had been sent to Namibia because he had previously distinguished himself in putting down anti-colonial uprisings in China and East Africa. The 'resistance' von Trotha hoped to uproot once and for all was not military resistance, but a resurgence of the social core of the Herero population, the 'old tribal organisations'.

[16] Translation from German by the author.

finally, in 1918, he withdrew to the picturesque atmosphere of Bonn-Poppelsdorf. During his residence in the city, von Trotha was 'celebrated in Bonn as an honorary guest' who 'was invited to elite rector's dinners at the University of Bonn' on several occasions, a distinct honour for local and visiting dignitaries at the time (Hagenberg-Miliu, 2024).

After his death on 31 March 1920, von Trotha was buried on the Cemetery of Poppelsdorf. His burial site is commensurate with the societal status he enjoyed: positioned at the cemetery's central square, ten metres from another major memorial that commemorates fallen soldiers of the past two world wars as 'our heroes', and that is routinely honoured with formal wreaths by the city's administration.[17]

Lothar von Trotha's grave, just steps away from this World War memorial, comes across as inconspicuous. His is a family grave—neither a memorial to honour him as a prominent member of the Imperial German forces, nor a cautionary historical site reflecting his role as perpetrator of the 20th century's first genocide. At the cemetery's side entrance, however, the administration's glass-covered notice board contains an informal notice: 84 words on an unassuming, small piece of paper, bent in the humidity, fixed to the vitrine with two shabby grey magnets, at the bottom of a collage of notices on opening hours, honorary grave locations and instructions for the handling of candles and flowers. Though undated, unsigned, and untitled, it references a project from the City of Bonn. While it appears to originate from the municipality, its informality, location, and materiality deprives it of the natural authority it could have had:

> On this cemetery lies the grave of Lothar von Trotha (1848–1920). As the Governor and Commander-in-Chief of the 'German Protection Force' in the then German South West Africa (now Namibia), he was responsible for the deaths of tens of thousands of children, women, and men of the indigenous Herero and Nama. His war of extermination is considered the first genocide of the 20th century. In his later years, he lived intermittently in Bonn, where he also died. As part of a planned municipal project 'Active Remembrance Culture', the (post)colonial heritage in Bonn will also be addressed.[18]

When consulting the cemetery administration during one of my research visits in January 2023, they appeared unaware of the notice. The reference to the municipal project, however, eventually led me to the responsible department at the City of Bonn. It also led me back to retrace a protracted local discussion that, at the end, made it into the national press: in 2020 the then un-contextualised nature of the von Trotha grave became subject of a public debate between the local activist group 'Bonn Postkolonial'—which called for a 'critical comment' on the grave—and the City of Bonn. Their exchange of letters (Bonn Postkolonial, 2020) revealed entrenched positions on both sides, reaffirming and rebutting without convergence. A nuanced or ultimately productive dialogue did not evolve, but a national newspaper, the conservative *Frankfurter Allgemeine Zeitung* escalated the case and asked 'How can a grave be 'critically commented' on?'.[19] In his own comment, the author rejects the modest proposal of a plaque, lamenting with hyperbole that 'at the grave, disputes fall silent: Postcolonial activists want to terminate this civilisational agreement' and 'discriminate against the death', arguing that the grave was now in the 'crosshairs' of a growing 'ethical radicalism of a historico-political iconoclasm' that had turned to 'moral barbarism' [sic] (Bahners, 2020).[20]

This is the (trans-)historical context—then and today—in which *transposition 904 (22°40'56.1"S 14°32'15.1"E •<-->• 50°43'05.1"N 7°04'57.6"E)*[21] took place in 2023, as part of my exhibition, in Bonn.

The Original Concept of transposition 904

From April to June 2023 *transposition 904* was presented as an inert sculptural intervention into the systemic collection of the Mineralogical Museum of the University of Bonn's Exhibition Hall 1 as part of my exhibition *We Build Our Languages Out Of Rocks—Transpositional Geologies* (Mikloweit, 2023c). During opening hours, the visitors were invited to activate *transposition 904* into a performance, carrying it out on their own at the grave of von Trotha. A performance script was provided (see in this volume, p.166ff) along with a simple, linen-clad archival box, custom-made by a local craftsman, displayed on a plinth made of translucent brown glass, placed within the museum's systemic collection (see in this volume, p.164-165). The box contained a folded piece of pure silk, printed with an image of the sandy heap of a nameless grave, one of the dozens I photographed while visiting the Nama & Ovaherero Genocide Monument 1904–1908 in Swakopmund, Namibia in October 2022 (see in this volume, p.162-163). The silk was measured to the size of von Trotha's grave slab. Accompanying it, the box contained four unclassified Tsumeb minerals, which I had extracted in collaboration with the spokesperson of the Tsumeb Municipality, Elli Stella Udda Imalwa, in October 2022: two minerals from inside the open pit mine—the former 'Green Hill'—and two from down on Level 1, to where we descended, slowly, carefully stepping along the pitch black drive-down that spirals for another 42 levels, winding ourselves with it into the humid and slippery Tsumeb mine, the extractive operation that the Imperial Germany's OMEG had started in 1907 (see in this volume, p.78, 163, 189).

The performance script guided the visitor from the premises of the museum to the cemetery. Upon arrival, the visitor was asked to unpack the box, extract the silk bearing the image of the unmarked sand grave, and to unfold it over von Trotha's tombstone until it was fully and precisely covered. To hold

[17] '*Unsere Helden* [Our Heroes] 1914–1918, 1939–1945': Originally inaugurated in 1930 to commemorate the First World War (Niesen, 2017), it now commemorates fallen soldiers of the First and Second World Wars. During a visit in January 2023, I noted wreaths placed at this memorial, one of them in pristine condition with the words '*Stadt Bonn*' (City of Bonn) printed in a typeset reminiscent of Old German-style script on one end, and the city's coat of arms on the other end of a yellow and red ribbon (Mikloweit, 2023a). The cemetery and city administration later confirmed that the wreath had been laid down by the cemetery's administration as part of their maintenance of the grave, in coordination with the City of Bonn's Office for Environment and Urban Green Spaces of the City (Telephone conversations with the Poppelsdorf Cemetery administration, and with the Commissioner for Sports and Culture on 17 July 2024 (Mikloweit, 2024). Regarding the areas of authority of Department III see further City of Bonn (2024a).

[18] Translation from German by the author based on photograph taken on site, 25 January 2023. (Mikloweit, 2023b)

[19] German original: '*Wie kann man eine Grabstätte „kritisch kommentieren"?*

[20] Translation from German by the author.

[21] See in this volume, p.160ff. In regards to the title: 22°40'56.1"S 14°32'15.1"E is the geo-location of the nameless grave which is part of the Nama & Ovaherero Genocide Monument 1904–1908 in Swakopmund. 50°43'05.1"N 7°04'57.6"E is the geo-location of the von Trotha grave in Bonn.

the delicate fabric in place, the visitor had to gently place the four unclassified Tsumeb minerals on the flat tomb stone to weigh down and align each of the four corners of the two graves. For this fleeting moment, the nameless grave layered and concealed the perpetrator's name. Von Trotha's grave itself was transformed: 117 years later it was linked to the site of the nameless graves of his victims and to imperial Germany's colonial geology.[22] The transposed nameless grave, temporarily set into place by the four minerals, connects these places and events—through time and space—with the Mineralogical Museum of the University of Bonn's collection.

As a final act, standing in front of the grave, I invited the visitor to silently read or recite a short work by the late Namibian poet Simon Zhu Mbako (1950-1994) entitled 'We Come and Die as Numbers' (Mbako, 1989),[23] before removing the unclassified Tsumeb minerals and the silk, stowing them back into the box, and returning to the exhibition space at the museum, leaving nothing behind at von Trotha's grave.

During this short performance—un-invasive, reversible, calm, with an act marked by the respect, piety, and dignity that the thousands of graves in Namibia were long denied—the site of von Trotha's grave remained undisturbed, and afterwards unchanged, bar the fallen leaves that may have been caused to stir. Formally, this constituted even less of an intervention than laying down a wreath or a candle. Through this short ritual, the visitor had the agency to transform the inert sculptural intervention they were confronted with inside the museum into an event. Yet, ultimately they had to relinquish their just-gained agency, and face the sobering moment of reverting the performance back to its original state of inertia, thus leaving behind the grave of the originator of a genocide as 'uncommented on' as they encountered it.

The performance part of *transposition 904* itself, however, was designed to evoke, on a visceral level, a personal connection between an individual visitor, the German colonial past, and the museum collection's Tsumeb objects. As an active participant in this process, the visitor 'becomes one' with minerals and graves to form an impermanent, fleeting counter-memorial to German colonial history (on impermanence and counter-memorial see in this volume (p.129f) Malina Lauterbach). With this evocative experience in mind, the visitor was confronted anew with the ontological status quo of the museum collection's 'Objects of Nature' while returning the box to its plinth in Exhibition Hall 1, thus reverting the performance back into an inert sculptural intervention.

During planning of this work, with the support of counsel, I took meticulous care to ensure the intervention would not be in violation of the German Criminal Code, Section 168 (German Federal Ministry of Justice, 2021), which regulates the 'disturbance of peace of dead' in the German Criminal Code (StGB). This was in order to provide reassurance to a range of actors involved: the visitors, the exhibiting institution, (federal) public funding institutions that officially supported the project, and potential onlookers or local staff on site.[24] The formal permission of *transposition 904* by the City of Bonn's administration was therefore important. It would render the legal difference between a provocative act of civil disobedience—operating in a legal grey area and potentially exacerbating existing discursive divides from the 2020 debate highlighted earlier—and an officially recognised act of remembrance and recognition.

This recognition and remembrance could have operated on three levels:

(a) on a trans-national level by at least sending a symbolic gesture of contemporary respect and recognition to the descendant communities in Tsumeb, bearing in mind that Tsumeb Municipality, like the City of Bonn, was an official cooperation partner in the project; while,

(b) on the national and local level allowing individuals in Bonn to engage in a novel and meaningful way with, and reflect on, German colonial history and its intertwinement with institutional structures such as that of the Institute of Geosciences of the Bonn University, and to reflect on their own position as contemporary citizens and what their individual relation to these transhistorical structures may be; while,

(c) on the level of public debate, by potentially creating a new impetus for constructive public exchange in Bonn, prompting to leave behind the trenches of the 2020 debate.

Down the Cemetery Path

In order to arrange the necessary paperwork, I reached out to the City of Bonn, whose project 'Active Culture of Remembrance' had been referred to in the short note regarding von Trotha's grave at the cemetery. The project turned out to be operated by the Centre for Urban History and Memory Cultures, Department IV Sports and Culture of the City of Bonn, which had been designated by City of Bonn as the 'point of competence' in this subject matter. Its personnel are specifically qualified to evaluate and decide on formats that bring to life an active engagement with remembering and confronting national-socialist and colonial histories in Bonn. Their mission statement outlines:

> The project [Active Culture of Remembrance] represents a multi-voiced, dialogical, and indeed ambivalent culture of remembrance. [...] The focus of this project is on sites of memory in urban spaces (such as street names, square names, memorials, monuments, information signs, statues, cemeteries, etc.) that have connections to National Socialism and colonialism. [...] In the spirit of an 'active' culture of remembrance, the aim is not only to passively respond to issues of cultural memory but to set active impulses through various formats for a long-term and participatory engagement with themes of cultural memory. [25] (City of Bonn, 202-)

[22] Regarding 'colonial geology', see in this volume: Selby Hearth, p.62ff; Helmut Maier, p.70ff; Christopher R. Hill, p.82ff; Saima Nakuti Ndahangwapo, p. 92ff; Bastian Herbst, p.106ff.

[23] In the framework of the exhibition the visitor was to encounter this poem again in *transposition 741* (see this volume, p.208ff. For reading the complete poem see p.173-174).

[24] Particularly relevant here is Section 168 (2) (§ 168 II StGB) and its extension in regards to 'defamatory mischief' which refers to an act that is grossly improper and demonstrates a crude attitude, expressing disrespect towards the protected subject.

[25] The author's translation from German.

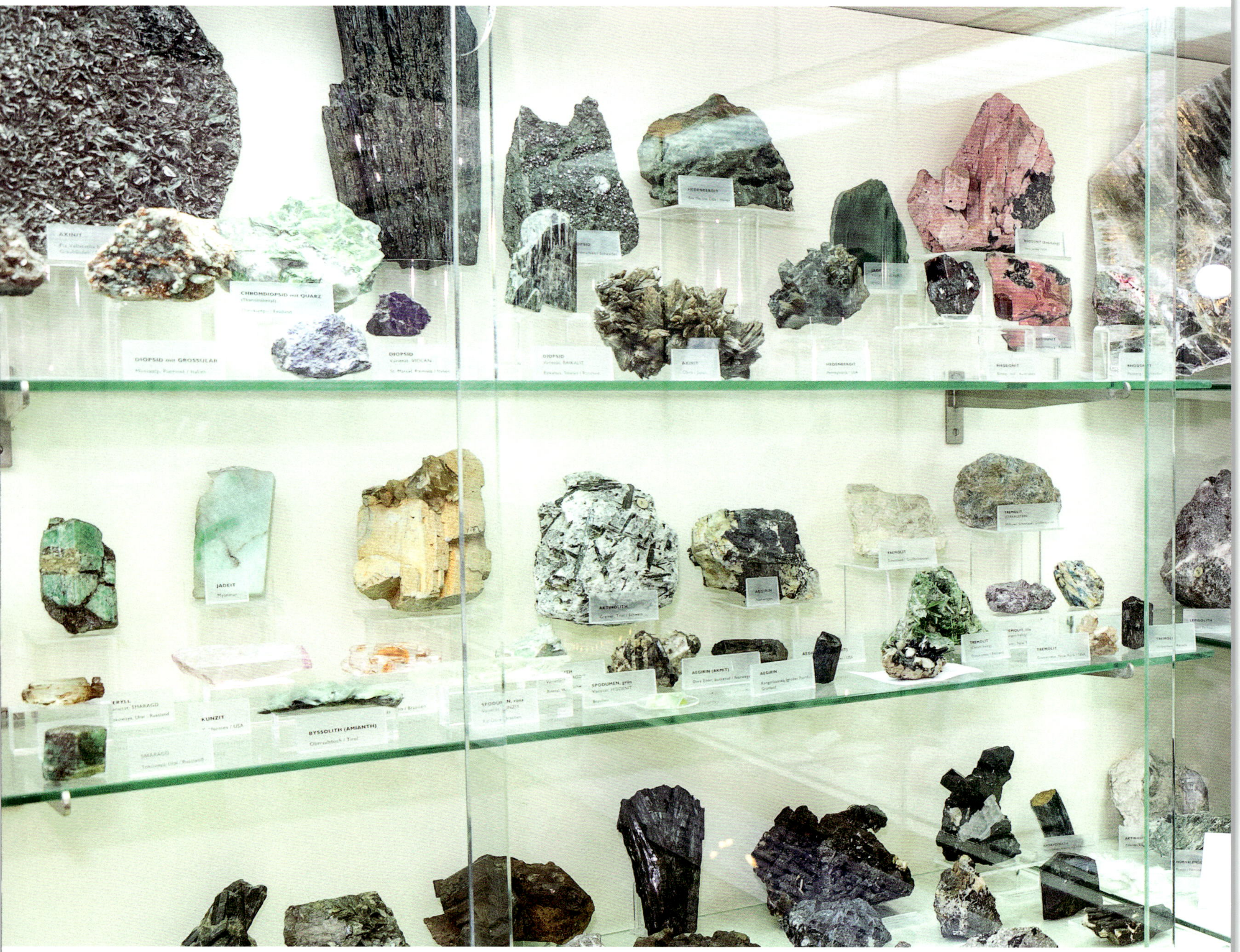
CHROMDIOPSID mit QUARZ
DIOPSID mit GROSSULAR
JADEIT
AKTINOLITH
KUNZIT
BYSSOLITH (AMIANTH)
SPODUMEN, grün
AEGIRIN (AKMIT)
AEGIRIN

Fig. 62

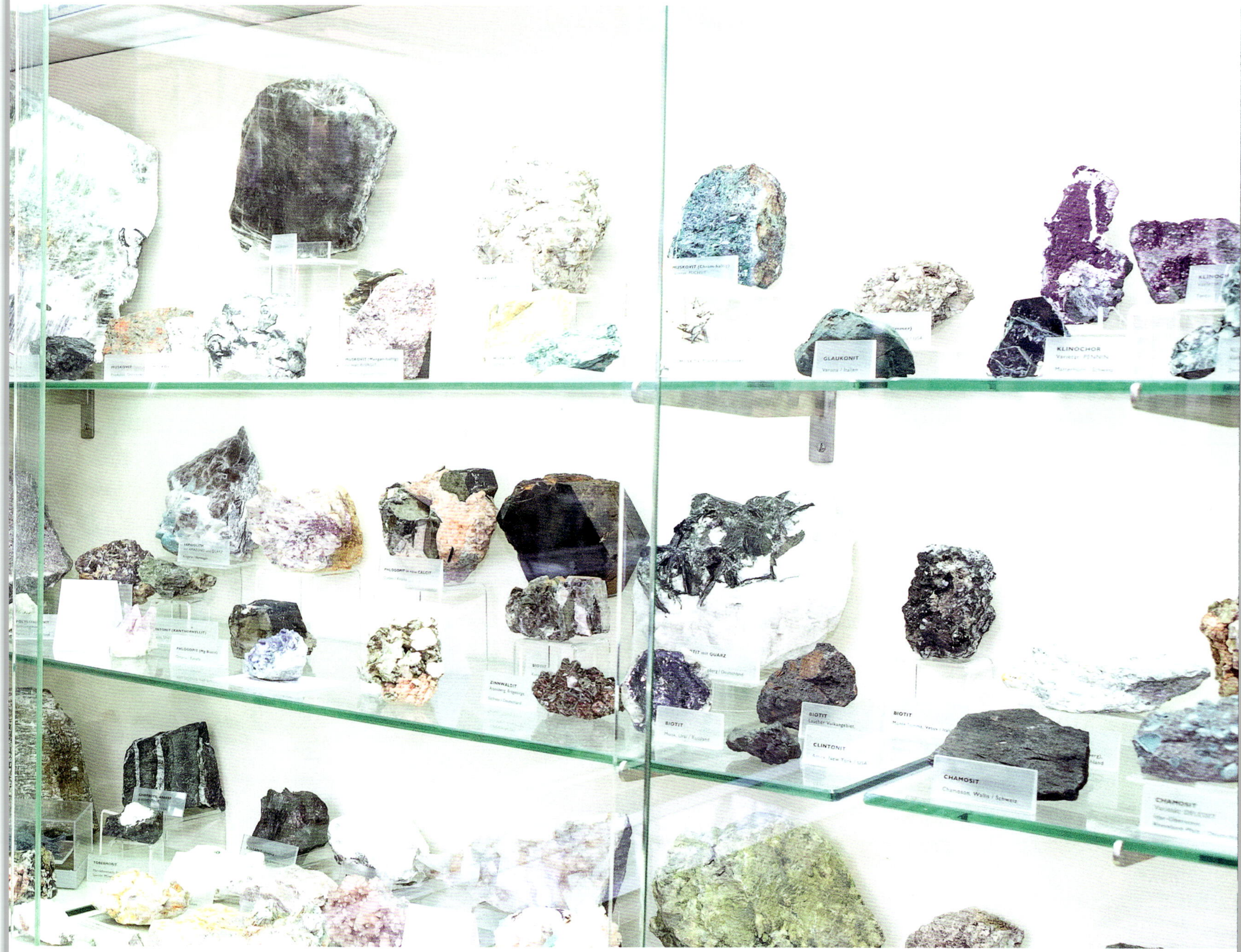
GLAUKONIT
KLINOCHOR
BIOTIT
CLINTONIT
CHAMOSIT
Chamoson Wallis / Schweiz

One of the three conceptual pillars of the project is the collaboration, and development of projects, with artists.[26] After my initial introduction to the project's research team in mid-February 2023, I began the process of seeking permission with them over the course of a month, and a range of phone conversations, meetings, and emails. In late March, the City of Bonn emailed confirming their interest in supporting t*ransposition 904* in the framework of their project 'Active Culture of Remembrance'. A meeting with the research assistant and the Head of Department was fixed, which was cordial and productive. The City's representatives agreed to a cooperation; its precise nature, and the specific form of its public communication to be developed in the process. Since the project was already sufficiently covered by public state and federal funding, I did not request any additional financial support. They agreed to liaise with the cemetery's management to formally legitimise *transposition 904*. Further we agreed that the City of Bonn would produce a document in the form of an official 'letter of permission'. This would become part of the performance kit to reassure both visitors and onlookers. Given the polarising debates three years earlier, this was—by both sides—understood to be a crucial precaution in case other visitors or the personnel of the graveyard would approach with interest, questions, or challenge.

At this point, neither of us foresaw that this work would eventually become a political issue. At the end, however, it did, and my work would be forced to deviate from its original conception that had initially been welcomed. In the process, the City of Bonn's inner workings would render visible self-inhibiting power structures within its administration, and eventually enter into a protracted phase of involuntary and voluntary deferral of administrative procedures. The following is hence an account not so much of an archetypical administrative process that is a fundamental requisite for cultural engagement in public spaces, but rather a first examination of the internal mechanism that appear to systemically inhibit the 'Active Culture of Remembrance' envisioned by the administration itself.

Around the opening of the exhibition at the Mineralogical Museum on 20 April, the City notified me that decisions had not yet been finalised and signalled that trans-departmental discussions were ongoing. On 27 April, I was informed that an approval had been denied by another department that the department of Culture had consulted. This was Department III, the Office for Environment and Urban Green Spaces, who to my understanding had to be involved as per their responsibility for the cemetery area. The grounds given in the informal correspondence referred to a vaguely proposed concern that a permission would unleash, in principle, a carte blanche for future artistic involvements of any grave in Bonn. The Head of Department III justified his decision internally with a reference to a newspaper article: the piece of Bahner's charged writing published in the *Frankfurter Allgemeine Zeitung* (2020).

Consequently, the process of granting a permission remained ongoing while the exhibition was on display. Having worked on a range of projects involving public stakeholders, I of course understood the necessity. Administrators are expected to remain objective and impartial, act with circumspection and proportionality, and take into account a realistic variety of perspectives on a given matter when preparing a decision. It is therefore mandatory not only to include affected or responsible departments and bodies for collaborative and transparent decision-making, but it is also reasonable to include and examine possible arguments for and against a proposed undertaking, especially when in light of recent public debates. It would, indeed, even be helpful to refer to examples of that debate, such as letters or comment pieces—including significant voices from the press or civil society—together with cases of precedent of previous decisions, recent evidence, or evaluation by experts.

It appeared, however, unorthodox to ground a refusal to approve a project on one single input citing a reference supporting only one side of the debate, especially when this was not a previous decision taken by the city as precedent. That this single input was, in fact, not an expert evaluation or outcome of an minuted internal discussion, but an *opinion* piece from the cross-regional print media—that is, a per se affect-oriented contribution to the debate—made this a one-sided base for reasoning. A potential bias inherent in the reasoning process would raise concerns in view of decision-making as set out in the *Administrative Procedure Act (VwVfG)* (German Federal Ministry of Justice, 2023), particularly with regard to the Department III's 'Competence of the Authority'.
Beyond reasoning, it was also an issue of weighting of arguments, and competent authority: it appeared that, effectively, the decision-input of the Office for Environment and Urban Green Spaces, Department III, equal in hierarchy to the neighbouring directorate's (City of Bonn, 2024c) Office for Sports and Culture, Department IV, which houses the Centre for Urban History and Memory Cultures, was given more weighting than the 'Competence of the Authority' of the entire Centre for Urban History and Memory Cultures—the very department initiated and assigned administrative competence by the City of Bonn to evaluate, develop and realise acts of 'indeed ambivalent culture of remembrance [...] on sites of memory in urban spaces (such as [...], cemeteries, [...]) that have connections to [...] colonialism.' (City of Bonn, 202-).[27]

In spite of their equality within administrative hierarchies, the Commissioner for Sports and Culture effectively succumbed to the Head of the green spaces department in a matter primarily concerning her own department's authority. There was no relevant information provided as to a possible internal

[26] Information according to a personal conversation of the artist with the Centre for Urban History and Memory Cultures' head of department on 24 May 2023.

[27] The author's translation from German.

questioning of this rejection of permission, nor a request to reconsider it, nor follow-up queries to explain or change the project.

The uncertainty that emerged from this protracted situation from 27 April 2023 lasted 46 days. Meanwhile, the overall exhibition itself had opened, and was on display. The performance—as a consequence of the City of Bonn's pending final decision—was included in the exhibition only in the form of its inert sculptural intervention. As it could not be performed officially by a general audience, it was effectively demoted to an act of civil disobedience (if performed nevertheless). The latter was never intended by the project, nor was it helpful for the visitors who were thus deprived of accessing a conceptually constitutive element of the exhibition: within the conceptual framework of the exhibition, *transposition 904* was the only instance where agency was granted to the spectator. All other 'transpositions' consciously relegated the audience to a passive and dominated position.[28] The performance was constructed to be the only, temporary release for the visitor from this domination by actively re-linking histories. As a consequence of the City of Bonn's stalling, the performance was not enacted by public visitors, but only by a group of professional visitors touring the exhibition: the Belgian artist and curator Patrick Carpentier, the British-Israeli artist Josh Baum, the British historian Chris Hill, and myself as artist and curator of the exhibition.

Another 28 days later, on 25 May 2023, I had a meeting with the Head of the Centre for Urban History and Memory Cultures; followed shortly after by email communication and a teleconference, which yielded no further information on an internal conclusion having been reached by the City. The exhibition then closed officially—and as planned—on 11 June 2023. A day later, on 12 June, I received an email by the Commissioner for Sports and Culture (City of Bonn, 2024d), who oversees the Centre for Urban History and Memory Cultures, declaring that the performance cannot take place as planned and that a permission had to be further evaluated:

> Currently, the City of Bonn cannot grant permission for the performance at the grave of von Trotta [sic] to take place in the cemetery. This is a precedent that requires detailed coordination within the administration. The administration will inform you of the results in due course. [...]
> (Commissioner for Sports and Culture, 2023)[29]

An amicable personal meeting with the Commissioner for Sports and Culture followed, in which we discussed the performance's formal and contextual details (Mikloweit, 2023d). I was reassured in person that my proposal was now handled as a 'precedence case', and even a potential future collaborative realisation was alluded to. A month later, on 9 July 2023, a further delay was announced of the planned trans-departmental meeting in which the case was supposed to be on the agenda. From there on, several notifications, including on my request, promised a final decision to be available 'in due course'. After further notifications of deferral, eventually the City of Bonn never issued a formal decision on my request for permission regarding the project they had initially welcomed.[30]

By indefinitely stretching its administrative procedures, the City of Bonn essentially prevented the performance piece from becoming a legitimised, temporary cultural activity of the exact 'Active Culture of Remembrance' the City was aiming for. This contravened the officially assigned 'Competence of the Authority' of their own experts: the Office for Environment and Urban Green Spaces effectively relegated the 'Competence of the Authority' of his colleagues at the Centre for Urban History and Memory Cultures, and acted against the recommendation of the Commissioner for Sports and Culture, who accepted this.[31]

At the end of the Cemetery Path

This lack of internal administrative leverage opens up questions as to the raison d'être and reach of the Centre for Urban History and Memory Cultures. How can a department for remembrance realise its projects within its publicly communicated sphere of authority, if it is not granted the power to do so within its own administration?

In an effort to resolve this impasse and close the file, I decided in a final attempt to approach the case from another angle: during our personal meeting in June 2024, I proposed to the Commissioner that I could open communication channels with Tsumeb Municipality, which I had established within the framework of the *Transpositional Geologies* project, to foster an active exchange and engagement between the two cities. In the course of the previous months of our correspondence, the City of Bonn's various actors had been informed about the fact that the Tsumeb Municipality was an official collaborator and cooperation partner in the exhibition in Bonn and the *Transpositional Geologies* project at large.[32]

A mutual engagement could have transformed what I came to see as a permission debacle into a new starting point, expanding the City's designated 'Active Culture of Remembrance' to include voices and perspectives of communities in Namibia on the genocide and the consequences of German colonialism. However, this proposal was never further pursued by the Commissioner for Sports and Culture. At the end, ideally it should not have to be an artist to make such proposals in the first place: restorative initiatives should originate from within a healthy, well-educated, innovative, and responsible city's administration and citizenry that is taking in public debates

[28] This dominated position in which the spectator experienced the exhibition alluded to the disempowered realities of Tsumeb mine workers (cross-reference to Kaluvi's piece) symbolically, without subjecting the workers—via the proxy of photographic representations of their bodies—to the predominantly white gaze of the Bonn visitors. They would have become subjects of such spectation had the workers been represented in the exhibition, for example, through photographic documents drawn from the colonial archives. I decided against such modes of representation, as they would have relegated the workers' histories to artistically aestheticised material, thereby merely perpetuating the colonial violence that the exhibition sought to address.

[29] The author's translation from German.

[30] Which is the status quo at the time of finishing this essay on 10 October 2024.

[31] The latter as head of Department IV has direct 'Competence of the Authority' as a superior only over the Centre for Urban History and Memory Cultures but not over the Department III's Office for Environment and Urban Green Spaces (City of Bonn, 2024c).

[32] I.e. *transposition 015*, *transposition 495*.

and is willing to deal with its own history—particularly if it prides itself on being a diverse and globally significant locale.

As a result, a carefully researched and designed public experience—part of a federally- and state-funded four-year project—did not take place in its proposed, modest form, despite an initially explicit welcome by a City that officially commits itself to an 'active culture of remembrance'; nor was its potential to contribute to the entrenched local debates recognised. This result raises concerns not only as to administrative decision-making,[33]or the potential restriction of artistic freedom.[34]

Essentially, what remains as a de facto outcome seemed to elude the City: Bonn, where 'von Trotha was celebrated [...] as an honorary guest' from 1906 to 1920 (Hagenberg-Miliu, 2024), has in 2023 prioritised the preservation of the final resting place of this colonial dignitary—who is recognised by the Federal Republic of Germany as responsible for the first genocide of the 20th century (GBRS, 2021; bpb, 2021)—over an even fleeting gesture of recognition and respect towards the living descendant communities of von Trotha's victims in Namibia.

[33] Administrative Procedure Act (*Verwaltungsverfahrensgesetz* [VwVfG]). (German Federal Ministry of Justice, 2023)

[34] Art. 5 Abs. 3 GG (Article 5 Paragraph 3 of the Basic Law (*Grundgesetz*) for the Federal Republic of Germany), which emphasises the freedom of artistic expression and scientific research. (German Federal Ministry of Justice, 2022)

Fig. 63

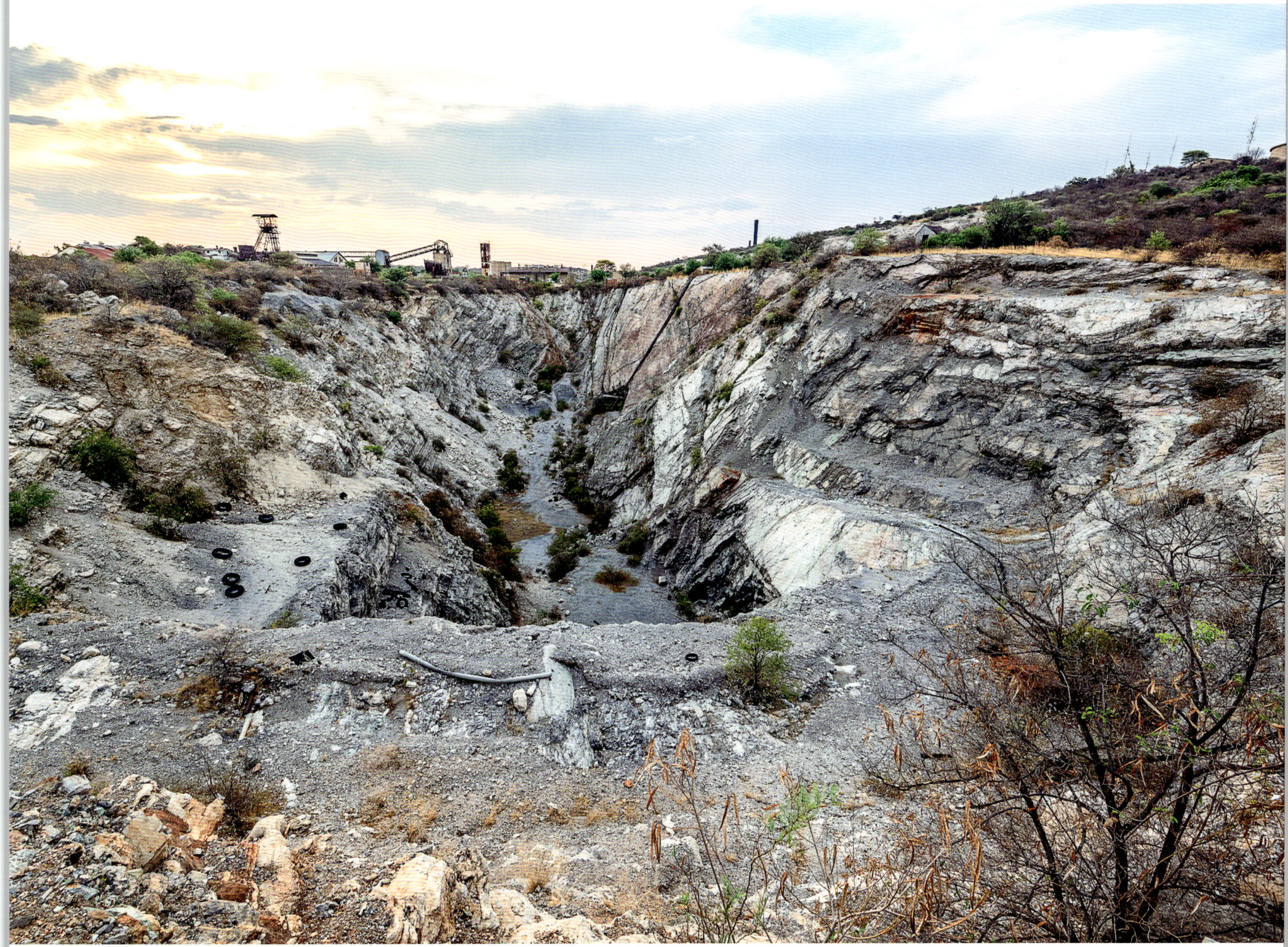

References

Bahners, P., 2020. Wie kann man eine Grabstätte „kritisch kommentieren"?. *Frankfurter Allgemeine Zeitung*, 10 September. [online] Available at: <https://www.faz.net/aktuell/feuilleton/debatten/postkoloniale-aktivisten-zum-grab-lothar-von-trothas-16945516.html> [Accessed 12 March 2022]

Bley, H. and Ridley, H., 1971. *South-West Africa under German Rule, 1894-1914*. London: Heinemann.

Blum, C., 2024. When cultures of remembrance clash: Contesting memorialisation in Namibia's urban spaces. *FGZ Leipzig, Research Institute Social Cohesion (RISC),* [online] 14 February. Available at: <https://fgzrisc.hypotheses.org/4757> [Accessed 07 Juli 2024].

Bonn Postkolonial, 2020. Offener Brief an die Stadt Bonn. [online] Available at: <hhttps://decolonize-bonn.de/offener-brief-an-die-stadt-bonn> [Accessed 05 Juli 2024].

Bundeszentrale für politische Bildung (bpb) / Federal Agency for Civic Education, 2021. Völkermord an Herero und Nama: Abkommen zwischen Deutschland und Namibia. *Bundeszentrale für politische Bildung Online*, [online] 22 June. Available at: <https://www.bpb.de/kurz-knapp/hintergrund-aktuell/335257/voelkermord-an-herero-und-nama-abkommen-zwischen-deutschland-und-namibia> [Accessed 24 July 2024].

City of Bonn, 202-. *Zentrum für Stadtgeschichte und Erinnerungskulturen: Projekt „Aktive Erinnerungskultur"*. [online] Available at: <https://www.bonn.de/vv/oe/_/Dez._IV/41/41-2/Zentrum-fuer-Stadtgeschichte-und-Erinnerungskulturen.php> [Accessed 07 Juli 2024].

____2024a. *University of Bonn.* [online] Available at: <https://international.bonn.de/business-and-science/science-research/university-of-bonn.php#:~:text=Bonn's%20university%20was%20founded%20in,Laureates%20in%20its%20recent%20history> [Accessed 07 Juli 2024].

____2024b. *UN in Bonn*. [online] Available at: <https://international.bonn.de/international-profile/united-nations-in-bonn/index.php> [Accessed 26 Juli 2024].

____2024c. *Dezernatsverteilungsplan des Verwaltungsvorstandes*. [pdf] 1 February 2024. [online] Available at: <https://www.bonn.de/medien-global/amt-10/Dezernatsverteilungsplan-Stand-01.02.2024.pdf> [Accessed 07 Juli 2024].

____2024d. *Frau Dr. Birgit Schneider-Bönninger*. [online] Available at: <https://www.bonn.de/vv/personen/Schneider-Boenninger-676.php> [Accessed 07 Juli 2024].

Commissioner for Sports and Culture, 2023. *Performance Grab v. Trotta* [sic]. [email] Message to S. Mikloweit. Sent 12 June 2023 17:15. [archive of the artist].

Chothia, F., 2022. Namibia pulls down German colonial officer's statue in Windhoek. *BBC news*, 23 Nov 2022. [online] Available at: <https://www.bbc.com/news/world-africa-63728105> [Accessed 02 August 2024].

Dedering, T., 1993. The German–Herero War of 1904: Revisionism of Genocide or Imaginary Historiography? *Journal of Southern African Studies*, 19(1), pp.80–88.

____1999. 'A Certain Rigorous Treatment of all Parts of the Nation': The Annihilation of the Herero in German South West Africa, 1904. In: M. Levene and P. Roberts, eds., 1999. *The Massacre in History*. New York, Oxford: Berhan Books, pp.205–222.

DPM (Dundee Precious Metals), 2024. *Tsumeb Specialty Smelter, Namibia*. [online]. Available at: <https://dundeeprecious.com/assets/other/tsumeb-specialty-smelter> [Accessed 02 August 2024].

Drechsler, H., 1966. *Südwestafrika unter deutscher Kolonialherrschaft: Die großen Landgesellschaften und Minengesellschaften (1885–1914).* Stuttgart: Franz Steiner Verlag.

____1980. *Let Us Die Fighting: The Struggle of the Herero and Nama against German Imperialism (1884-1915)*. London: Zed Press.

____1984. *Aufstände in Südwestafrika: Der Kampf der Herero und Nama 1904 bis 1907 gegen die Deutsche Kolonialherrschaft*. Berlin: Dietz.

Eberle, L. and Weiland, S., 2024. Beschwerde von Namibias Botschafter: AfD-Politiker Tritschler löst diplomatische Spannungen aus [Complaint from Namibia's ambassador: AfD politician Tritschler triggers diplomatic tensions]. *Spiegel Online,* [online] 25 July. Available at: <https://www.spiegel.de/politik/deutschland/nach-kranzniederlegung-in-namibia-afd-und-nrw-abgeordnete-sven-tritschler-loest-diplomatische-spannungen-zwischen-namibia-und-deutschland-aus-a-ca974484-1998-47dd-93f4-69899649419d> [Accessed 25 July 2024].

GAoEJ (Global Atlas of Environmental Justice), 2022. *Namibian Custom Smelter, Tsumeb, Namibia*. [online]. Available at: <https://ejatlas.org/conflict/tsumeb-smelter> [Accessed 02 August 2024].

German Bundestag Research Service (GBRS), 2021. *Einzelfragen zum Oberbefehlshaber der kaiserlichen „Schutztruppe" in „Deutsch-Südwestafrika" Lothar von Trotha*. [pdf] WD 1 - 3000 - 018/21, 20 July 2021. [online] Available at: <https://www.bundestag.de/resource/blob/855620/da4256f6d584139c527efb161b66d606/WD-1-018-21-pdf-data.pdf> [Accessed 16 July 2024].

German Federal Ministry of Justice, 2021. *German Criminal Code (Strafgesetzbuch – StGB)*. [online] Available at: <https://www.gesetze-im-internet.de/englisch_stgb/englisch_stgb.html> [Accessed 07 Juli 2024].

____2022. *Basic Law for the Federal Republic of Germany*. [online] Available at: <https://www.gesetze-im-internet.de/englisch_gg/englisch_gg.html#p0031> [Accessed 07 Juli 2024].

____2023. *Verwaltungsverfahrensgesetz (VwVfG)*. [online] Available at: <https://www.gesetze-im-internet.de/vwvfg/index.html> [Accessed 07 Juli 2024].

Green, R.H., Louw, J., eds., 1999–2024. History of Namibia. *Britannica*. [online] Accessible at: <https://www.britannica.com/place/Namibia/History#ref480264> [Accessed 30 May 2024].

Hagenberg-Miliu, E., 2024. Verantwortlich für Völkermord in Afrika: Wie Lothar von Trotha in Bonn als Ehrengast gefeiert wurde. *General Anzeiger Bonn,* 14 March. [online] Available at: <https://ga.de/bonn/bad-godesberg/voelkermord-an-herero-und-nama-lothar-von-trotha-lebte-in-bonn_aid-108367867> [Accessed 30 May 2024].

Hasheela, I. 2018. Contamination mapping and land use categorization for Tsumeb. *Communications of the Geological Survey of Namibia*, 19, pp.1–7. Windhoek: Ministry of Mines and Energy.

Hearth, S., 2021. The 'World's Greatest Mineral Location': Hai||om, Ndonga, Herero, and the Early Colonial Histories of Tsumeb, Namibia. *Earth Sciences History*, 40(2), pp.433–460.

Holmes, A., 2024. *TG research trip: Swakopmund Genocide Memorial.* Interviewed by S. Mikloweit. [email] Berlin, Brussels, 27 July 2024.

Kondarev, G., 2015. Exporting Toxic Pollution from Europe to Namibia. *CEE Bankwatch Network*, 19 November. [online] Available at: <https://bankwatch.org/blog/exporting-toxic-pollution-from-europe-to-namibia> [Accessed 07 Juli 2024].

Magoley, N., 2024. 'Verhöhnung der Opfer': AfD-Vize legt Kranz ans Grab deutscher Kolonial-Täter in Namibia ['Mockery of the victims': AfD vice-president lays wreath on the grave of German colonial perpetrators in Namibia]. *WDR online*, [online] 19 June. Available at: <https://www1.wdr.de/nachrichten/landespolitik/afd-tritschler-kranz-namibia-100.html> [Accessed 24 July 2024].

Mikloweit, S., 2022. Research field notes. [notebook] *TG Archive.* Notebook 1. Brussels, Berlin: Transpositional Geologies.

____2023a. Images of P. Terkatz WW1, WW2 memorial at Poppelsforfer Cemetery. [jpg] *TG Archive*. TG-T904-ra_pc-01, TG-T904-ra_pc-02. Brussels, Berlin: Transpositional Geologies.

____2023b. Images of Poppelsforfer Cemetery LvT notice. [jpg] *TG Archive*. TG-T904-ra_pc-04 _DSC2395. Brussels, Berlin: Transpositional Geologies.

____2023c. *We Build Our Languages Out Of Rocks—Transpositional Geologies.* [exhibition] Mineralogical Museum of the University of Bonn, Germany. 20 April 2023 – 11 June 2023.

____2023d. Research field notes. [notebook] TG Archive. Notebook 1. Brussels, Berlin: Transpositional Geologies.

____2024. Calling notes 17-07-24 City of Bonn. [pdf] *TG Archive*. TG-T904-ra_pc-03_cn-17-07-24CoB. Brussels, Berlin: Transpositional Geologies.

Mindat.org, n.d. *Tsumeb Mine (Ongopolo Mine)*. [online] Available at: <https://www.mindat.org/loc-2428.html> [Accessed 09 Juli 2024].

Mbako, S. Z., 1989. We Come and Die as Numbers. In: S. Z. Mbako, ed. 1989. *Tell Them of Namibia: Poems from the National Liberation Struggle*. London: Karia Press.

Niesen, J., 2017. Peter Terkatz: Bildhauer (1880–1954). *Portal Rheinische Geschichte des LVR*. [online] Available at: <https://www.rheinische-geschichte.lvr.de/Persoenlichkeiten/peter-terkatz/DE-2086/lido/57c93acda13043.92640220> [Accessed 16 Juli 2024].

NIOH (National Institute for Occupational Health), 2013. *Preliminary Report on the Survey of Namibian Customs Smelter Workers Tsumeb, Namibia*. Johannesburg, Windhoek: National Institute for Occupational Health, Ministry of Environment and Tourism of Namibia.

Oesterle, K., n.d. *Das Lied vom guten Kameraden: Die heimliche deutsche Hymne. Auszug eines Artikels von Kurt Oesterle*. [online] Available at: <https://www.100-jahre-erster-weltkrieg.eu/fileadmin/redaktion/Micro_Weltkrieg/Downloads/Das_Lied_vom_guten_Kameraden.pdf> [Accessed 02 August 2024].

Petersen, S., 2022. Great-grandson says Von François 'did a lot for this country'. *The Namibian*, 30 Oct, p.1–2. [also online] Available at: <https://www.namibian.com.na/great-grandson-says-von-francois-did-a-lot-for-this-country> [Accessed 02 August 2024].

Popov, D., Kohrs, G. K. B., and Bacheva-MacGrath, F., 2016. *Dirty Precious Metals: Dumping European toxic waste in Tsumeb, Namibia*. Namibia, Bulgaria: CEE Bankwatch Network, Earthlife Namibia, and Za Zemiata – Friends Of The Earth Bulgaria. [pdf] [online]. Available at: <https://bankwatch.org/publication/dirty-precious-metals-dumping-european-toxic-waste-in-tsumeb-namibia> [Accessed 02 August 2024].

SAHO (South African History Online), Anon., 2024. *German Lieutenant-General Lothar von Trotha lands in Swakopmund, South West Africa*. [online] Available at: <https://www.sahistory.org.za/dated-event/german-lieutenant-general-lothar-von-trotha-lands-swakopmund-south-west-africa> [Accessed 30 May 2024].

Van Wyk, B., 2023. 'They died for us': The people tending to the graves of Namibia's colonial genocide victims. *Minority Africa,* 10 January 2023. [online] Available at: <https://minorityafrica.org/they-died-for-us-the-people-tending-to-the-graves-of-namibias-colonial-genocide-victims> [Accessed 02 August 2024].

Wallace, M., 2013. *A History of Namibia: From the Beginning to 1990*. New York: Oxford University Press.

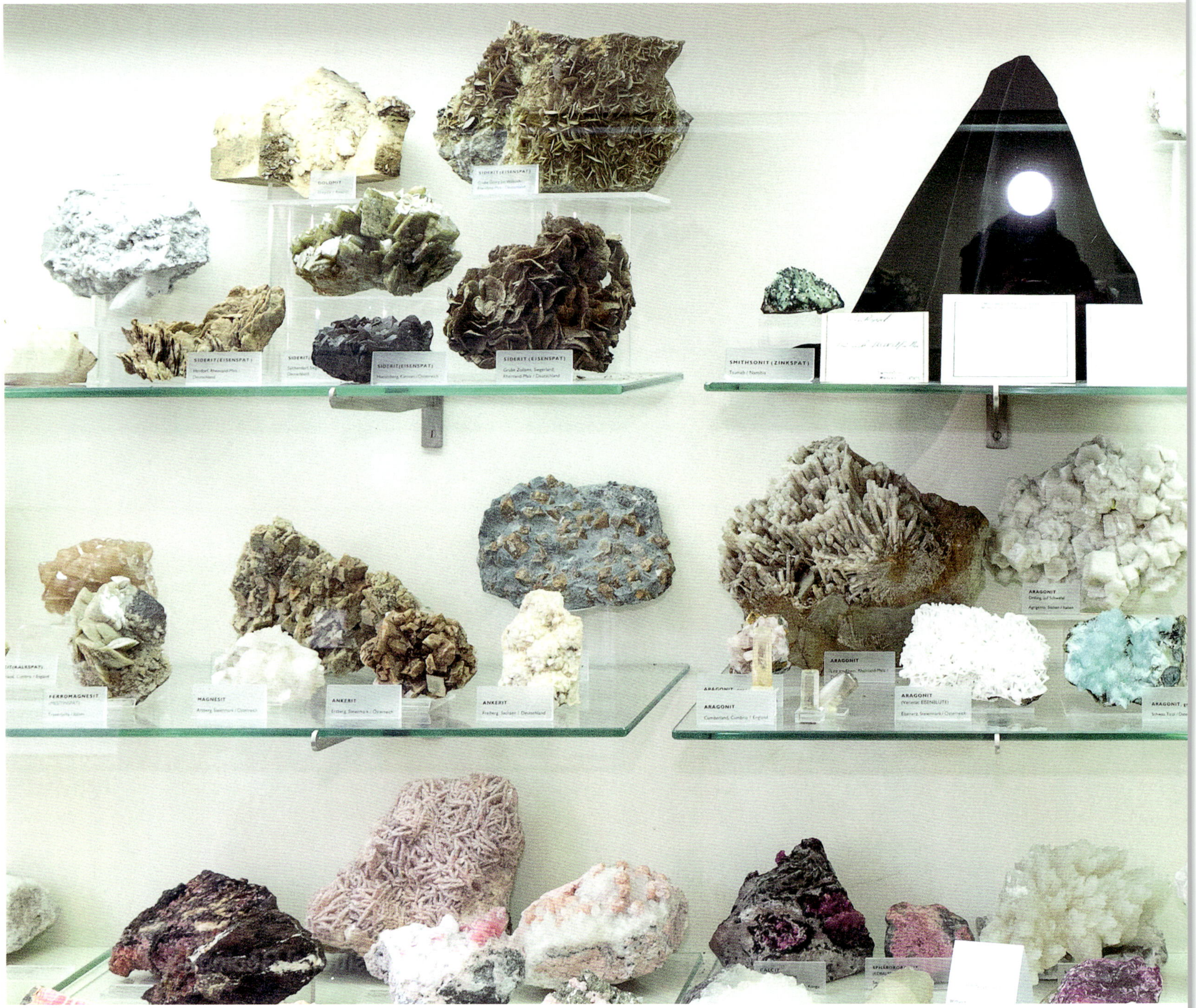
SIDERIT (EISENSPAT)
SIDERIT (EISENSPAT)
SIDERIT (EISENSPAT)
SMITHSONIT (ZINKSPAT)
FERROMAGNESIT
MAGNESIT
ANKERIT
ANKERIT
ARAGONIT
ARAGONIT
ARAGONIT

Fig. 64

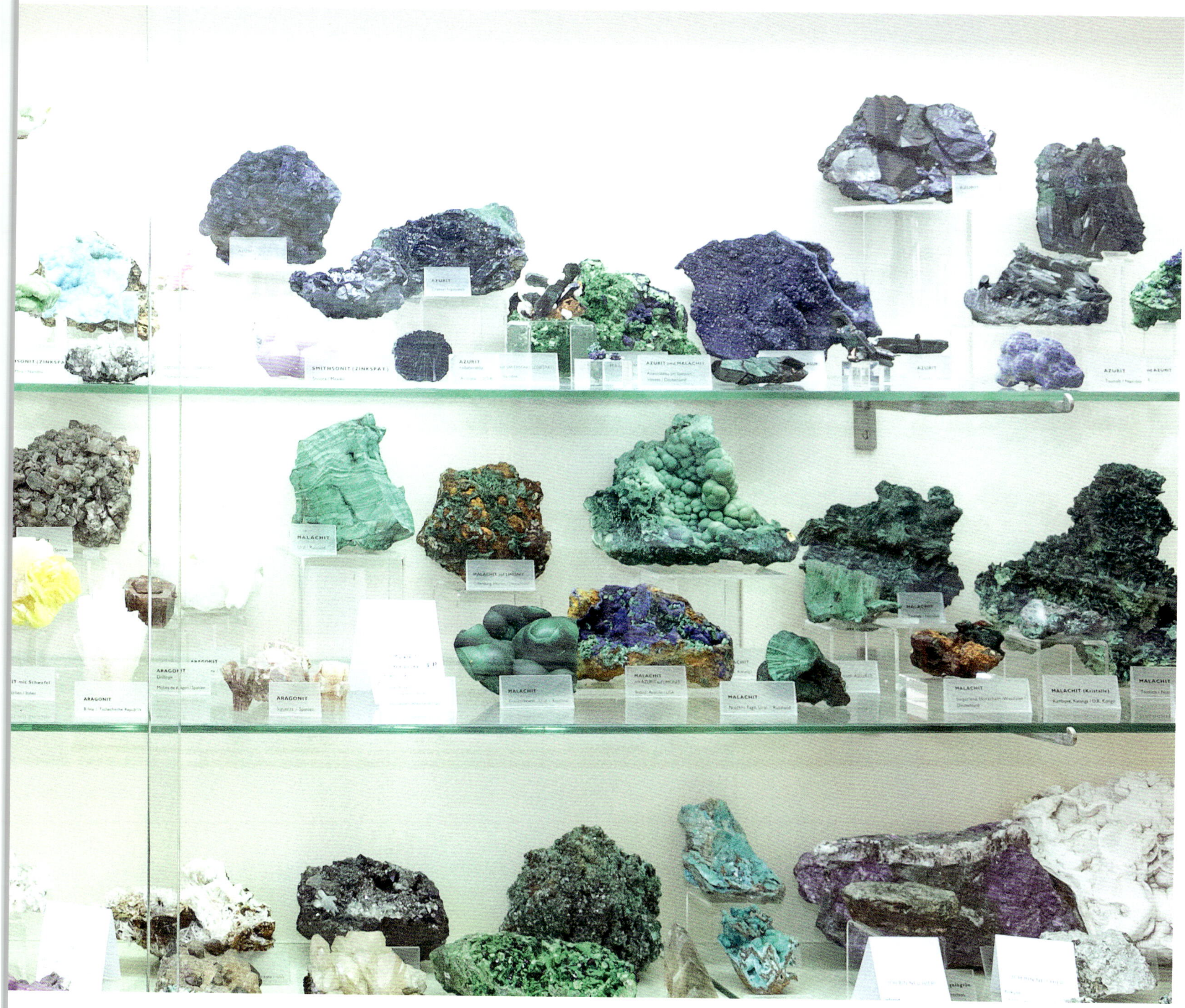
SMITHSONIT (ZINKSPAT)
AZURIT
AZURIT und MALACHIT
AZURIT
AZURIT
MALACHIT
ARAGONIT
ARAGONIT
ARAGONIT
MALACHIT
MALACHIT
MALACHIT
MALACHIT (Kristalle)
MALACHIT

TG-T441
transposition 441
(Black No More, MMdUB Version)

Artist: Sascha Mikloweit
Year: 2023
Location: Mineralogical Museum of the University of Bonn, DE, Exhibition Hall 1, Systemic Collection
Type: Sculptural Intervention
Dimensions: 40 x 32 x 28 cm
Material: Systemic Collection, Mineralogical Museum of the University of Bonn; Collection Object #50215 (location: Exhibition Hall 1, 30 top); object labels early 1900s to 2022; black velvet

Veiled collection object of the Mineralogical Museum of the University of Bonn, number 50215 from the Tsumeb Mine (Namibia), complete with its three museum archive labels. The oldest label (left) dates approximately from the early 1900s, the centre one from the 1950s, and the rightmost from 2022. (this volume, p.288) Number 50215's initial label marks it as originating from 'DSW' (*Deutsch-Südwestafrika*, or German South West Africa), the historical German term (1884–1915) for its former colonial territories, which are now Namibia. The assimilated object is part of a collection amassed by German industrialist and I.G. Farbenindustrie AG / IG Farben founder Friedrich Carl Duisberg (1861–1935). The Duisberg Collection has been housed at the Mineralogical Museum of the University of Bonn, under a permanent loan agreement with Bayer AG (one of the successor companies of IG Farben) since April 1954.

Fig. 65

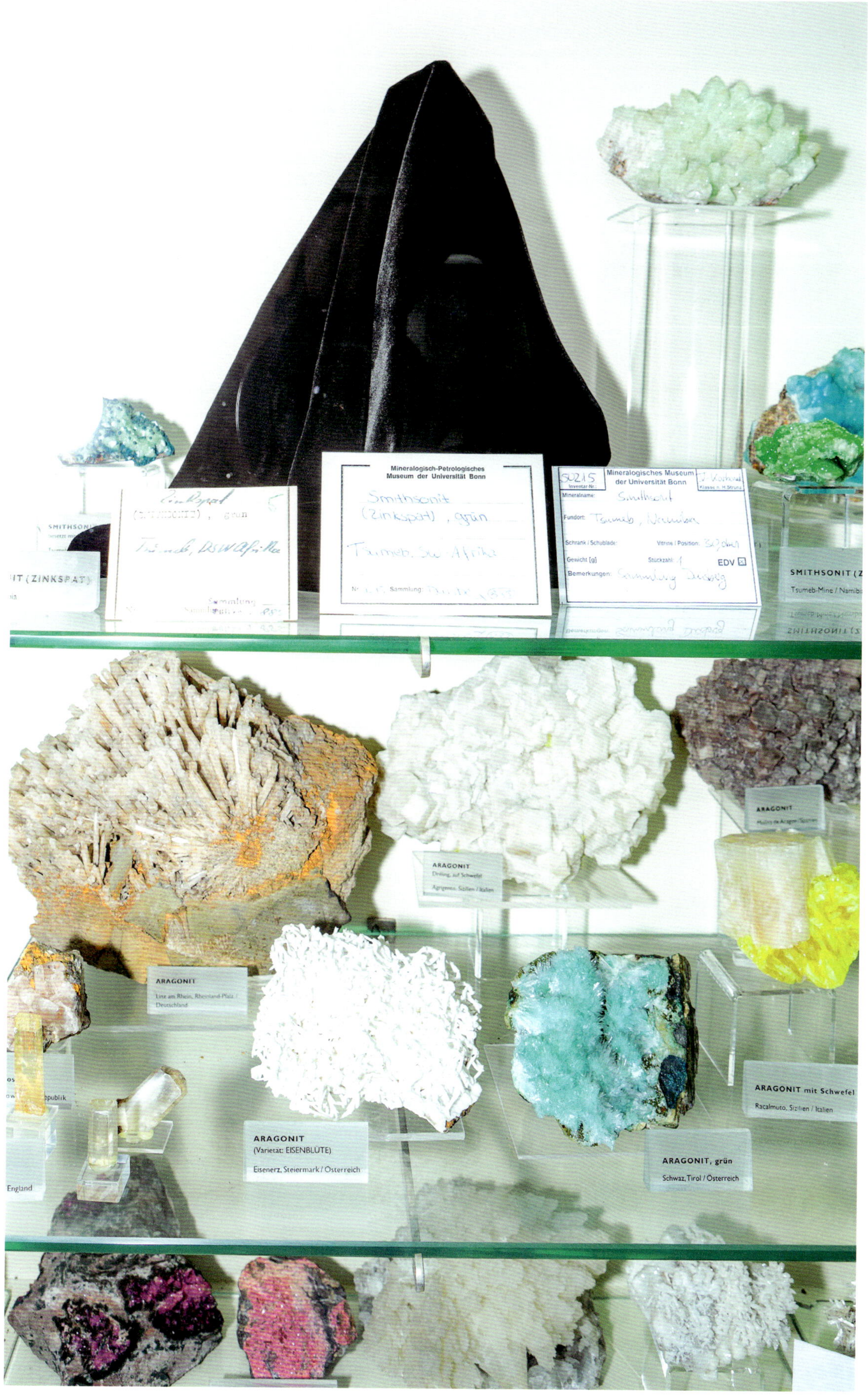
Mineralogisch-Petrologisches
Museum der Universität Bonn
Smithsonit
(Zinkspat), grün
Mineralogisches Museum
der Universität Bonn
SO215
Inventar-Nr.:
Klasse n. H.Strunz
Mineralname: Smithsonit
Fundort: Tsumeb, Namibia
Schrank / Schublade:
Vitrine / Position:
Gewicht [g]
Stückzahl: 1
EDV
Bemerkungen:
SMITHSONIT (Z
Tsumeb-Mine / Namibi
ARAGONIT
Agrigento, Sizilien / Italien
ARAGONIT
Linz am Rhein, Rheinland-Pfalz / Deutschland
ARAGONIT
(Varietät: EISENBLÜTE)
Eisenerz, Steiermark / Österreich
ARAGONIT, grün
Schwaz, Tirol / Österreich
ARAGONIT mit Schwefel
Racalmuto, Sizilien / Italien

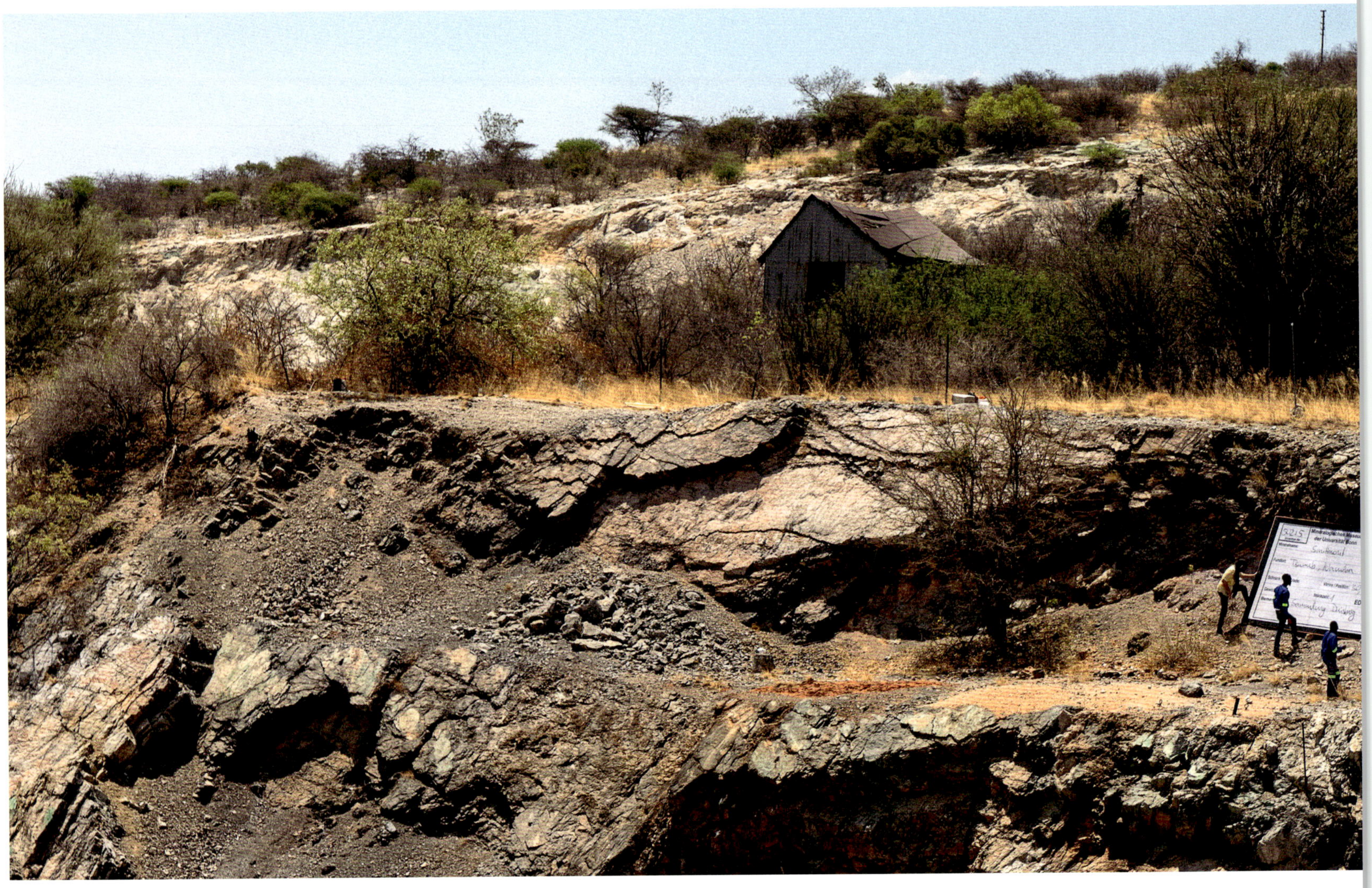

TG-T086
transposition 086
(Tsumeb Relay)

Artist: Sascha Mikloweit
Collaborator: August Kapolo, Rakker Festus, Festus Alushiya, David David
Year: 2023
Location: Former Otavi Mining and Railways Company (OMEG) open pit mine, Tsumeb, Otjikoto, Namibia: -19.2390120, 17.7165468
Type: Performance, Permanent Installation, Image—Object (I—O)
Dimensions: 356 x 252 x 4 cm (each; four billboards)
Duration: 05:00:00h (Performance)
Material: Former OMEG open pit mine; labels of MMdUB collection object #50215; O—I_MMdUB-50215_L_1, 2, 3; German 18/1 billboard prints; wood; cardboard

transposition 086 (Tsumeb Relay) re-links the coloniality of the Mineralogical Collection in Bonn to its colonial origins. Initially, these billboards were intended to be part of the Bonn exhibition, where they would have been placed in the public space of the city, thus expanding the archive of the Mineralogical Museum into the public sphere. During the planning phase, I proposed the work to several institutions: the Bonn Minster (a grand church at the central plaza), which delayed a decision for weeks and ultimately refused; the Academic Art Museum (part of the Bonn University Museums), which after prolonged negotiations, chose to use the space for advertising purposes; and Bonn University itself, where I proposed the billboards be installed in front of the Electoral Palace (the university's main building) on the Hofgarten lawn, a popular area for students and citizens to relax in the summer. The university's directorate only confirmed the possibility of realising the work one day before the opening of my exhibition in Bonn, making production not feasible. This situation, however, led to the development of *transposition 441* (Black No More, MMdUB Version) (see this volume, pp. 192–195) and I decided to take the billboard prints I had produced, but not

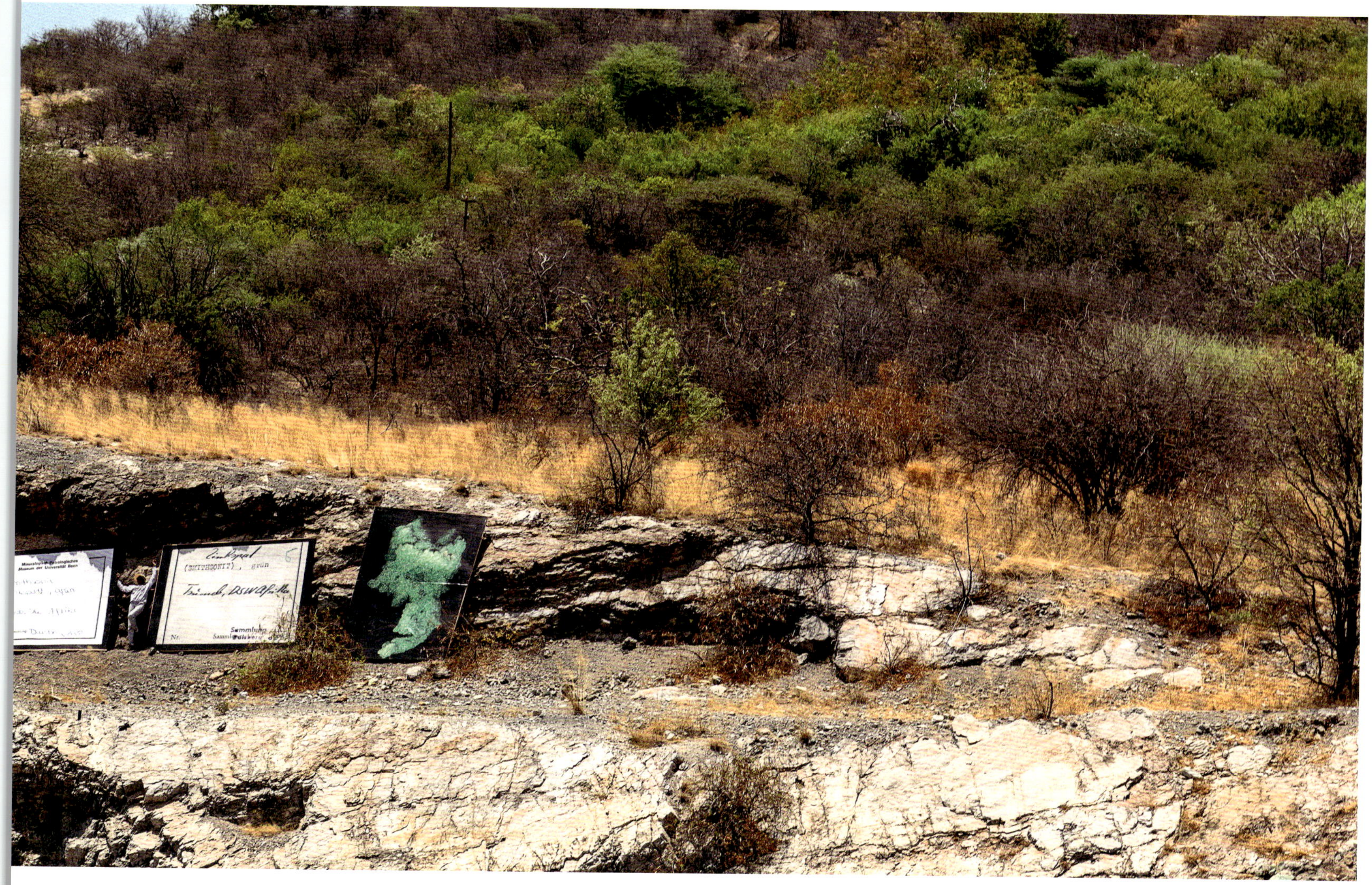

Fig. 66

Fig. 67 [pp.198–199]

Fig. 68 [p.200]

Fig. 69 [p.200]

Fig. 70 [p.201]

Fig. 71 [p.201]

used, to Tsumeb to potentially contextualise them at the site of the former OMEG mining operation.

During my Tsumeb residency and exhibition preparations in late October 2023, Elli Stella Udda Imalwa (Tsumeb Municipality) introduced me to the Community Skills Development Centre (COSDEC Tsumeb), where I had the pleasure of meeting local artist August Kapolo. This encounter sparked a collaboration that led to this performance and permanent installation. August invited three young carpenters, apprentices at COSDEC into the collaboration: Rakker Festus, Festus Alushiya, and David David. On 28 October 2023, a five-hour performance began in the morning, in which the five of us built the four billboards, carried them into the open-pit mine, and installed them eventually on a reinforced fence at the northern edge of the former 'Green Hill'—now the vast OMEG open-pit mine. (This location was just 100 metres east of where the sediment for *transposition 015* (see this volume, pp. 228ff.) was extracted and returned.) The following day, the fence gave way, and the billboards tipped over, facing the sky. I initially attributed this to the 'force of nature,' most likely the wind. However, when a dear friend visited the installation roughly three weeks later, he found the billboards completely turned around, now upside down. The metal rods that reinforced the fence were bent, as though they had been flicked like toothpicks. Given the strength required to flip the 10-square-meter billboards (which were strongly fixed to the fence), the 'force of nature' explanation began to seem unlikely. When Johannes Giebel (Institute of Applied Geosciences, TU Berlin) visited the site in March 2024 during a field trip with his students from the Technical University of Berlin, and sent updated pictures, it became clear that this was no accident. The billboards had turned a third time, from face-side-down to face-side-up, and had been clearly smashed by blunt force, as evidenced by the chipped edges of the fragmented materials.

Furthermore, the Marula tree we planted during *transposition 015* just 100 metres west, simply disappeared entirely. Whether this is connected to locations in Tsumeb, where the German Reich flag is still proudly displayed, remains speculative.

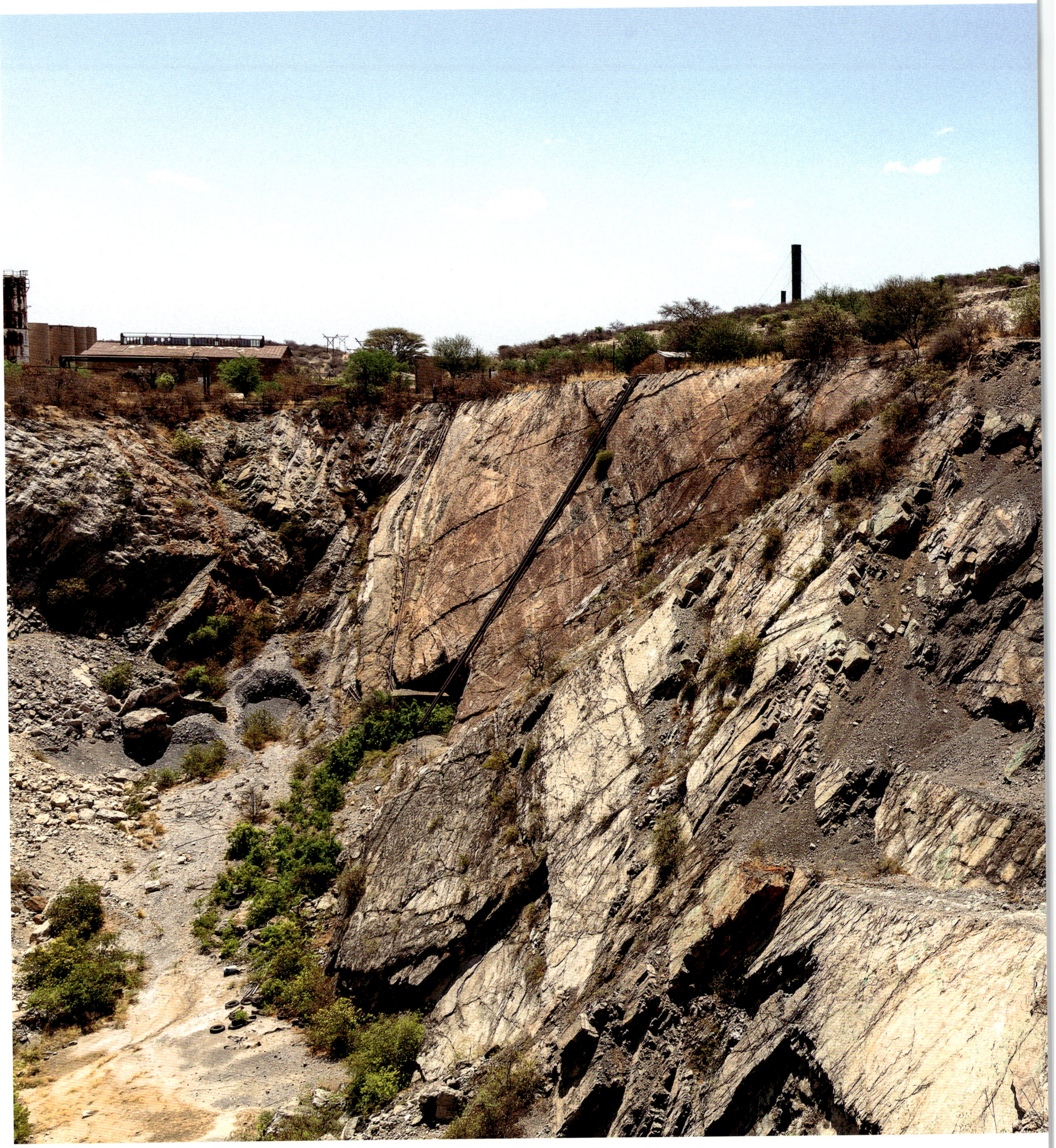

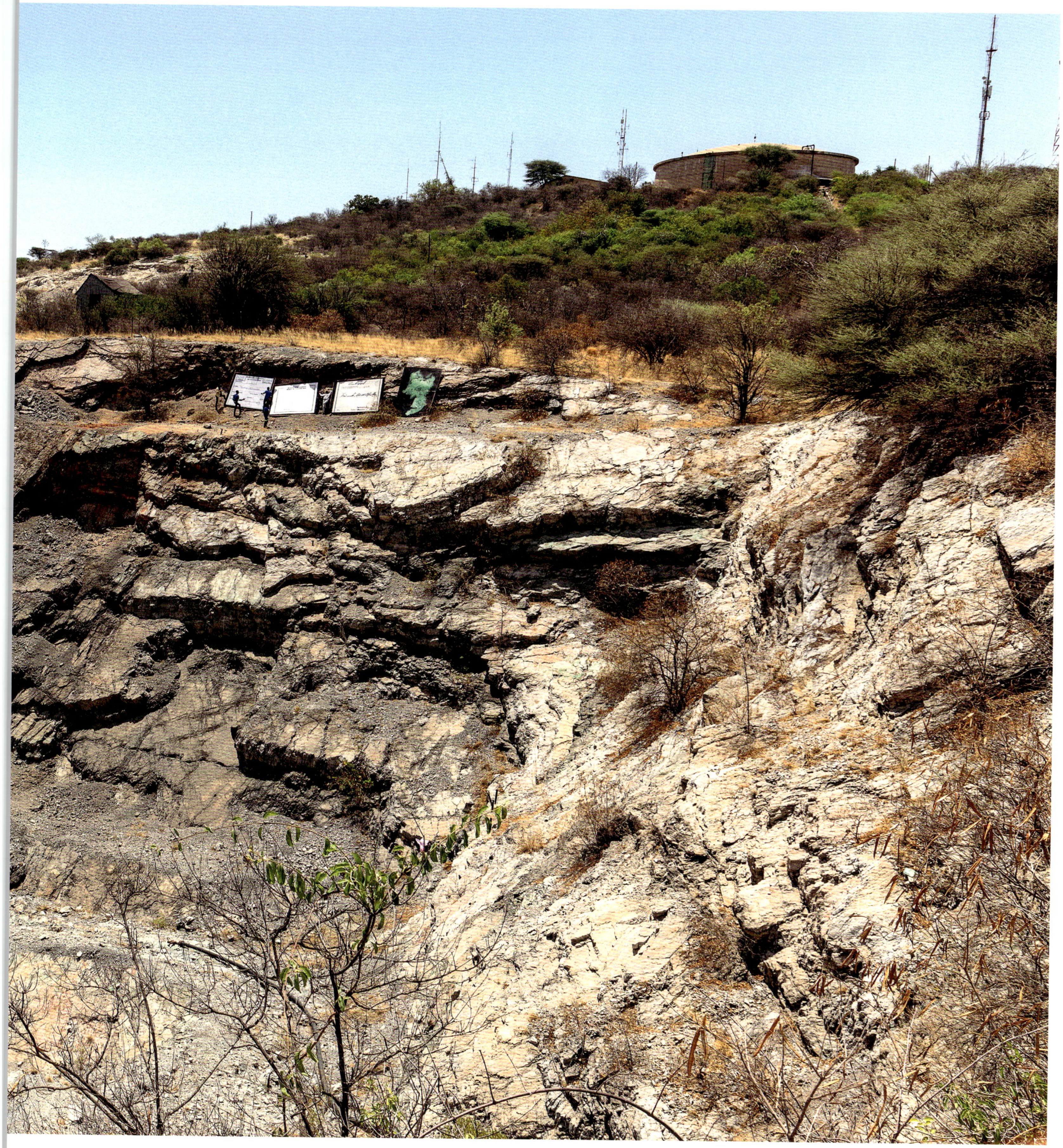

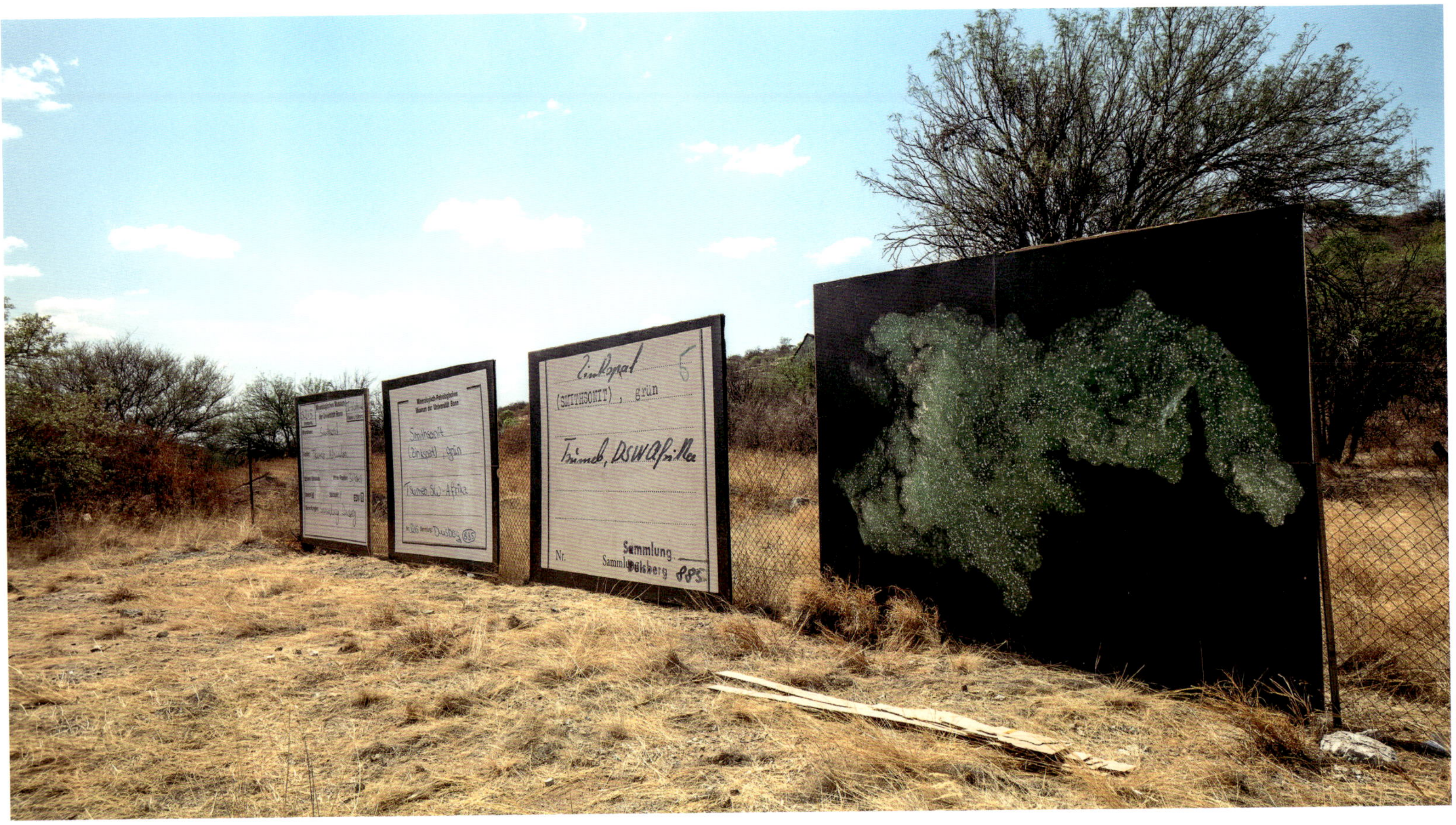
Sammlung

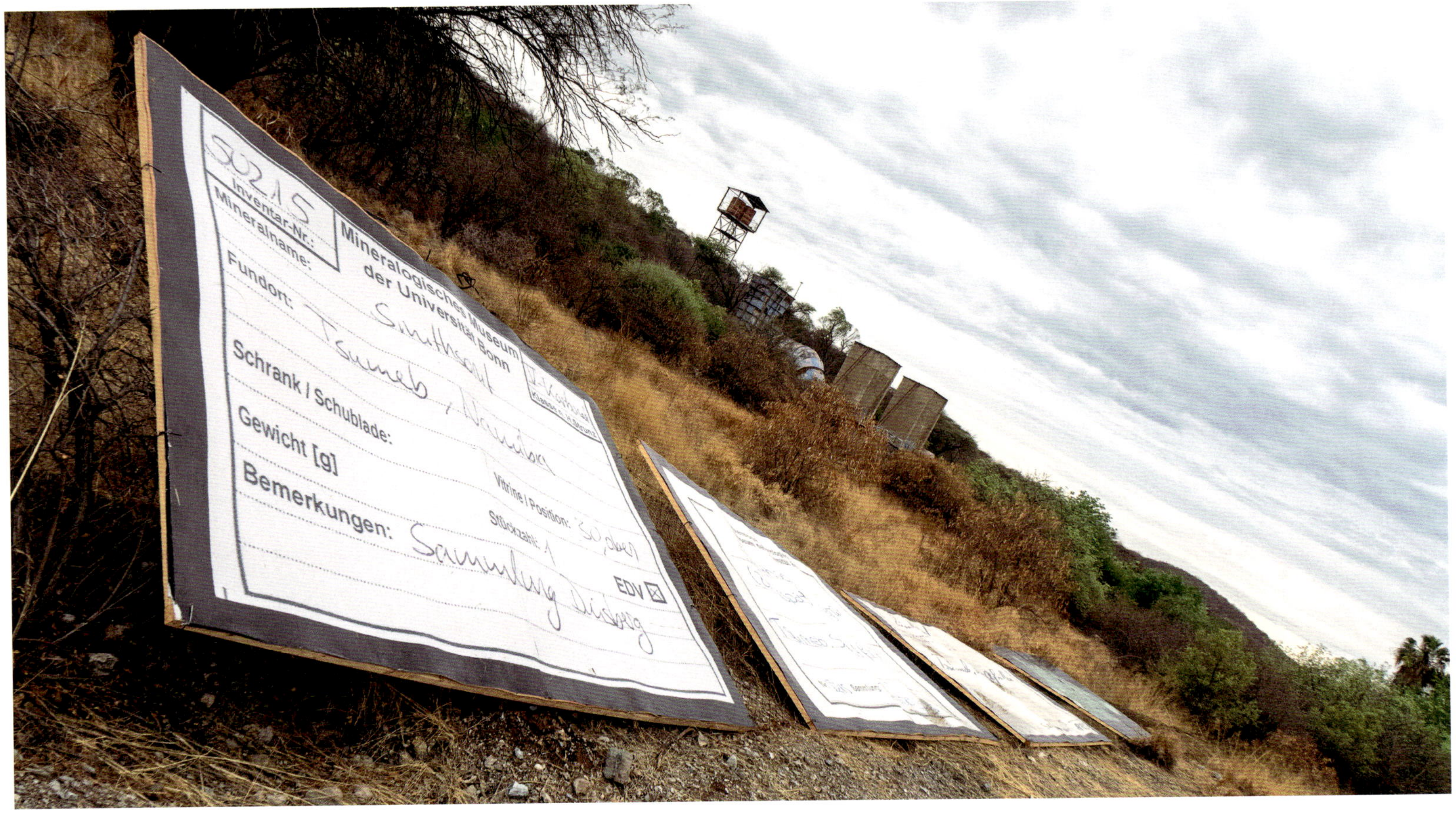
Mineralogisches Museum
der Universität Bonn
Inventar-Nr.:
Mineralname:
Fundort:
Schrank / Schublade:
Gewicht [g]
Bemerkungen:
Vitrine / Position:
Stückzahl:
EDV

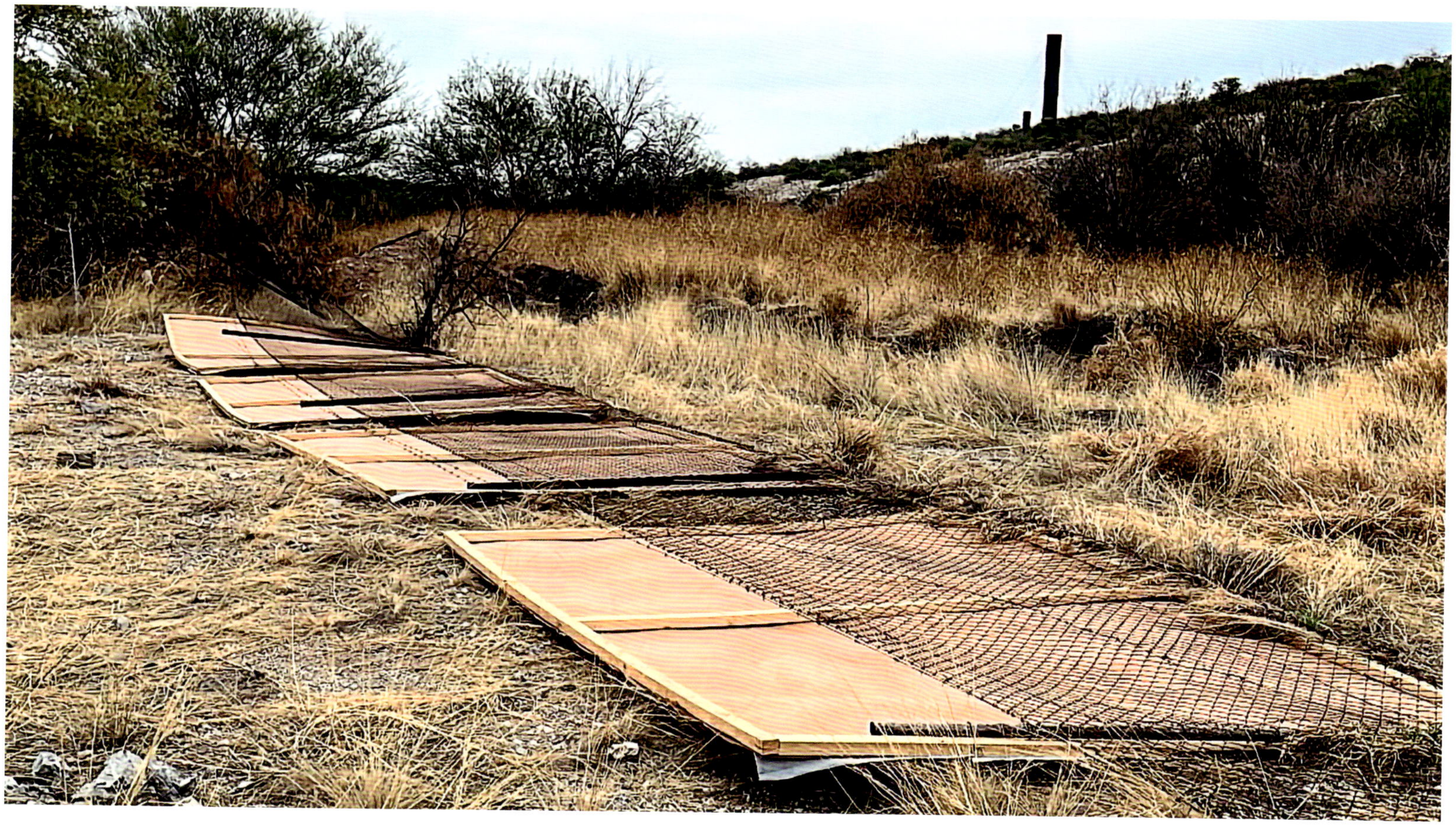

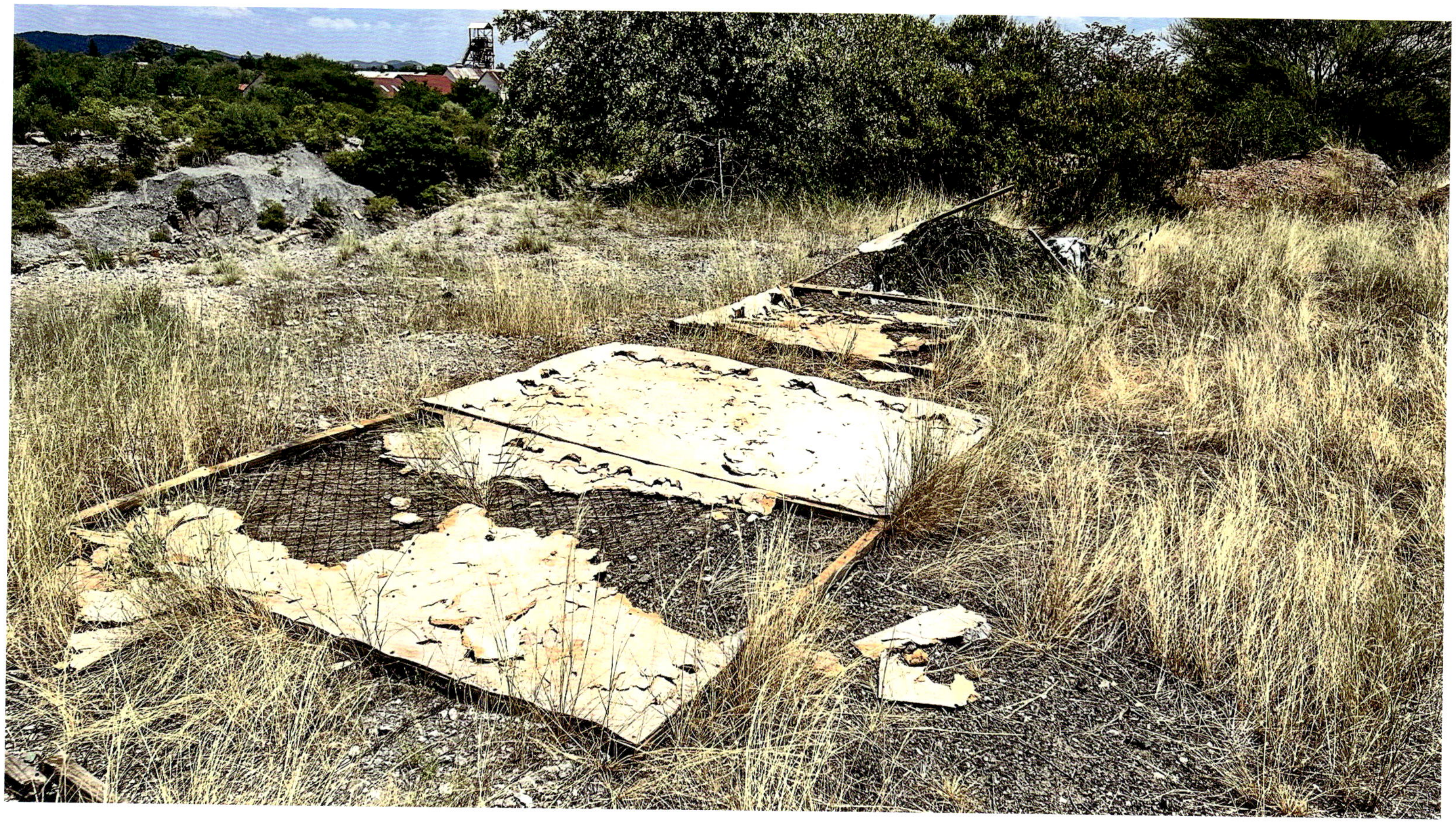

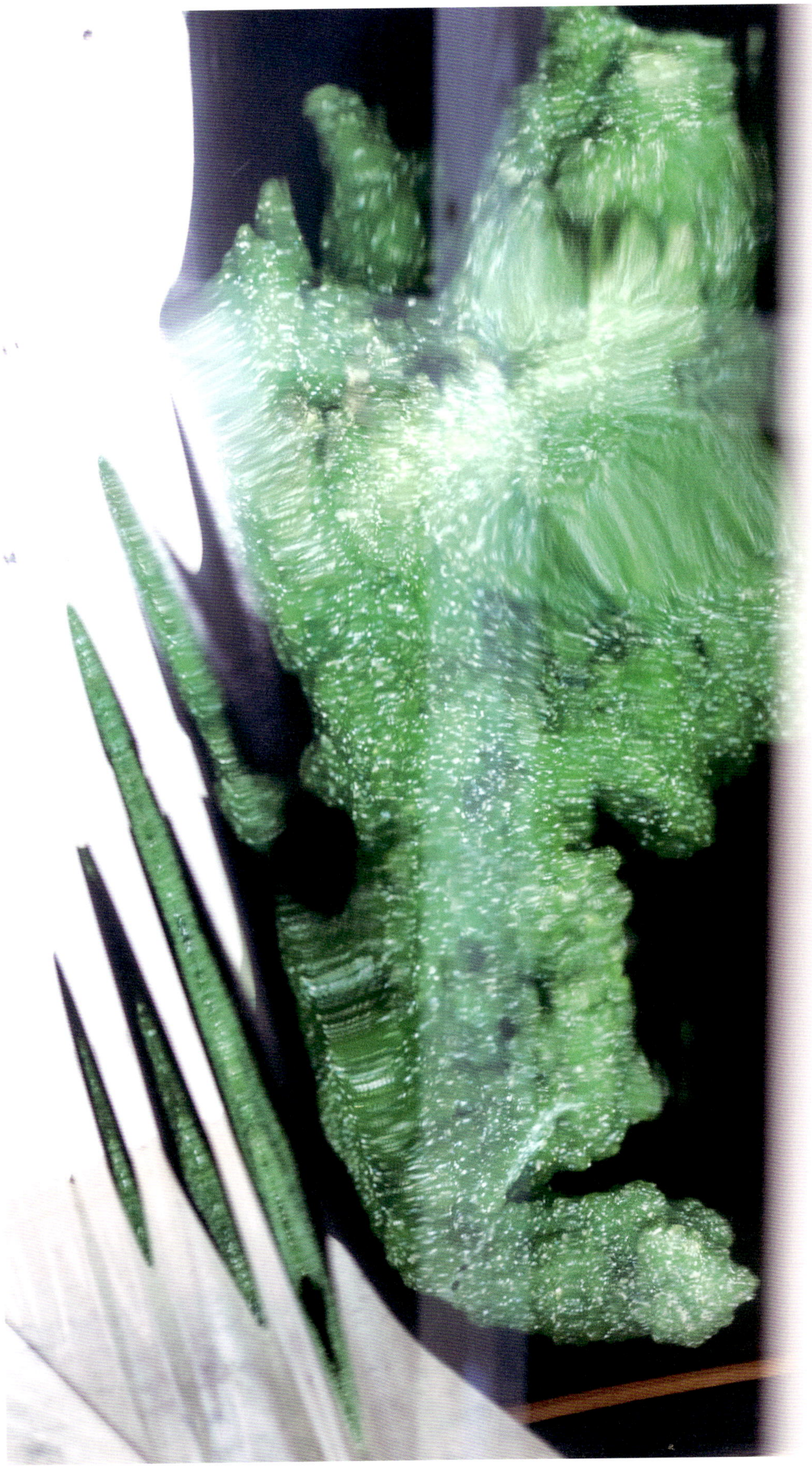

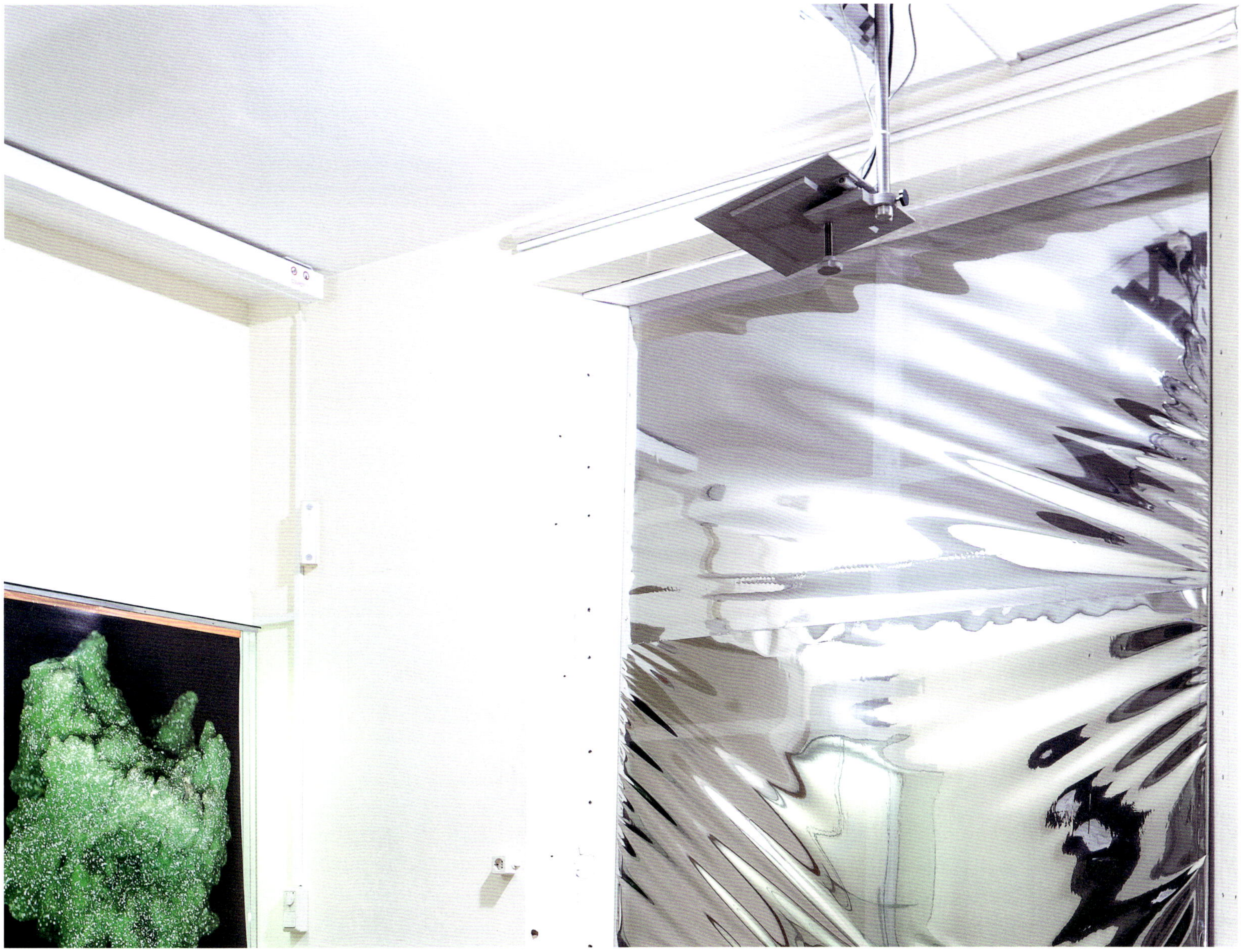

TG-T756
transposition 756
(Just Another Unexpected Guest)

Artist: Sascha Mikloweit
Year: 2023
Location: Mineralogical Museum of the University of Bonn, DE, Exhibition Hall 2, Special Exhibitions
Type: Sculptural Intervention
Dimensions: 182 x 112 x 3 cm (each; four panels)
Material: Special Exhibitions Hall, Mineralogical Museum of the University of Bonn; mylar mirror film; Molton; wood

The view out of the four windows of the Mineralogical Museum of the University of Bonn's Special Exhibitions Hall (Hall 2, see this volume, p.57) was obstructed by four mylar mirror film panels. The result was a distorted reflections of the institutional space itself, and of the persons visiting this space. The mylar mirror film was reused from the exhibition 'Just Another Unexpected Guest' which took place in 2012 in the framework of the 7th Liverpool Biennale at Static Gallery, Liverpool, GB (http://mikloweit.net/resources/spbr_P1_net2020.pdf).

Fig. 72

Fig. 73

TG-T504
transposition 504
(I—O_MMdUB-50215)

Artist: Sascha Mikloweit
Year: 2023
Location: Mineralogical Museum of the University of Bonn, DE, Exhibition Hall 2, Special Exhibitions
Type: Sculptural Intervention, Image—Object (I—O)
Dimensions: 208 x 130 x 43 cm
Material: Special Exhibitions Hall, Mineralogical Museum of the University of Bonn; museum vitrine; collection object #50215, O—I_MMdUB-50215; copper tape

Photographic print of O-I_MMdUB-50215, derived from collection object number 50215 of the Mineralogical Museum of the University of Bonn, mounted with copper tape onto a vitrine emptied specifically for this installation in the museum's Special Exhibitions Hall.

The print depicts a smithsonite mineral—the one veiled in *transposition 441* (this volume pp.192–195)—that was mined from the German colonial OMEG Mine (Otavi Mining and Railways Company (OMEG), later the TCL Mine) in Tsumeb approximately around 1900-1905. Subsequently it went into the possession of the German industrialist and I.G. Farbenindustrie AG / IG Farben founder Friedrich Carl Duisberg (1861–1935). During the artist's time at the institution in Bonn, the scattered records provided allegedly no information on how the object made its journey from the mine to the collection of Duisberg.

Bayer AG (one of the successor companies of IG Farben) outsourced Carl Duisberg's collection under a permanent loan contract to the Mineralogical Museum of the University of Bonn, Germany in the 1950s: it is still on display there at present.

Fig. 74

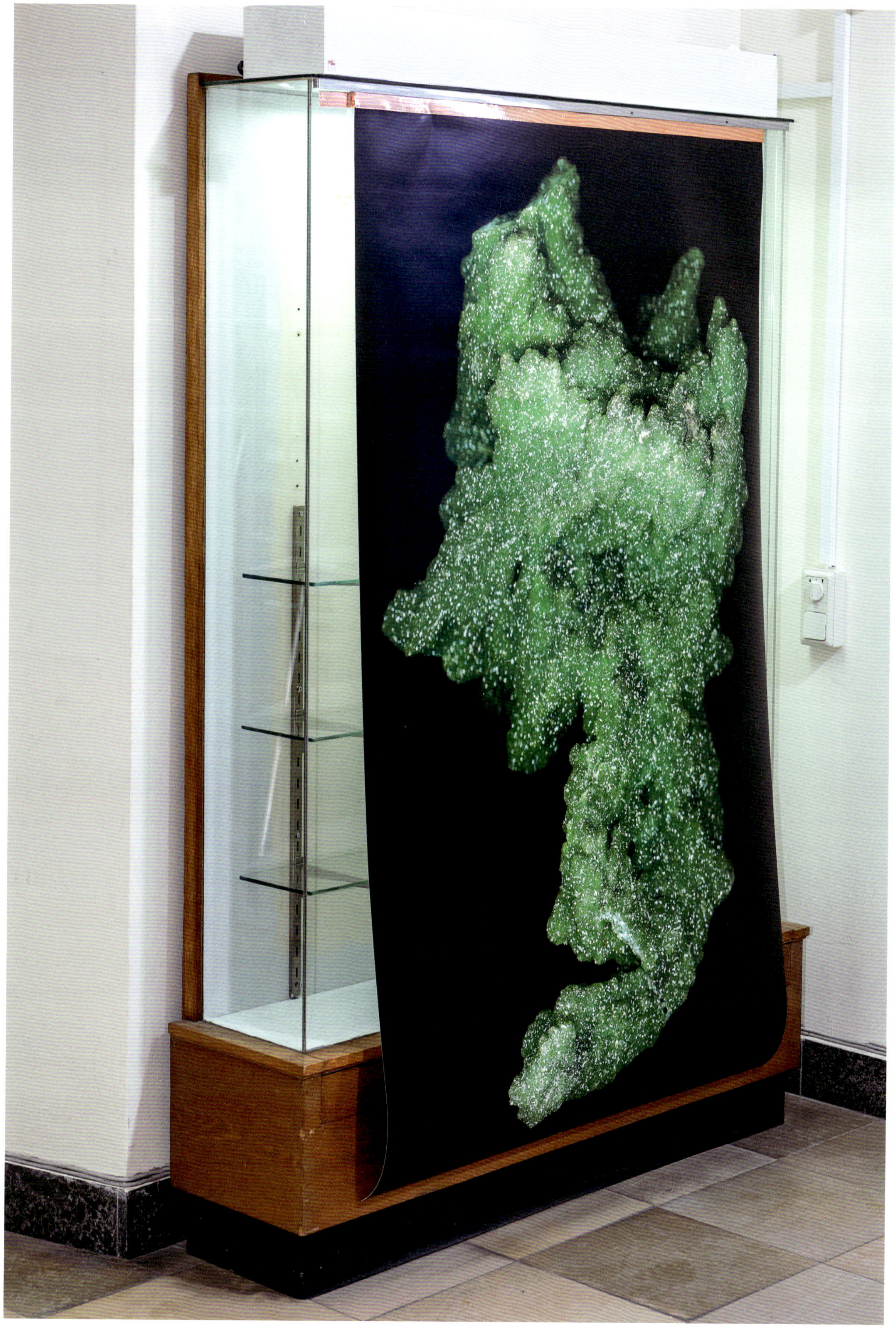

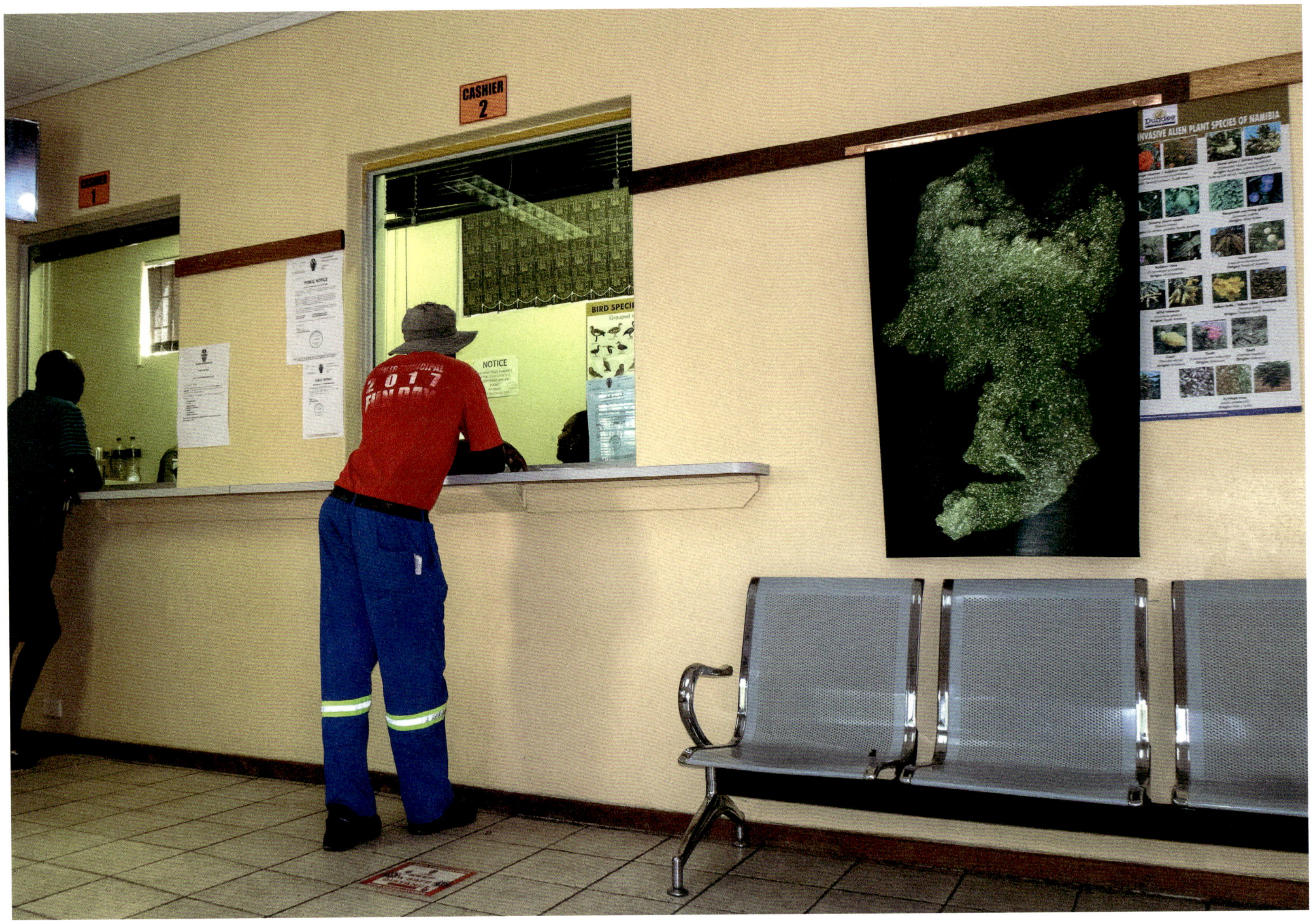
CASHIER
2
NOTICE
INVASIVE ALIEN PLANT SPECIES OF NAMIBIA

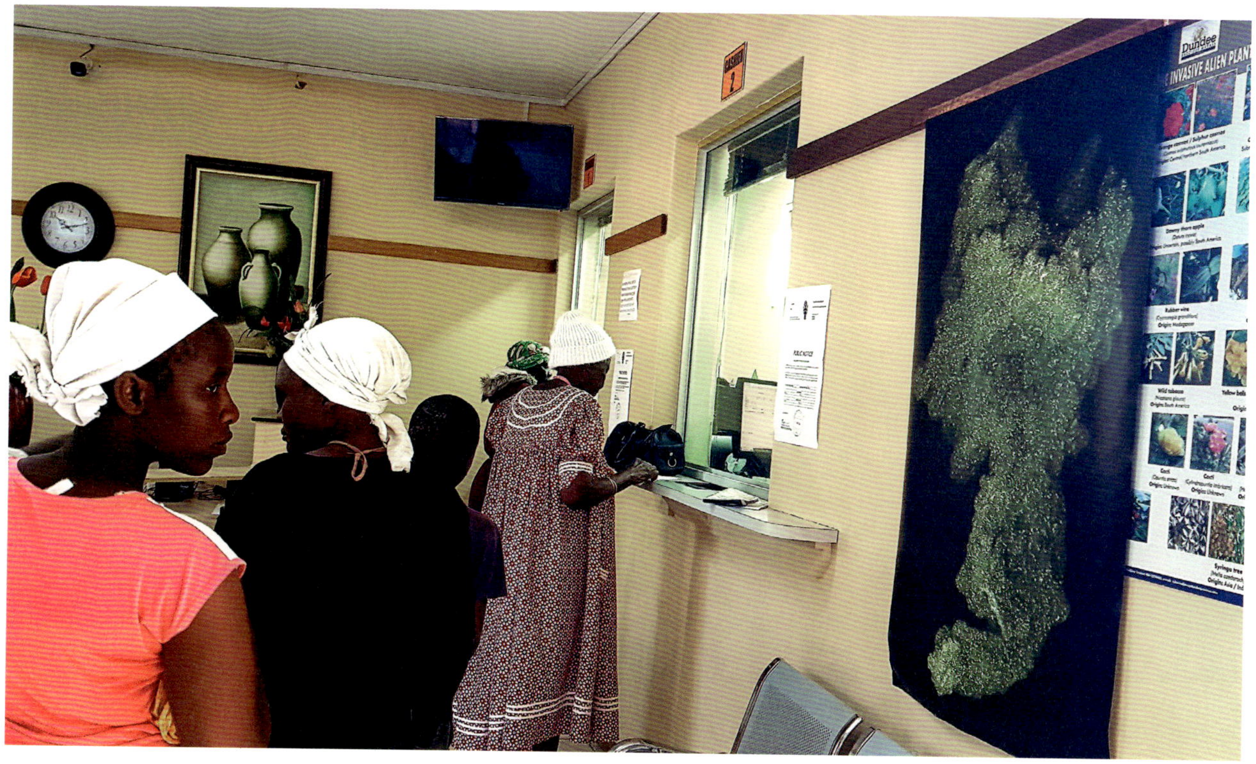

TG-T995
transposition 995
(I—O_MMdUB-50215@TM-NA)

Artist: Sascha Mikloweit
Year: 2023
Location: Tsumeb Municipality entrance hall, Tsumeb, NA
Type: Immersive installation, Image—Object (I—O)
Dimensions: 1500 x 450 x 310 cm, 122 x 74.2 x 4 cm
Material: Tsumeb Municipality entrance hall; O—I_MMdUB-50215; pigment print on baryta paper; Dundee Precious Metals Inc. 'Invasive Alien Plant Species Of Namibia' info board; copper tape

Photographic print of O—I_MMdUB-50215, an image obtained from collection object number 50215 of the Mineralogical Museum of the University of Bonn, mounted with copper tape, overlapping a Dundee Precious Metals Inc. corporate public relations campaign board (then the owner of the Tsumeb Specialty Smelter) that addresses its target audience with the message: 'Learn About Some Of The Invasive Alien Plant Species Of Namibia.' This permanent installation merges with the entrance hall of the Tsumeb Municipality main building to create an immersive installation.
The print depicts a smithsonite mineral—the one veiled in *transposition 441* (this volume pp.192–195)—that was mined from the German colonial OMEG Mine (Otavi Mining and Railways Company (OMEG), later the TCL Mine) in Tsumeb approximately around 1900-1905. Subsequently it went into the possession of the German industrialist and I.G. Farbenindustrie AG / IG Farben founder Friedrich Carl Duisberg (1861–1935). During the artist's time at the institution in Bonn, the scattered records provided allegedly no information on how the object made its journey from the mine to the collection of Duisberg.

Fig. 75

Fig. 76

Bayer AG (one of the successor companies of IG Farben) outsourced Carl Duisberg's collection under a permanent loan contract to the Mineralogical Museum of the University of Bonn, Germany in the 1950s: it is still on display there at present.

TG-T741
transposition 741
(We Come and Die as Numbers)

Artist: Sascha Mikloweit
Year: 2023
Location: Mineralogical Museum of the University of Bonn, DE, Exhibition Hall 2, Special Exhibitions
Type: Sculptural Intervention, Scratching
Dimensions: 126 x 226 x 50 cm
Material: Poem; glass; vitrine of the Special Exhibitions Hall, Mineralogical Museum of the University of Bonn

Fig. 77

Fig. 78 [p.210]

The poem *We Come and Die as Numbers* by the late Namibian poet Simon Zhu Mbako (1950–1994) was scratched—with a diamond—into a glass panel of an emptied exhibition vitrine in the Special Exhibitions Hall of the Mineralogical Museum of the University of Bonn. The poem was originally published in: Mbako, S. Z., 1989. *We Come and Die as Numbers*. In: Mbako, S. Z., ed. *Tell Them of Namibia: Poems from the National Liberation Struggle*. London: Karia Press.

WHEN WE
BY THE
WE ARE SENT
WHEN WE ARE
WE ARE DENIED
FOR IT IS NOT
WHEN THE
WE ARE SENT
AS NUMBERS.

E SUCKED DRY
ROUS SPEED OF WOR
O THE RESERVATIONS
ICK OF POISONOUS
MPENSATION —
RTH NUMBERS
DUST EATS AWA
OUR GR

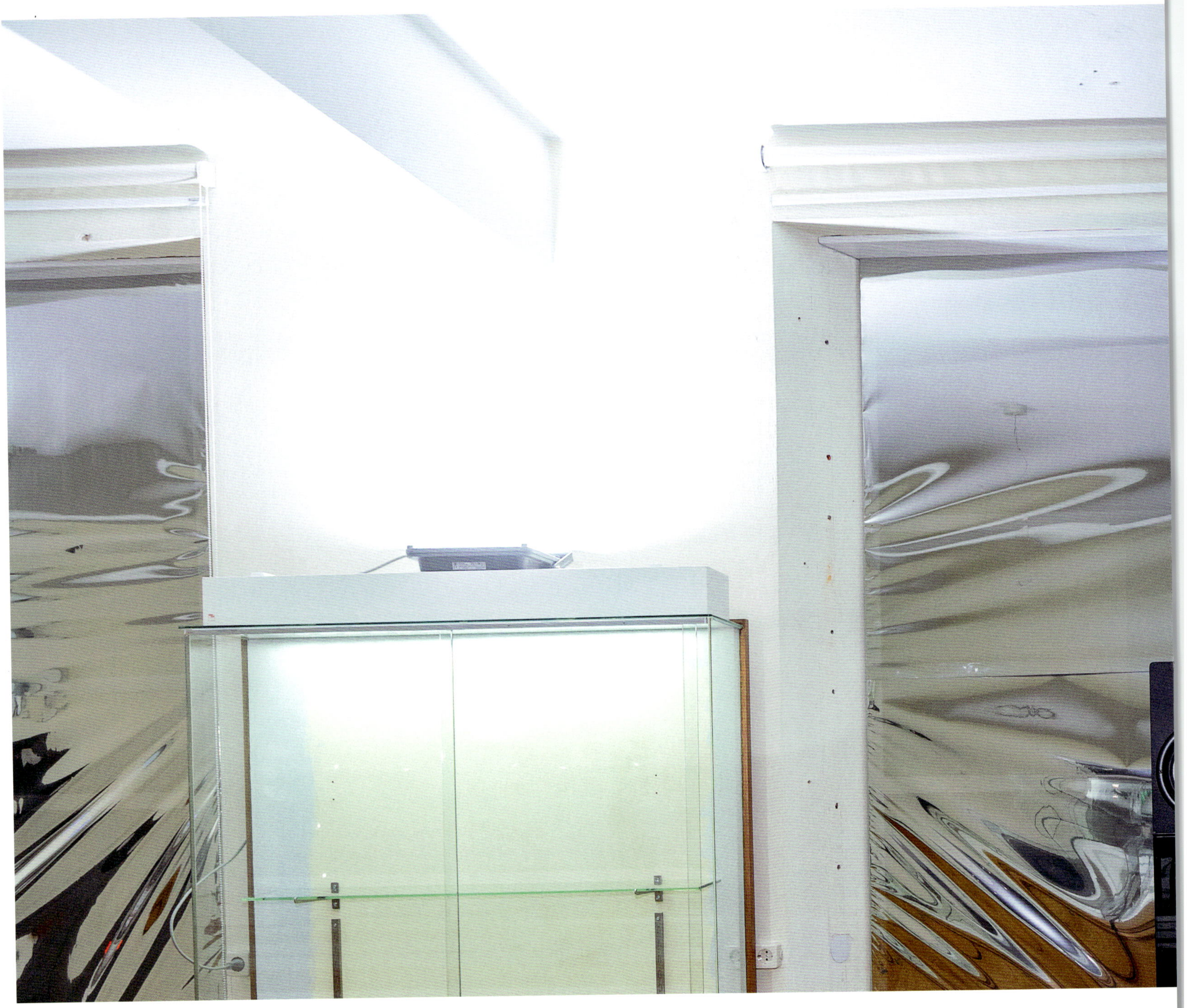

Fig. 79

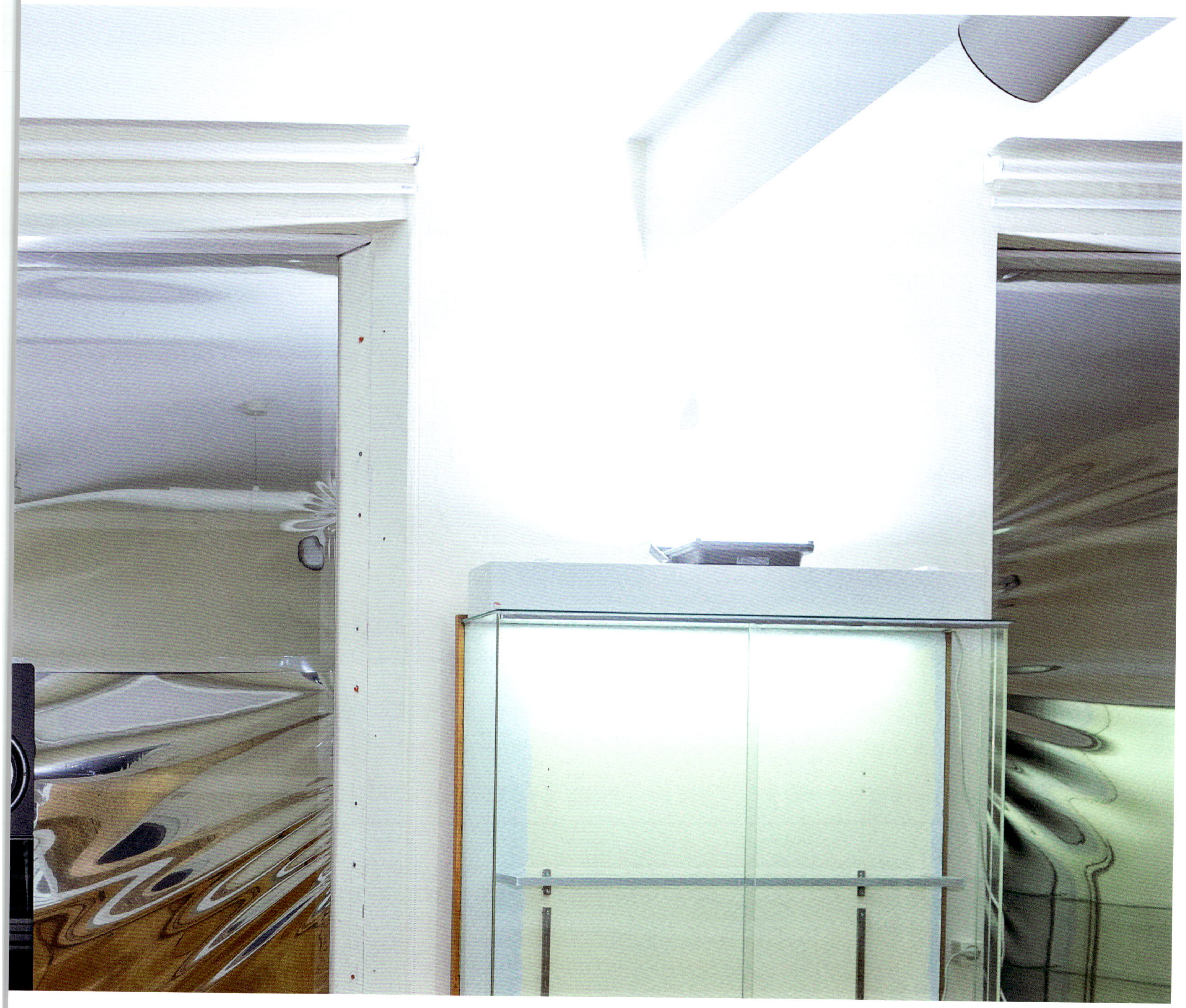

TG-T368
transposition 368
(The Future Sound of Destruction)

Artist: Sascha Mikloweit
Collaborator: Paul Philipp Heinze
Year: 2023
Location: Mineralogical Museum of the University of Bonn, DE, Exhibition Hall 2, Special Exhibitions
Type: Sound Installation
Dimensions: 84 x 25 x 41 cm (each element, overall dimension variable)
Duration: 00:05:00, loop
Material: Sound recordings from the 'Crushing Room', Raspberry Pi, sound card, speaker, glass

Sound installation based on recordings from the basement of the Institute of Geosciences, Department of Geochemistry/Petrology's 'Crushing Room'. The sounds of two machines—the 'Torpedo Crusher' and the *Scheibenschwingmühle* (vibrating disc mill)—were recorded while running empty. These recordings were modulated into a four-channel sound installation on a five-minute loop, with 20 seconds of silence before each restart. The soundscape produced its most intense effect acoustically when standing in front of the exhibition vitrine designated as *transposition 741* (this volume, pp. 208–211; see also graphic on p. 57), creating a sensation of being acoustically pushed toward the vitrine. On 5 June 2023, these machines in the 'Crushing Room' were used to grind collection object #50592 (originally from the colonial Tsumeb mining operation) to dust. This performance became part of transposition 015 (this volume, pp. 228–245).

Fig. 80

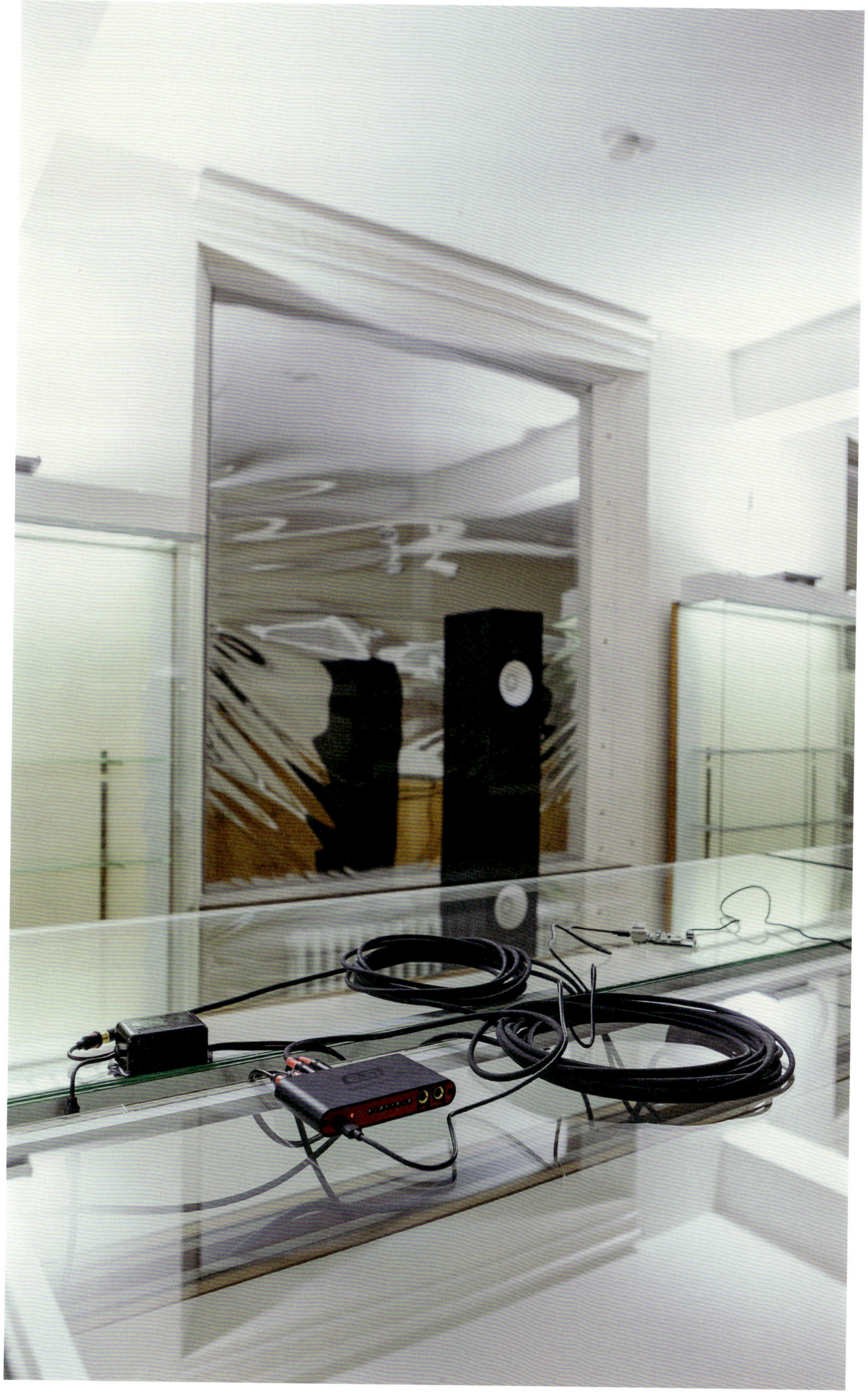

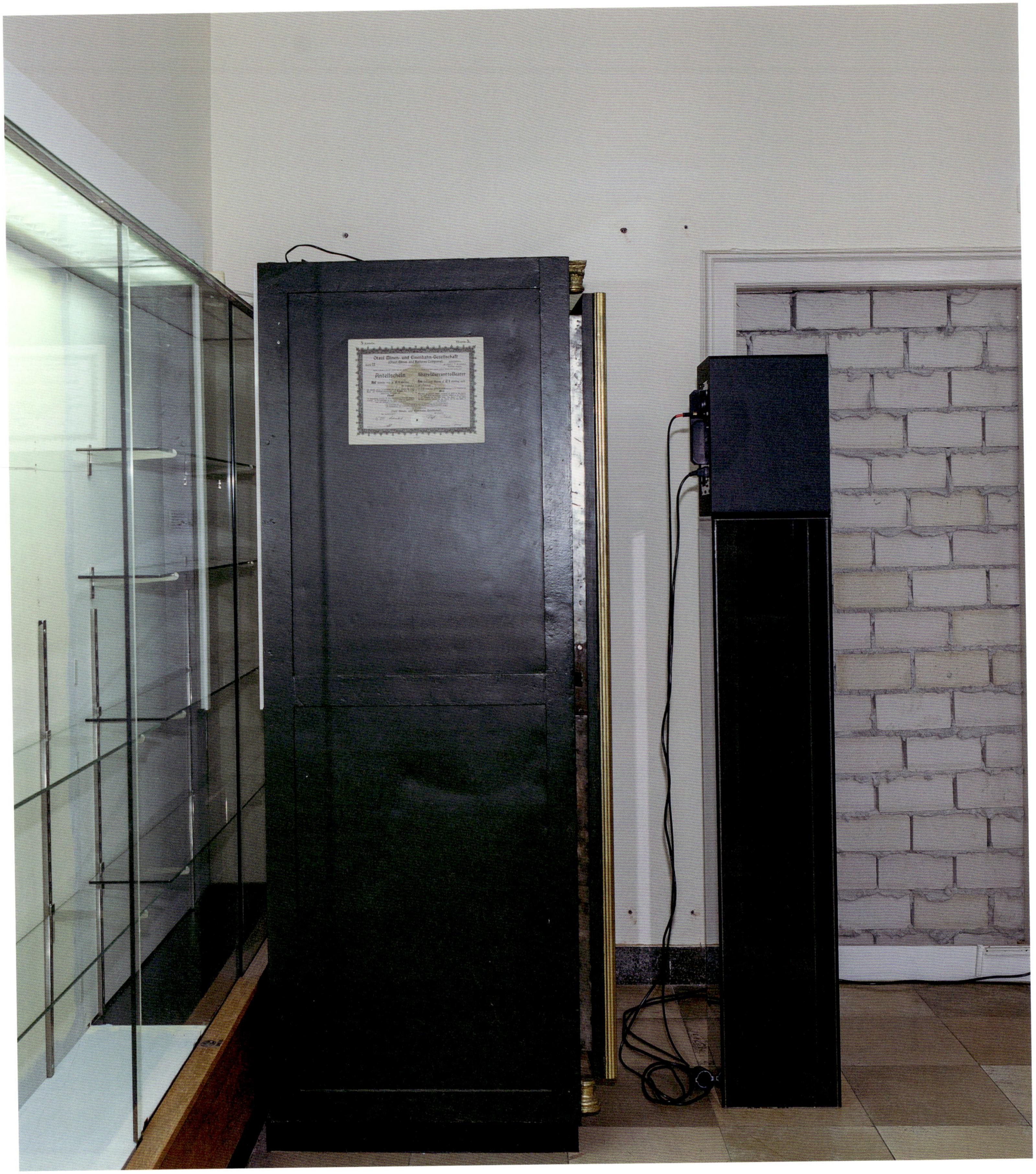

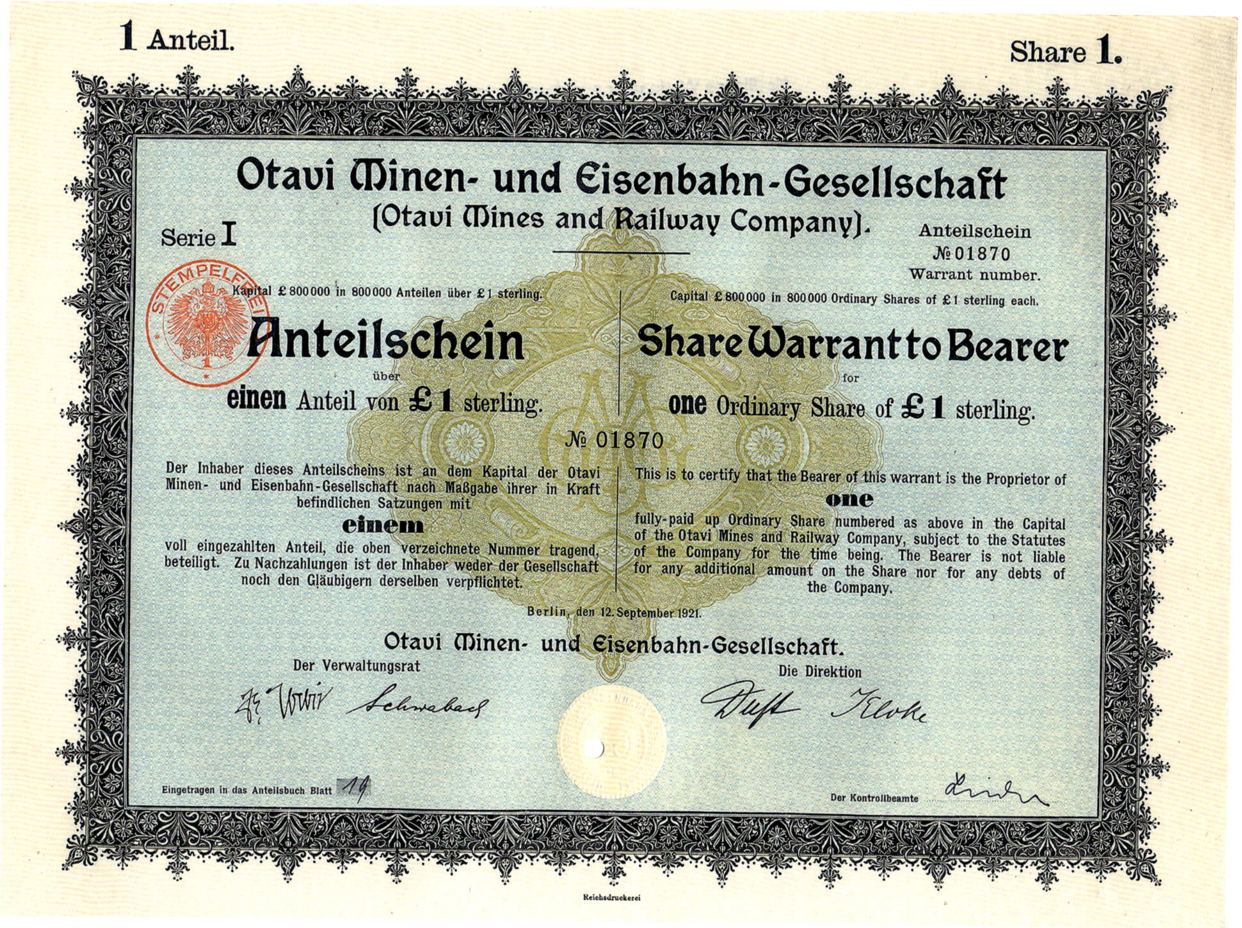
1 Anteil.

Share 1.

Otavi Minen- und Eisenbahn-Gesellschaft
(Otavi Mines and Railway Company).

Serie I

Anteilschein
№ 01870
Warrant number.

STEMPELFREI

Kapital £ 800 000 in 800 000 Anteilen über £ 1 sterling.

Capital £ 800 000 in 800 000 Ordinary Shares of £ 1 sterling each.

Anteilschein
über
einen Anteil von £ 1 sterling.

Share Warrant to Bearer
for
one Ordinary Share of £ 1 sterling.

№ 01870

Der Inhaber dieses Anteilscheins ist an dem Kapital der Otavi Minen- und Eisenbahn-Gesellschaft nach Maßgabe ihrer in Kraft befindlichen Satzungen mit
einem
voll eingezahlten Anteil, die oben verzeichnete Nummer tragend, beteiligt. Zu Nachzahlungen ist der Inhaber weder der Gesellschaft noch den Gläubigern derselben verpflichtet.

This is to certify that the Bearer of this warrant is the Proprietor of
one
fully-paid up Ordinary Share numbered as above in the Capital of the Otavi Mines and Railway Company, subject to the Statutes of the Company for the time being. The Bearer is not liable for any additional amount on the Share nor for any debts of the Company.

Berlin, den 12. September 1921.

Otavi Minen- und Eisenbahn-Gesellschaft.

Der Verwaltungsrat

Die Direktion

Eingetragen in das Anteilsbuch Blatt 14

Der Kontrollbeamte

Reichsdruckerei

TG-T665
transposition 665
(OMEG)

Artist: Sascha Mikloweit
Year: 2023
Location: Mineralogical Museum of the University of Bonn, DE, Exhibition Hall 2, Special Exhibitions
Type: Sculptural Intervention
Dimensions: 243 x 100 x 121 cm
Material: Otavi Mining and Railways Company (OMEG) share No 01870, Series 1, Berlin, 1921.; safe;
Special Exhibitions Hall, Mineralogical Museum of the University of Bonn

Original Otavi Mining and Railways Company (OMEG) share attached to an antique safe.

Fig. 81

Fig. 82

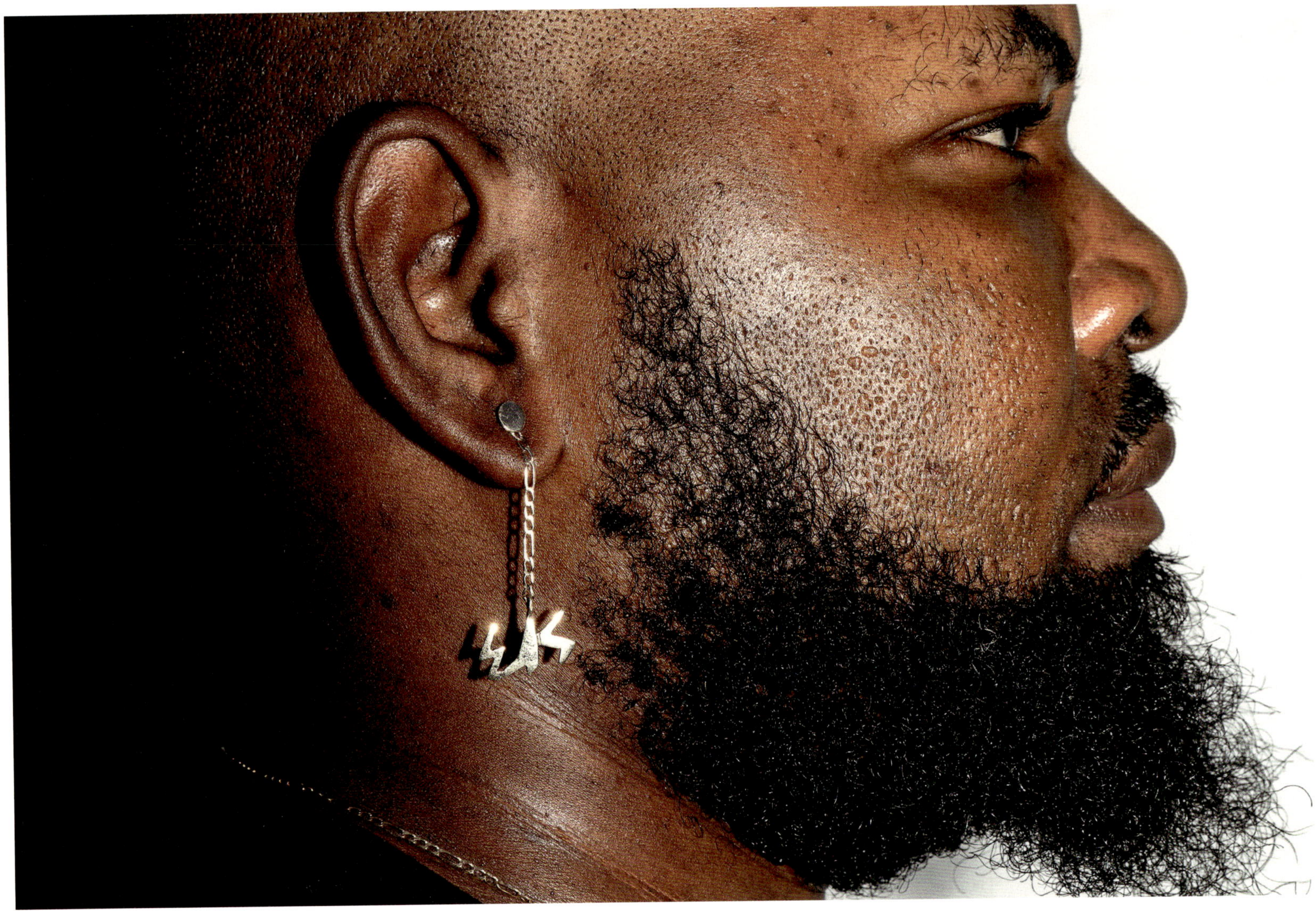

TG-T610
transposition 610
(US / THEM)

Artist: Sascha Mikloweit
Collaborator: Richardlee Shoombe Jewellery Design, Manuel Pandalis Photography
Year: 2023
Location: Tsumeb Municipality, Tsumeb, Otjikoto, NA
Type: Jewellery (Prototype)
Dimensions: 5 x 2 x 0.1 cm, 5 x 3 x 0.1 cm
Material: Gold, B2Gold, Otjikoto Mine, Namibia
(see also this volume, pp.222–225)

During our first research visit to Namibia in 2022, two observations became central to this work: 1) The majority of private properties, especially—but not exclusively—in affluent white areas, were secured with elaborate high-voltage electric fencing; and 2) in most conversations during our three-week stay, the question was not *if* but rather *when* our conversation partners would shift to a terminology of 'Us' and 'Them,' applied in a divisive, othering manner. These two experiences became the foundation of this jewellery piece. While Namibia's high-voltage electric fences physically divide the world into 'Us' and 'Them,' the true origin of such divisions lies in our minds—where hopes, abilities, limitations, dogmas, ideologies, ignorances, and fears have forged an exclusive, skewed version of humanism for the few. These earrings, with their form and material, encapsulate this division: a tangible reminder of the invisible lines we draw, both around us and within us.

Fig. 83

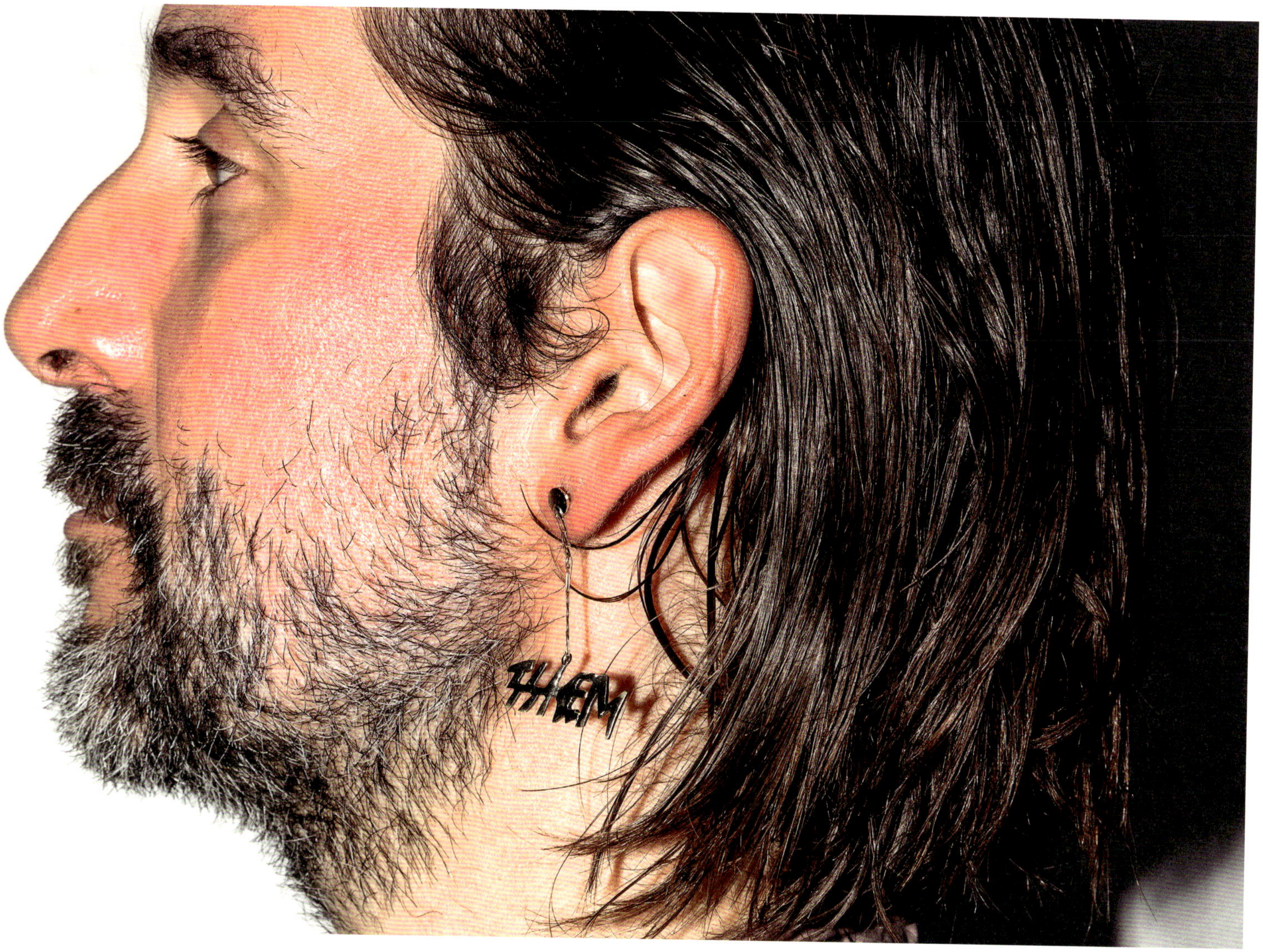

When I returned one year later for my residency and exhibition planned with the Tsumeb Municipality, I had the pleasure of meeting the jewellery designer Richardlee Shoombe. Together, we decided to bring this concept to life as a collaborative piece—an object that is both a symbol and a practice of healing in the face of History. This collaboration then evolved to include German-Greek photographer Manuel Pandalis, who captured portraits of Richardlee and me wearing the earrings, gazing at one another—a moment of reflection and confrontation.

In the context of *Transpositional Geologies*, however, this piece is not solely limited to a human-centric reading. Its material's journey from ore extraction to refined product extends the 'Us / Them' divide, evoking the Enlightenment-era separation of Nature and Culture. In this expanded perspective, the jewellery piece serves as a reminder of the ideological and material structures that shape human relations with the Earth—a relationship in which the exploitation of its riches without restraint has been, and is, legitimised through logics of superiority and subordination.

Fig. 84 [pp. 220–221]

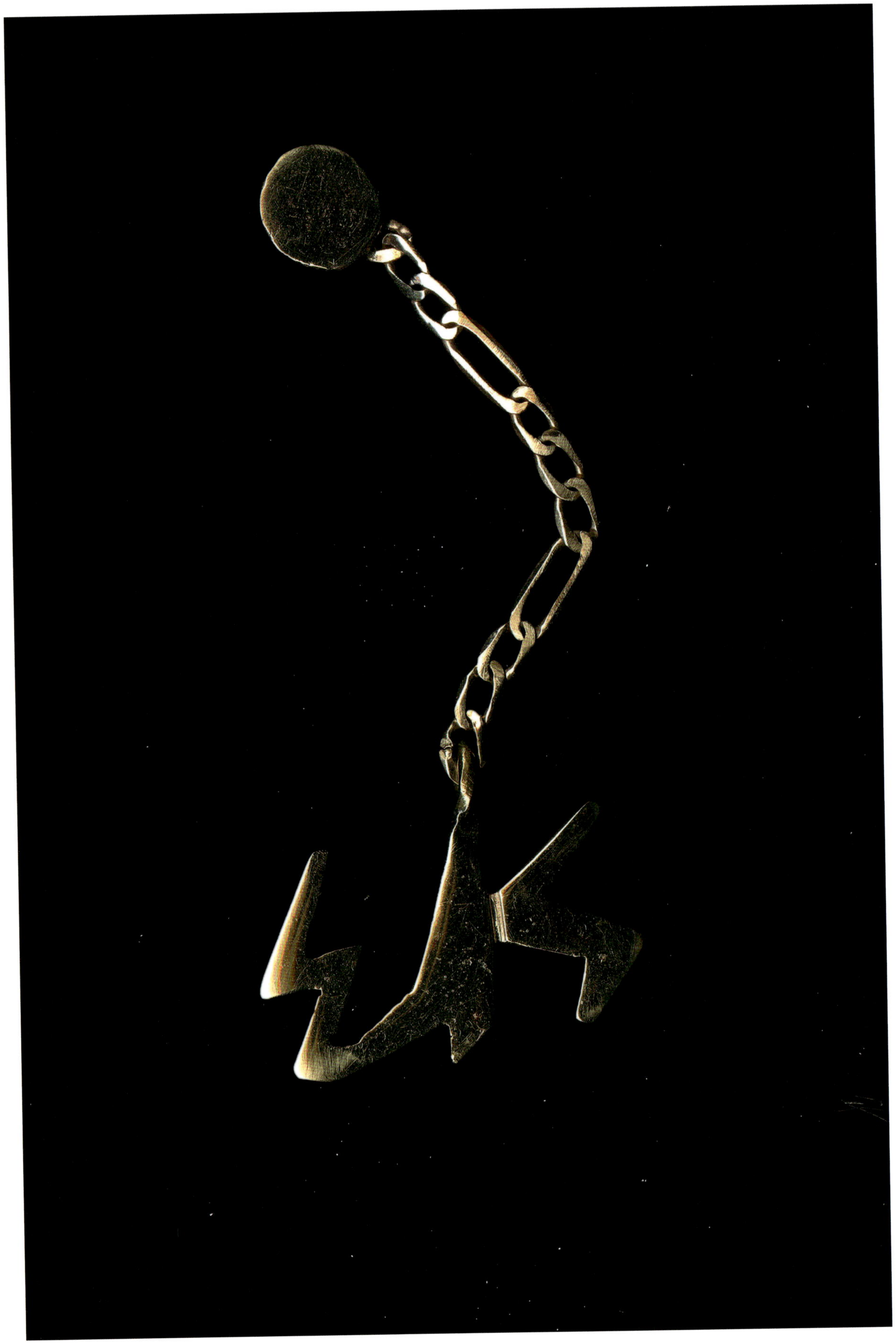

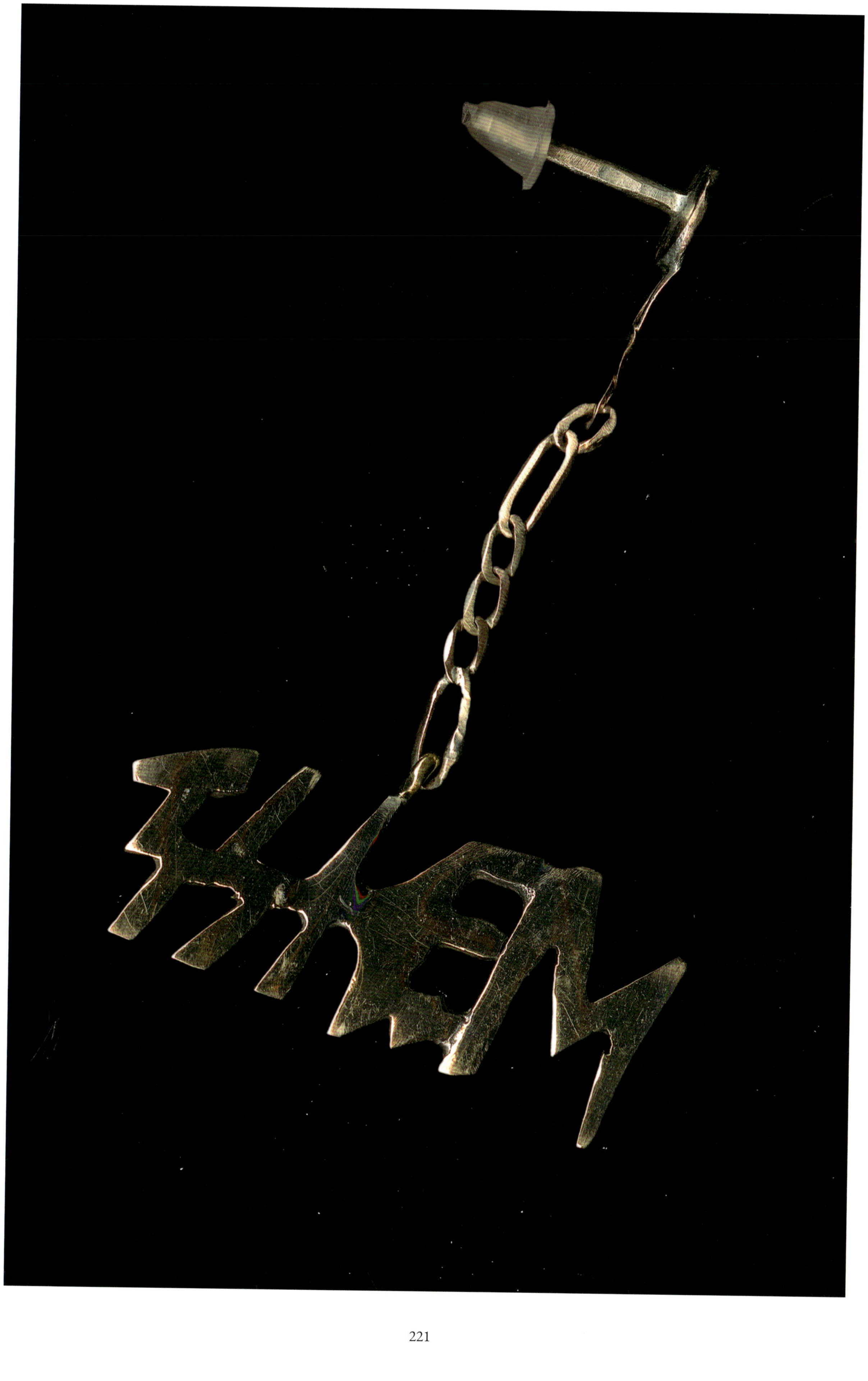

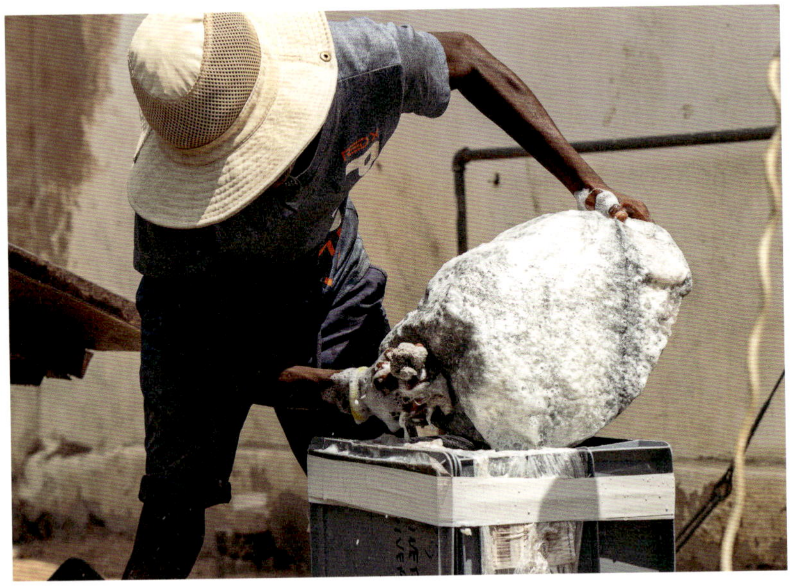

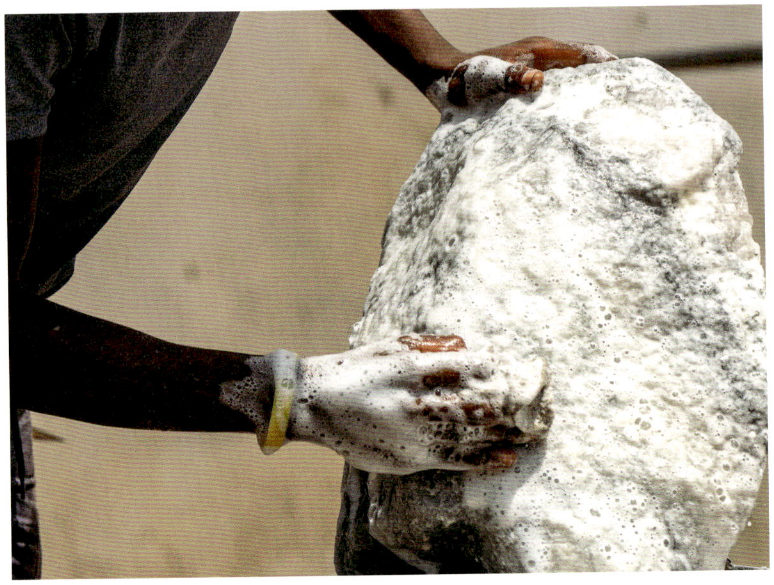

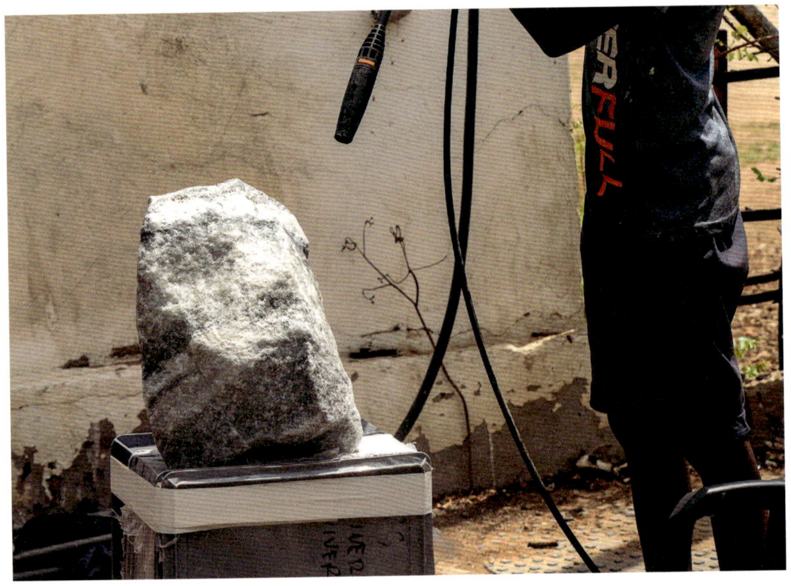

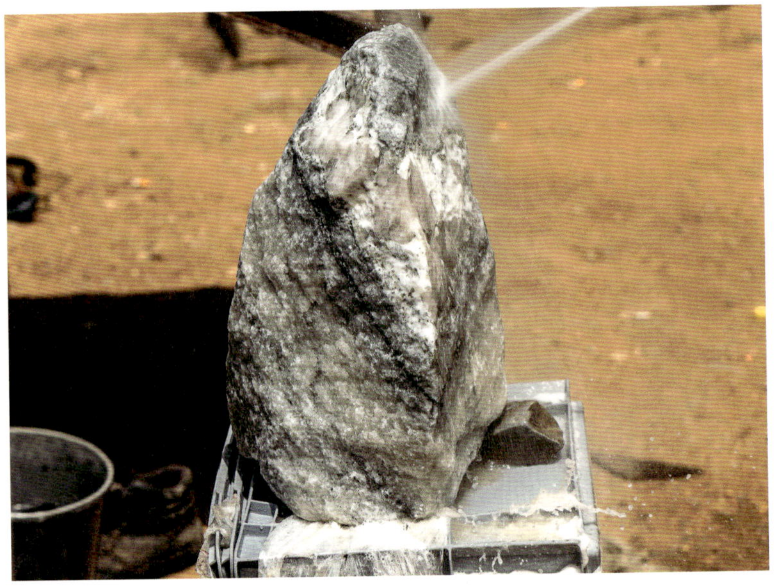

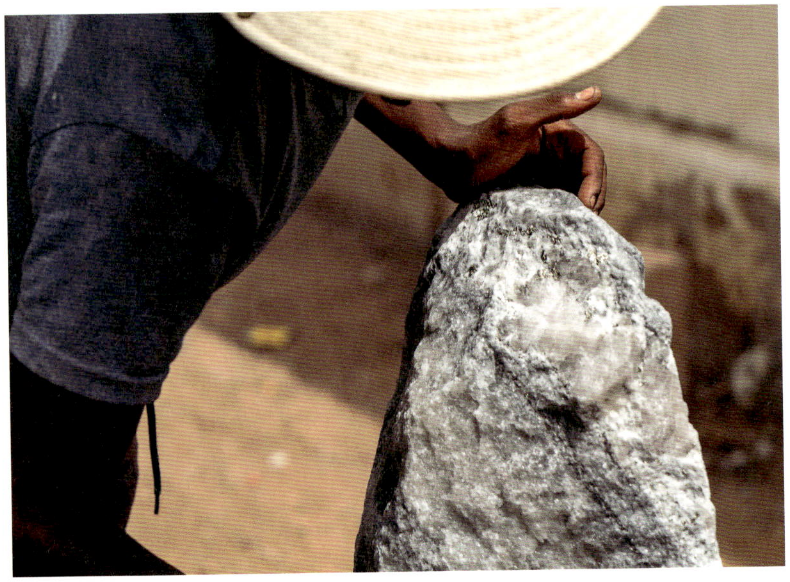

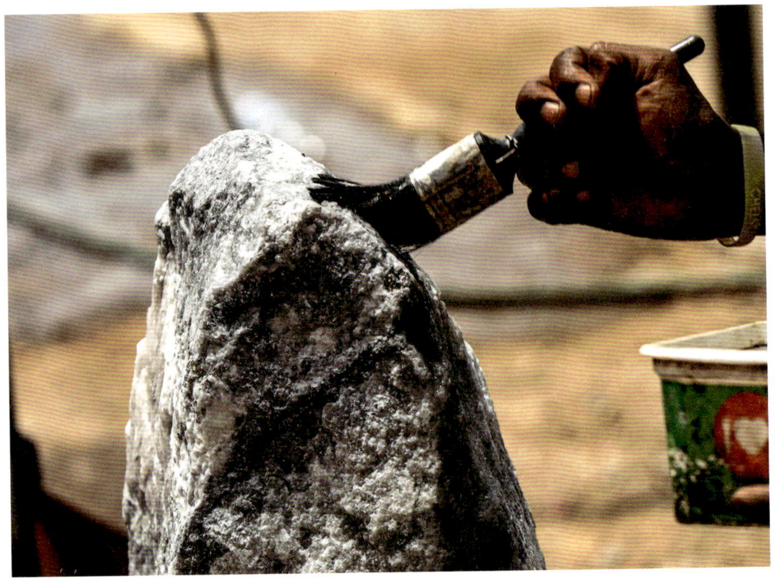

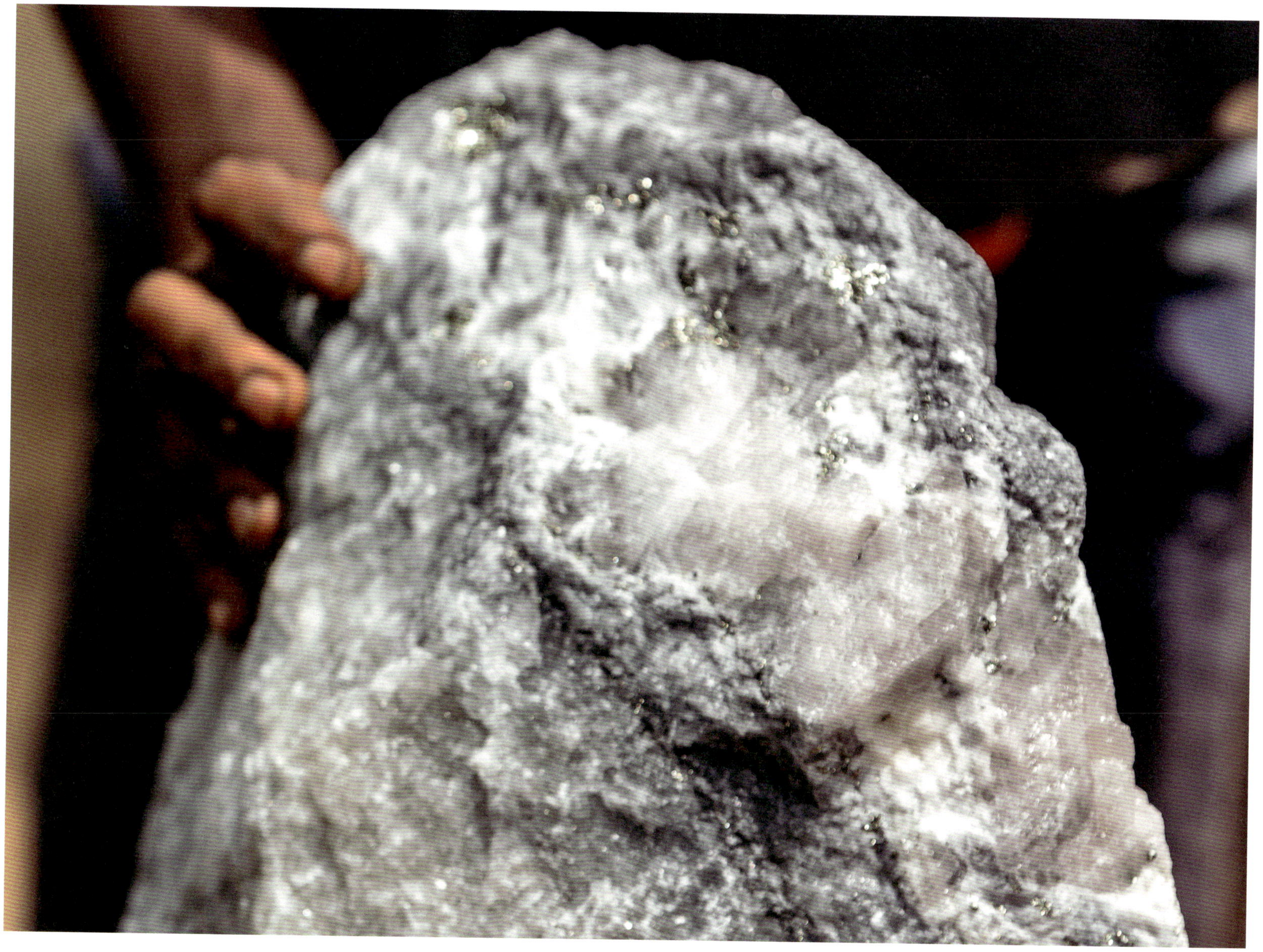

TG-T521
transposition 521
(Does Not Contain B2Gold)

Artist: Sascha Mikloweit
Collaborator: Medala's Car Wash , Manuel Pandalis
Year: 2023
Location: Medala's Car Wash, and Steinbach Beergarden, Tsumeb, Otjikoto, NA
Type: Performance, Process Sculpture, Administrative Negotiation, Administrative Documents, Photography
Dimensions: 46 x 32 x 28 cm
Duration: 06:00:00h (performance)
Material: B2Gold Otjikoto Mine, Namibia; B2Gold marble tailings fragment; shampoo; liquid car wax

Fig. 85–90

Fig. 91

During our research visit to the B2Gold mine (Otjikoto, Namibia, about 120km SSW of Tsumeb) on 21 October 2023, we encountered an incidental detail that became the starting point for this work: this vast open-pit mining operation, which managed to exhaust its targeted resource—gold—in just about 10 years, produces marble tailings as a waste product. These tailings are deemed unsuitable for commercial trade due to the immense detonation forces used in the mining process, which leave the marble's structure brittle and unfit for controlled processing.

When we inquired whether it would be possible to obtain two small pieces of this marble, we were informed that, under current Namibian law, it is illegal to remove even commercially irrelevant marble tailings from the designated territory of the B2Gold mine. The law presumes that such tailings might still contain traces of gold. Possession of these tailings, therefore, constitutes a criminal act in Namibia

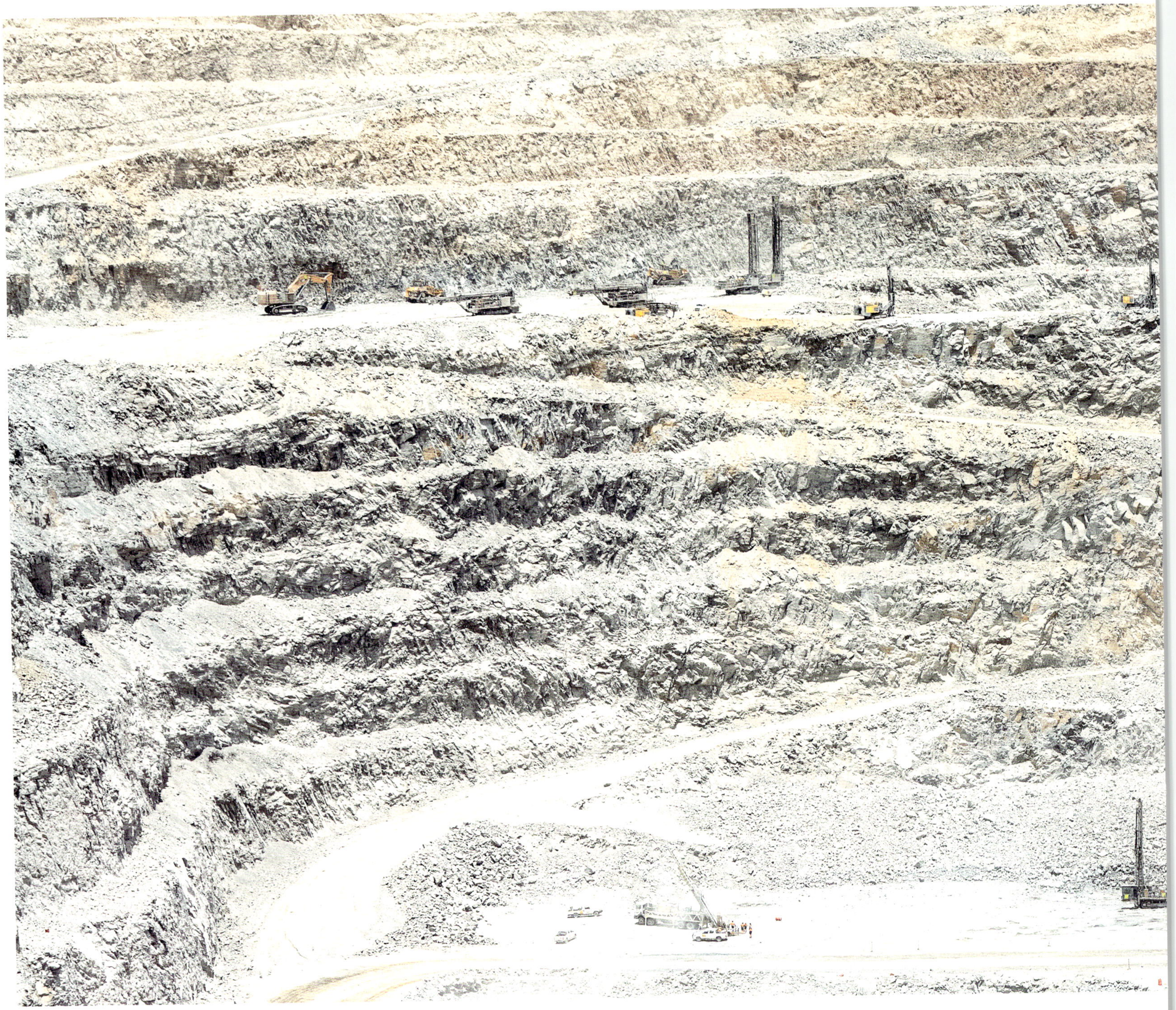

unless they have been inspected and certified by a geologist to contain no gold residues, in which case ownership becomes legitimate outside of the mine's premises.

In the days following our visit, we were able to negotiate to obtain two pieces of marble tailings, and B2Gold agreed to put them through the legally-required inspection and certification process. Three days later, the pieces were delivered to Tsumeb, accompanied by the required paperwork. One piece was passed on to the Namibian sculptor Kambezunda Ngavee, while the other became the process sculpture *transposition 521*.

Jenny, the owner of Medala's Car Wash, located directly across the road from Richardlee Shoombe's jewellery workshop (this volume, pp. 218–221), immediately expressed interest in getting involved. We agreed to create a performance piece that took place over six hours on 2 November 2023. Once per hour, the B2Gold marble specimen received the standard treatment offered to car wash customers: shampooing, high-pressure cleaning, and hot wax finishing. At the end of the six treatments, the marble was delivered to, and exhibited at, Steinbach Beergarden as part of *transposition 554 (Before Thought Hardens Into Language II)* (this volume, pp. 226–

Fig. 92

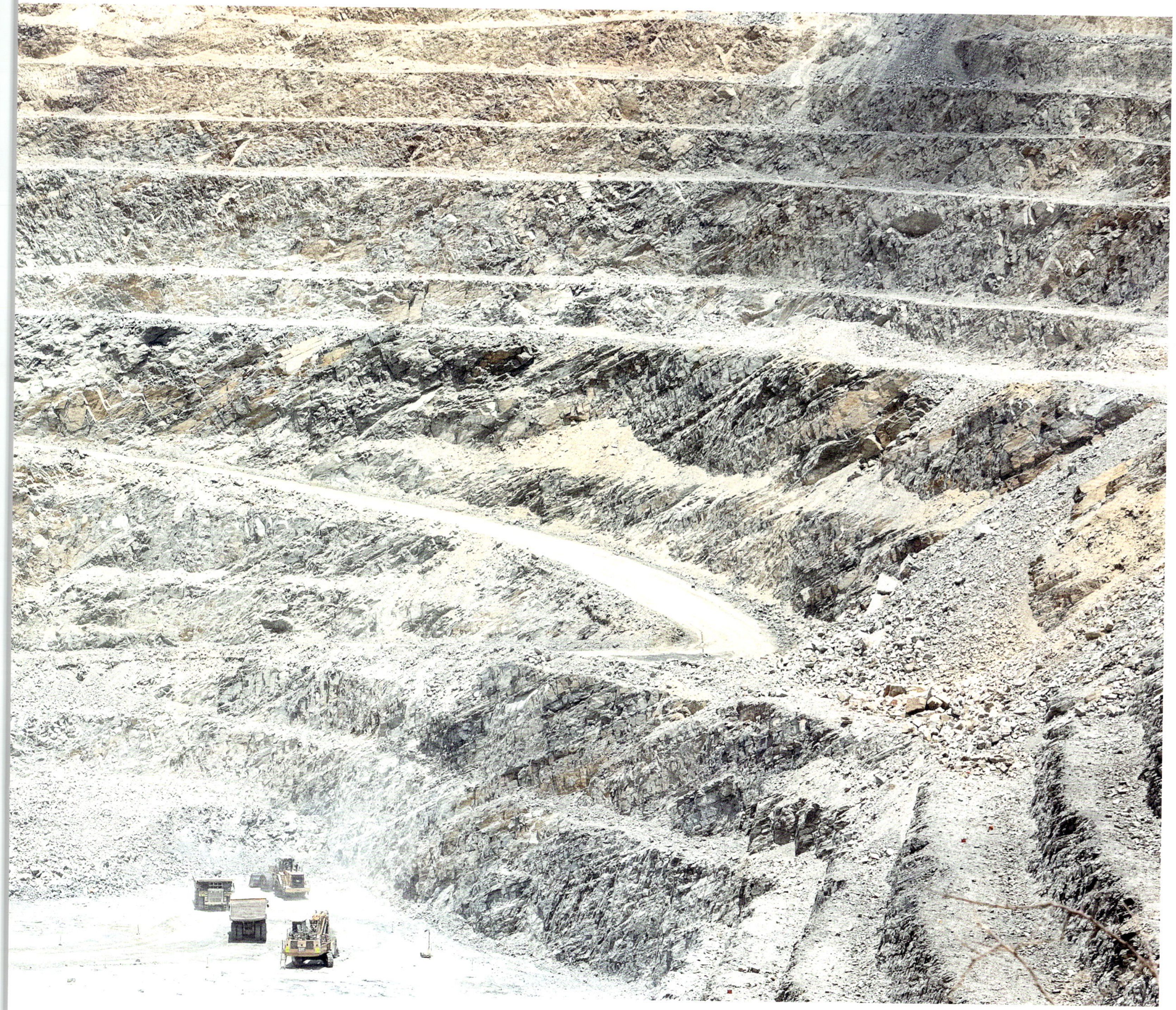

227). Currently, the marble is in intermediary storage at the Ministry of Mines and Energy, National Earth Science Museum, Geological Survey of Namibia awaiting export.

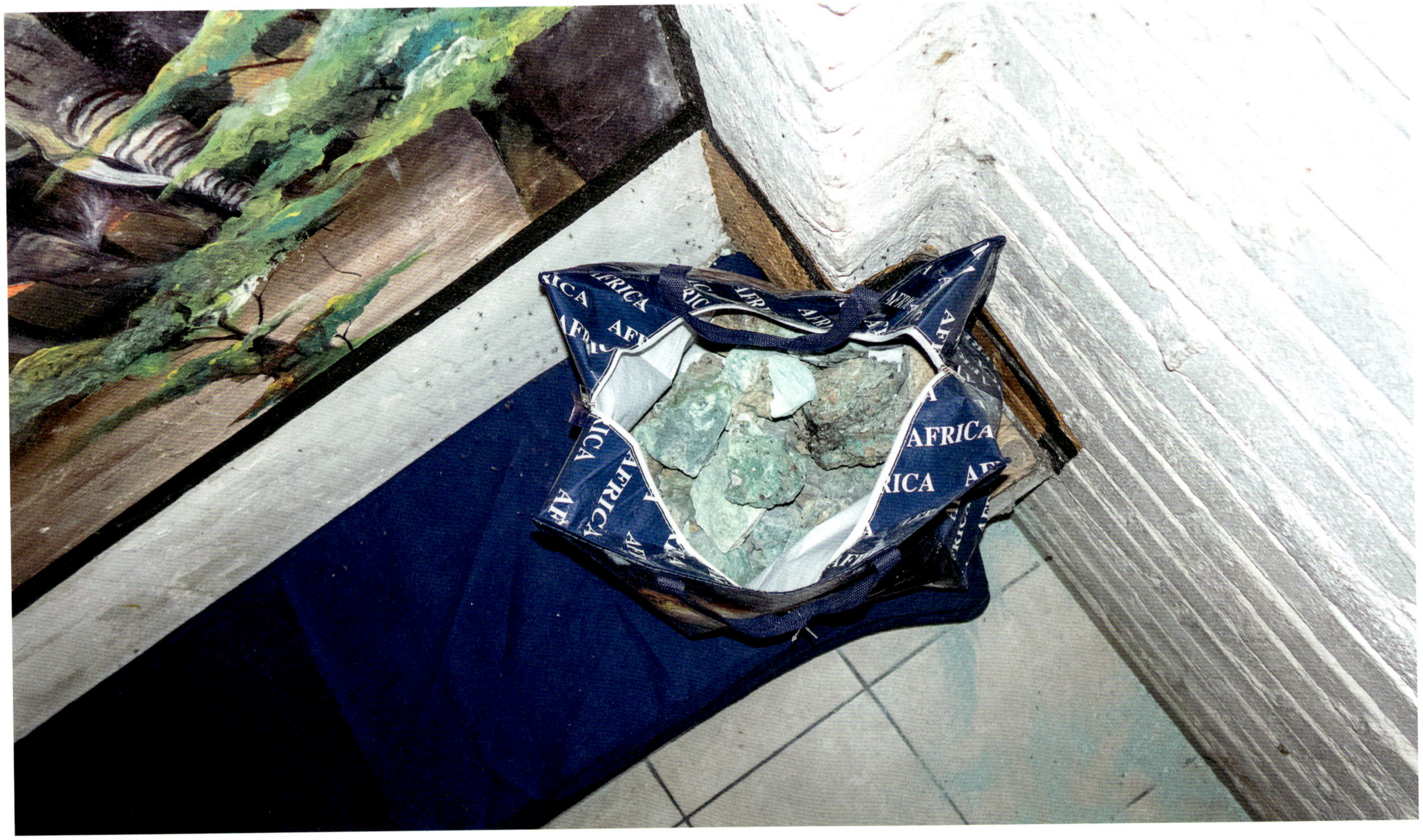

TG-T554
transposition 554
(Before Thought Hardens Into Language II)

Artist: Sascha Mikloweit
Collaborator: Tsumeb Municipality, Elli Stella Udda Imalwa, Francesca Steinbach / Steinbach Beergarden, Richardlee Shoombe, Manuel Pandalis, Hot Air Balloon Resort former TCL Mine Recreation Club \ CEO Bar, August Kapolo, Rakker Festus, Festus Alushiya, David David
Year: 2023
Location: Tsumeb Municipality, Former Otavi Mining and Railways Company (OMEG) / TCL open pit mine, Steinbach Beergarden, former TCL Mine Recreation Club \ CEO Bar: Tsumeb, Otjikoto, NA
Type: Experimentally evolving transhistorical transcultural relay, Exhibition, Immersive Installation, Sculptural Intervention, Process Sculpture, Experimental Film
Material: Tsumeb Municipality; former Otavi Mining and Railways Company (OMEG) / TCL open pit mine; Steinbach Beergarden; former TCL Mine Recreation Club / CEO Bar; Tailings from the Dundee Precious Metals Inc. Namibian Custom Smelter; Tsumeb minerals; various theory texts; *transposition 086, transposition 015, transposition 995*, *transposition 512*, *transposition 610*, and *transposition 001*

This exhibition was part of an experimentally evolving, transhistorical, and transcultural relay I organised in Tsumeb, Otjikoto, Namibia, from 19 October to 30 November 2023. The project took place in the format of an artist residency and exhibition, hosted by the Tsumeb Municipality across four different locations. The exhibition comprised *transposition 086* (this volume, pp.196–201), *transposition 015* (this volume, pp.228–245), *transposition 995*, *transposition 521* (this volume, pp.222–225), *transposition 610* (this volume, pp.218–221), and *transposition 001* (this volume, pp.252–265), together with several spontaneous, small, untitled sculptural interventions integrated into the infrastructure of Steinbach Beergarden, two of which are documented here: an 'Africa' shopping bag printed with a 'touristy' image of an elephant, filled with minerals I collected inside the former Otavi Mining and Railways Company (OMEG) / TCL open pit mine (above), and a line of black tailings on one of the beer garden's tables that I collected from the Dundee Precious Metals Inc. Namibian Custom Smelter in Tsumeb (right). Large-scale prints of the two photographs showing Richardlee and myself staring at each other, wearing the US / THEM jewellery that came out of our collaboration with Manuel Pandalis for *transposition 610* (this volume, pp.218–219) were pasted onto the entrance wall of the beer garden, facing Tsumeb's main commercial road.

Fig. 93

Fig. 94

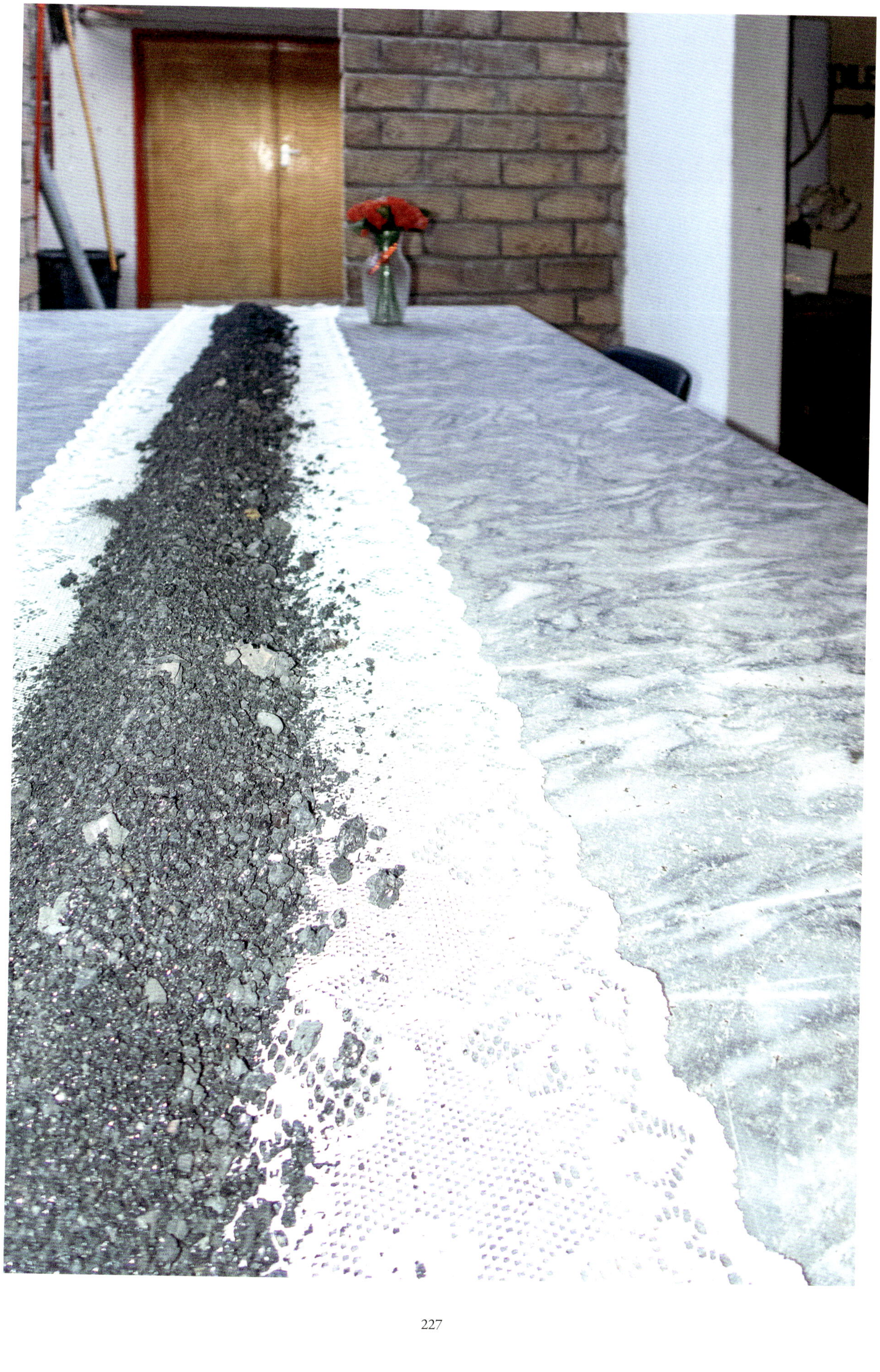

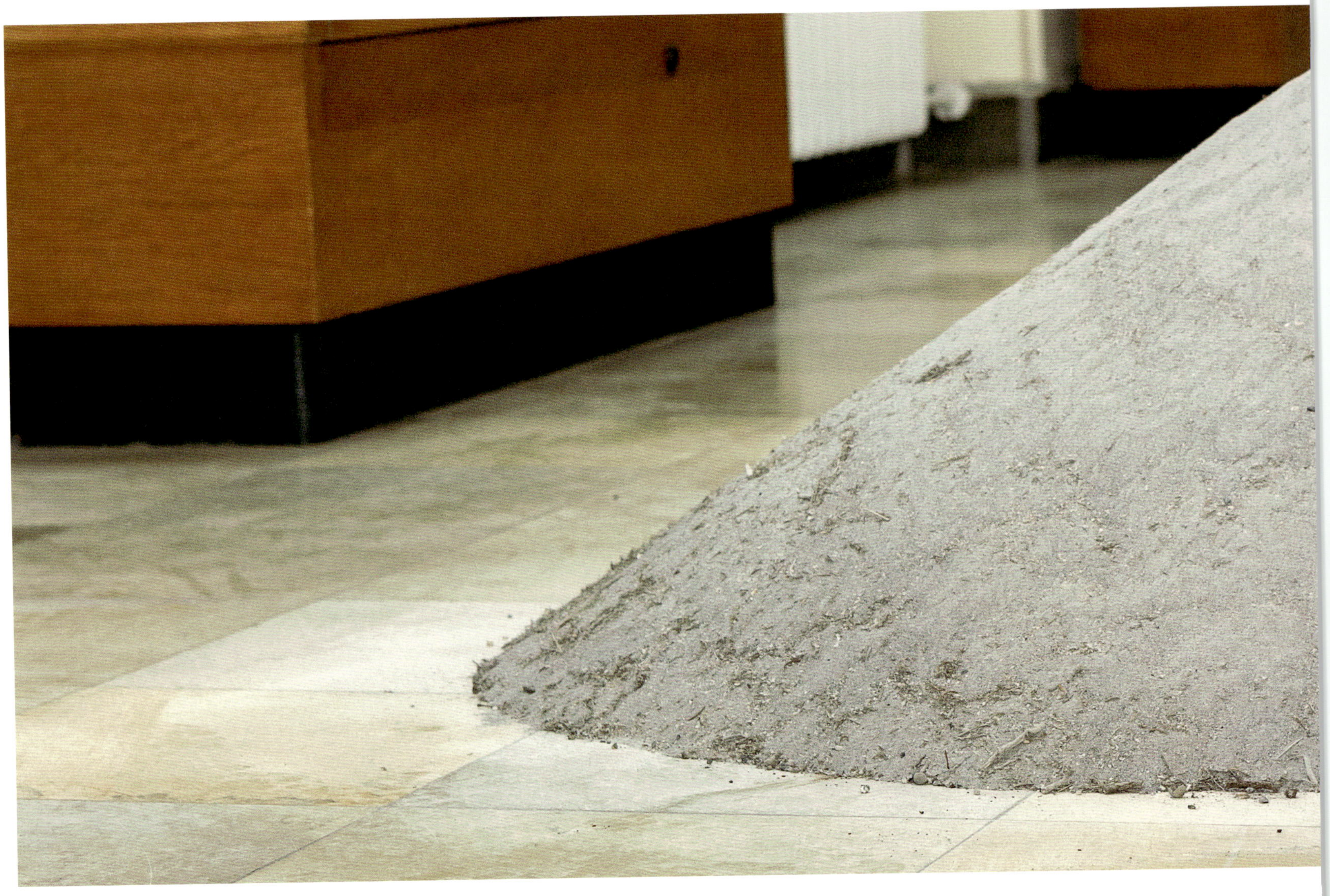

On the impermanence of Sascha Mikloweit's transposition 015 (re—existence)

Malina Lauterbach

Abstract: Drawing from James E. Young's analysis of 'counter-monuments', the author suggests thinking of Sascha Mikloweit's *transposition 015 (re—existence)* in the context of the project's implications of impermanence, auto-destruction and collectivity. Located in two very different semi-public sites: The Mineralogical Museum of the University of Bonn in Germany and the crater of a colonial open pit mine, the former 'Green Hill' in Tsumeb, Namibia, *transposition 015 (re—existence)* does not fit the conceptual form of a counter-monument. Nevertheless, Mikloweit, with his multi-layered artistic practice, initiates a collaborative act of 'epistemic disobedience' in Germany and Namibia, leading to the symbolic transformation of mineral relations embedded in the repercussions of geo-colonial histories. As such, Mikloweit not only follows histories of extraction, but undermines the very nature of artistic remembrance as an act of material manifestation.

Keywords: Remembrance, geo-coloniality, counter-monument, auto-destructive, impermanence, 'Green Hill', Tsumeb, epistemic transformation, collectivity, translocation

In 1992, James E. Young published the article 'The Counter-Monument: Memory against Itself in Germany Today', examining the concept of a 'counter-monument' by the artists Jochen Gerz and Esther Shalev-Gerz, in relation to Germany's remembrance of the Holocaust. In 1986, the Gerzes had installed a 12m-high lead pillar against fascism in Hamburg-Harbug. Near its base, one could read a temporary inscription in German, French, Russian, Hebrew, Arabic, Turkish and English:

> We invite the citizens of Harburg and visitors to the town, to add their names here to ours. In doing so, we commit ourselves to remain vigilant. As more and more names cover this 12 meter tall lead column, it will gradually be lowered into the ground. One day, it will have disappeared completely and the site of the Harburg monument against fascism will be empty. In the end, it is only we ourselves who can rise up against injustice. (Gerz, 1986)

As the pillar would be covered with the citizen's memorial graffiti, it would be partially lowered into the ground, leaving only its top plate flush with the pavement. Analysing how the disappearing graffiti pillar counters the traditional form of a monument, Young points out its disruptive qualities:

> [I]ts aim is not to console but to provoke; not to remain fixed but to change; not to be everlasting but to disappear; not to be ignored by its passersby but to demand interaction; not to remain pristine but to invite its own violation and desecration; not to accept graciously the burden of memory but to throw it back at the town's feet. (Young, 1992, p.277)

Fig. 95

Apparently, Sascha Mikloweit's *transposition 015 (re—existence)* does not fit the conceptual form of a counter-monument in public space as framed by Young. It is not intended to evoke the viewers' interaction and it is located in two very different semi-public sites: The Mineralogical Museum of the University of Bonn in Germany and the crater of a colonial open pit mine, the former 'Green Hill' in Tsumeb, Namibia, located in an industrial area, which is partially abandoned, partially sublet. Nevertheless, it is productive to think with the counter-monument's implications of impermanence, collectivity, and *auto-destruction* to navigate through Mikloweit's practice. The idea of *Auto-Destructive Art* was already introduced in the late 1950s and early 1960s by German-born artist and political activist Gustav Metzger as a critique of the destructive capabilities of technology, capitalism and the threat of nuclear war (Wilson, 2008, pp.177–178). By creating art that self-destructs, Metzger drew parallels to the self-destructive tendencies of modern civilisation (Metzger, 1960).

To characterise *transposition 015 (re—existence)*, I start by mentioning the artist's involvement in entangled spheres of 'geo-coloniality' in Germany and Namibia. As such, the work consists of a series of actions by the artist involving people holding institutional and political agency in both countries, as well as conducting two collaborative actions with the descendants of the historically victimised people of the Tsumeb Municipality. During this process legitimisation and access are provided for geological matter to pass through hands, borders, institutions, storages, and phases: 300kg of sediment from the colonial historical site, the former 'Green Hill' in Tsumeb, Namibia, being exported to Germany, where it is mixed with a pulverised colonial mineral from the Mineralogical Museum of the University of Bonn and placed in the artists' intra-institutional exhibition 'We Build Our Languages Out Of Rocks—Transpositional Geologies' as a pile-like earth sculpture. At the same time, the remainders of the museums' pulverised colonial mineral (object box, historical and contemporary object labels, and label referencing the art project) are being integrated into *archive cabinet #92 drawer #8* (the original archival storage place of the object) as a 'new' evolved epistemic object. What follows the German transition is the enriched sediment's return to the exact extraction site at the edge of the crater of the colonial open pit mine. The matter now being mineralised with the pulverised colonial epistemic object from the Mineralogical Museum of the University of Bonn re-connects symbolically with its double-sensed site of extraction in Tsumeb, Namibia. Marking the exact place, a Marula tree is being planted on top, and for the next weeks, the tree is being looked after and watered. One day, it vanishes under enigmatic circumstances, leaving only the dusty remains of *transposition 015 (re—existence).*

Through his translocation of matter, Mikloweit enters the realm of contemporary transnational power dynamics, wherein threads of German–Namibian colonial histories are continuously preserved and perpetuated. Reading the artist's project documentation almost feels like going through a screenplay with an ensemble of characters being orchestrated to perform bureaucratic arrangements in the aftermath of colonial rule. And here lies a crucial aspect of Mikloweit's interventionist practice: through extensive research and engagement in social relations he activates someone else's involvement/action/power outside of that person's professional *ratio*. Gaining permission to export and re-import 300kg of sediment from the colonial open pit mine for geo-scientific purposes, which, as a matter of fact, has no geo-scientific value. Gaining permission for the destruction of a museum specimen from the mineral collection, which conventionally is to be archived, studied and preserved. By entering these ambivalent relations, *transposition 015 (re—existence)* performs a collaborative transnational operation of epistemological disobedience, enacting a state of constant transformation. The 'fleeting presence' (Young, 1992, p.295) of *transposition 015 (re—existence)* also entails states of auto-destruction, e.g. pulverisation of the epistemic object of the Mineralogical Museum of the University of Bonn or re-connecting the mineralised matter with earth at the extraction site. However, the most unintentional but significant expression of the work's impermanence seems to be signified through the disappearance of the Marula tree—a figuration of living matter, which could have been grown, connecting past and future. Like the counter-monument's resistance to commemoration, what is left at the crater of the former 'Green Hill' are the dusty remains of the artist's symbolic gesture—a 'trace of its makers, not of the memory itself' (ibid., p.279).

References

Gerz, J., 1986. *Mahnmal gegen Faschismus / Monument Against Fascism*. [online] Available at: <https://jochengerz.eu/works/monument-against-fascism> [Accessed 20 June 2024].

Metzger, G., 1960. *Manifesto Auto-Destructive Art*. [online] Available at: https://designmanifestos.org/gustav-metzger-auto-destructive-art-manifesto> [Accessed 20 June 2024].

Wilson, A., 2008. Gustav Metzger's Auto-Destructive/Auto-Creative Art. *Third Text*, 22 (2), pp.177–194.

Young, J. E., 1992. The Counter-Monument. Memory against Itself in Germany Today. *Critical Inquiry*, 18 (2), pp.267–296.

Fig. 95

Fig. 96 [pp.232–233]

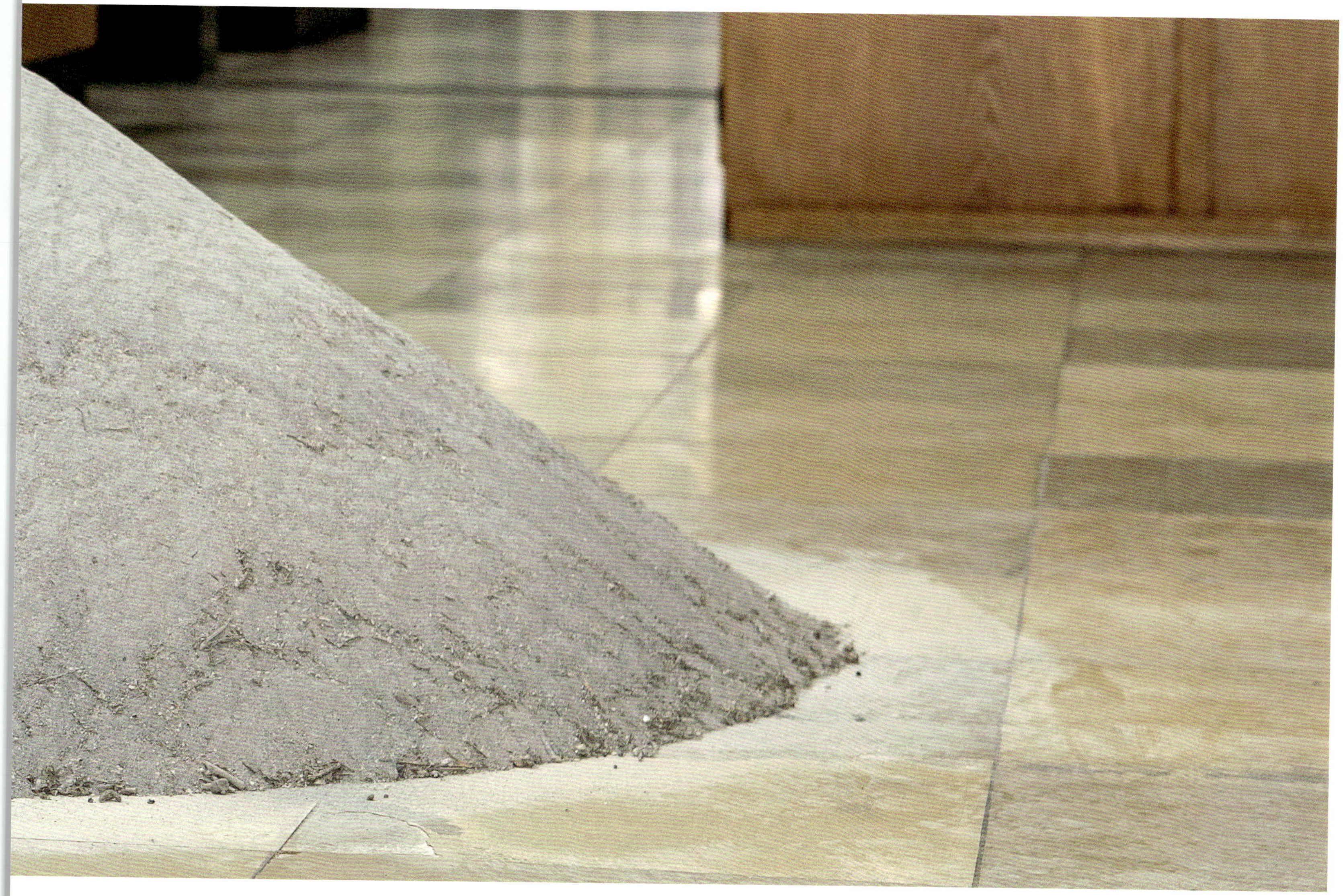

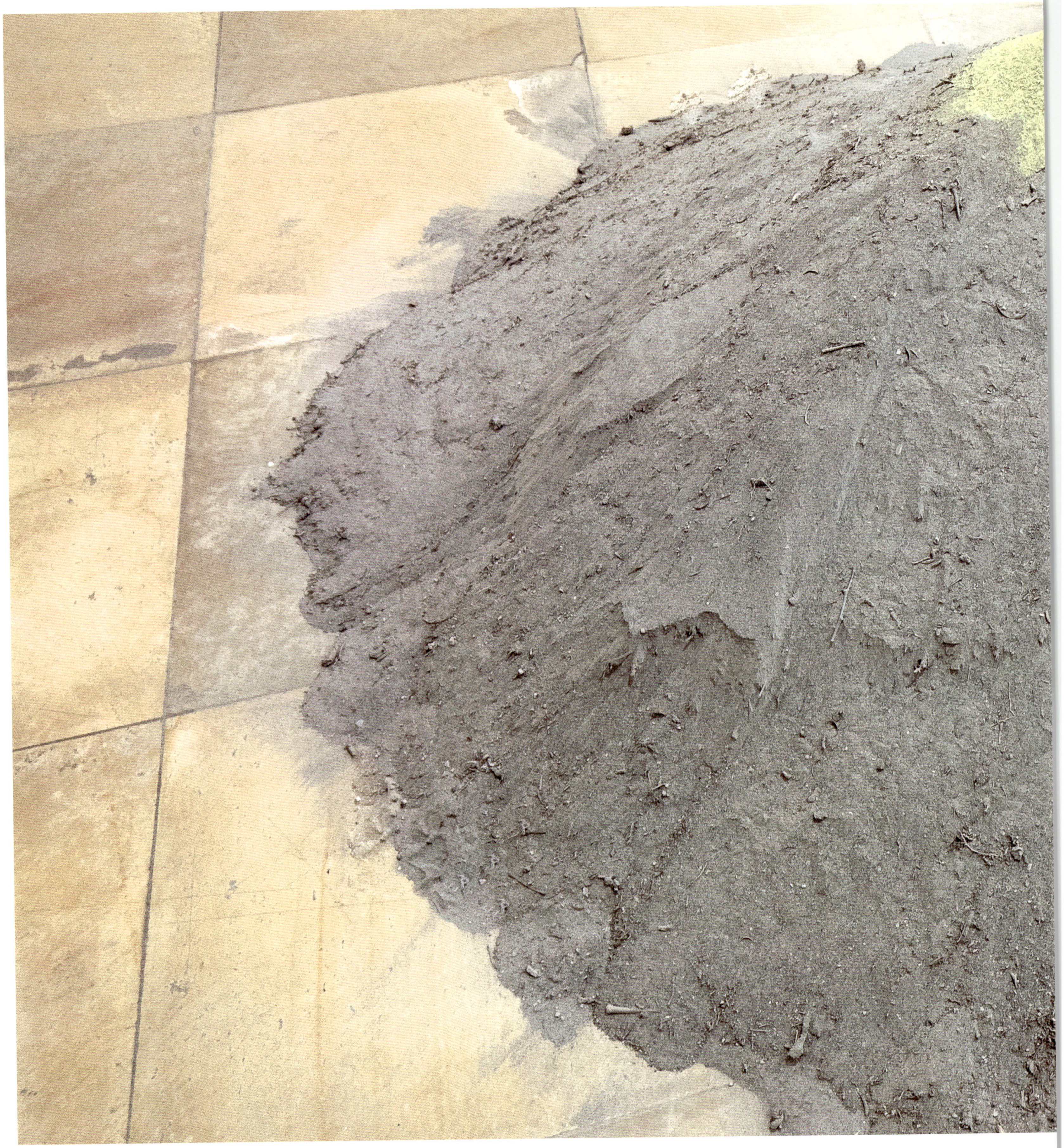

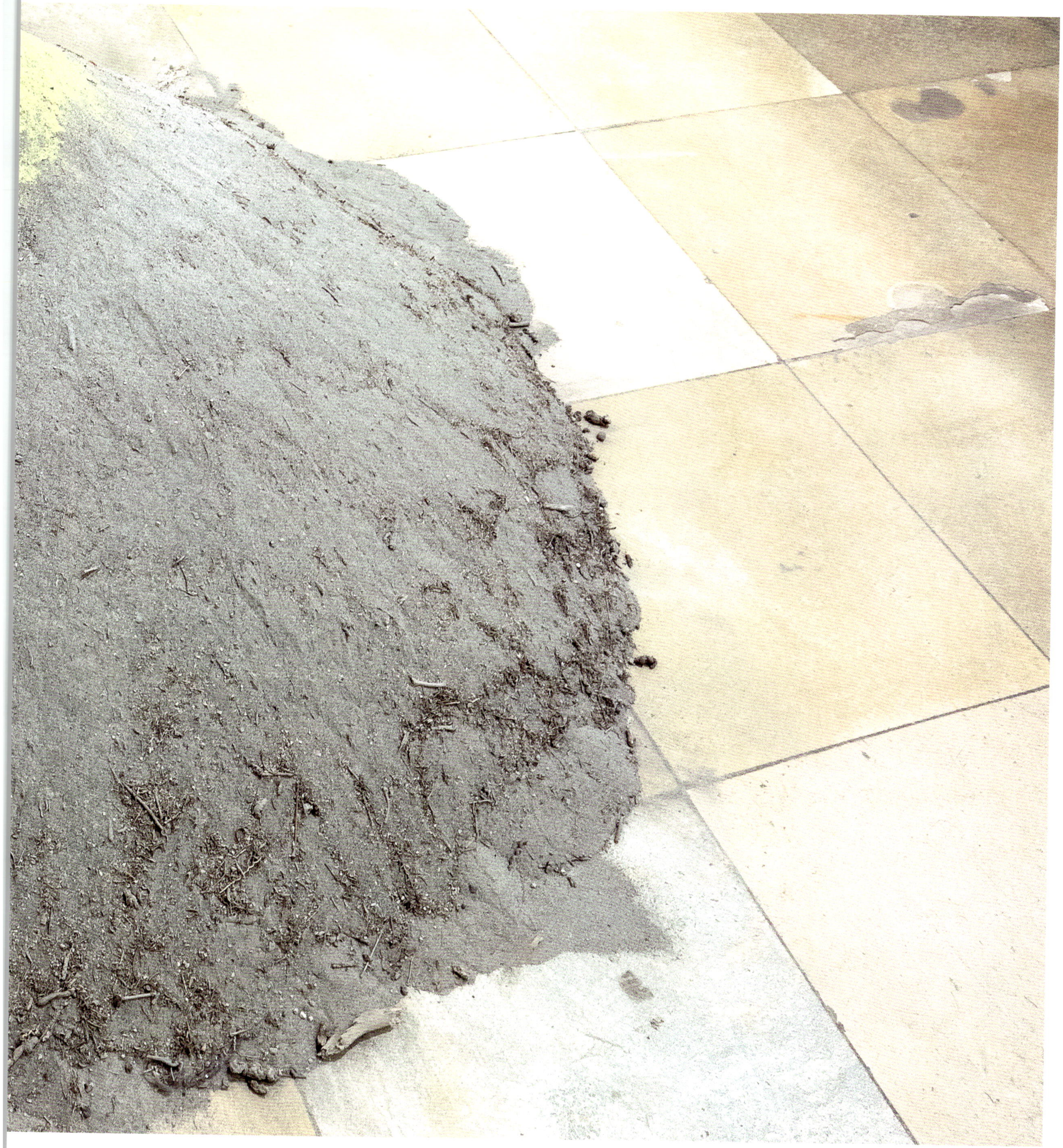

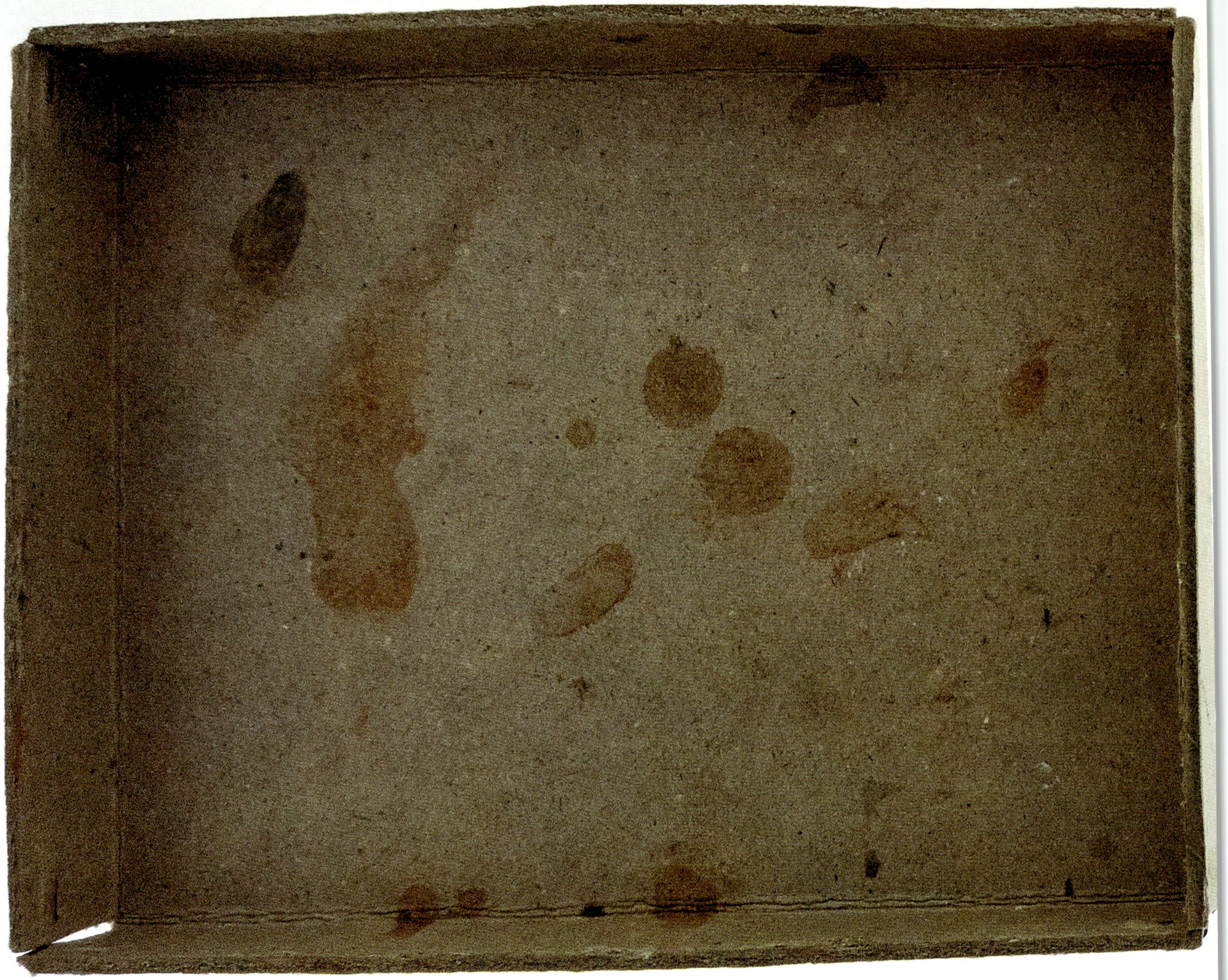

Fig. 97

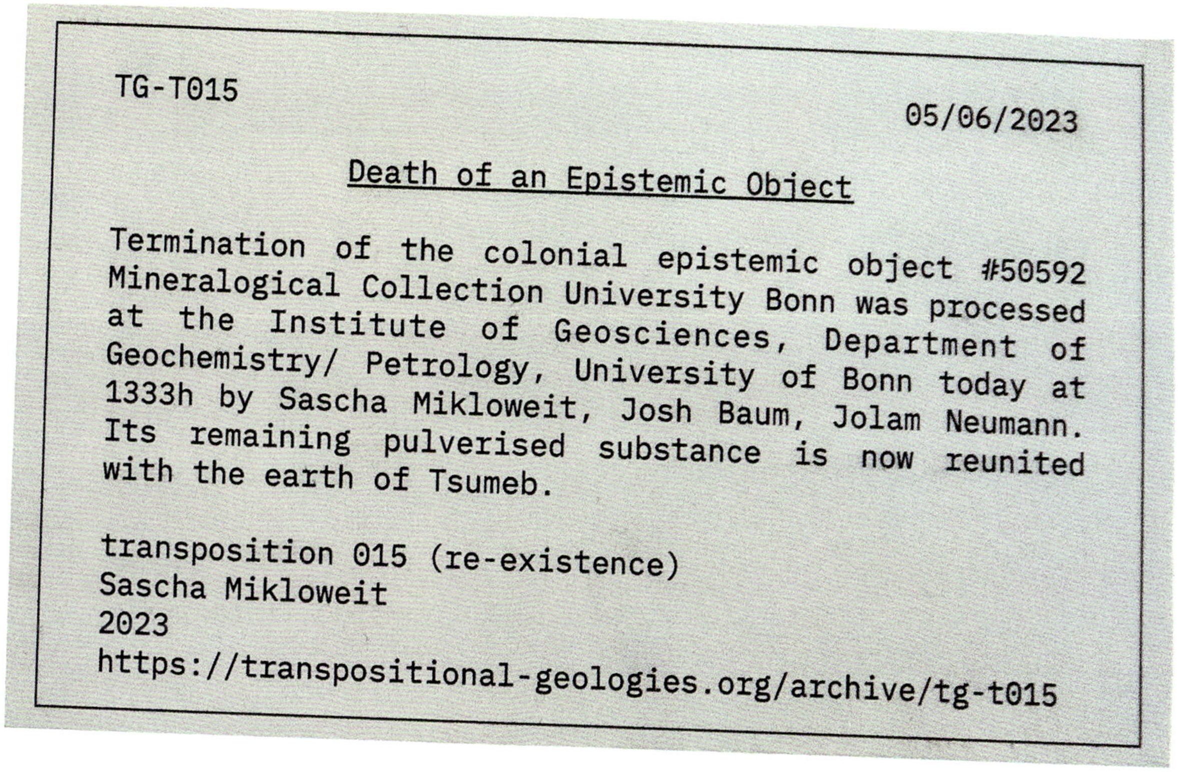

TG-T015

05/06/2023

Death of an Epistemic Object

Termination of the colonial epistemic object #50592 Mineralogical Collection University Bonn was processed at the Institute of Geosciences, Department of Geochemistry/ Petrology, University of Bonn today at 1333h by Sascha Mikloweit, Josh Baum, Jolam Neumann. Its remaining pulverised substance is now reunited with the earth of Tsumeb.

transposition 015 (re-existence)
Sascha Mikloweit
2023
https://transpositional-geologies.org/archive/tg-t015

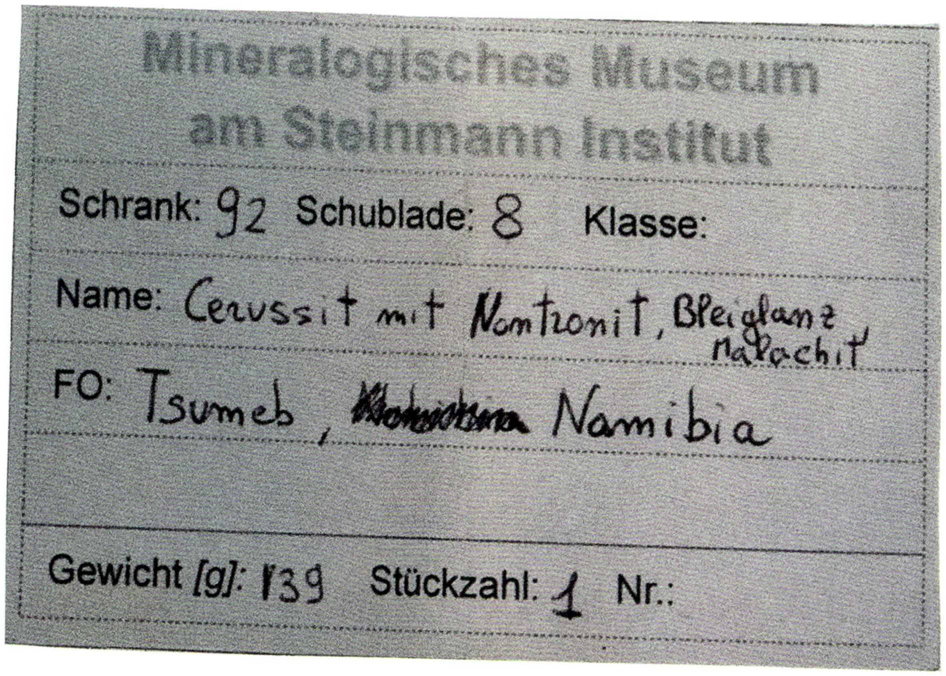

Mineralogisches Museum
am Steinmann Institut

Schrank: 92 Schublade: 8 Klasse:

Name: Cerussit mit Nontronit, Bleiglanz, Malachit

FO: Tsumeb, [illegible] Namibia

Gewicht [g]: 139 Stückzahl: 1 Nr.:

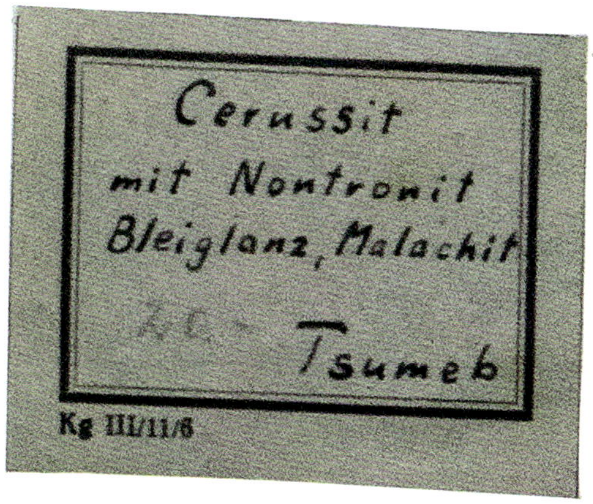

Cerussit
mit Nontronit
Bleiglanz, Malachit
Tsumeb

Kg III/11/6

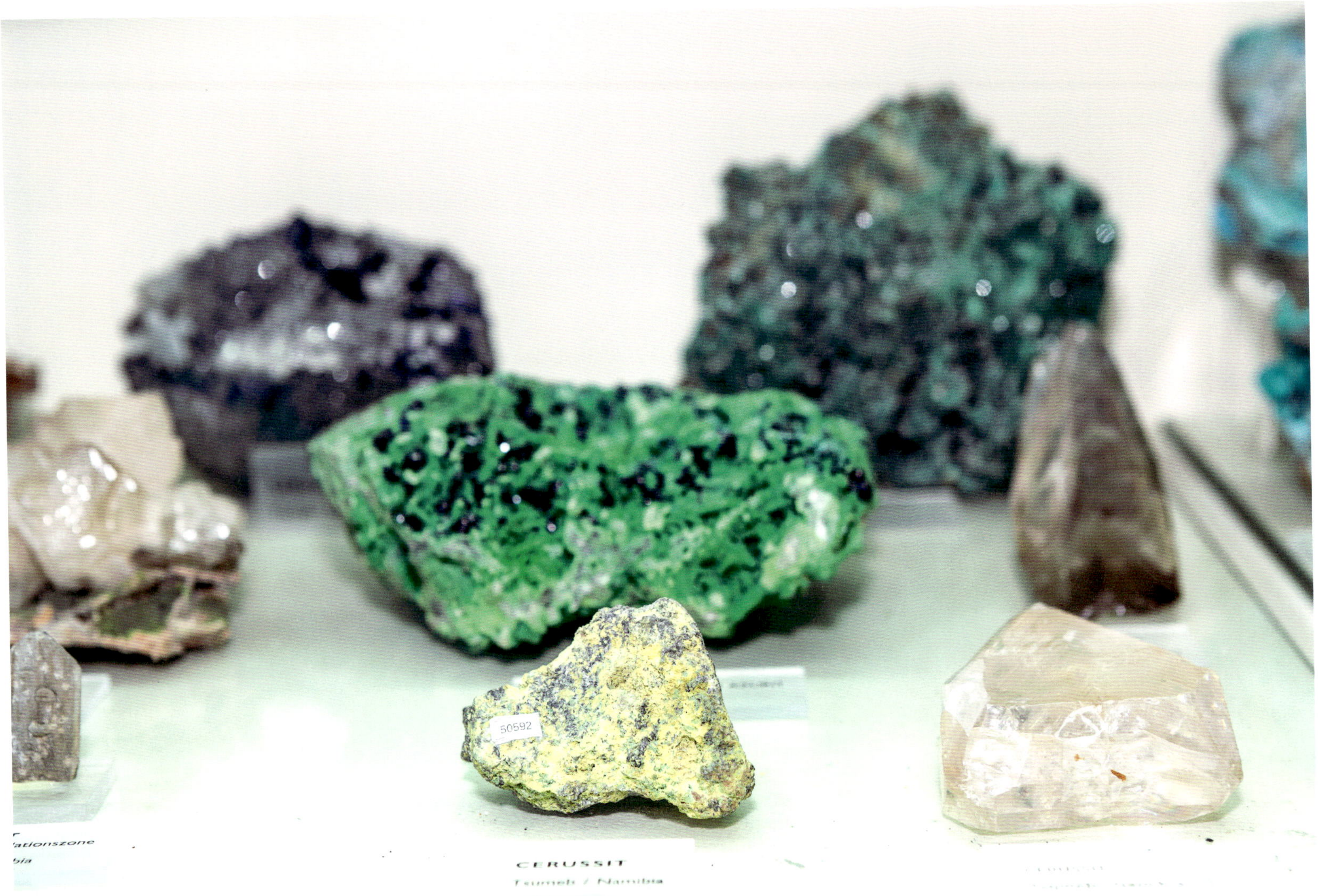

TG-T015
transposition 015
(re—existence)

Artist: Sascha Mikloweit
Collaborator: Dr R. Johannes Giebel (Director of Mineralogical Collections TU Berlin, DE), Elli Stella Udda Imalwa (Spokesperson for the Municipality of Tsumeb, NA), Dr Anne Zacke (Director of Mineralogical Museum of the University of Bonn) , Dr Helke Mocke (Director of National Earth Science Museum, Geological Survey of Namibia), Anna Williams (Administrative Officer at the National Earth Science Museum), Jolam Neumann (Phd student at the Institute of Geosciences University of Bonn), Josh Baum (Artist, UK / IL)
Year: 2021–2024
Cooperation: Goethe Institute Namibia, German Embassy Windhoek, Namibia; Ministry of Mines and Energy, Namibia; Tsumeb (private) Museum, and Tsumeb (private) Mineralogic & Mining Museum, Namibia.
Involved institutions (in order of process): Mineralogical Museum of the University of Bonn, Germany; Mineralogical Collections, Technical University of Berlin; Goethe Institute Namibia; German Embassy Windhoek, Namibia; Ministry of Mines and Energy, Namibia; Ministry of Education, Arts and Culture, Director of National Heritage and Culture Programmes, Namibia; Ministry of Education, Arts and Culture, Directorate of Arts, Namibia; Ministry of Mines and Energy, National Earth Science Museum, Geological Survey of Namibia; Tsumeb (private) Museum / Tsumeb (private) Mineralogic & Mining Museum, Namibia; Dundee Precious Metals Inc., Namibian Custom Smelter, Tsumeb, Namibia; Tsumeb Muinicipality, Namibia.
Location: Former Otavi Mining and Railways Company (OMEG) / TCL open pit mine; Ministry of Mines and Energy, National Earth Science Museum Geological Survey of Namibia; Mineralogical Museum of the University of Bonn, Germany; Tsumeb Mine, Namibia
Type: Performance, Opaque Relational Practice, Process Sculpture, Aesthetic Systems Analysis,
Administrative Negotiation, Administrative Documents

Fig. 98

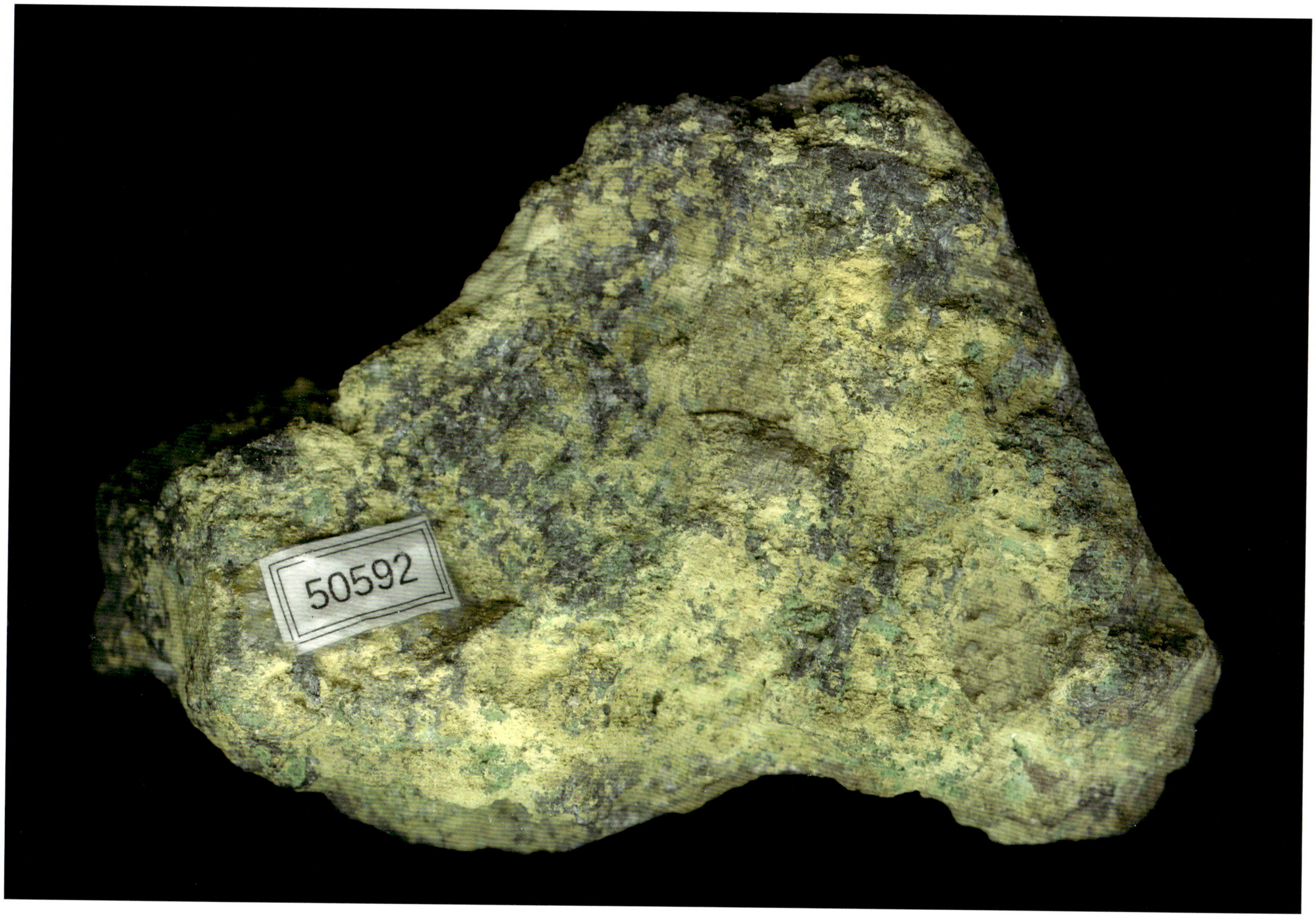

Dimensions: Morphodynamic
Duration: 3 years
Material: 300kg sediment, Tsumeb Mine, Tsumeb Namibia; scientific mineral specimen from Tsumeb with colonial heritage (Mineralogical Museum of the University of Bonn's collection object #50592); administrative processes; institutions; export / import procedures Namibia / Germany

Fig. 99

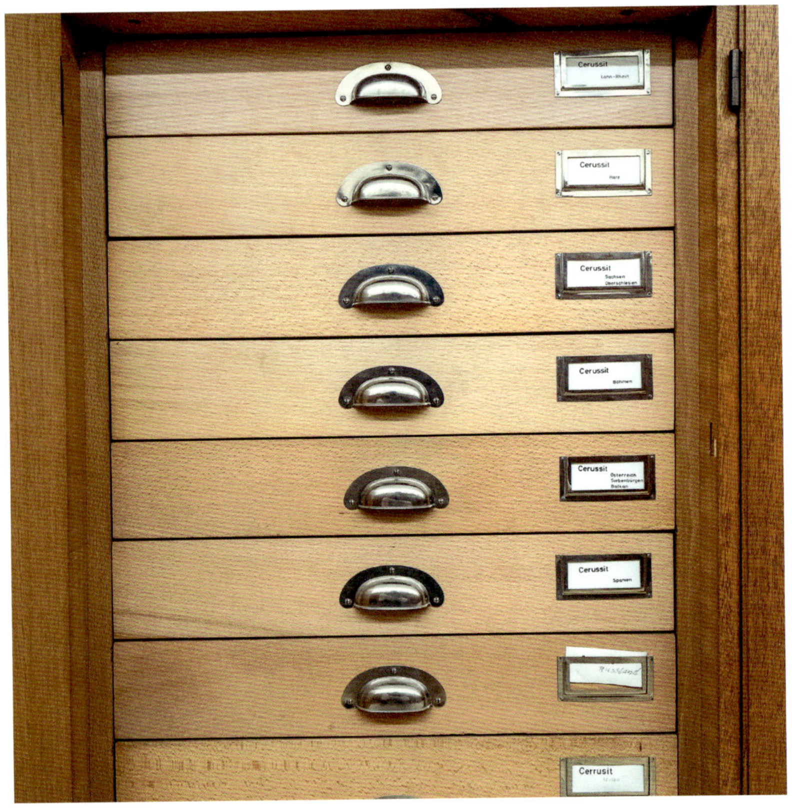
Cerussit
Cerussit
Cerussit
Cerussit
Cerussit
Cerussit
Cerussit

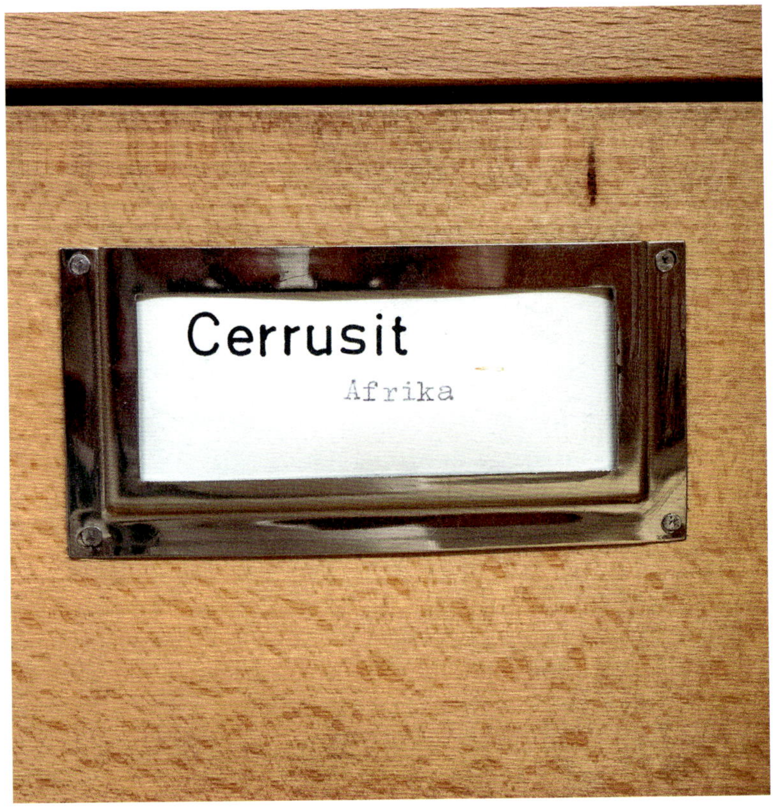
Cerrusit
Afrika

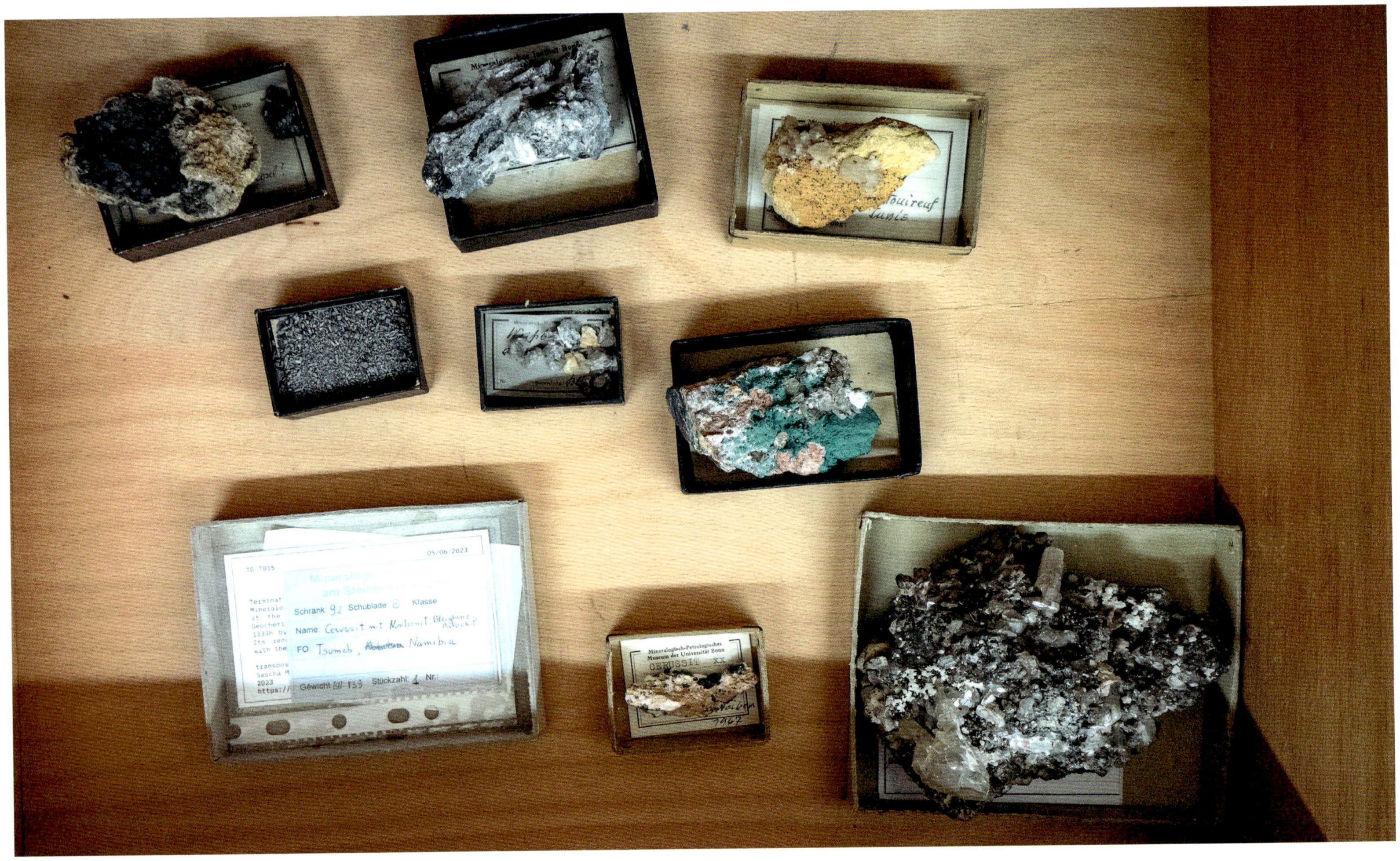
Schrank 92 Schublade 8 Klasse
Name
FO
Gewicht Stückzahl Nr.

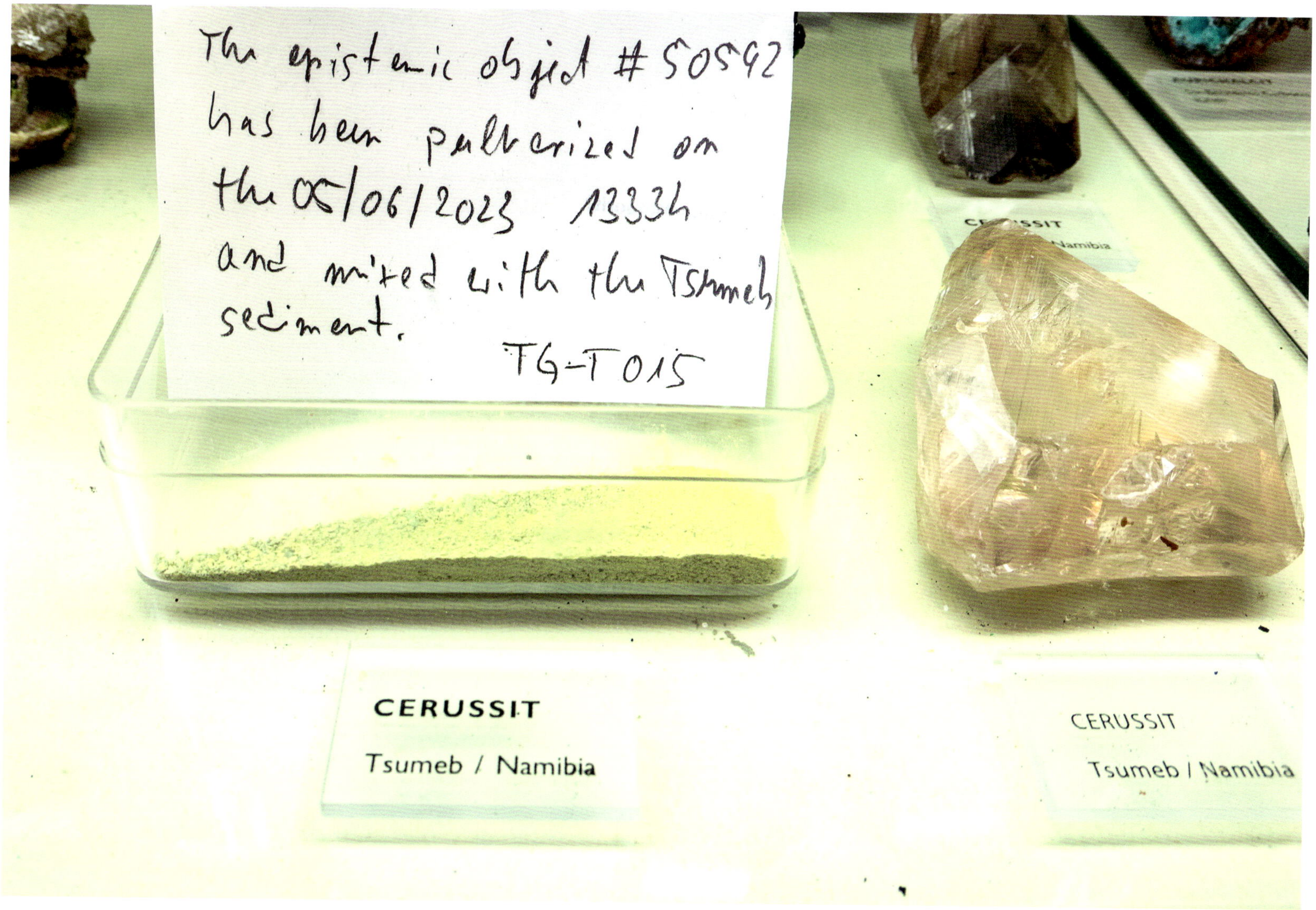

[*Artist's Note: This is a condensed version of this multifaceted work, offering a glimpse into its broader narrative and conceptual framework. Although 18 pages are dedicated here, a full unpacking of 'transposition 015' would require considerably more space and merits a separate publication format and case study, which is currently in preparation.*]

This morphodynamic collaborative work explores European-African histories of dominant modes of thought and existence, unfolding from a German-Namibian archetype.
In October 2022, 300kg of earth with no commercial value were extracted in a collaborative, performative act in Tsumeb, Namibia, involving representatives of the Tsumeb Municipality and artist Sascha Mikloweit.
The extraction took place on the northern edge of the open-pit mine, at the very site where the German colonial Otavi Mining and Railway Company (OMEG), founded in 1905 in Berlin, began large-scale copper mining in 1906, erasing
Fig. 100–102 the 'Green Hill'. This followed their completion of a 567km railway line (constructed from 1897–1906) connecting the
Fig. 103 Tsumeb Mine in the Otavi Mountains with Swakopmund on the Atlantic coast—an operation tied to the context of the Herero and Nama War, 1904–1907 (for context, see this volume, pp. 180ff).

Following 18 months of logistical research and administrative negotiations (from May 2021 to November 2022), the project revealed the complex dynamics of agency and power structures that govern not only the flow of materials but also knowledge and cultural heritage. Ultimately, Namibian authorities granted export permission for the sediment to Germany, where it was displayed at the Mineralogical Museum of the University of Bonn as part of Mikloweit's exhibition, *We Build Our Languages Out of Rocks—Transpositional Geologies* (April–June 2023).

In June 2023, during this exhibition, a Tsumeb mineral from the museum's mineral collection (object #50592), one with a colonial heritage, was destroyed—at least from a museological and human-centred perspective—by being ground into powder in the Institute of Geosciences' basement 'Crushing Room' (a performance video is available at: *https://*

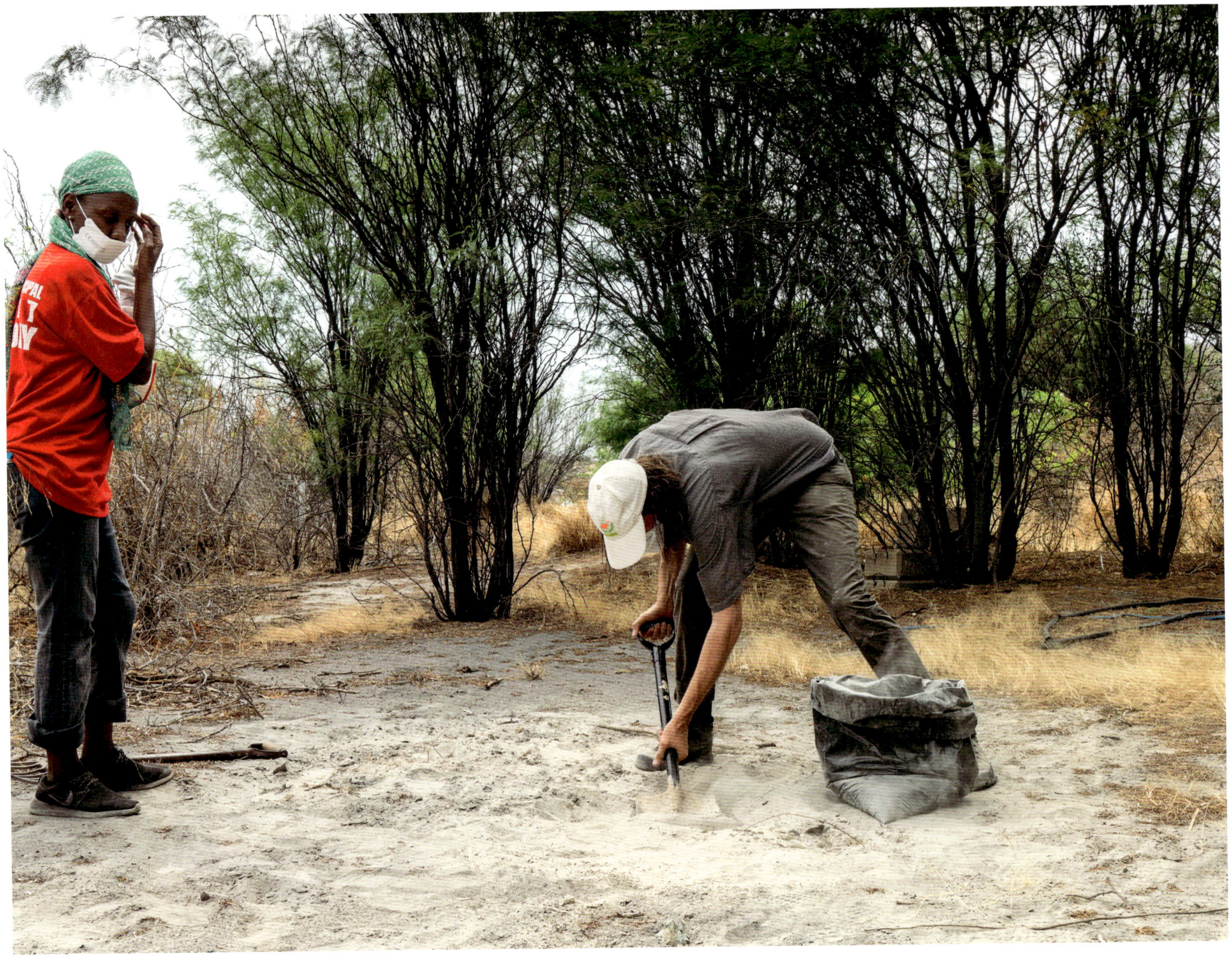

transpositional-geologies.org/archive/tg-t015). In a further act, this pulverised mineral was united with the imported Tsumeb sediment, symbolically restoring it to its pre-scientific, pre-human origin.

For a blink of geological time, within the framework of the Bonn exhibition, first the Tsumeb sediment alone, and later the sediment combined with the destroyed mineral's powder, underwent an ontological shift, becoming fine art—imbued with all the racial and political ideologies inherent in this Western concept (Lloyd, 2019, this volume, p. 59; Shiner, 2001, ibid., p. 60)—and materialising as *sculpture*.

After the exhibition, the sediment mixture was re-imported to Namibia in October 2023, where it was returned in a collaborative, performative act with representatives of the Tsumeb Municipality at the very site of its extraction. During this act, a young Marula tree was planted in the sediment to nourish itself with this remnants of Western knowledge (November 2023). The tree later disappeared, and evidence suggests it was uprooted by unidentified citizens of Tsumeb sometime before mid-March 2024.

Fig. 104

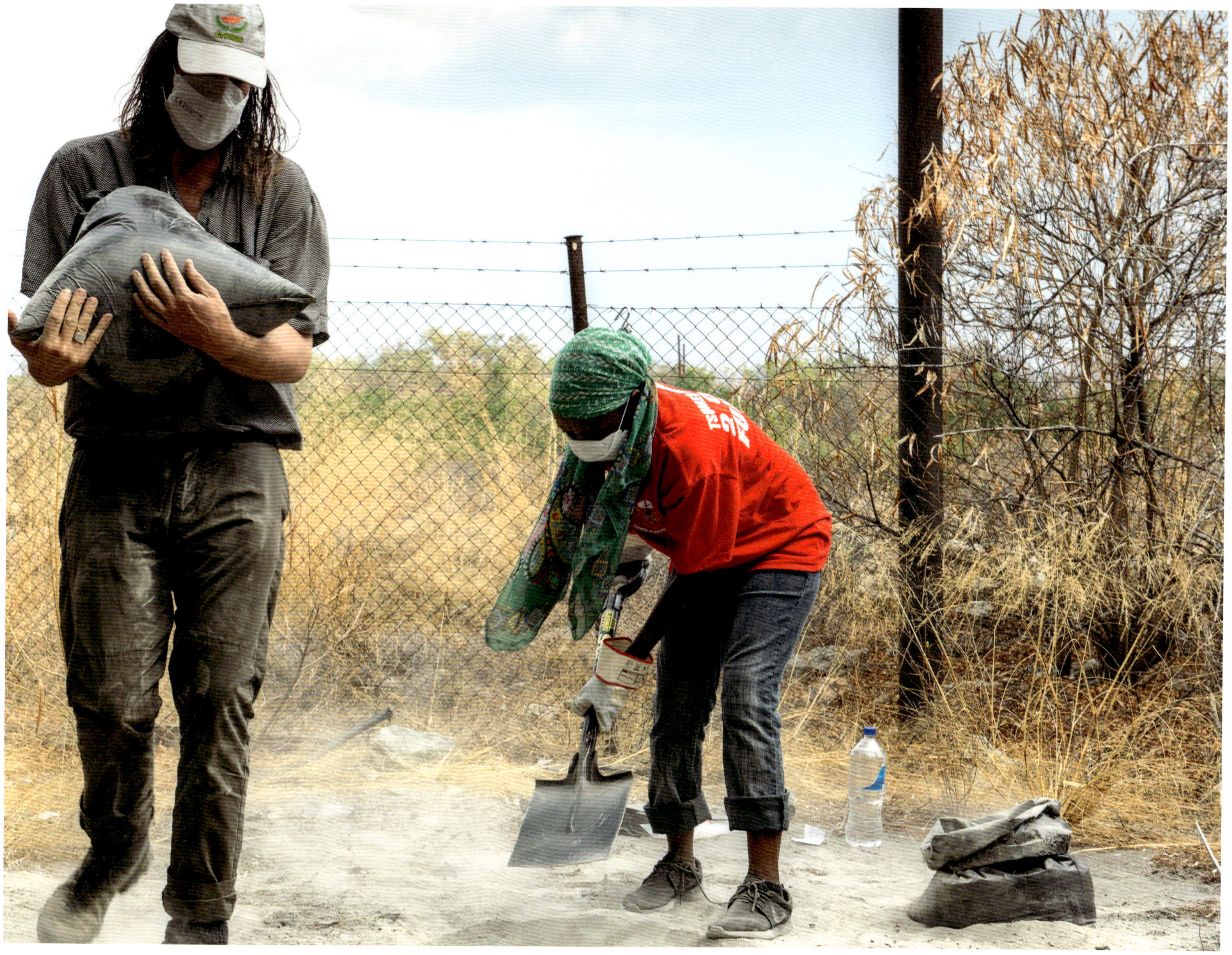

Fig. 105

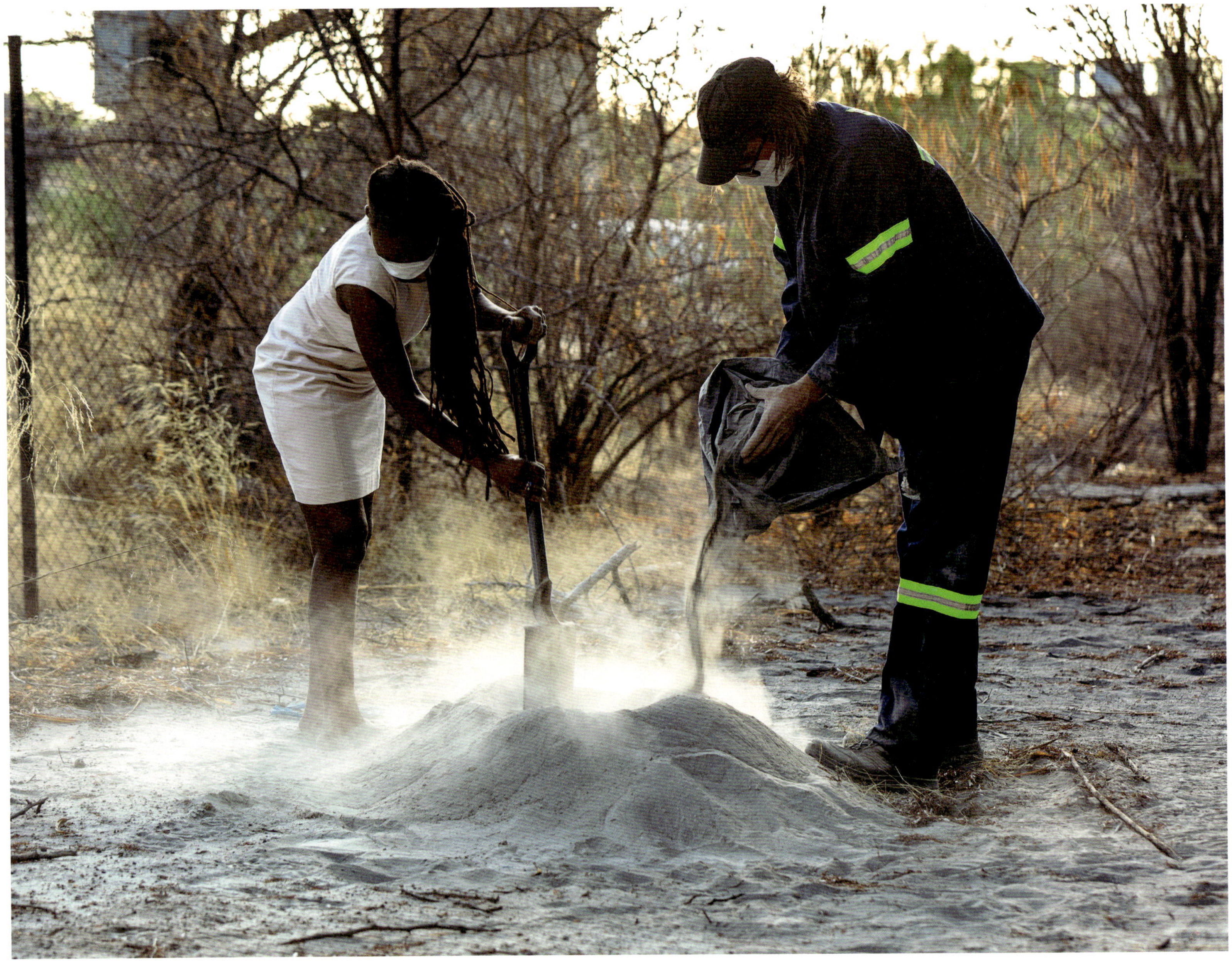

Fig. 106

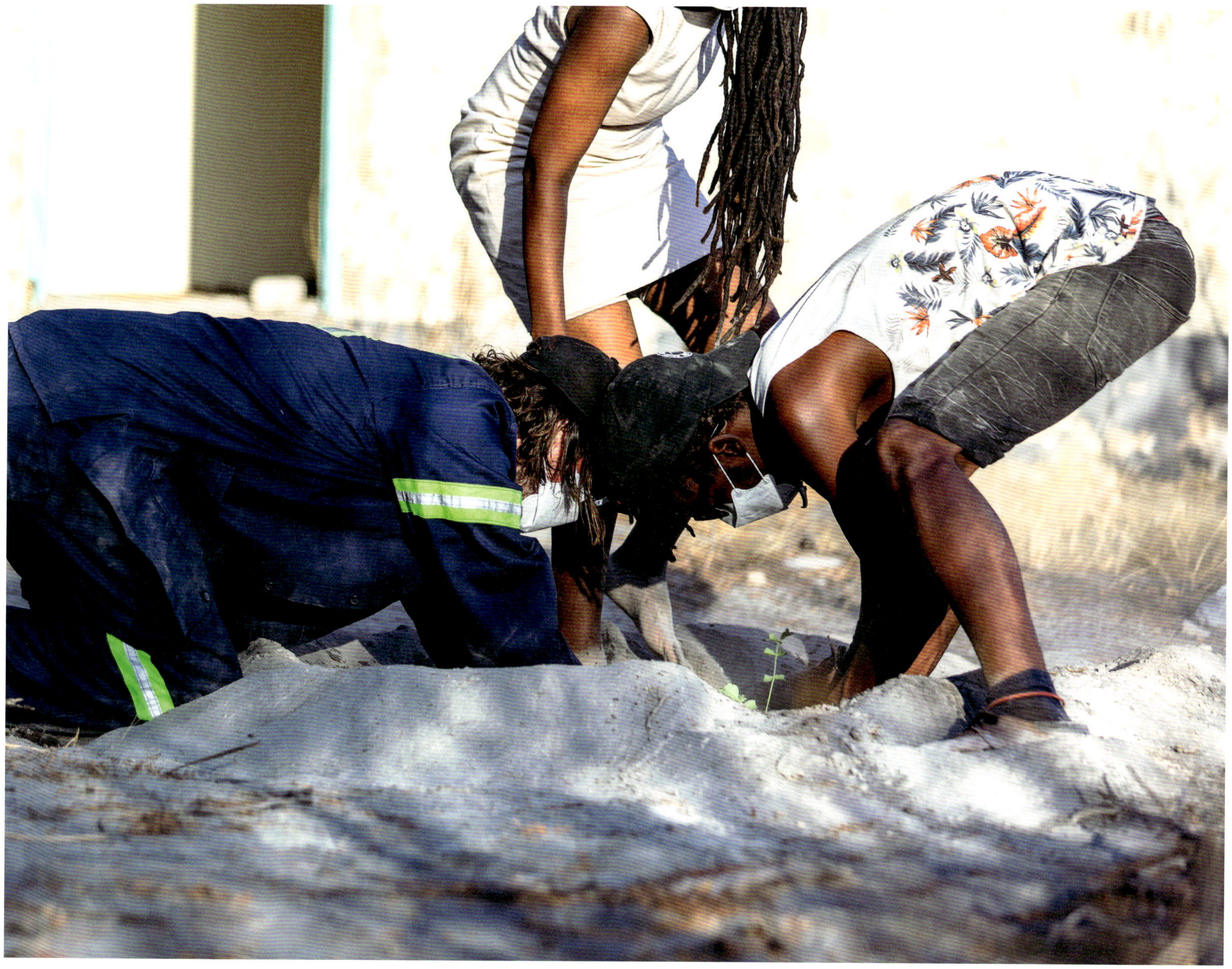

Fig. 107

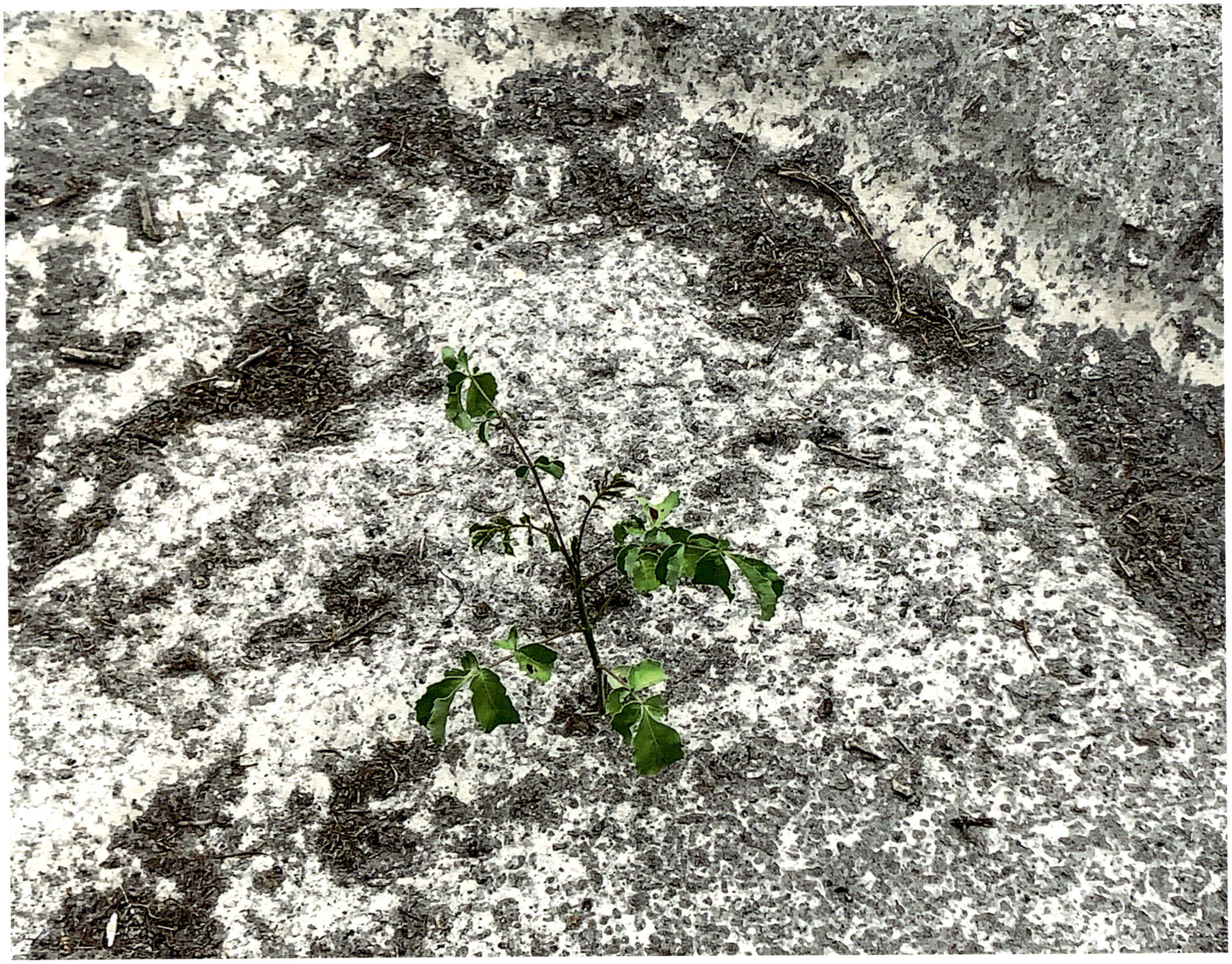

Fig. 108

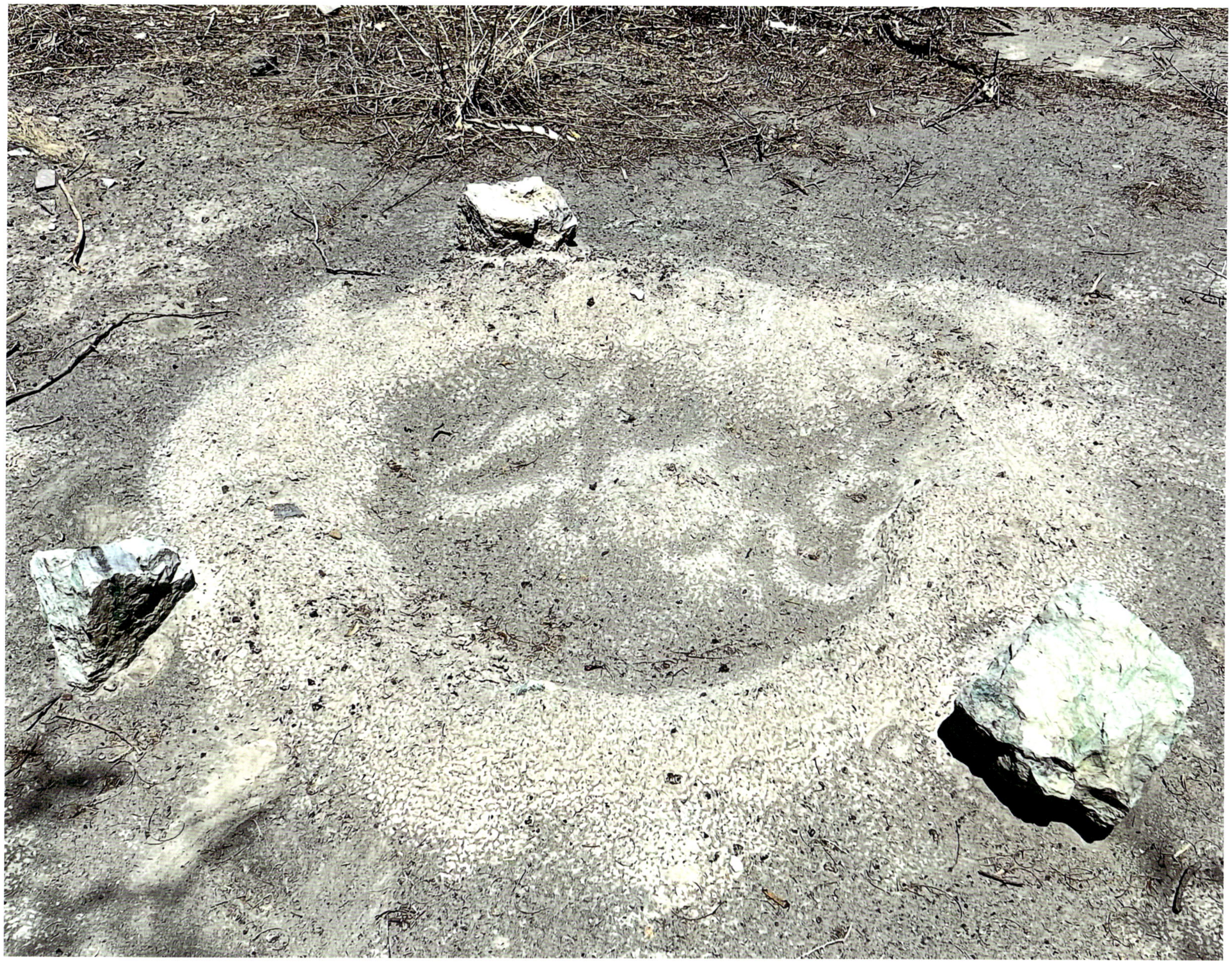

Fig. 109

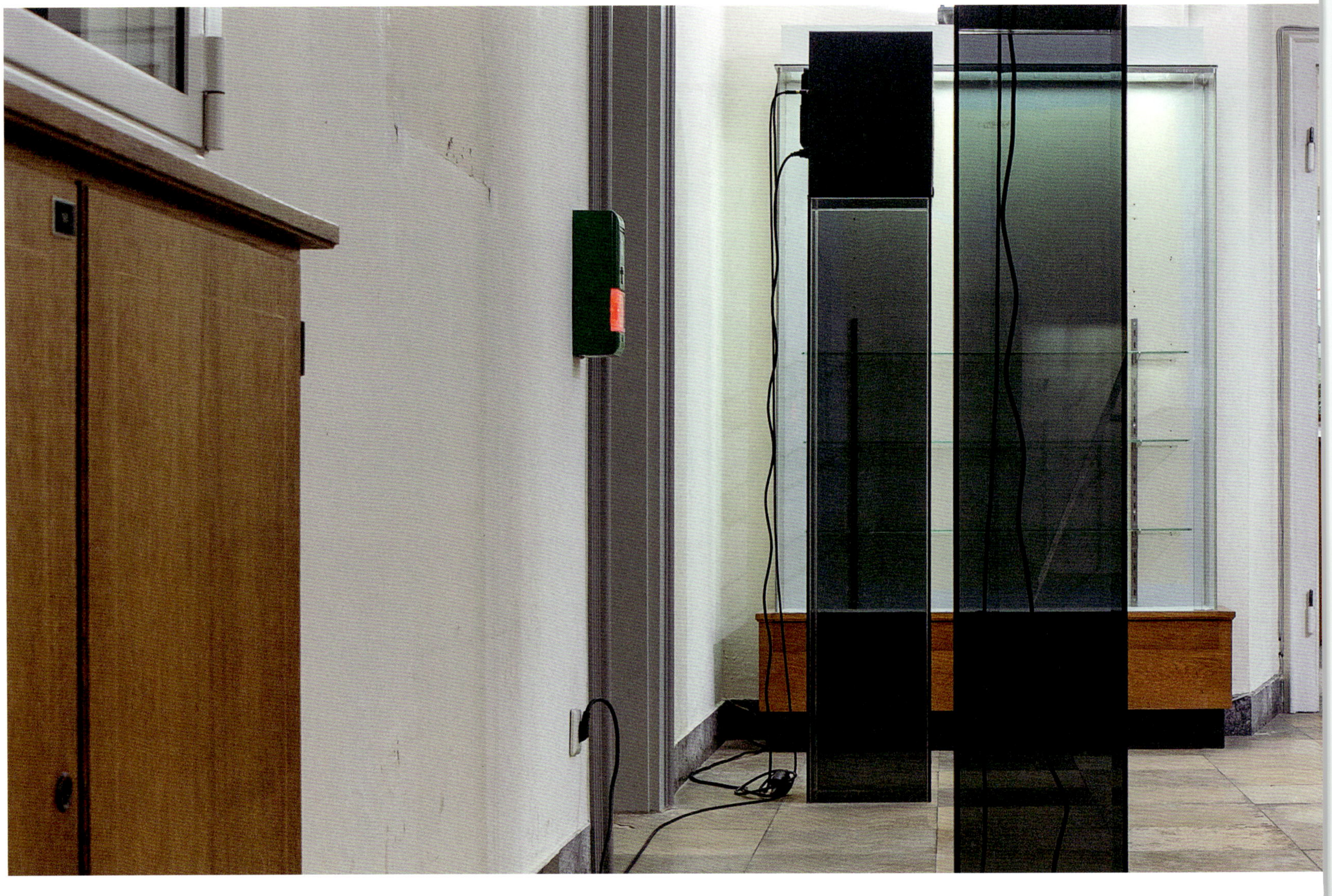

TG-T241
transposition 241
(Before Thought Hardens into Language I)

Artist: Sascha Mikloweit
Year: 2023
Location: Mineralogical Museum of the University of Bonn, DE, Exhibition Hall 1 & 2
Type: Exhibition, Immersive Installation, Sculptural Intervention, Process Sculpture
Dimensions: 330 x 480 x 1558 cm (Hall 2); 134 x 25 x 81 cm, 40 x 32 x 28 cm (Hall 1)
Material: Systemic Collection and Special Exhibitions Hall, Mineralogical Museum of the University of Bonn; sound; 6500K /13.500 lm light; *transposition 441*, *transposition 904*, *transposition 015*, *transposition 504*, *transposition 756*, *transposition 741*, *transposition 064*, *transposition 665*, *transposition 368*

transposition 241 was the central module of the exhibition *We Build Our Languages Out Of Rocks—Transpositional Geologies*, that took place at the University of Bonn ffrom 20.04.2023 –11.06.2023:

After a three-year artist residency at the Mineralogical Museum of the University of Bonn, Sascha Mikloweit presented an experimental exhibition at the Poppelsdorfer Schloss in 2023. The show was the first of its kind in the institution's 200-year history, connecting various locations throughout Bonn with the Poppelsdorfer Schloss at its gravitational centre.

Mikloweit related the university's mineralogical collection to different times and spaces in history and geography. By expanding the scope of scientific collections, his practice reframes our understanding of the minerals—un-earthing their complex and often violent pasts. The exhibition offered a unique opportunity for visitors to explore the interconnectedness of colonialism, science, and extractivism through the lens of mineralogical specimens.

Fig. 110

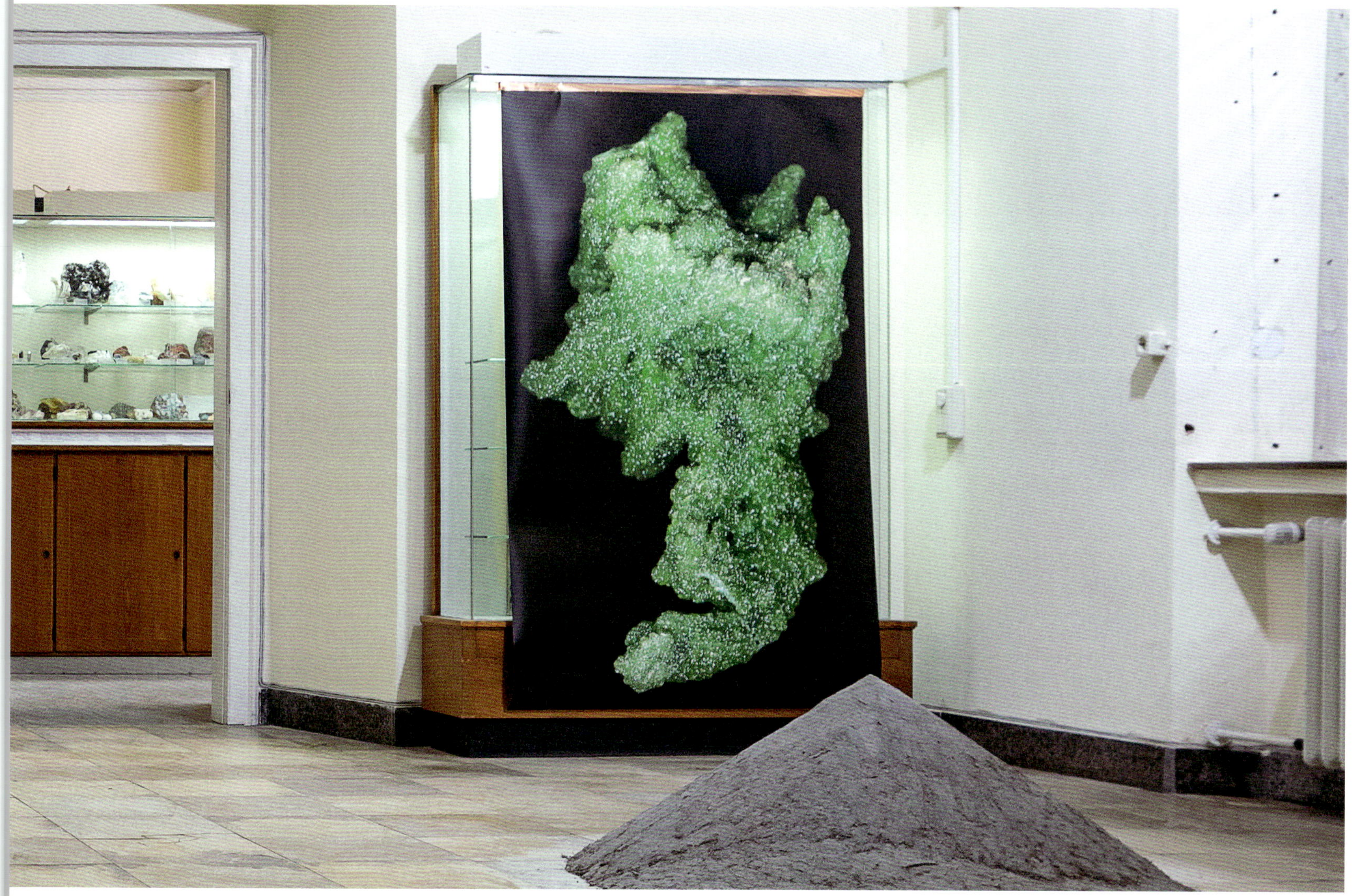

We Build Our Languages Out Of Rocks—Transpositional Geologies combined immersive installation, process sculpture, situational performance, experimental film and a series of lecture interventions welcoming guest speakers from Africa, North America, and Europe.

The central exhibition, *transposition 241 (Before Thought Hardens into Language I)*, questioned the institution's self-evident appearance by introducing two sculptural interventions into the visual density of the Systematic Collection—*transposition 441* (this volume, pp. 192–195), and *transposition 904* (ibid., pp. 160–175)—alongside an immersive, juxtaposed installation in the adjacent Special Exhibitions Hall of the Mineralogical Museum of the University of Bonn. To start, Mikloweit entirely emptied this latter space, leaving only the institutional architecture and its archive and display infrastructure.

The addition of five ultra-bright lighting sources intensified the materiality of the interior space, creating an alienating, blinding, cold white atmosphere for the visitor. The immersive installation was further constituted by the introduction of *transposition 015* (ibid., pp. 228–245), *transposition 504* (ibid., pp. 204–205), *transposition 756* (ibid., pp. 202–203), *transposition 741* (ibid., pp. 208–211), *transposition 064* (ibid., pp. 156–159), *transposition 665* (ibid., pp. 216–217), and *transposition 368* (ibid., pp. 212–215) into the space.

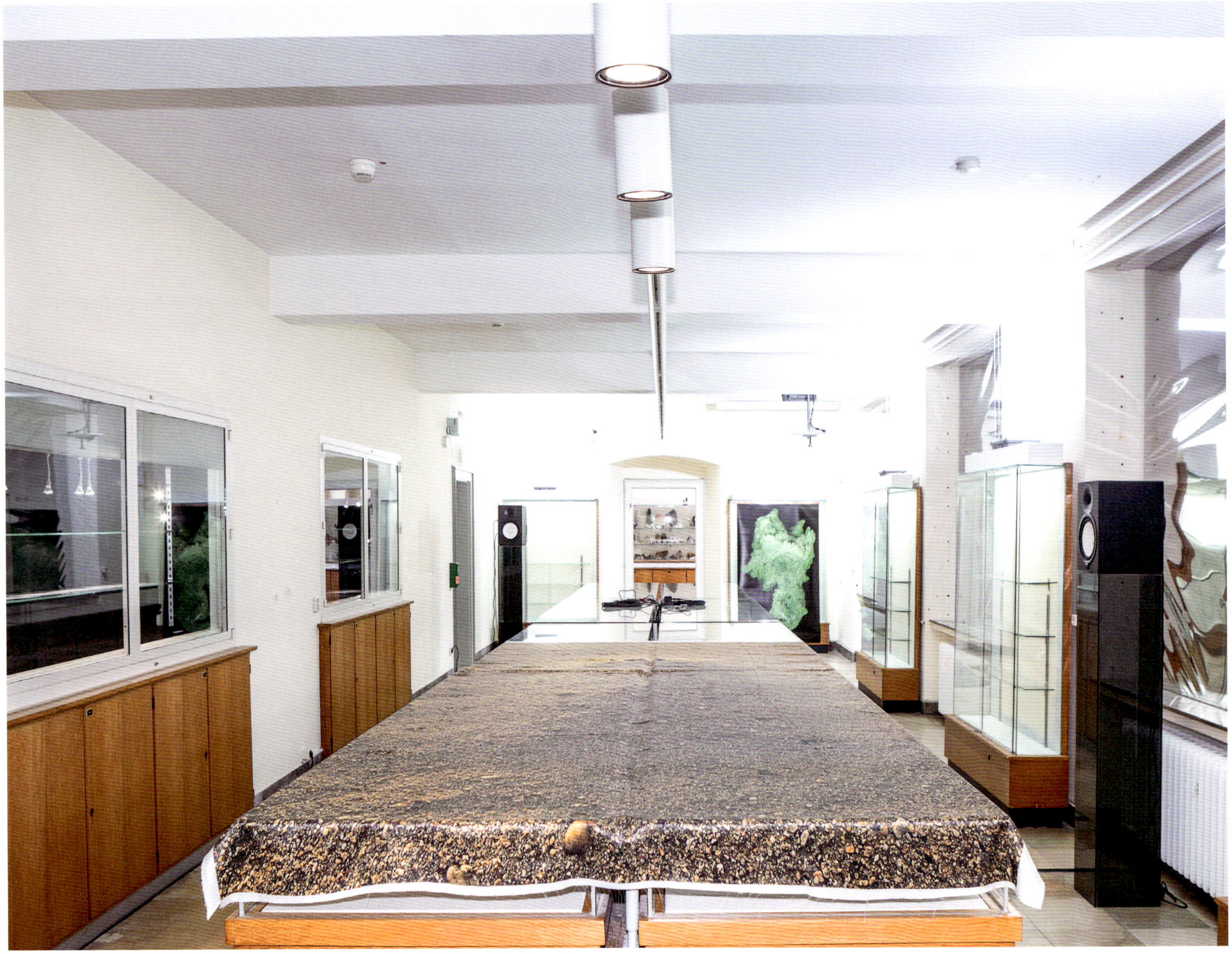

Fig. 111

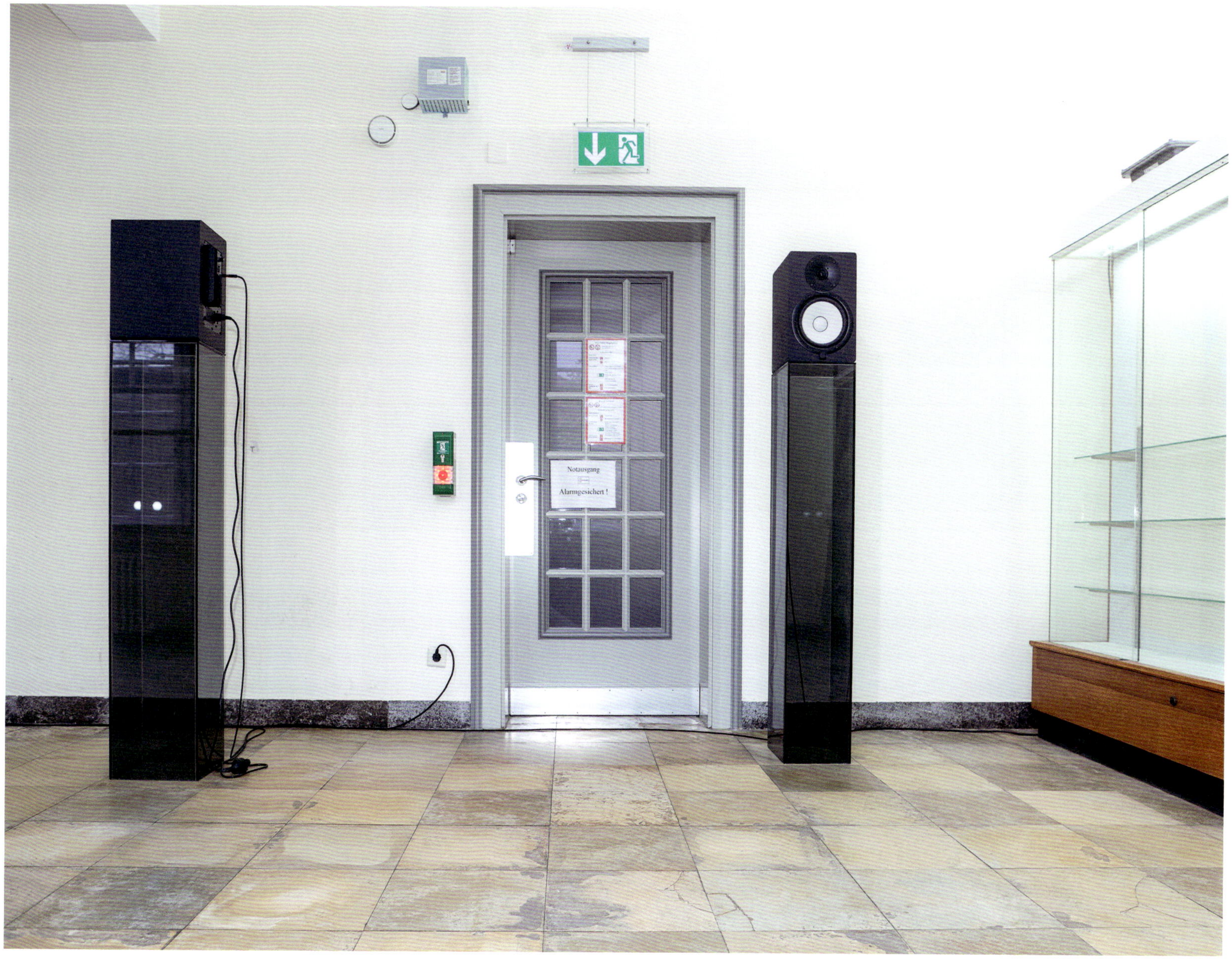

Fig. 112

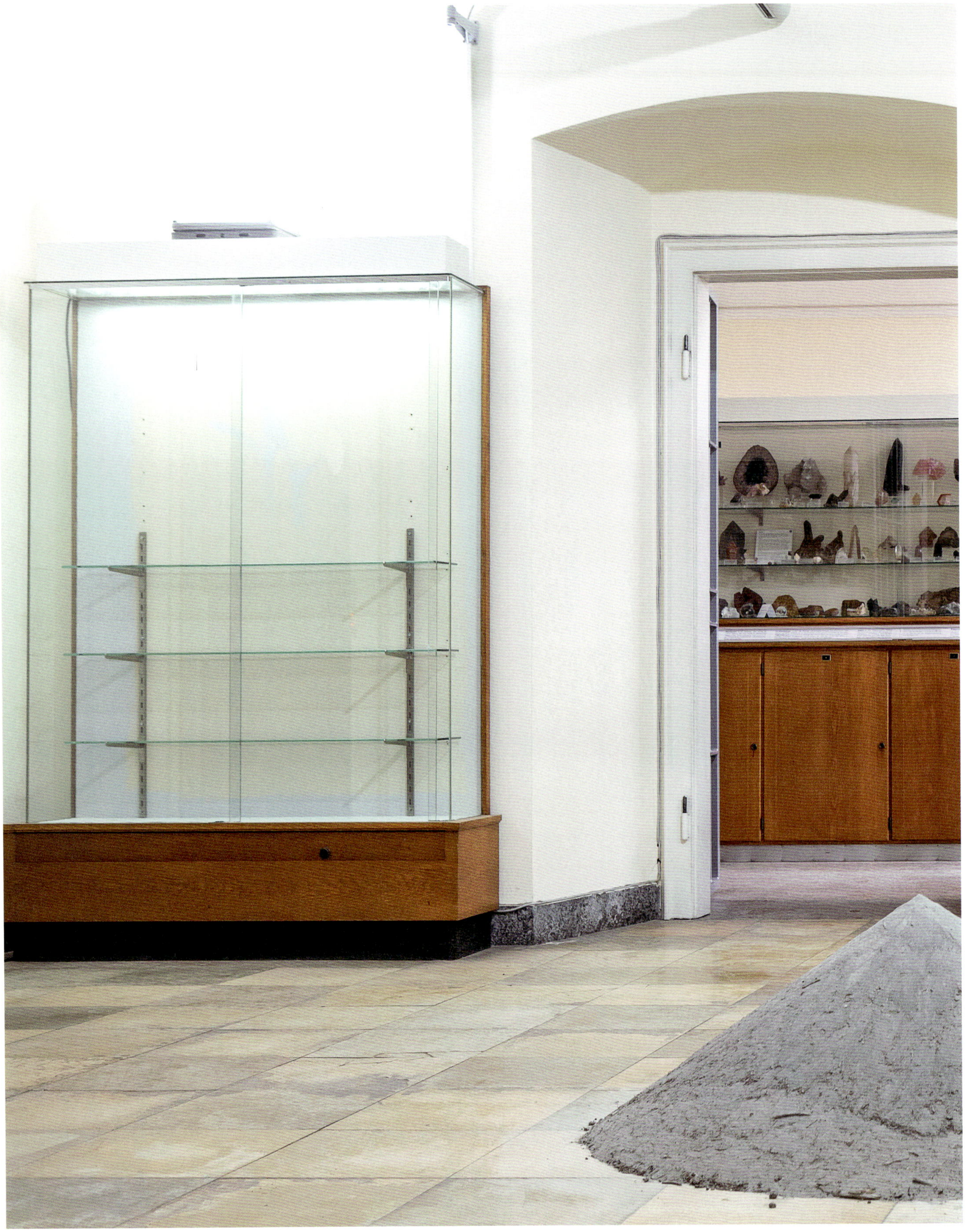

Fig. 113

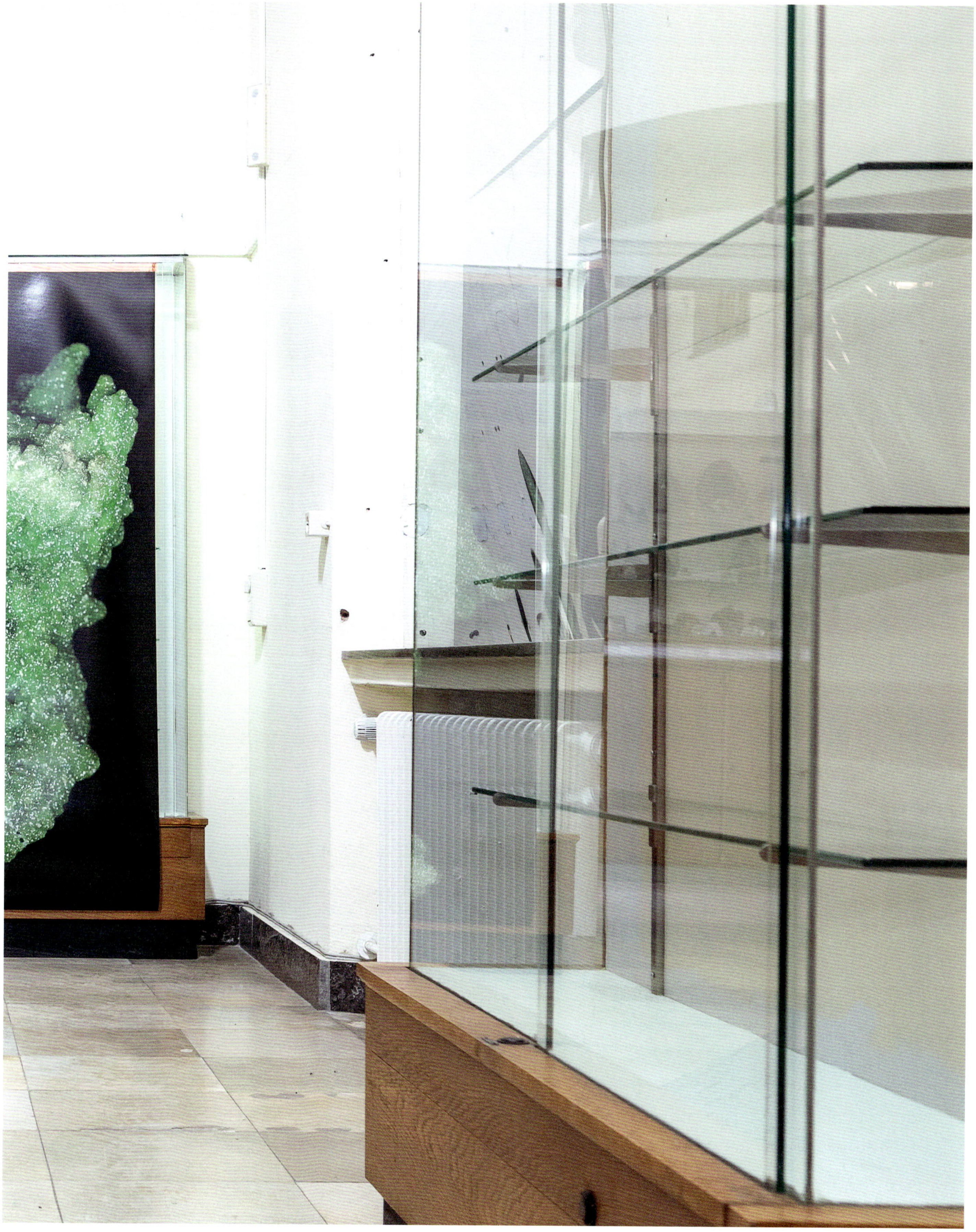

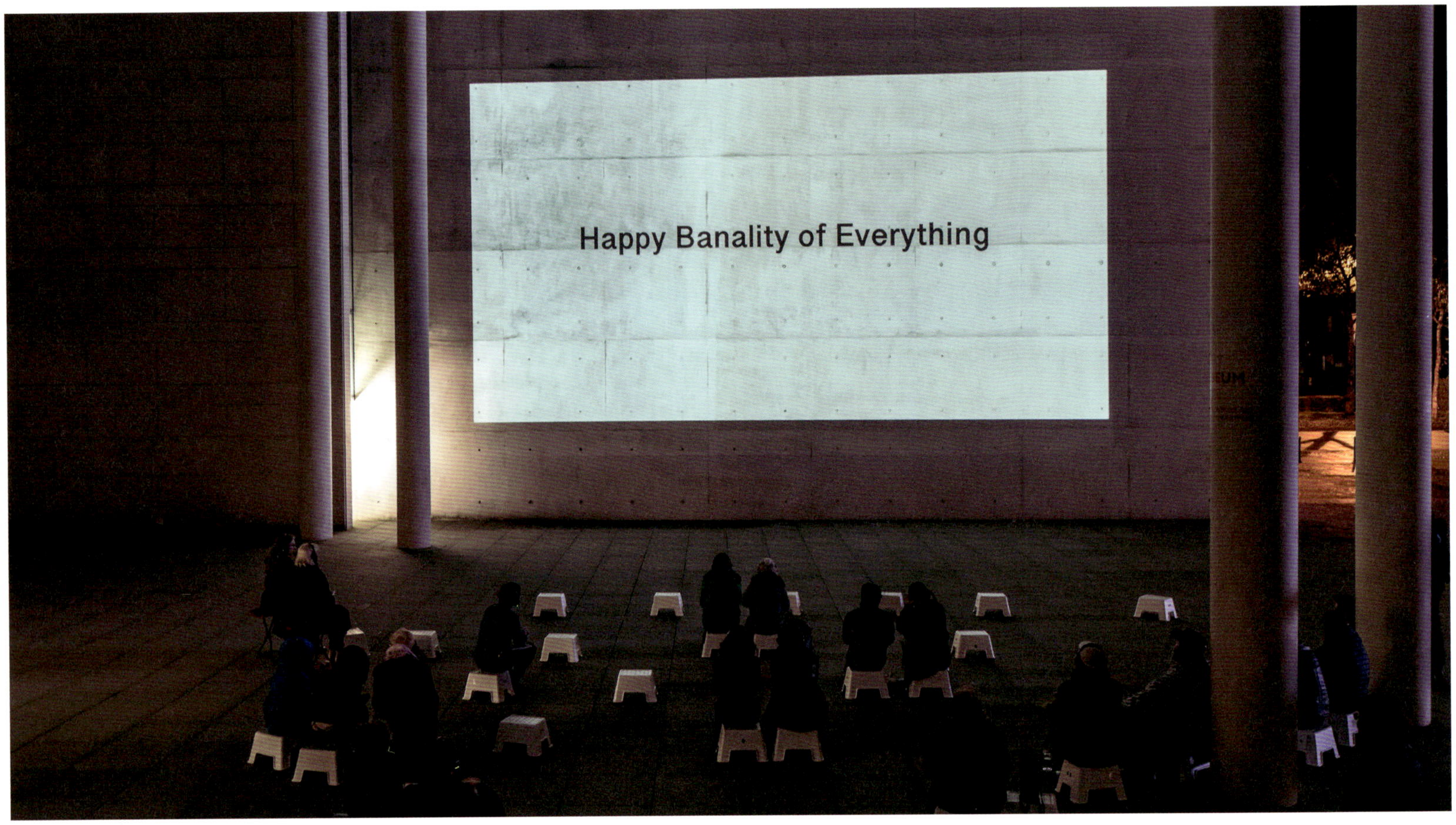
Happy Banality of Everything

TG-T001
transposition 001
(Happy Banality of Everything)

Artist: Sascha Mikloweit
Soundtrack: Irakli Kiziria
Animation: Holger Risse
Year: 2022
Type: Experimental Film
Duration: 00:46:01h
Resolution: 4k
Material: Selected objects from the collection of the Mineralogical Museum of the University of Bonn; field recordings from mining operations;
Richard Wagner—Der Ring des Nibelungen (1876);
Richard Wagner—Parsifal (1882);
Anton Bruckner—Symphony No. 8 (Second Version, 1890);
Anton Bruckner—Symphony No. 3 in D minor (1888/89)

Synopsis by Peripatetic Thinking

The film *transposition 001 (Happy Banality of Everything)* emerged from the *Transpositional Geologies* project's four-year art residency hosted by the Museum of Mineralogy of the University of Bonn. The collection houses 60,000 minerals, all meticulously named, and catalogued according to Strunz's classification system, which follows their chemical composition and crystal structure.

The starting point for the production was artist Sascha Mikloweit's vast archive of 'Object-Images' generated from the collection using a modified flatbed-scanner. After a nine-month collaboration between Mikloweit, the animation expert Holger Risse and the composer Irakli Kiziria, the film emerged as a poetic re-engineering of the scientific collection driven by the trio's desire to convey the untold violent *History* inscribed in these minerals. The effect is at once seductive and profoundly unsettling.

As an act of epistemic disobedience, the film defies the constitutive narrative of *Western* classification systems, re-narrating the collection as a continuum of de-classified textural and morphological events, lavish in their visual wealth and haunted by *Man's* insatiable material appetite. The effect is compounded by the beguiling, pulsating soundtrack. Built in parts from classical compositions of Anton Bruckner and Richard Wagner—referencing the Imperial German colonial *imaginary* of its self-assumed cultural superiority—these fragmented classical sounds appear as acoustic ghosts, along with environmental recordings from mining operations, embedded in techno-industrial sound-scapes. The result offers a spectre of *extractivism*: vague streets of desolate cities, broken components of drifting machinery, fragments of poisoned sky above a river of uninhabitable magma. *Man's* objects of desire, lusted after and attained at all cost, are liberated from scientific nomenclature; and in their digital transposition, they are given a new foreboding voice.
Freed from the fetishism of the museum's vitrine, the film presents the calm dignity and endless promise of undefined material: gleaming like metal, opaque as bark, crinkled to dark plastic foil. *transposition 001* evokes the great wealth of technological invention which inheres in raw materiality: the incomparable hardness of a diamond, the magnanimous ductility of copper, the weapon lurking in uranium, and the unique paramagnetics of gadolinium which cause our smartphones to vibrate.

Fig. 114

Fig. 115 [pp.254–255]

Fig. 116 [pp.256–257]

Fig. 117 [pp.258–259]

Fig. 118 [pp.260–261]

Fig. 119 [pp.262–263]

Fig. 120 [pp.264–265]

The film is part of a larger project which has taken Mikloweit to the former German colonial copper mine of the Otavi Mining and Railways Company, located in present day Namibia, to physically dig and transpose 300kg of sediment, together with the Tsumeb Municipality, to the Mineralogical Museum of the University of Bonn in Germany.

Imminent in Mikloweit's work is the violence necessary to amass institutional collections, and the epistemic violence implicit in their function as benign teaching devices—for generations of geologists. Following a smithsonite mineral back from the collection to its geographical origin in Tsumeb, has lead Mikloweit to a place where there once stood the 'Green Hill' of chrysocolla and malachite, whose secret the Hai||om community closely guarded. The Hai||om would trade copper ore with the Ndonga community, who were the keepers of the knowledge of smelting and crafting the ore—most likely utilising vacant termite mounds as furnaces. All that remains today of the 'Green Hill*'* and this gracious exchange of knowledge is a defunct colonial open pit mine, burrowed deep into the Earth—a counter monument to *Man's* insatiable global appetite for metal.

Given that all advances of technology are already always complicit with complex histories of material extraction, all *transpositions* in this project involving digital production cycles are likely to have used equipment containing Namibian minerals themselves. From within this entanglement, the *Transpositional Geologies* project at large and the film in particular remain open and reflexive meditations on the relationships between colonial pasts and techno futures, their knowledge productions, *Western imaginaries,* and our places within them.

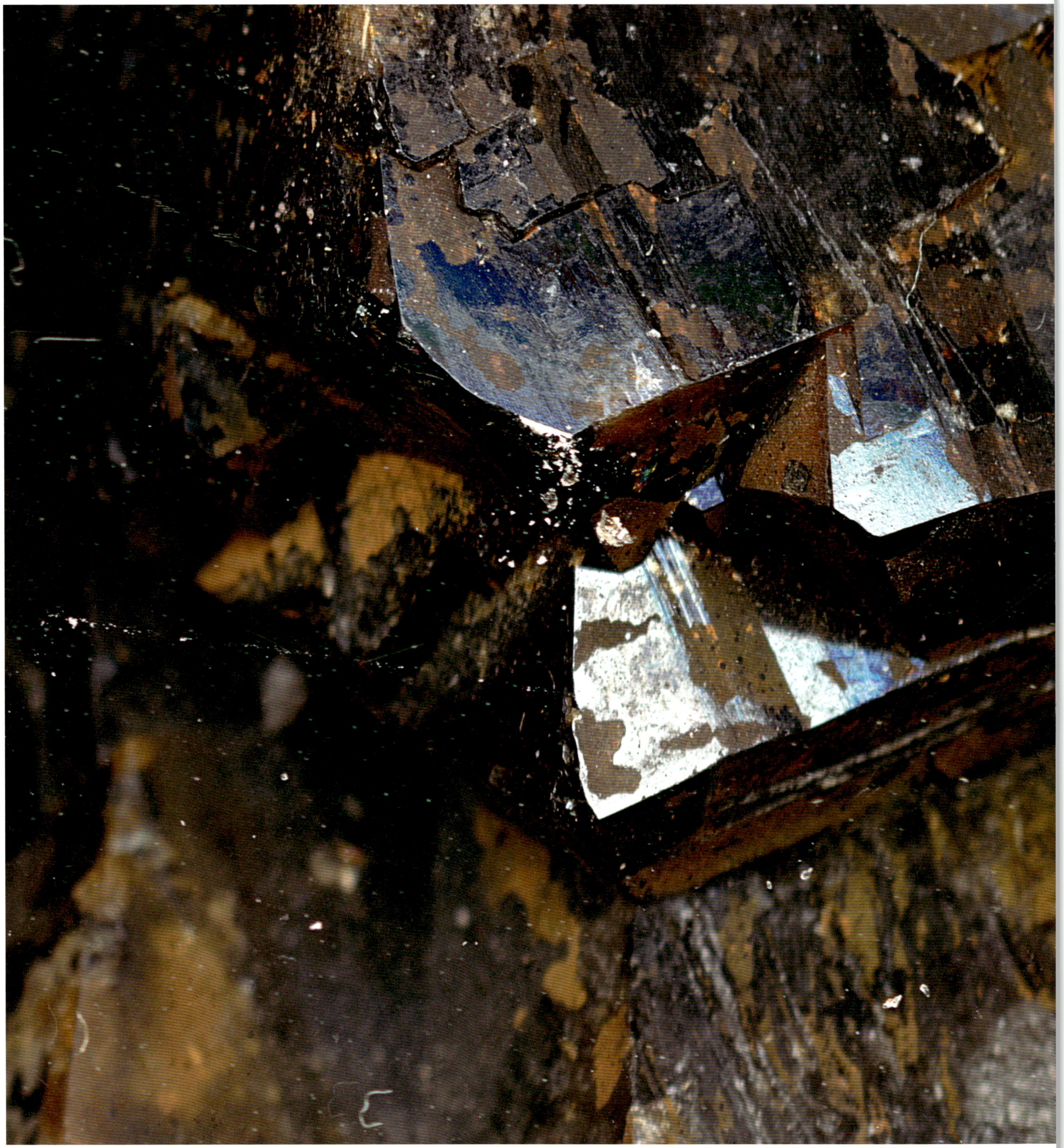

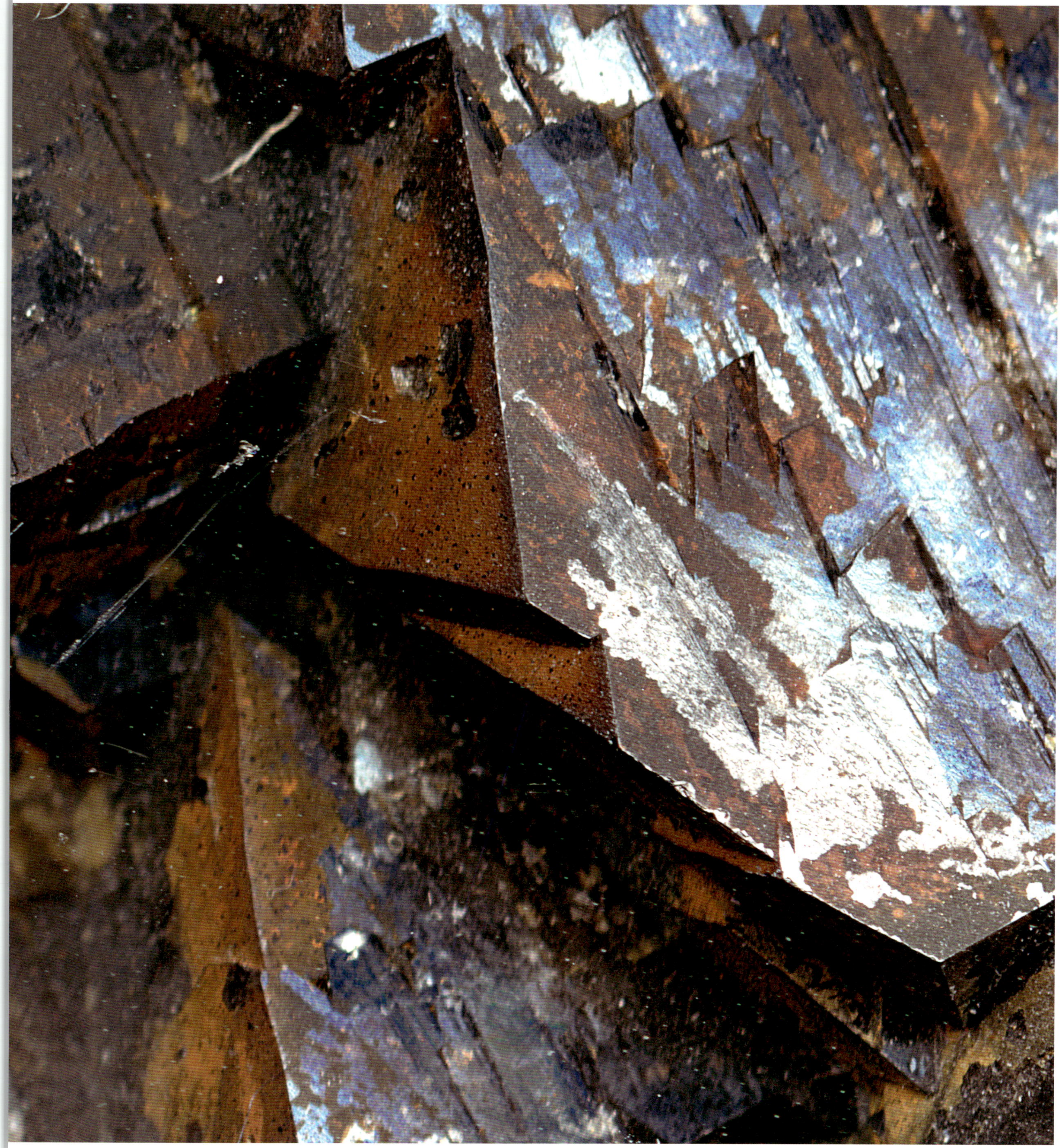

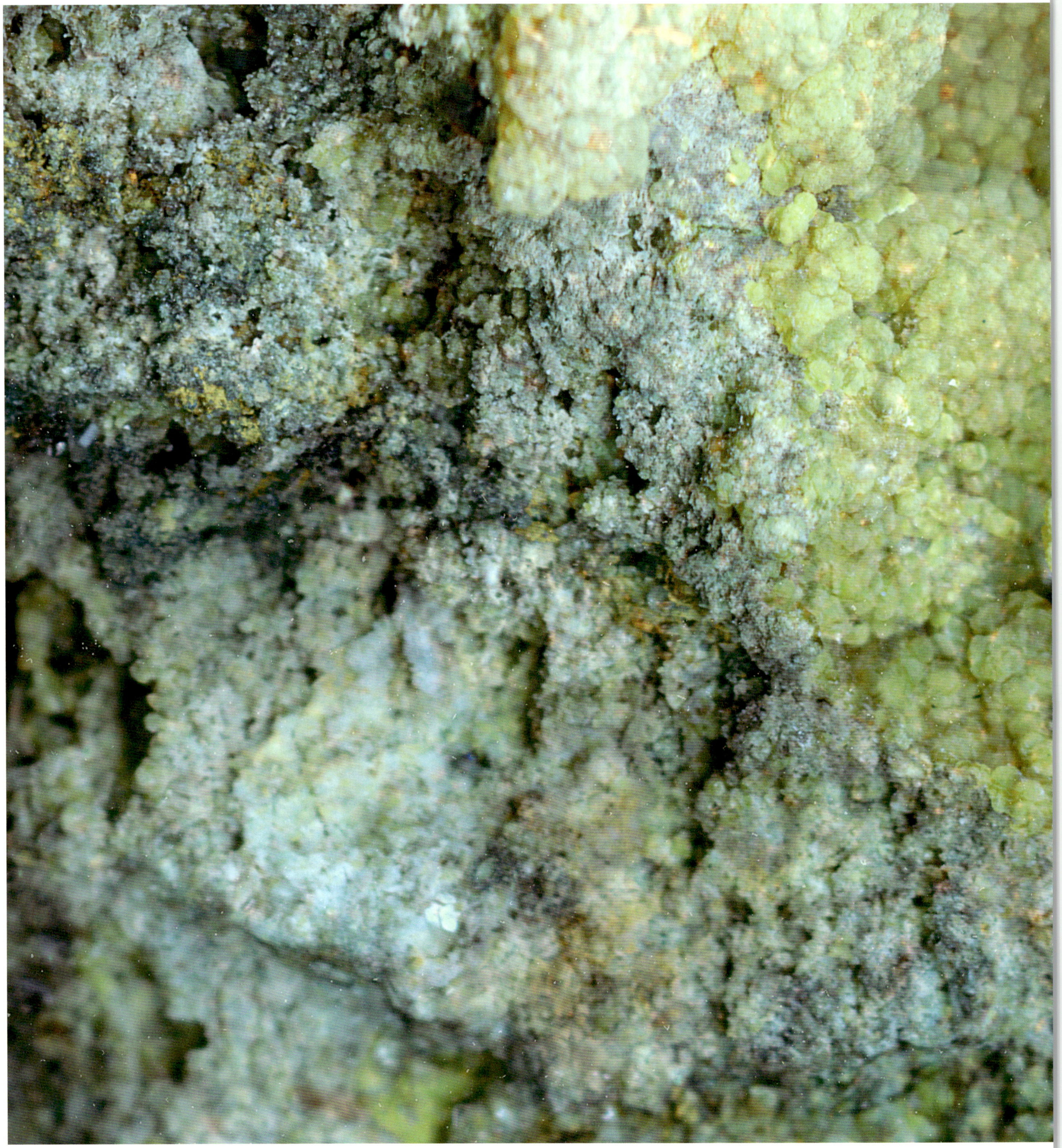

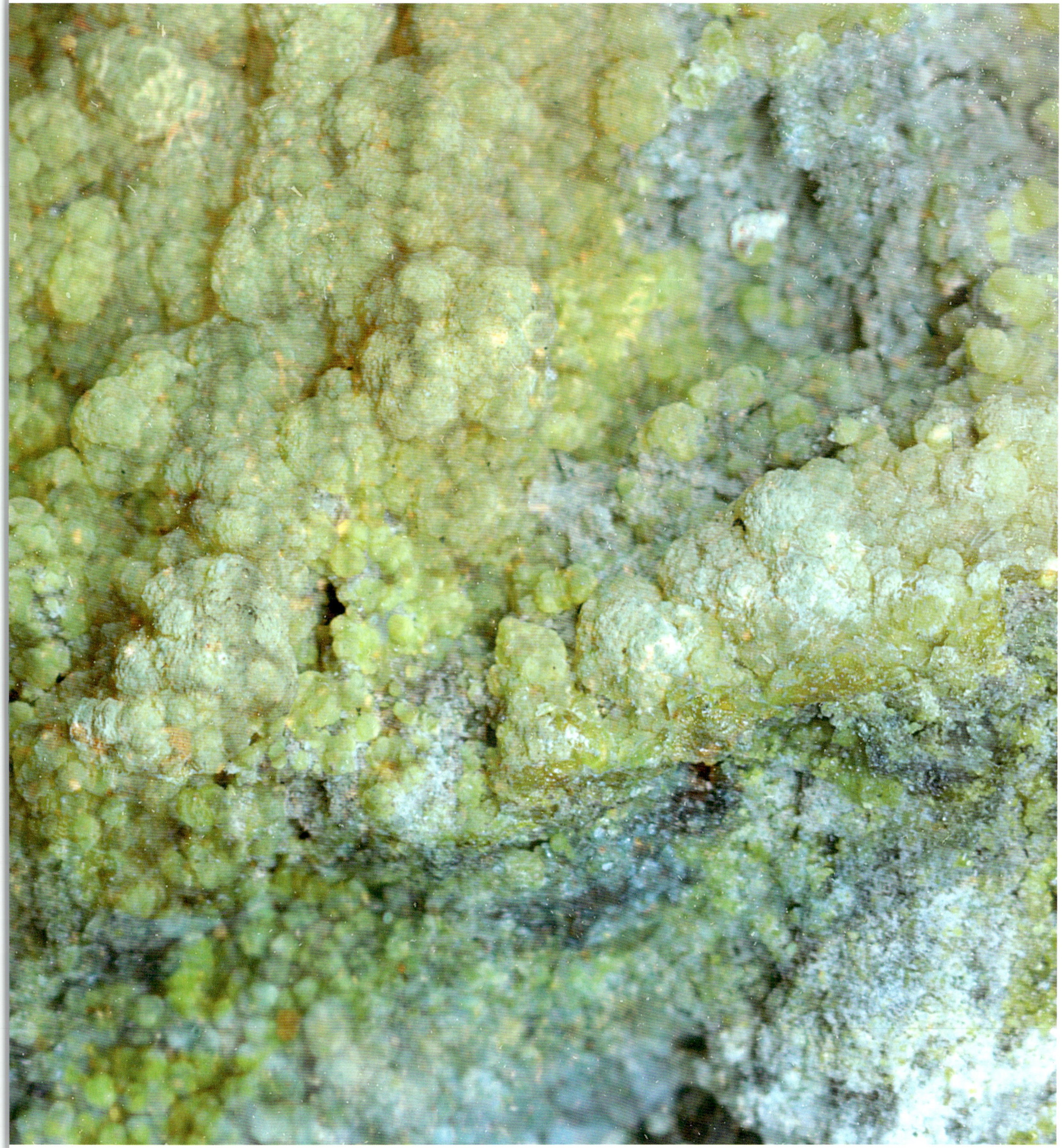

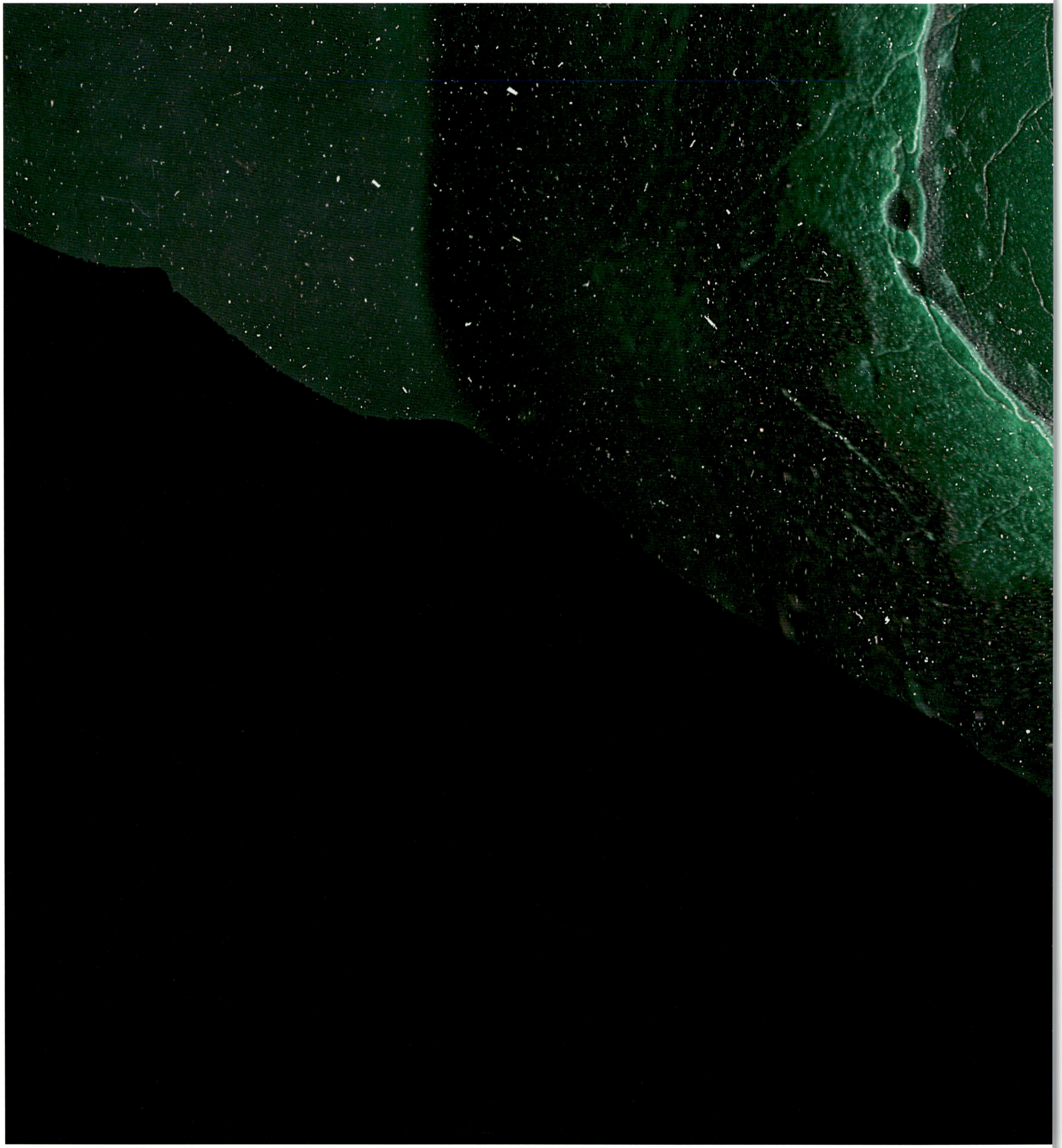

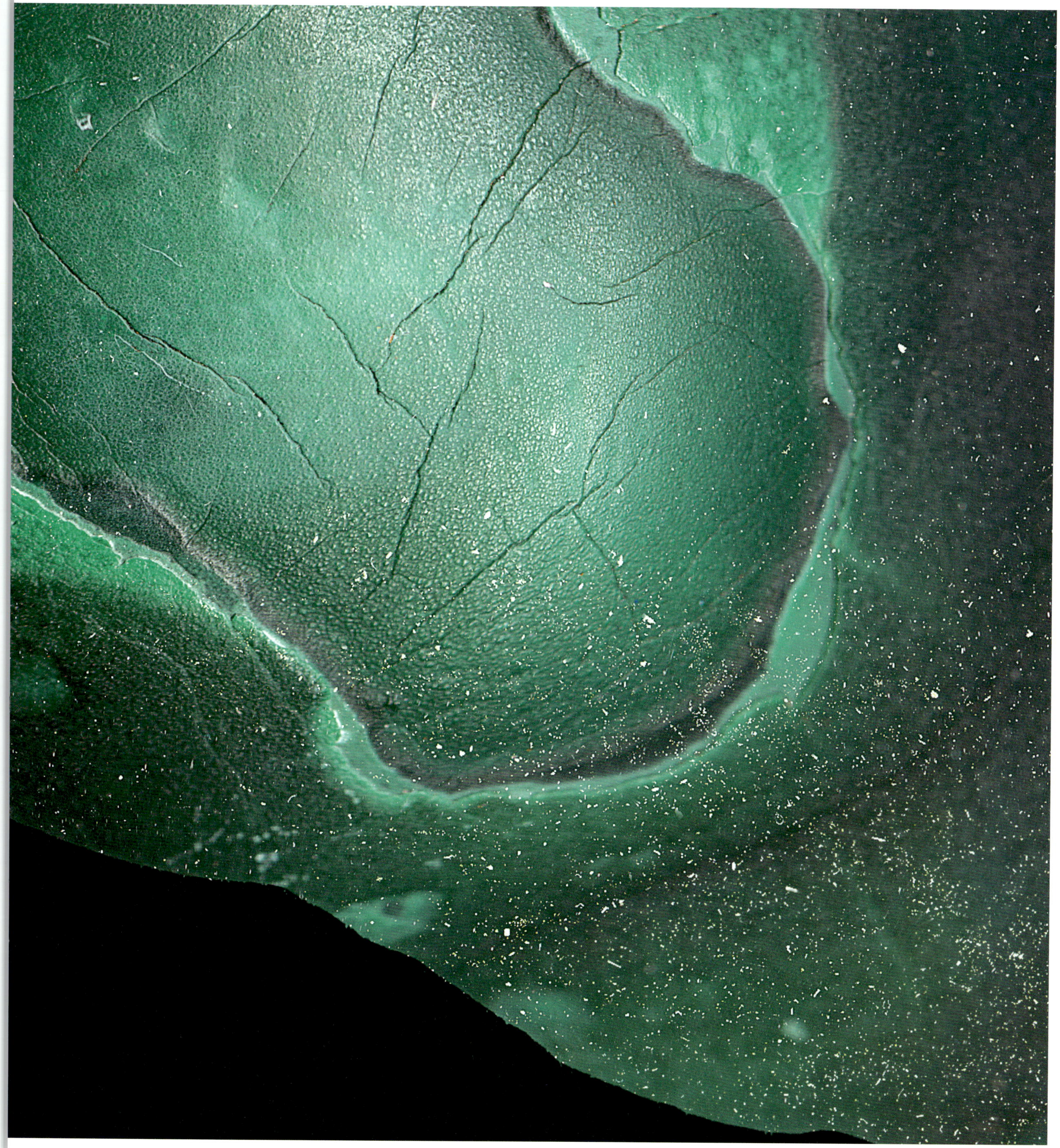

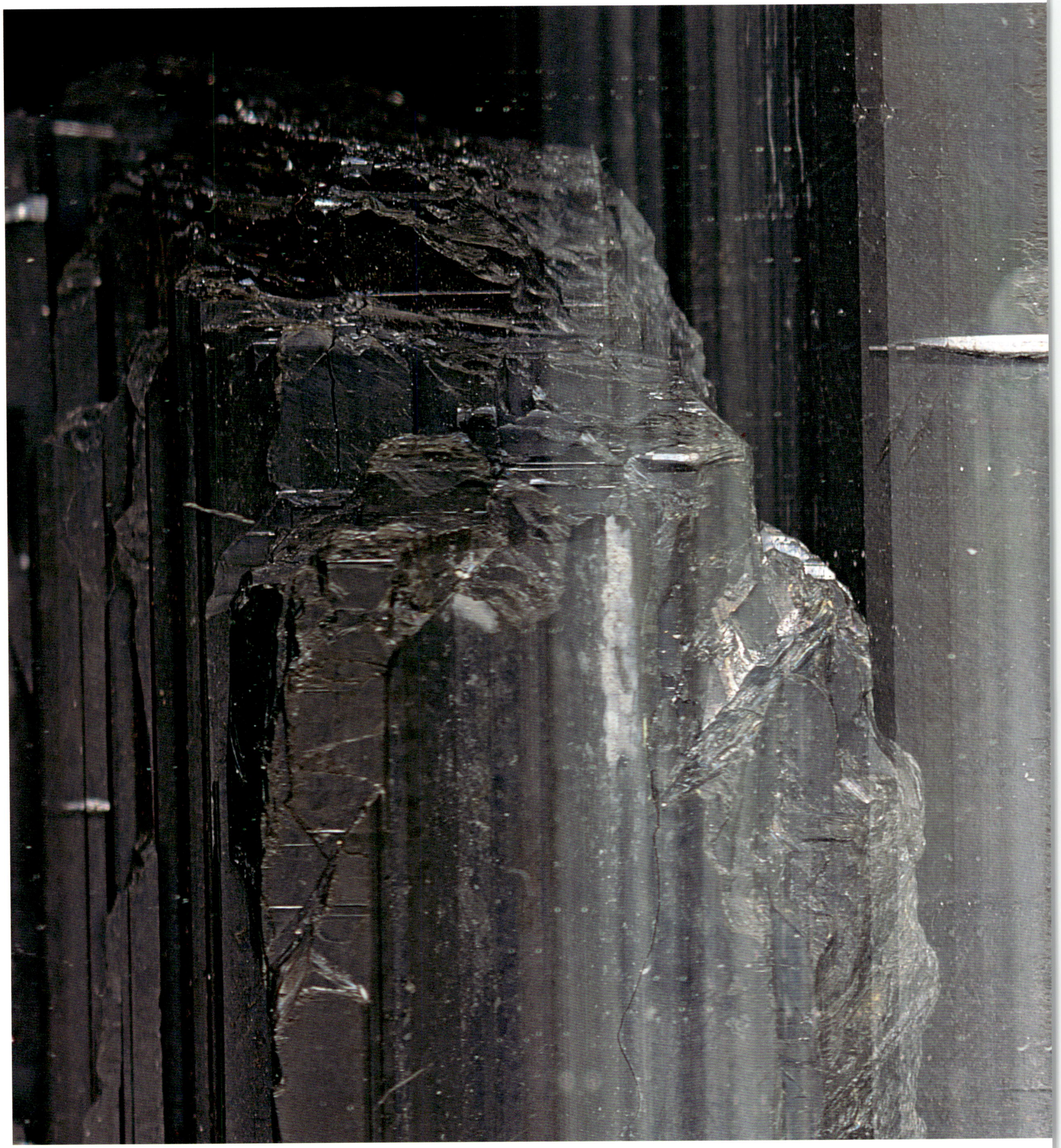

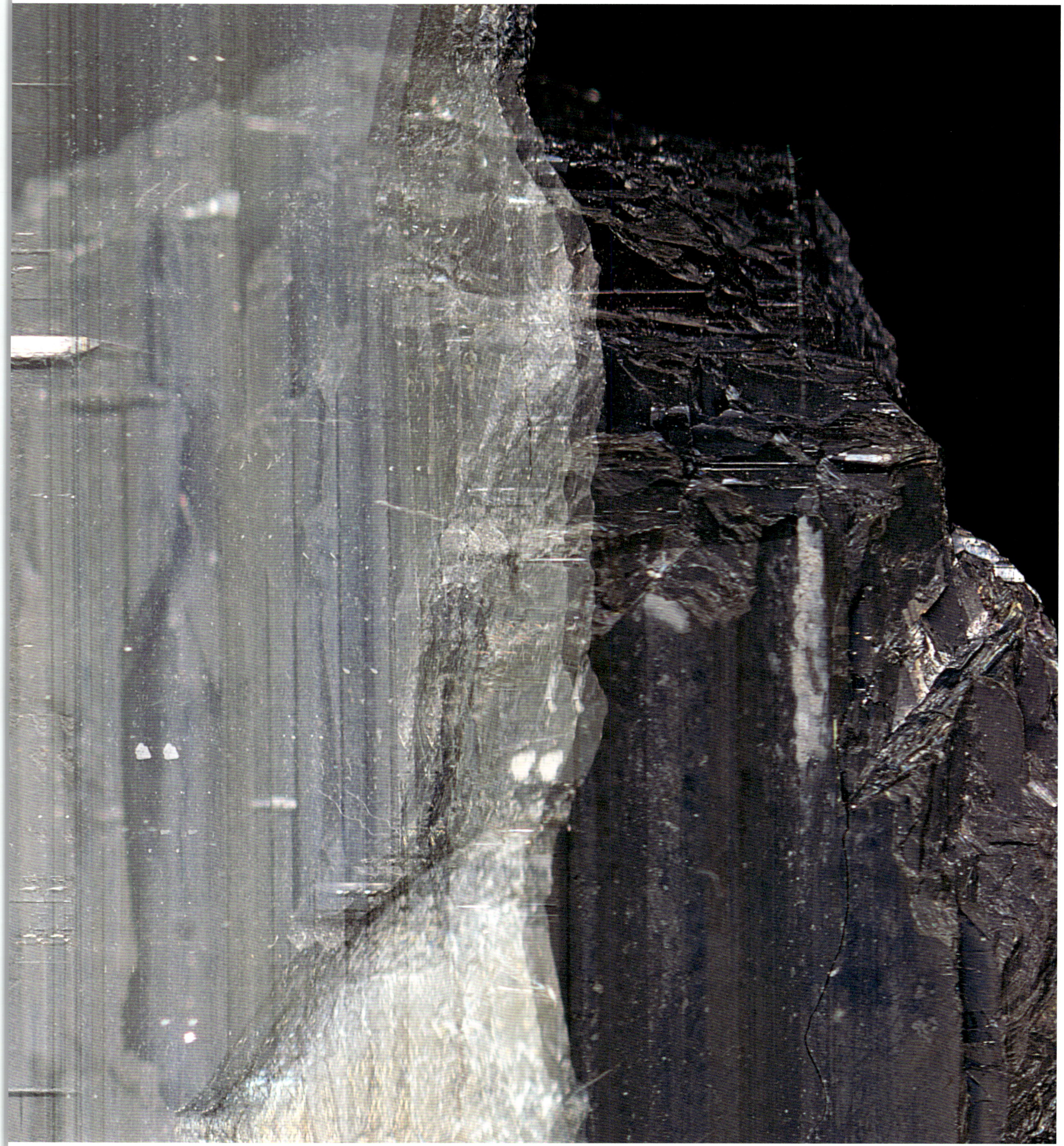

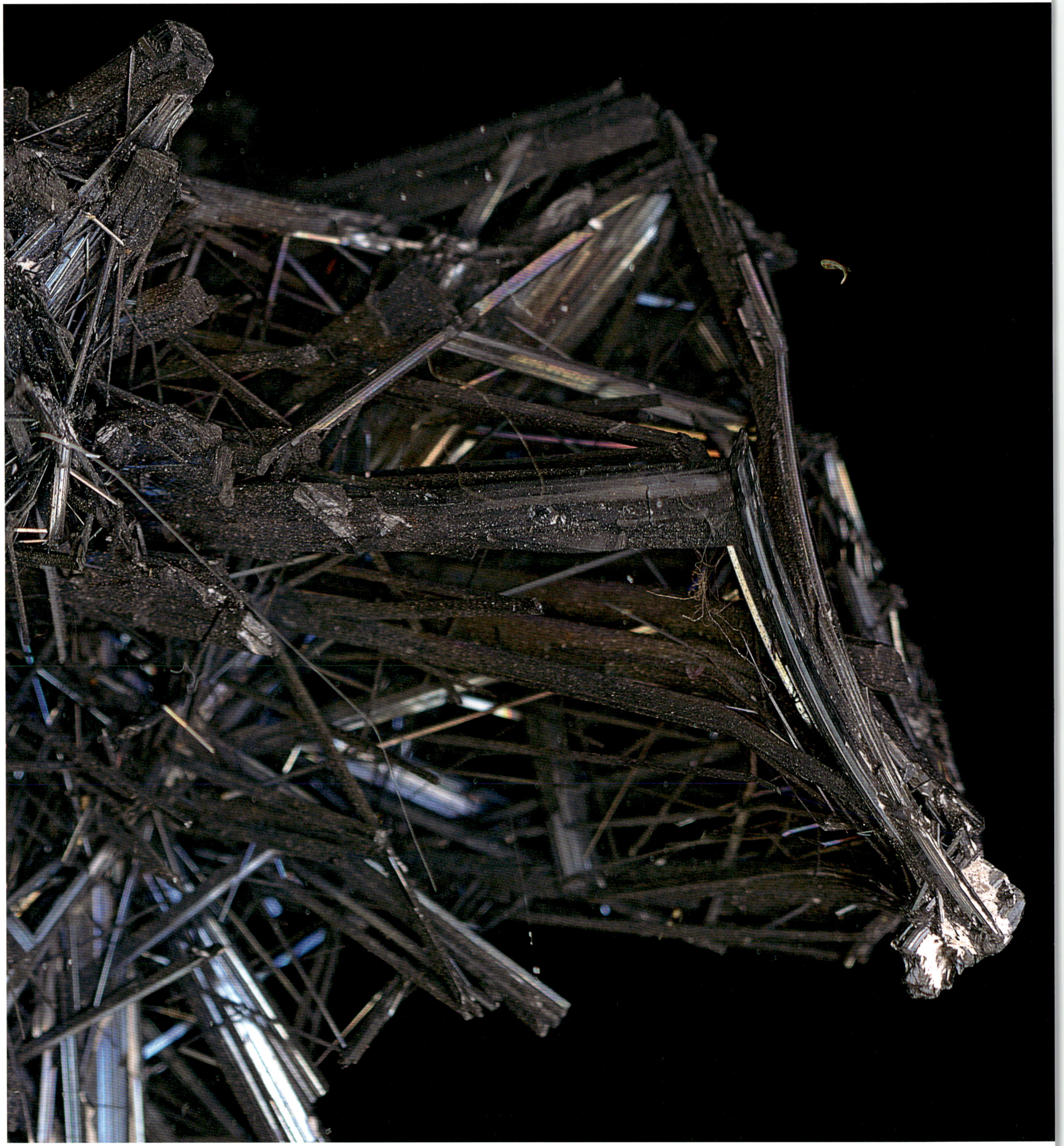

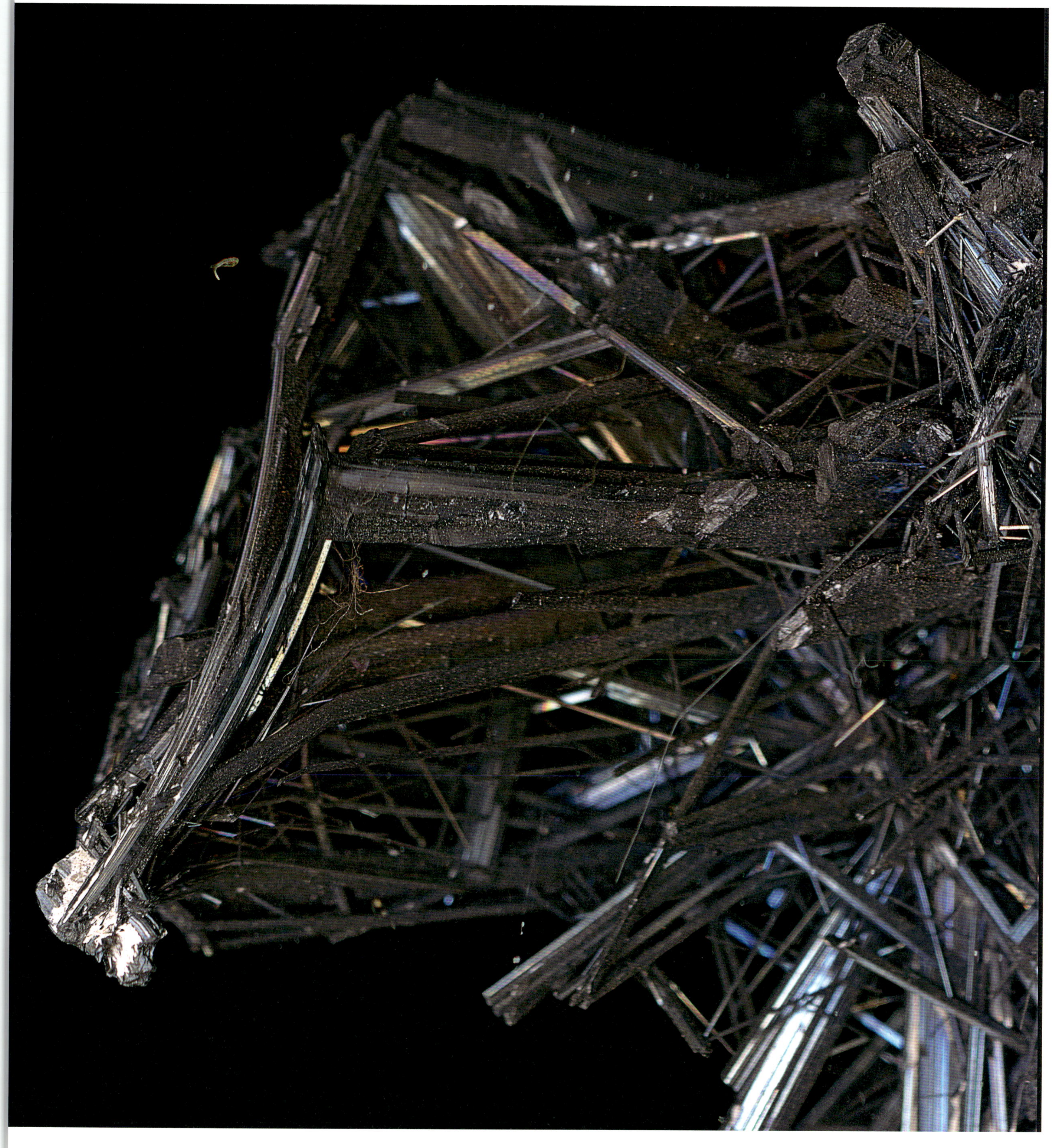

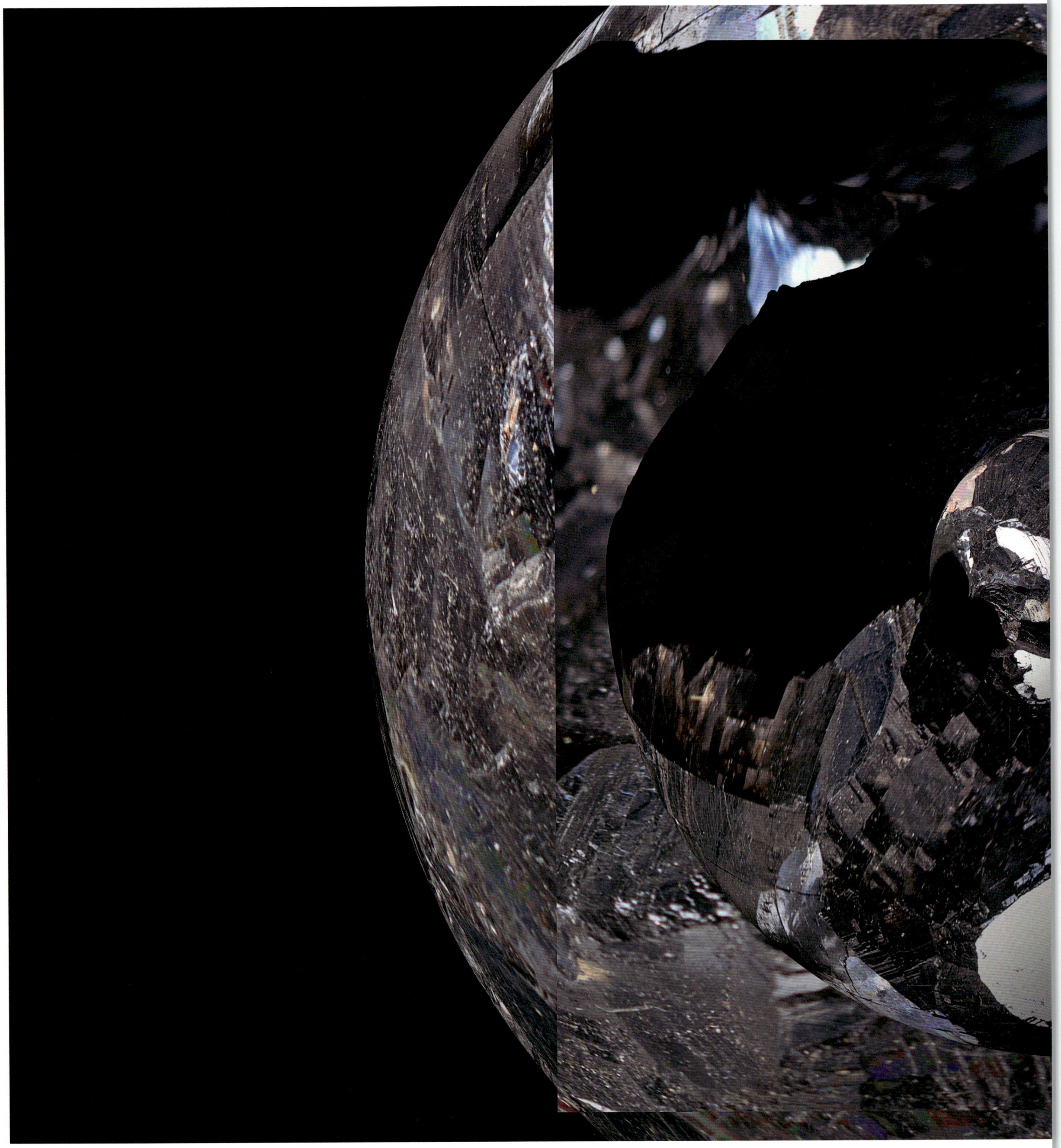

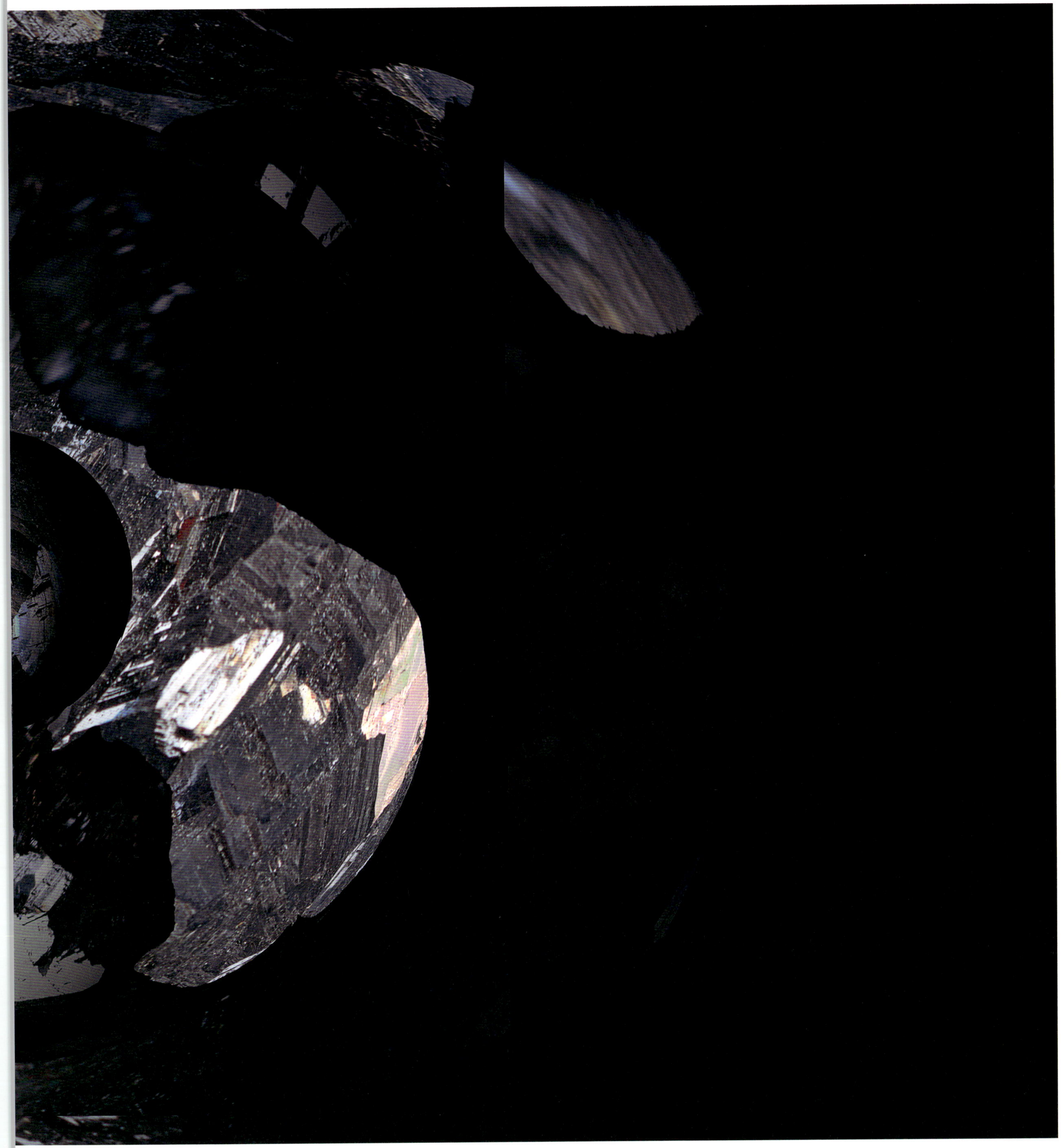

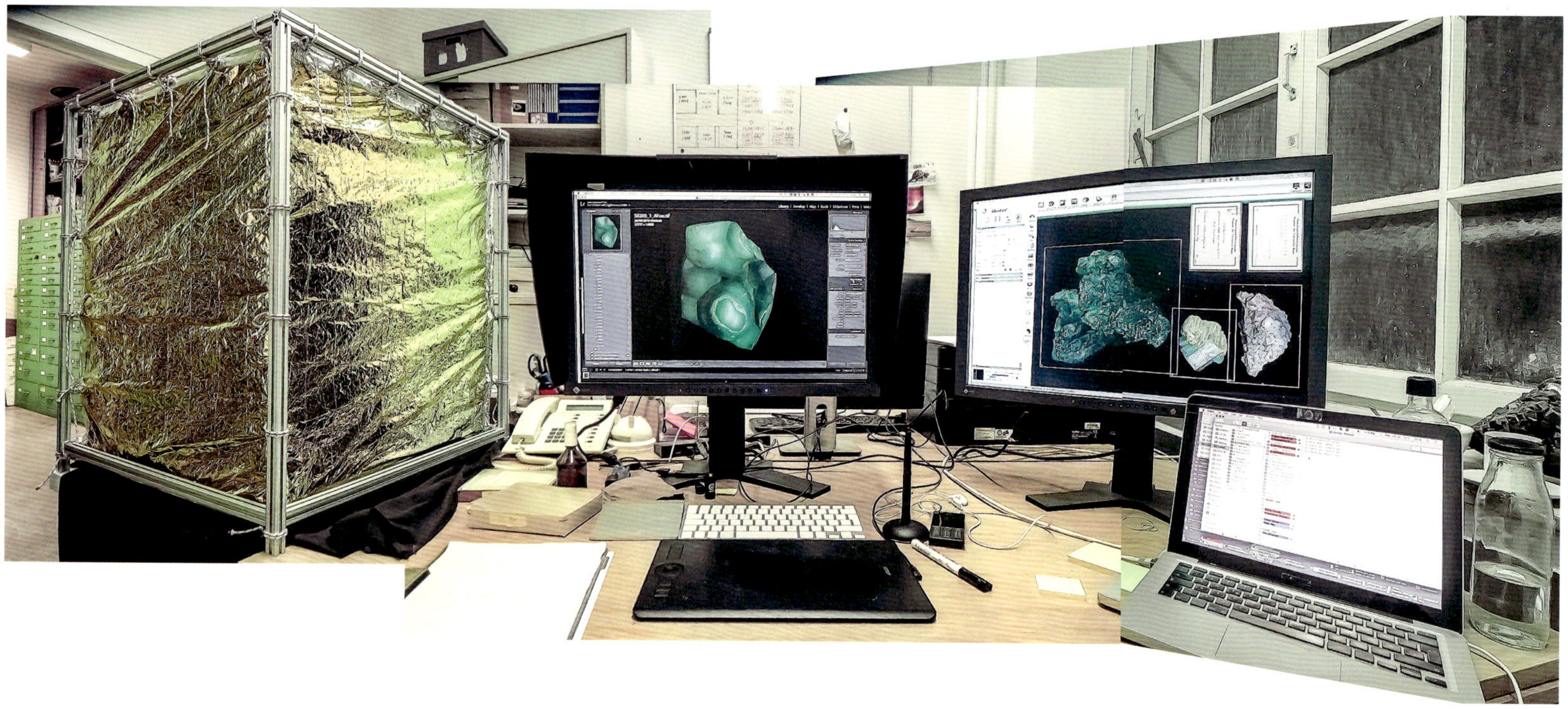

A Non-Violent Explosion in the Taxonomic Schloss (Towards a Neo-Baroque)

Paul O'Kane

Abstract: In this text, art writer Paul O'Kane creates a speculative description and evaluation of a series of projects carried out and evolving under the attention of Sascha Mikloweit. The writing brings Mikloweit's latest work into alignment with previous projects, as O'Kane strives to reveal continuities and a process of growth and development, in terms of scale, ambition, and idiosyncrasies in Mikloweit's practices. Taking 'explosion' as an initial theme, the writer finds references in the ideas of Georges Bataille. The technology of scanners becomes embroiled with the epistemologies of the museum and vitrine, as O'Kane moves towards a Deleuzian 'neo-baroque' model, which might be capable of accommodating a certain explosive formlessness that might be said to compel Mikloweit's enquiries. Screen might 'become vitrine' for a 21st-century world, in which only a baroque model provides sufficient complexity within which to play with such profound pre- and post-historical references.

Keywords: explosion, neo-baroque, baroque, taxonomy, Deleuze, Bataille, scan, screen, vitrine, sublime

All is explosion, every 'thing' an event. Even this writing, appearing to take place in lines of ordered words, resonates in ways that spread in many directions at various speeds. Writing is an explosion, a diaspora, spraying out, emitting from the event of its starting point. How do we shape and control this energy, this writing, while also 'giving it its head'—as drivers of harnessed horses sometimes say. The solution is not to impose a known form out of a fear of chaos, but to allow an unanticipated form and content to arise from the material, from the event, from the energy of the writing. Another solution might be to simply become more comfortable with chaos and more familiar with formlessness (Deleuze and Guattari, 1986; 2004).[1]

The apparent formlessness of an explosion is something that can nevertheless be valued, interpreted, analysed. The enduring popularity of firework displays might remind us that a certain affective thrill or *jouissance* can derive from the making and witnessing of explosions made to contrive complex patterns. And the most powerful explosions of all, like those produced by 20th century atomic warfare, have left their form, image and shape imprinted on our cultural retina.

Meanwhile, a 21st century, already shaped by terrorism, with its suicide bombers monstrously extending the legacy of 1970s airplane hijackers, amplifies the explosion to a newly grim status, wherein it comes to challenge representation—of both the political and artistic kind. Something like this was implied by the *avant garde* composer Stockhausen when he referred to the event we call '9/11' as 'The Devil's Masterpiece'—elsewhere described as 'the mother of all images', and also well illustrated by Johan Grimonprez's postmodern classic video essay 'Dial H-I-S-T-O-R-Y' (1997).

Fig. 121

Dissident Surrealist, Georges Bataille, listed 'spit' among a series of examples he gave of what he called 'the formless' (Bois and Krauss, 1997). Spit, too, is an explosion, as are those everyday parts of speech known as 'plosives'. Here I briefly digress into an anecdote wherein I recall meeting the artist Alex Schady in London in the 1990s, when his practice was, like that of many of us at that time, influenced by Bataille. Schady made a work in which he spat on to sheets of paper or handkerchiefs, then sewed around the resulting stain to give the marks a form and permanence that might proffer meaning in something like the way the ancients once deciphered entrails or 'read' residual tea leaves.

For Bataille, and perhaps for Schady too, the apparent formlessness of explosive spit might have both profane and sacred implications.[2] Spit invokes dirt, ill health, abjection, and yet its multiple and complex trajectories might equally remind us of infinity, eternity, or transcendent escape, like a baroque religious ceiling painting that seems to spiral up and away into the forever. Spit might even invoke the so-called 'Big Bang' theorised by scientists to have initiated the universe.

That which reminds us of all that is beyond human scale and control has elsewhere been aesthetically accommodated within Romantic theories of 'the sublime'. But Romantic art might also share unexpected affinities with the Catholic baroque art referred to above, though only as much as it shares with spit, sneezes, fireworks and bombs, as well perhaps as thunderstorms and other forms of horror-inducing or vertigo-inducing experience.

[1] Here the figure of James Joyce, and his adoption by Deleuze & Guattari (1986; 1988; 2004) comes to mind.

[2] See also the art of Gustav Metzger, Cornelia Parker, and others.

A Controlled Explosion in London – Scanners & Vitrines

I first encountered the artist Sascha Mikloweit and his practices in London circa 2009. At that time, he had acquired the remains of a briefcase which, left unattended and suspected of being a terrorist device, had been subjected to a 'controlled explosion' at a major London railway station by an anti-terror unit of the British Transport Police. Without going into detail about its contents and the subsequent narratives that the artist derived from the same, what was initially most striking about Sascha's use and analysis of his 'unattended object' was his specific use of high-quality, large-scale scans and digital printing, deployed to amplify and transmit an appropriate image, and used as a way of capturing, recording, and sharing this energetic event and its residual material. Sascha has recently reflected on this choice of technological representation and explored the idea that scanning is a particularly relevant vehicle for representing a thing once regarded as an object (an object left unattended) now transformed by explosive forces into an event.

Unlike previous generations of photographic processes utilising cameras, lenses, shutters and distance, the flatbed scanning pursued by Sascha requires objects to be placed directly upon or pressed up against a glass sheet. The image is then created not by briefly clicking open a shutter and aperture, but by calmly gliding over or under it, like a hawk, glider, or drone, while reading (or, as Sascha prefers, 'inventing') what it 'sees' as line after line of data.[3] The results tend to engage the audience with the special qualities of the digitally scanned image while celebrating what we might call its 'super superficiality', i.e., a superficiality spectacularly redoubled and affirmed.[4] But while the resulting seductive and dramatic scan imagery might be valued as 'merely' aesthetic (or comparable with that 'flatness' and shallowness embraced and affirmed by Andy Warhol or described by Douglas Crimp with regard to Robert Rauschenberg), there may yet be hidden depths and complexities to be gleaned here.

Anyone who has made 'analog' photographic negatives, and subsequently scanned them, will know the pleasure of having thus seemed to bring the past into the present while translating one kind and quality of image into another. As a result, the past is made newly malleable, preserved and ordered by the scanning process, but is also potentially freed of its actuality by means of its virtualisation. The scanned, digital image is no longer subjected to the incarceral conditions of the physical vitrine and of the paper-based photographic print. In fact, the vast realm of digital, screen-based images might today be justifiably regarded as one great image and one great vitrine, any part of which can be accessed by almost anybody, anywhere, simultaneously, countlessly, via any hand-held, digital, screen-based device. And here, amid the massive murmurations of virtual clouds and belligerent 'barrages' of Big Data, the neo-Catholic imagery of the neo-baroque once again calls out for comparison.

Through his use of scanners, Sascha draws our attention to a historical procession of taxonomic technologies, enhancing their visibility while rendering them relativised, questionable, peculiar, and particular—ready for comparison with other, prior and future technologies. The various processes involved in the realisation of his work, while sensual, spectacular and initially transmitted in striking aesthetic terms, is also a form of epistemology, a particular way of knowing, of making sense of the world, of an object, of an event.

Perhaps we can think of other examples of objects, events, or ideas that we come to know and understand first by 'exploding' them, then scanning and presenting the remainder (see our reference to 'entrails' above), by first destroying their integrity, shattering their identity, their 'thing-hood', in order to read and know them anew?

It is hard to resist here aligning Sascha's process with the (anti-) philosophical process known as deconstruction, which, we might say, destroys in a non-violent manner, and only to construct. Or perhaps we could say that deconstruction 'constructs violently' (and deconstruction's founding [albeit anti-foundational] guru Jacques Derrida has spoken of 'doing violence to one's thought'). While explosions appear destructive, painful, harmful, if we regard everything as explosion and all as diasporic, we might be able to shrug off the 'violence' that is seemingly inherent to the explosion—a *moral* association that we automatically award it.[5]

Today, any post-enlightenment, post-human, and post-modern project that can deconstruct and read history 'against the grain' provides a form of explosion that reaches out and ripples through the given and inherited order of things. It nevertheless appears to treat the *status quo* non-confrontationally, we might say peacefully, with infinite delicacy, even while reducing it to tiny components—even to dust (Benjamin, 1968, aphorism VII, p.257).[6] In this way we emulate F. W. Nietzsche's proto-deconstructive notion of 'philosophising with a hammer', albeit now interpreting this hammer not necessarily as a large, crude, weighty device, deployed with malice and maximum force, but as the kind of hammer deployed by a jeweller, watchmaker or a doctor seeking responses from a patient's knee.[7]

Ariane 5

Sascha Mikloweit extended his London-based enquiries by later working with the European Space Agency (ESA), again using his scanning process, but this time to engage with debris resulting from the destroyed payload of an Ariane 5 rocket which exploded shortly after take-off on 4th June 1996.[8] Here, a scanning process that once again tends to

[3] See also Benjamin's Chinese curios (1979, pp.49–50).

[4] Here, 'affirmation' and 'nay-saying' are references to Nietzsche, who seemed to have strived in his thought to affirm as much as he possibly could, if not, everything; meanwhile disparaging and satirising a tendency and tradition that he called 'nay-saying'.

[5] Invoking Nietzsche's essay On Truth and Lies in an Extra-Moral Sense (Kaufman, 1976, pp.42–47).

[6] Hence my 'non-violent explosion' of the title. See also David Campany's curation A Handful of Dust (2017) at London's Whitechapel Gallery.

[7] Nietzsche, along with Heidegger, influenced the 'deconstruction' of Jacques Derrida. Elsewhere Nietzsche writes 'It is the stillest words which bring the storm. Thoughts that come on doves' feet guide the world', perhaps emphasising a certain gentleness he deemed crucial for thought, despite his reputation for iconoclasm (Nietzsche,1969, p.168). We might also consider Twilight of the idols (1915), another of Nietzsche's concepts and titles, by means of which we are reminded that any conflation of art and destruction is bound to conjure-up consideration of the history of iconoclasms that have always attended the history of art and ideas.

[8] Ariane flight V88 was the failed maiden flight of the Arianespace Ariane 5 rocket, vehicle no. 501, on 4 June 1996. It carried the Cluster spacecraft, a constellation of four European Space Agency research satellites. The launch was in the Guiana Space Centre, also called Europe's Spaceport. It is a European spaceport to the northwest of Kourou in French Guiana, an overseas territory of France in South America. The artist clarifies that the material he had access to was Cluster mission satellite debris, residue of four satellites that were the payload of Flight 501 / V88 of the Ariane 5 rocket type, which are archived at the European Space Operations Centre (ESOC), Darmstadt, Germany.

Fig. 122

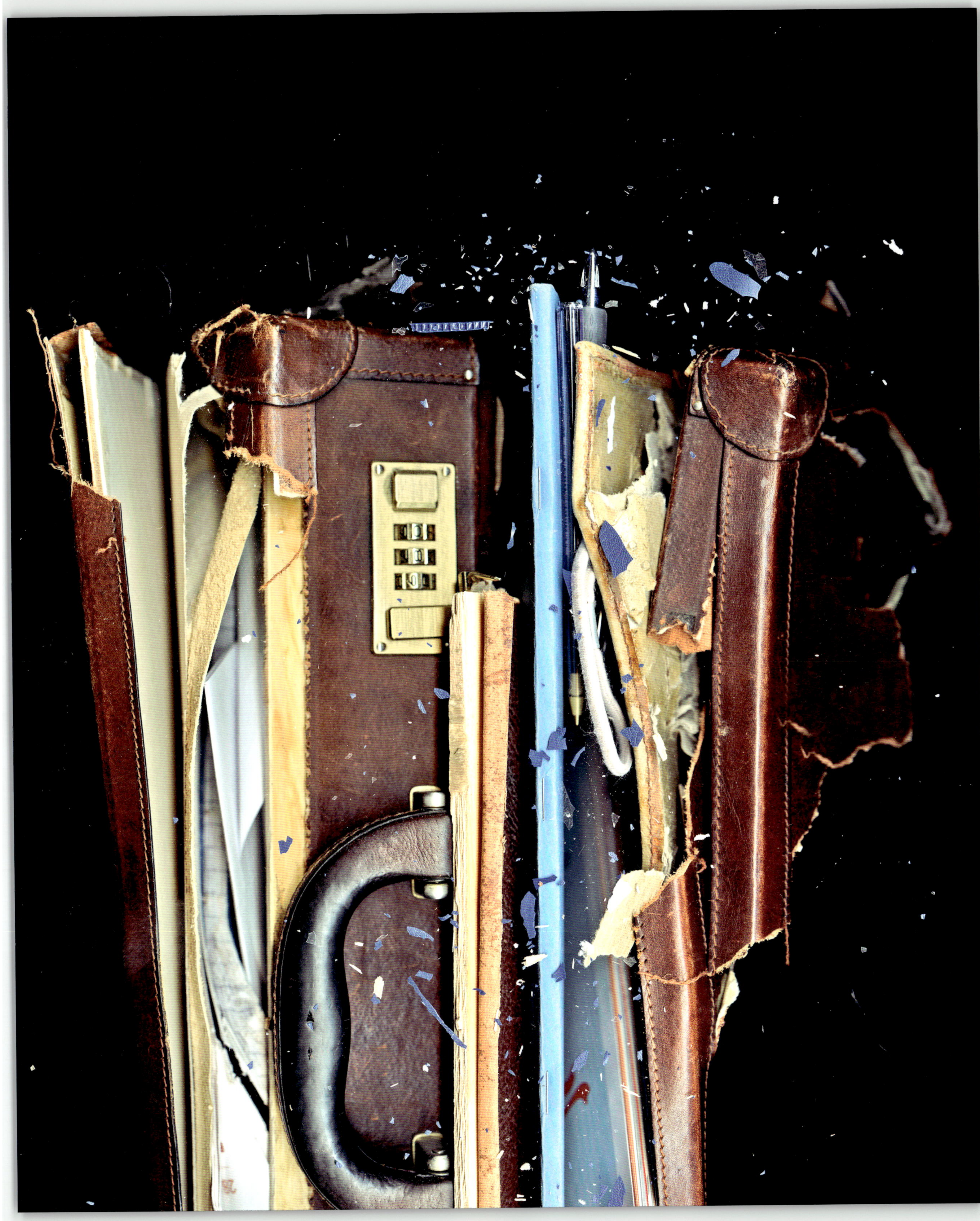

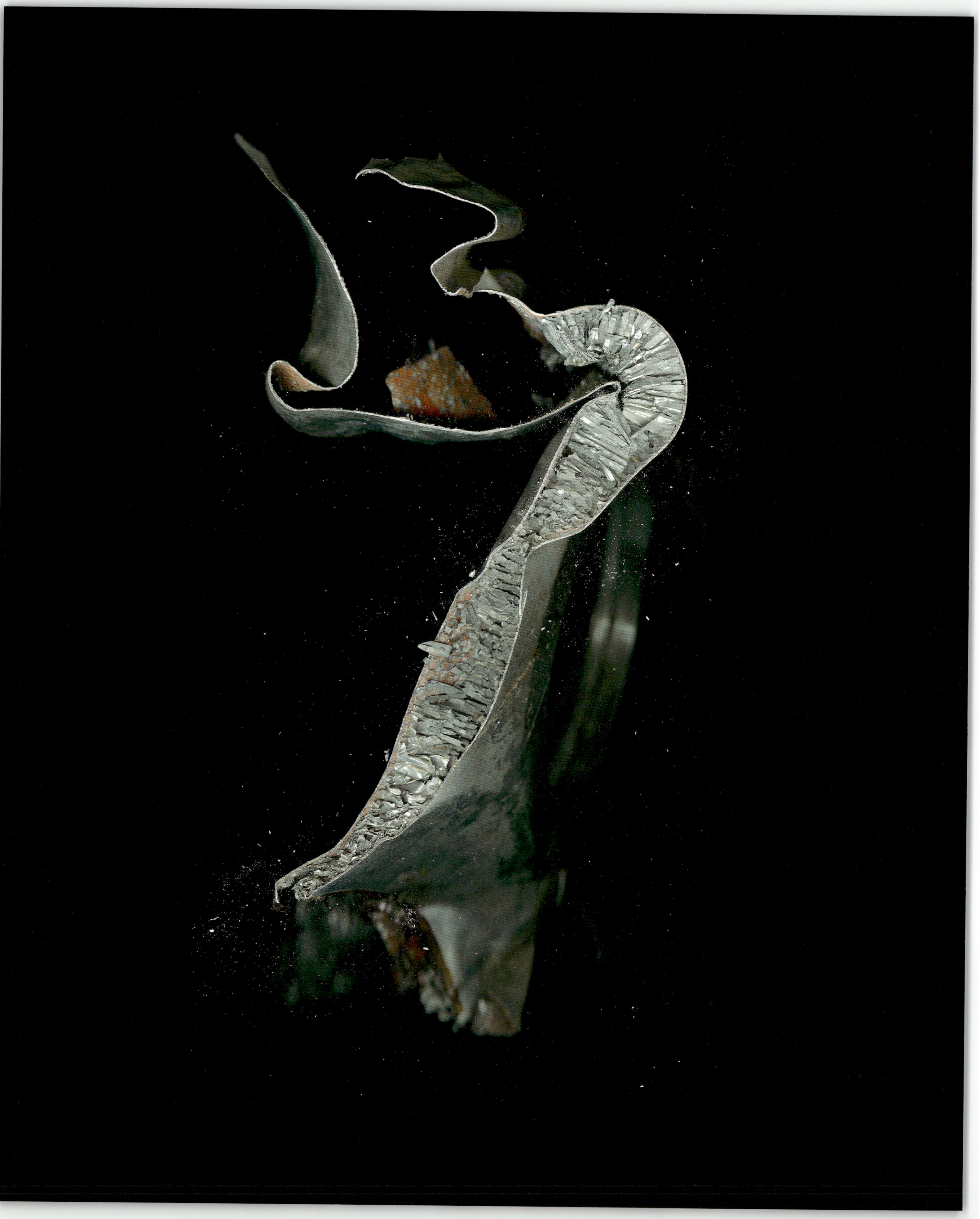

isolate an object against the cosmic-looking black void of its background, produced images of buckled, fried and dented objects, again 'seen' by the relatively new archival apparatus of the scanner, and absorbed into its particular taxonomic realm that renders all things not only digital and virtual, but eventual.

The debris of the so-called 'space age' has here come to signify failure, a lost future, and a disappointed adventure that is yet to find an appropriate way to write its history. If we call our current epoch the 'digital age' we might also claim that, for us 'screen becomes vitrine', as three-dimensional objects are scanned in increasingly sophisticated ways, ultimately to be viewed by means of a glass or plastic screen, rather than displayed in glass cases and lofty museum halls (O'Kane, 2019).[9]

[9] For the origin of this phrase and concept see O'Kane (2019).

Poppelsdorfer Schloss – Radical History and Prehistory

Given the above, it is perhaps appropriate that now, as I write in 2023, I find Sascha entrusted to work with a precious mineralogical collection in what we might call a 'traditional' European museum, there engaging with equally traditional vitrines and archives. Sascha is once again working with objects that are (like the exploded briefcase that set these themes in motion) perhaps 'left unattended', but not yet suspected of any potential threat. Rather it is the artist himself perhaps who appears threatening to the established order and to a taxonomic regime that holds certain power structures together and in place.

Today, when we look at the monumental systems of classification that determined what came to be known as enlightenment and empire we do so critically, sceptically, as well as with a certain shame, hostility and resentment. These systems set up a violent and imposing relationship between man and man and between man and the world, first by claiming to 'discover' and presuming to collect, then by taking pains to order and measure (according to its own measure, of course) everything as other than and other to itself—and also other to the human. As Walter Benjamin (1968) suggests in his *Theses on the Philosophy of History*, all such apparently caring curatorial control is also potentially criminal:

> There is no document of civilisation which is not at the same time a document of barbarism. And just as such a document is not free of barbarism, barbarism taints also the manner in which it was transmitted from one owner to another.

As above, a toxic taxonomic regime does not only restrict itself to the relationship between the human and the supposedly non-human realm, but between men and other men, most notably between those (on one hand) who by means of a certain convenient ignorance have elevated themselves to a position of supremacy, and (on another) others thereby condemned to a less than, non-, or barely human condition (Arendt, 1958; Levi, 1987; Agamben, 2007). This regime, closely allied with and motivated by commerce, enabled the unprecedentedly profitable practice of the Atlantic slave trade to fund what came to address itself as the 'first world's' (or 'leading nations'') consequent excesses and cultural 'achievements'—including seats of learning, majestic architecture, palatial homes, mighty museums and triumphant tomes, all of which become flagships and signifiers of enlightenment, while concealing their dependence on the iniquitous operations of empire, and on what Paul Gilroy (1993, pp.1–16) has called 'capitalism with its clothes off'.

Fig. 123

To the thoughts, thinkers and theorists referred to here, some of whom might be implicated in various affirmations of explosivity, we could add the practices of Robert Smithson (1996), whose land-art, in manoeuvring minerals *en mass*, entered or emerged from a dialogue with George Kubler, another unique thinker who seems to have pursued the aim of writing an art history in geological time (Kubler, 1962). Such a prehistory might begin to allow us to free ourselves from the kind of Euro-centric, phal-logo-centric, modern-centric, empire-centric, anthropo-centric history of art that is, admittedly, still informing lines of thought in this writing, even as they attempt to explode themselves in an act of auto-deconstruction.

Kubler's implications suggest a history of art that is no longer concisely and conveniently contained by our currently established museology and historicisation, its particular concrete forms and constructions. A more 'geological' history could release us into a wider, wilder, more formless field of art history or prehistory involving base materials, energetics, and events of which the Earth, and by implication, the universe, is constantly being formed.

The word 'radical' is perhaps overused in common cultural discourse, but its original meaning, with regard to 'roots' (see e.g. the humble *rad*ish) is appropriate here as, via Smithson and Kubler, we come upon a kind of 'dig', signalling a geological and radical history of art that is conscious of its innate, immanent and original explosivity, its constant velocity, its latent, vibrating potential to escape containment and order. This radicalism can be exerted even as and when we are asked to look at its products and 'discoveries' ordered in conservative rows on gallery walls, in museum cases, or in purpose-built hyper-illuminated white-cube galleries where it seems that nothing can be hidden. This radicalism can be exerted in palatial, panelled museums, the cultural castles (*Schlösser*) of curating that tend to impress us even before we climb their vertiginous entrance steps and pass between their lofty columns.

Towards a Neo-Baroque

As we have mentioned above, 'the sublime' theorises an aesthetic thrill, emerging from a brush with fear or a glimpse of the immeasurable, an encounter with the formless void where human perspectives no longer protect and assure the human. The baroque meanwhile provides a certain and particular form to non-human scale and formlessness. Its imagery involves a tumbling, fluid expression once claimed by the philosopher Gilles Deleuze to be encapsulated in the endless complex potential and dynamism of the apparently simple fold. The universe *qua* fold is not an assembly of discrete, separated objects, or things with fixed forms, but a folding and unfolding continuum (17th century baroque philosophers and theologians also refer to a *plenum*). The fold, as paradigm, enables a non-exclusive and holistic vision, allowing for what the Christian bible (1 Corinthians, 15: 28) calls the 'all in all' (and for what Deleuze sometimes calls 'one in the other') (Deleuze and Guattari, 1988, p.477; Deleuze, 1997, pp.57, 165, 236, 250).[10]

Today, as modern societies and 'leading nations' begin to confront and contemplate the divisive, destructive, and short-sighted approaches that have been taken to explore and exploit the earth and its surroundings, the neo-baroque's non-exclusive and holistic vision, in which all is enfolded in all else, can encourage a greater, perhaps more holistic consideration for the whole within the particular. A contemporary artist, informed by this model, finds themselves not only unprecedentedly free to explore any process, period, style, and agenda (the legacy of liberative modernism) but able to approach current issues in the context of the greatest expanses of time and space, in a postmodern reclamation of the premodern; of the ancient; the prehistoric; the traditional; the cosmic; and the religious.

A traditional and conservative form of taxonomy might tend towards the production of a still image of our mobile universe (a means by which questions are halted by knowledge) and thus lead us, through still-image making, to mistakenly believe that we have contained and controlled the universe as known and knowable objects and 'things'. For too long, it seems, photographic technologies have beguiled us with an illusory stillness that is misleading and unrepresentative of experience. But for Sascha Mikloweit, what he calls his 'Image-Objects', obtained through scanning the outcomes of explosions, are always tools to 'mobilise, to re-animate history'.

Even if we take up Andre Malraux's enthusiasm for the infinite and permissive sounding idea of still photography (in its entirety) as a *Museum Without Walls*, we nevertheless suspect that still photography (despite the influence of the 'moving image') has tended to consolidate and confirm erroneous perspectives on otherness by falsely capturing it as immobile and thereby pacified (Malraux, 1967). See, for example, the tradesmen and tribesmen photographed by Irving Penn (1974) within the consistent context of his portable portrait studio, or consider Edward Steichen's (1986) idealistic post-WW2, *Family of Man* project.

To taxonomically freeze and represent the great formless explosion that all ultimately is and that we also are, is not only a great conceit of still photography but also that of a would-be encyclopaedic enlightenment. It is something that anti-philosopher and radical historian Gilles Deleuze sought to disrupt in his own way by digging behind, before and beyond enlightenment, perhaps most notably in the direction of his advocacy of a 'neo-baroque' paradigm. Deleuze wrote about the way in which cinema creates new possibilities for thought and philosophy, but he also sees—in a pre-enlightenment, less rational 17th-century baroque paradigm—a way of encompassing, acknowledging, working with and relinquishing control over the truer and greater complexity with which an enquiring 21st-century mind is now confronted when considering a universe where chaos is affirmed as a more complex form of order (Deleuze, 1990; 2001; Voltaire, 2001).

Deleuze's neo-baroque paradigm is event-based as opposed to object-based or thing-based; it is *in*tensive rather than extensive; it is malleable, pliable (see the French *pli* for 'fold'), even fluid and pale in its most rococo manifestations (Deleuze, 1990, p.16; 2001, pp.27–38, pp.86–94). And this might account for the constant energetic motion, of the rhythms and vibrations, speeds and hesitations entertained by molecular physics, that gives an impression of distinct events manifested by relative speeds and rhythms.

Meanwhile, Deleuze encourages us to see ourselves as events, even suggesting that we regard ourselves as akin to: 'a wind, a season, or 5 o'clock' (Deleuze, 1991, pp.94–95). If we investigate the infinitely and eternally folded and enfolded realm of Deleuze's model of the neo-baroque, we might perceive therein the possibility and capability of encompassing a greater complexity than that which enlightenment presumed and promised to master, and which it ultimately reduced and restricted.

Before, after, beyond and notwithstanding enlightenment, all is explosion and diaspora, of which we are each and all a part, as events within events within events [sic]; events that will not be corralled and contained for long, and certainly not forever, but which territorialise only on their way to a subsequent deterritorialisation, a further fold; events that exist (or should we say *per*sist) one-in-the-other, all in all, without exclusion or exclusivity.[11]

As diasporic and explosive, we are all and we are one, all in one, the whole in one, fused and united by our mutual becoming, our common difference and constant differentiation. There is no separation of self-and-object, human-and-nature, living and lifeless matter, only one great

[10] Regarding 'all in all' see: "And when all things shall be subdued unto him, then shall the Son also himself be subject unto him that put all things under him, that God may be all in all." (The Holy Bible, King James Version, 1769)

[11] Here Michel Foucault's Preface to his book The Order of Things comes to mind as a withering, and laughter-inducing critique of order per se, and in which the laughter perhaps arises from a kind of vertiginous fear (Foucault, 1974).

Fig. 124

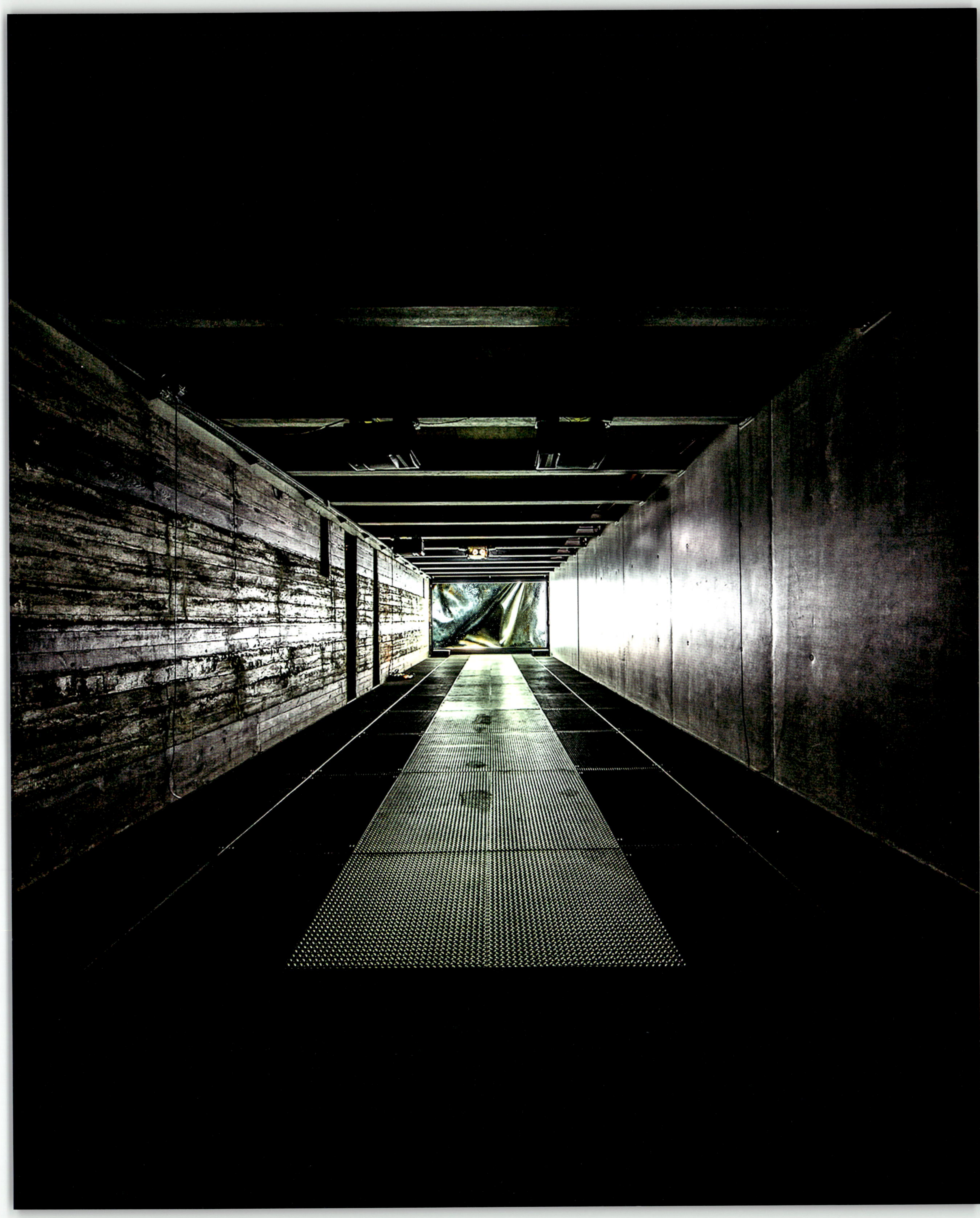

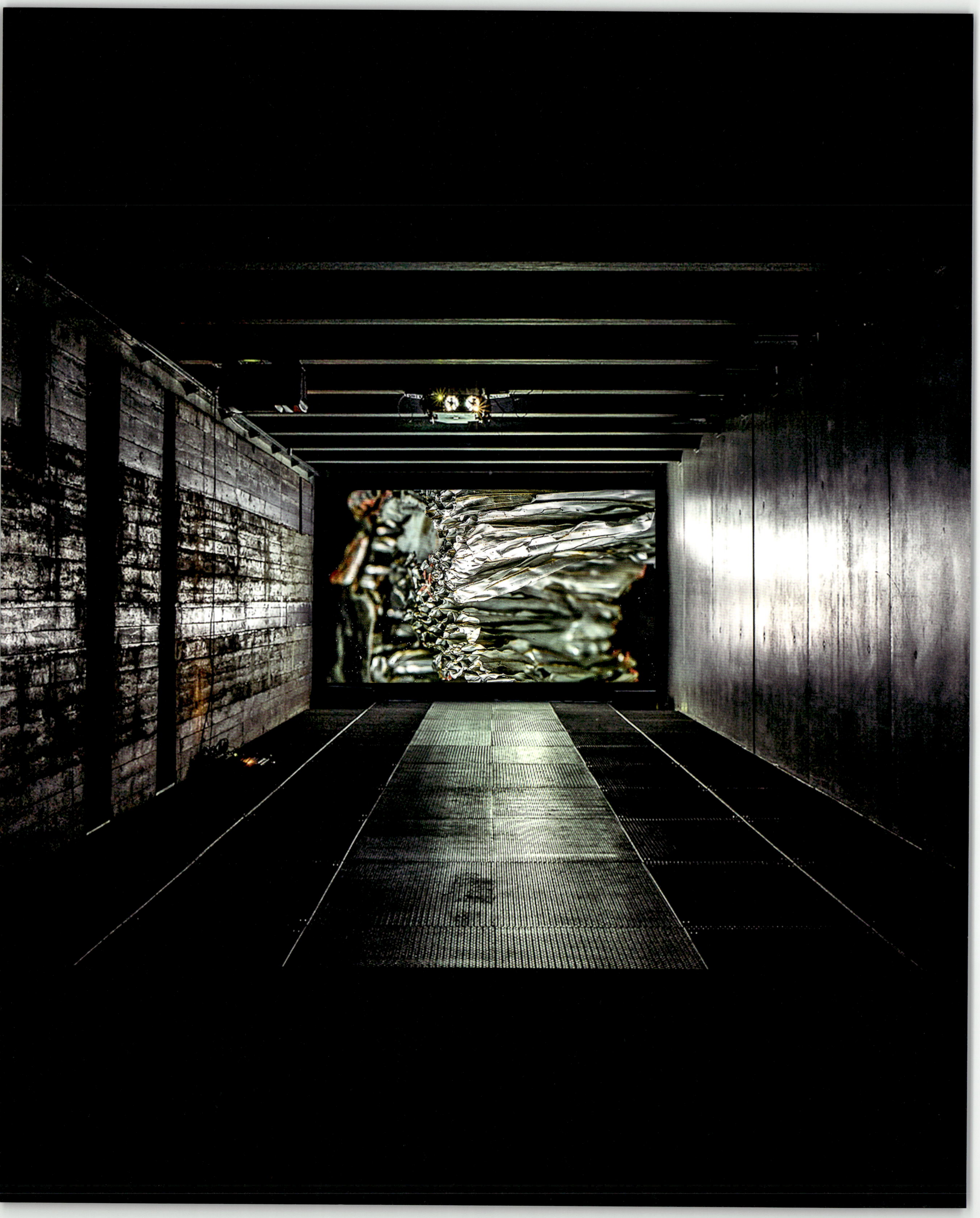

energetic and explosive enfolding and unfolding within which apparent individuations are merely hesitations, flights, manifestations of relative speeds and varying trajectories resulting in temporary formations, all rippled through by a dynamic force.

The neo-baroque may offer a model by means of which relatively crude, simplistic and repetitive forms of order, while reducing fear and inducing a sense of safety, give way to a brave and radical taxonomic expansion, a fluid liberty according to which we no longer shy from un-control or the potential loss implied by relinquishment. Loss and gain no longer relate to a *plenum* in which all is in all and where one is enfolded in other.

The neo-baroque is *not* the sublime, but like the sublime it acknowledges and affirms a kind of vertigo, induced by adventure, curiosity, and enquiry; a kind of dizziness brought on by keeping all in question and by associating the least of events with the greatest of questions. The neo-baroque might just lie within the dynamic and energetic force that drives the multiple explosive projects of Sascha Mikloweit, and that equally drives this writing—itself an explosion.

To conclude with a question (and all questions, we seem to have shown here, are themselves explosive and diasporic, opening-up ever-widening paths): what might be the duty, the task, the responsibility of an artist today when strangely charged to respond, when confronted by a prestigious mineralogical collection in the impressive context of a grand taxonomic *Schloss*? Perhaps that duty is a responsibility to explode all that can be exploded, to repatriate safely ordered objects, not to their pre-historic origins but to their most explosive *ur* condition, showing that the museum never completely captures things, as 'things' but merely induces a still and silent coma, a hesitation or delay of things (O'Kane, 2009).

The artist's duty, as a *counter*-curator or *anti*-curator, might then be to awaken and free things from their 'thing-hood', releasing them from any temporarily tamed stasis and known form, liberating things from their status as evidence, and from their organisation as objects of knowledge, allowing them to move, fragment, dissolve, to lose their apparently fixed forms while returning them to questionability, reinstating their trajectories and becomings so that they can be renewed, revisited, and perhaps re-read—like the detritus of an explosion, like entrails, tea leaves, koans, runes and riddles, providing a form of 21st-century augury that is as futural as it is ancient.

Fig. 125

Bibliography

A Handful of Dust, 2017. [exhibition] Whitechapel Gallery, London. 7th June 2017 – 3rd September 2017.
Agamben, G., 2007. *Sovereignty and Life*. Stanford: Stanford University.
Arendt, H., 1958. *The Human Condition (Charles R. Walgreen Foundation Lectures Series)*. Chicago: University of Chicago Press.
Ballard, J. G., 1973. *Crash*. London: Jonathan Cape.
Benjamin, W., 1968. Theses on the Philosophy of History. In: *Illuminations*. H. Arendt, ed. 1968. Translated from German by H. Zohn. New York: Schocken, pp.253–264.
____1979. *One Way Street and Other Writings*. Translated from German by E. F. N. Jephcott and K. Shorter, 2000. London, New York: Verso.
Bois, Y. A. and Krauss, R. E., 1997. *Formless – A User's Guide*. New York: Zone.
Deleuze, G., 1990. *Negotiations*. New York: Columbia University Press.
____1991. A Philosophical Concept. In: E. Cadava, ed., 1991. *Who Comes After the Subject?*. London: Routledge.
____1997 (1968). *Difference and Repetition*. London: The Athlone Press.
____2001. *The Fold, Leibniz and the Baroque*. London: The Athlone Press.
Deleuze, G., Guattari, F., 1986. *Kafka, Towards a Minor Literature*. Minneapolis, Minn. University of Minnesota Press.
____1988. *A Thousand Plateaus – Capitalism & Schizophrenia*. London: The Athlone Press.
____2004. *Anti-Oedipus: Capitalism and Schizophrenia*. London: Continuum.
Dial H-I-S-T-O-R-Y. 1997. [film] Directed by Johan Grimonprez. France: Centre Georges Pompidou.
Foucault, M., 1974. *The Order of Things: An Archaeology of the Human Sciences*. London: Tavistock Publications.
Gilroy, P., 1993. *The Black Atlantic: Modernity and Double-Consciousness*. London: Verso.
James, K., 1769. The Holy Bible, King James Version. Cambridge Edition: 1769. Available at: <https://www.kingjamesbibleonline.org/book.php?-book=1+Corinthians&t=1&chapter=15&verse=28> [Accessed 3 December 2023].
Kaufman, W., ed., 1976. *The Portable Nietzsche*. Viking Portable Library. Harmondsworth: Penguin.
Kubler, G., 1962. *The Shape of Time – Remarks on the History of Things*. New Haven: Yale University Press.
Levi, P., 1958. *If This is a Man – The Truce*. Translated by S. J. Woolf, 1987. London: Abacus.
Malraux, A., 1953. *Museum Without Walls*. Translated by S. Gilbert and F. Price, 1967. London: Secker & Warburg.
Nietzsche, F. W., 1883. *Thus Spake Zarathustra - A Book for Everyone and No One*. Translated from German by R. J. Hollingdale, 1969. London: Penguin Books.
____1915. *The Twilight of the Idols, or, How to Philosophise with the Hammer, the Antichrist, notes to Zaratustra, and Eternal Recurrence*. Edinburgh: T. N. Foulis.
O'Kane, P., 2009. *A Hesitation of Things*. PhD. Goldsmiths College, University of London. Available at: <https://okpaul.com/wp-content/up-loads/2015/02/PaulOKane-phd-hesitation-of-things.pdf> [Accessed 3 December 2023].
____2019. Technologies of Romance: Looking For "Object Love" In Three Works of Video Art. *Science Museum Group Journal*, Autumn 2019(12). Available at: <https://journal.sciencemuseum.ac.uk/article/looking-for-object-love/#abstract> [Accessed 3 December 2023].
Penn, I., 1974. *Worlds in a Small Room as an Ambulant Studio Photographer*. London: Studio Vista.
Smithson, R., 1996. The *Collected Writings*. Berkeley: University of California Press.
Steichen, E., 1986. *The Family of Man: The Greatest Photographic Exhibition of All Time*. New York: The Museum of Modern Art.
Virilio, P., 1993. *The Art of the Motor*. Translated by J. Rose, 1995. Minneapolis: University of Minnesota Press.
Voltaire, F., 1759. *Candide or, Optimism*. Translated by R. M. Adams, 2001. New York: W. W. Norton & Company.

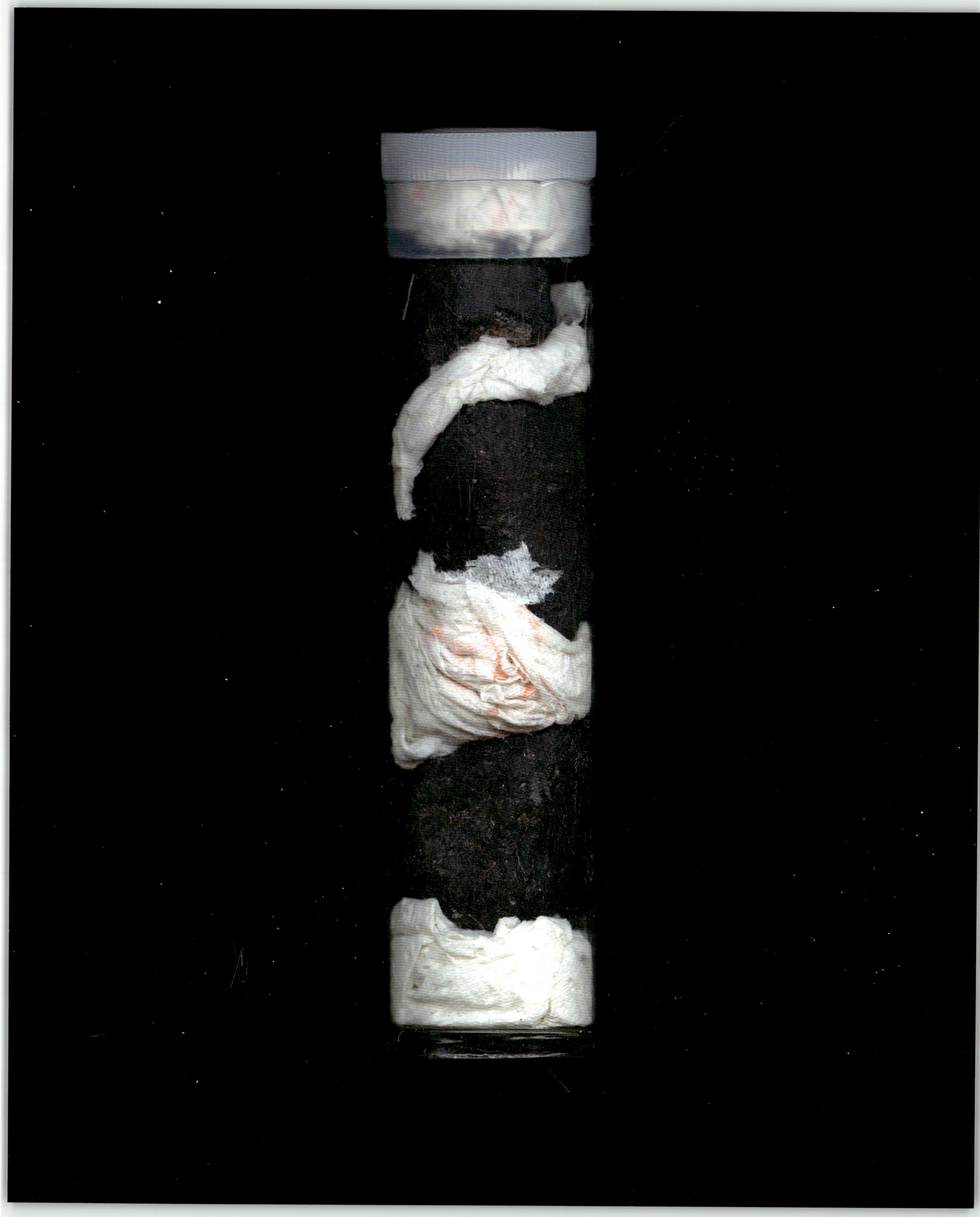

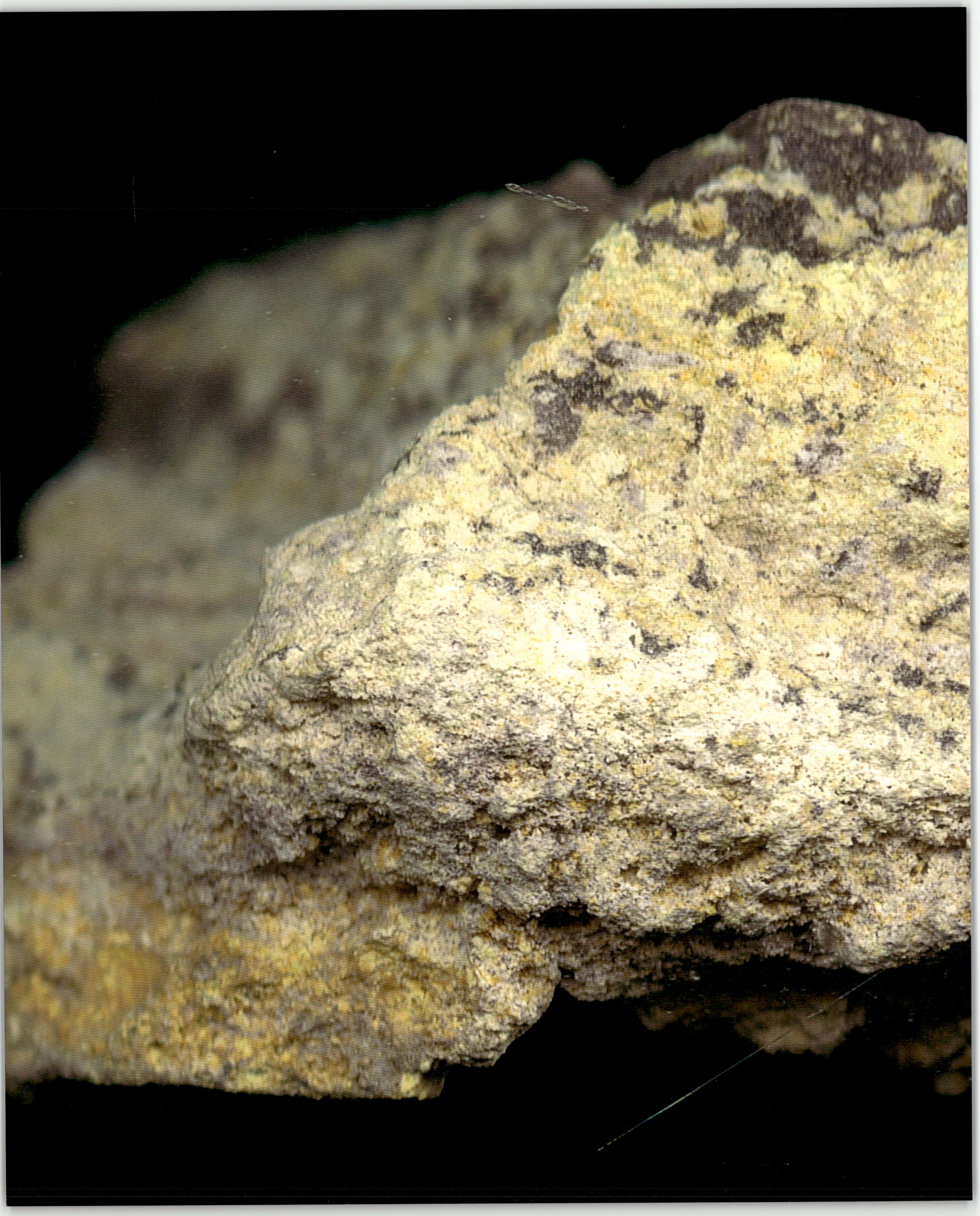

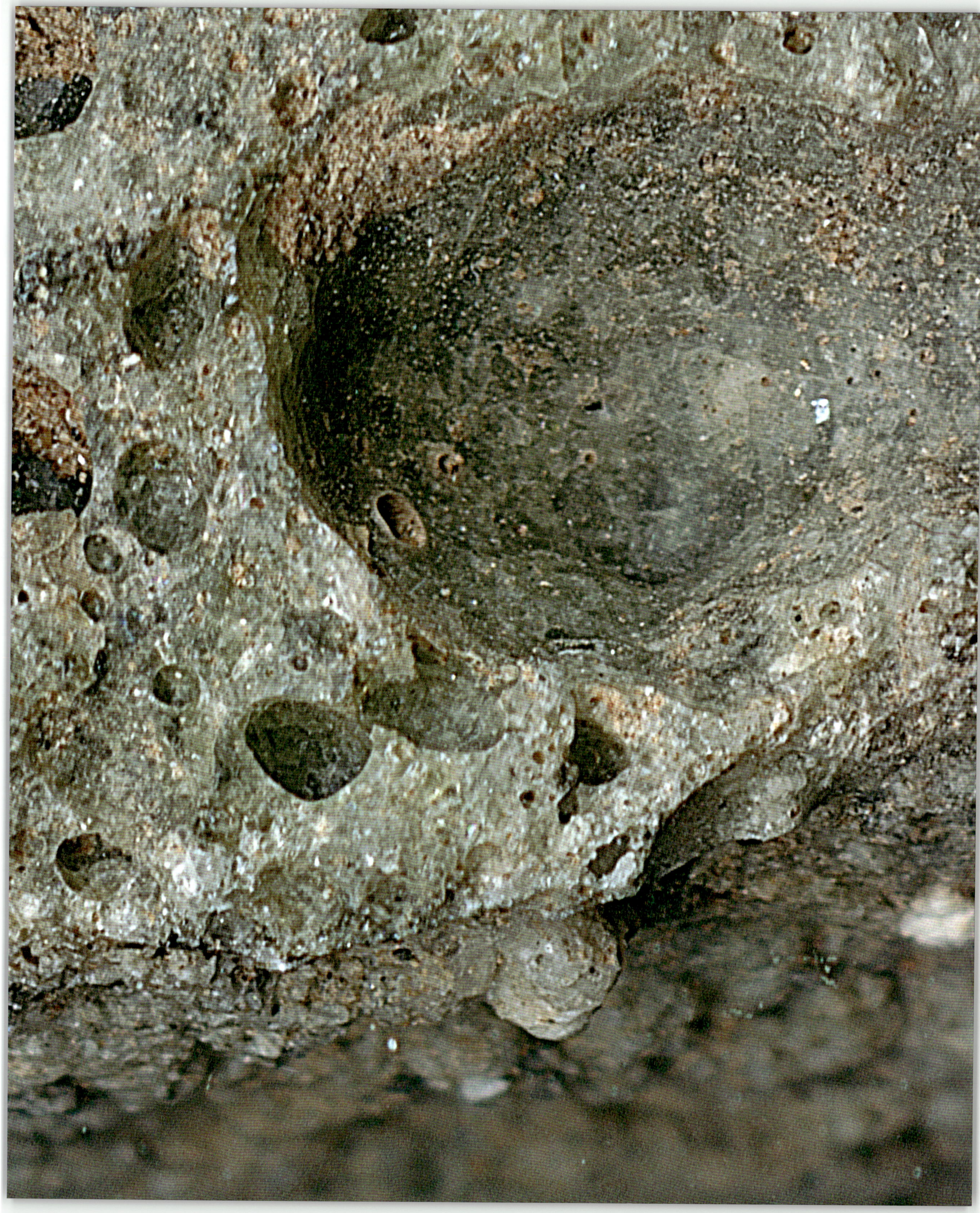

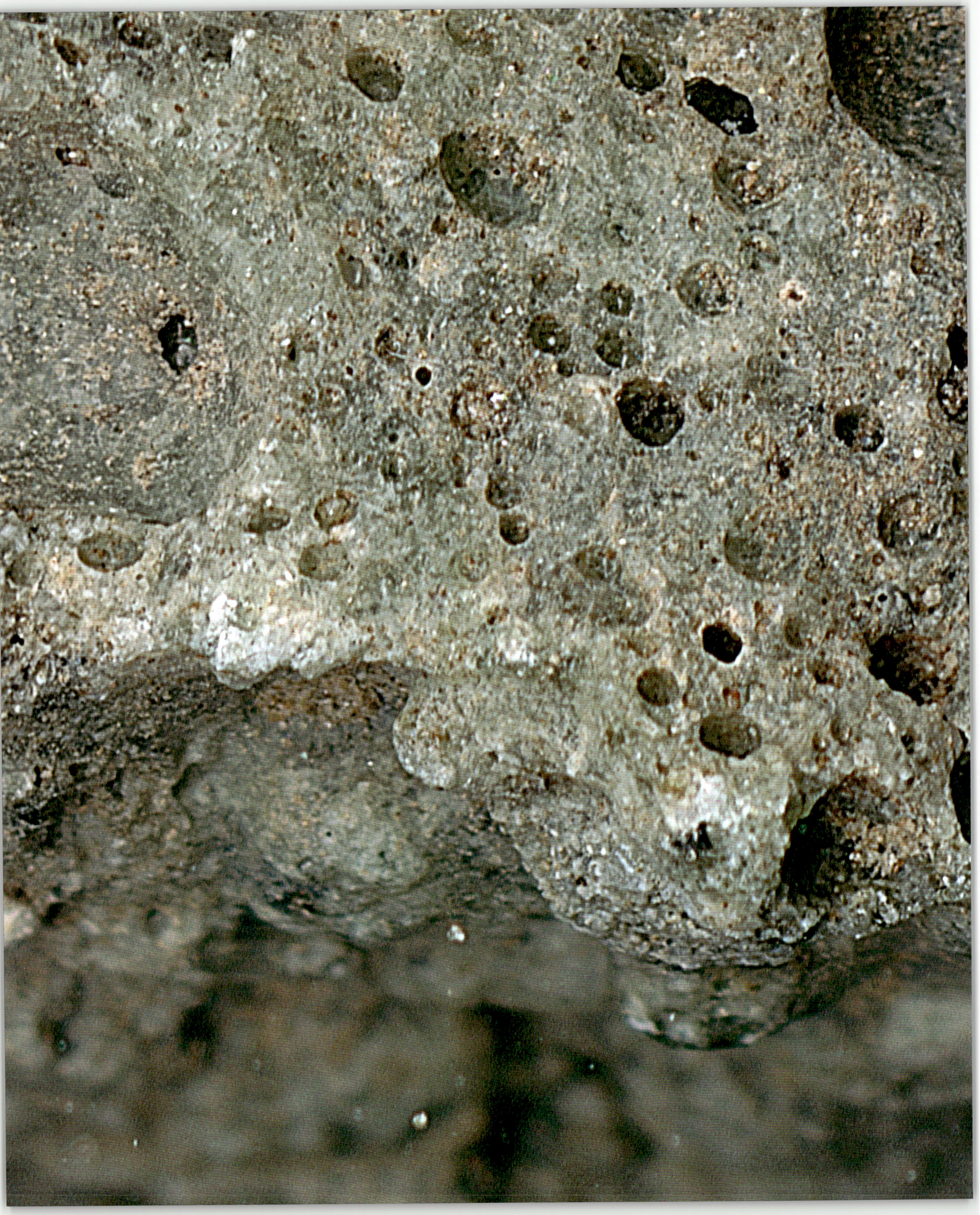

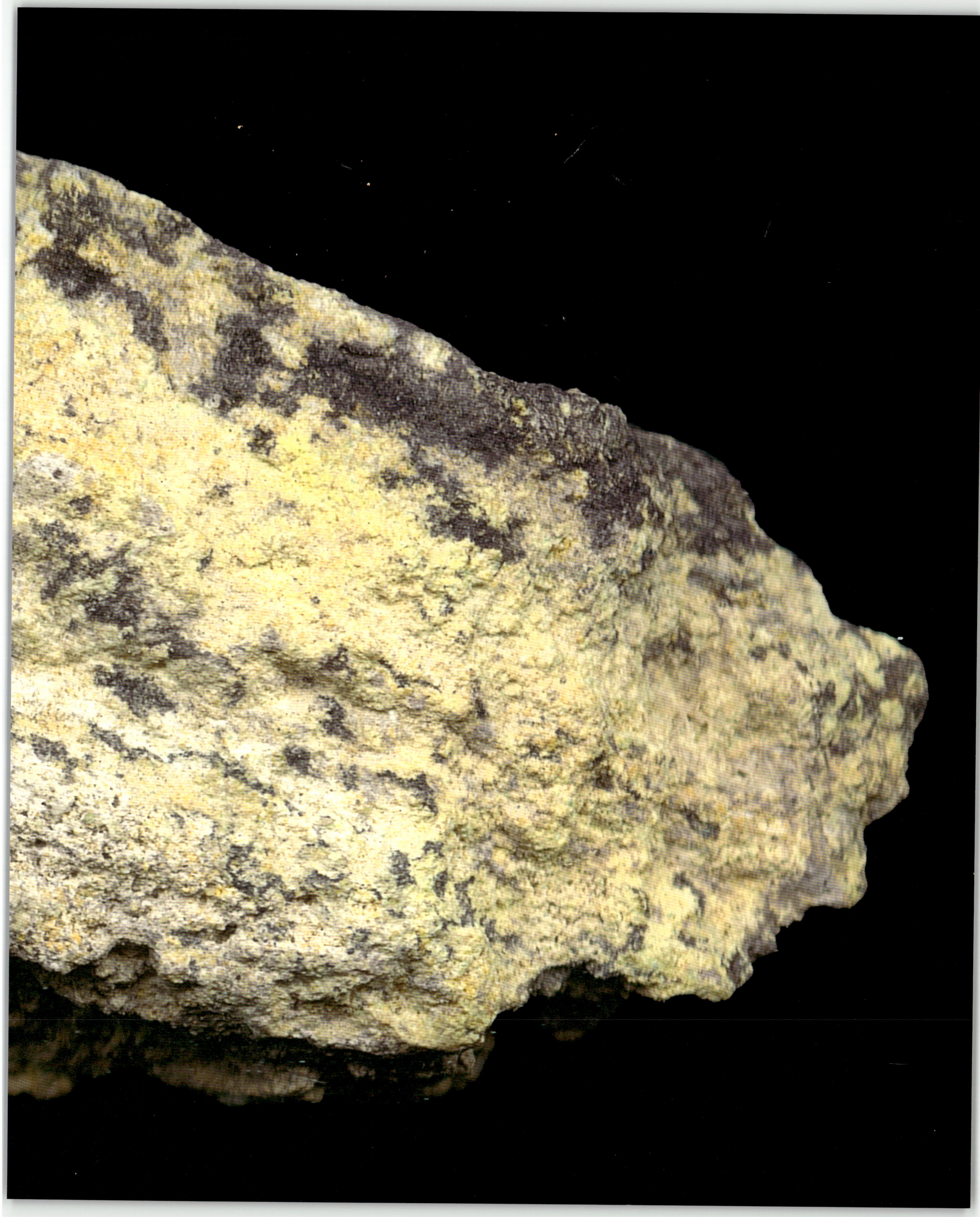

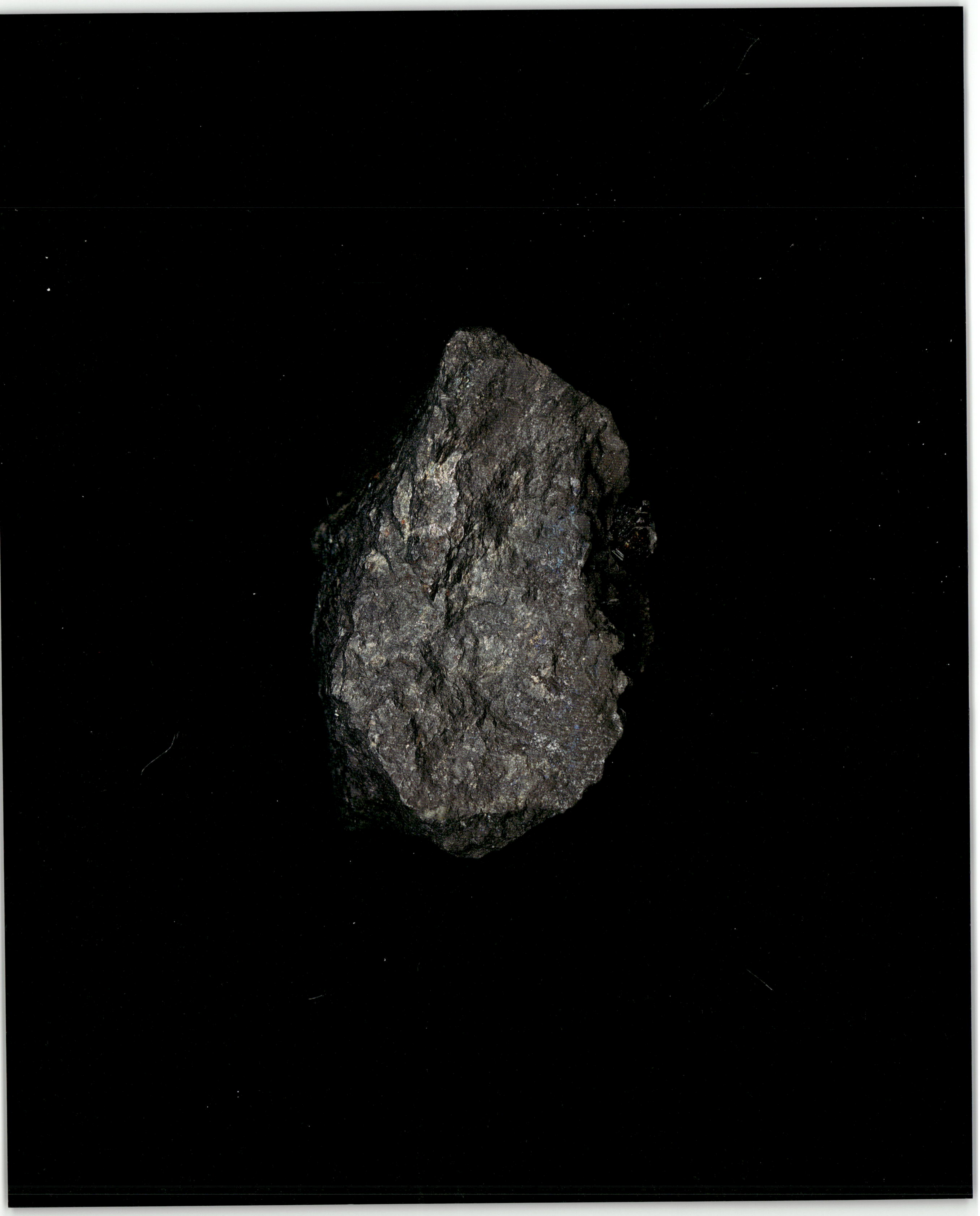

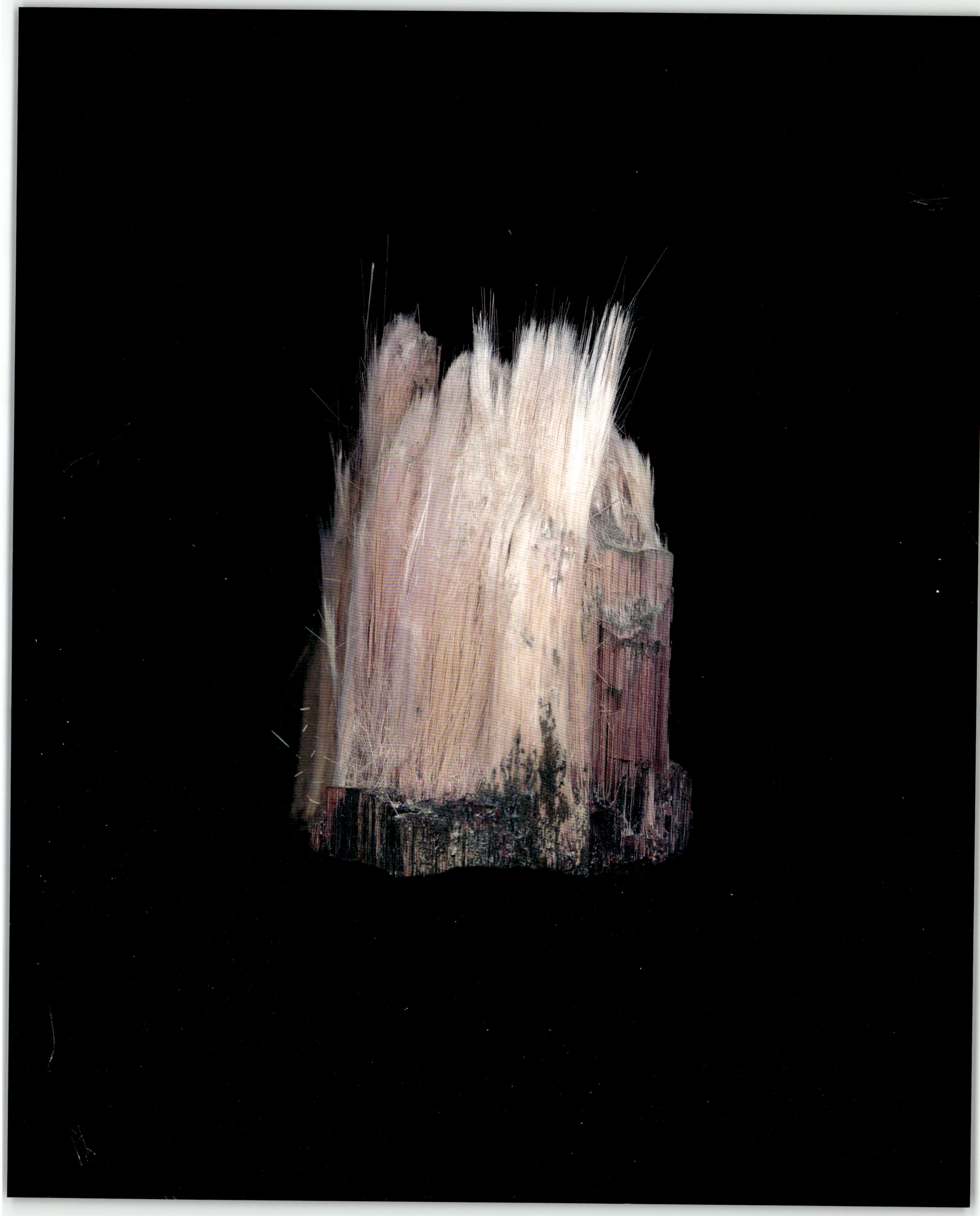

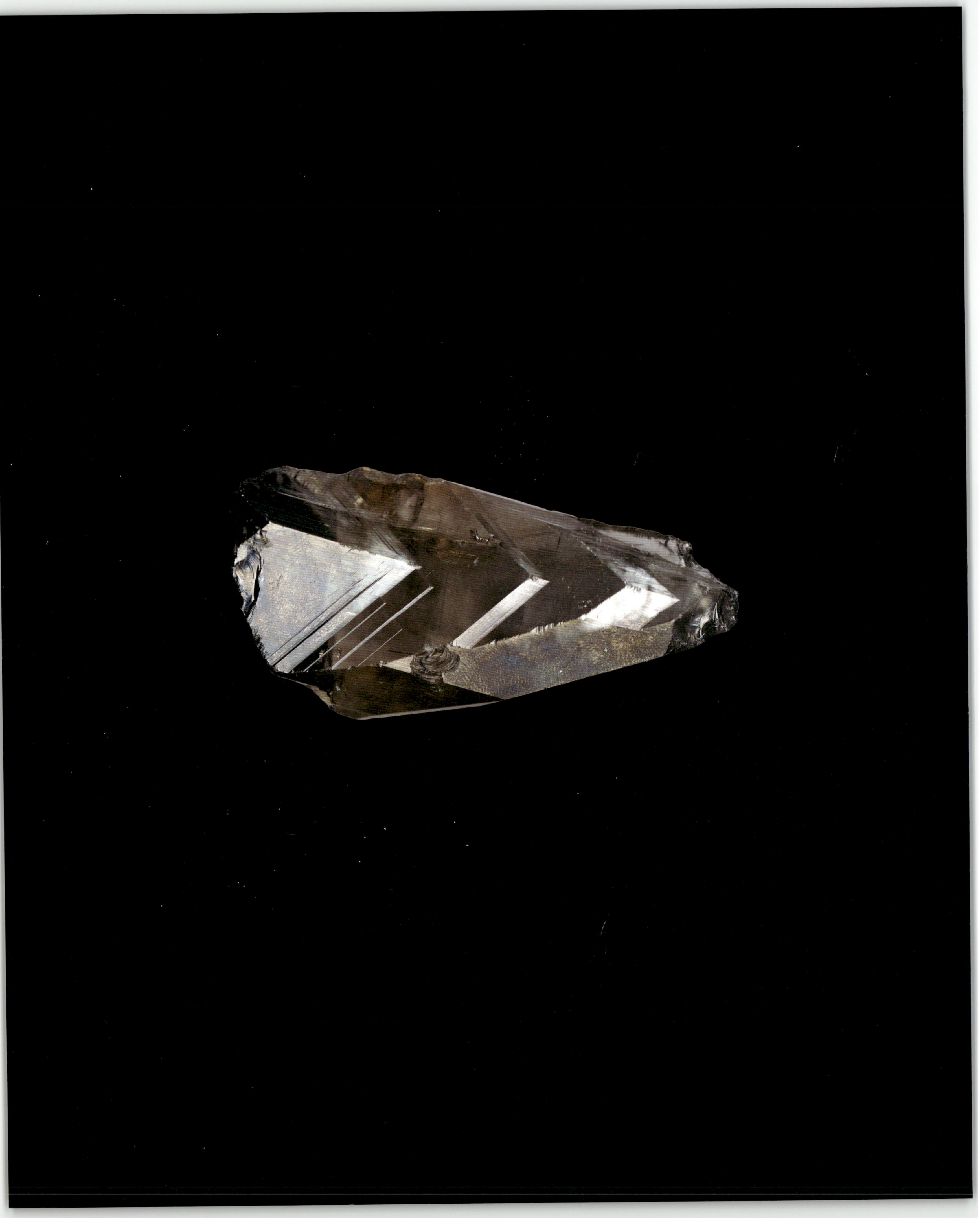

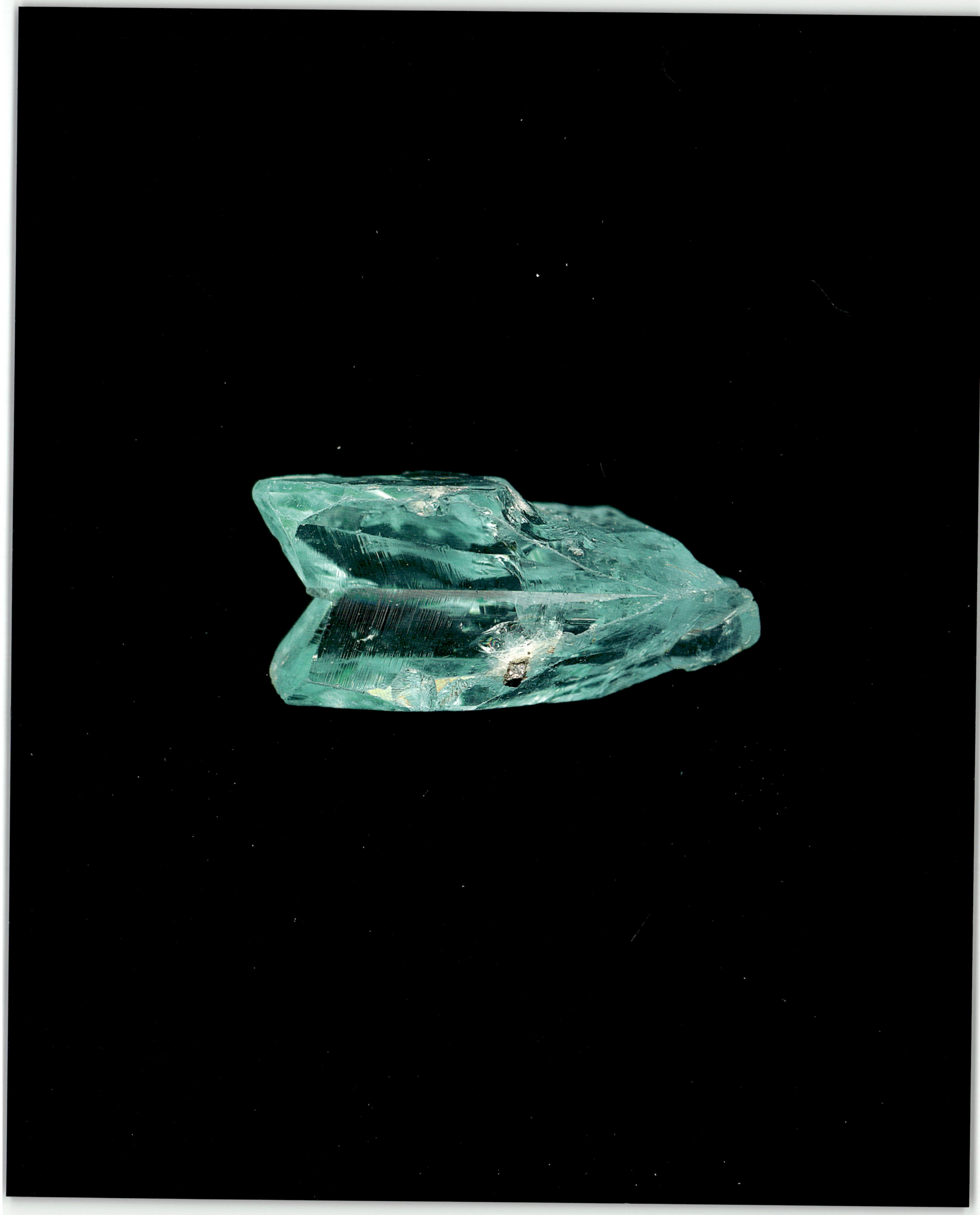

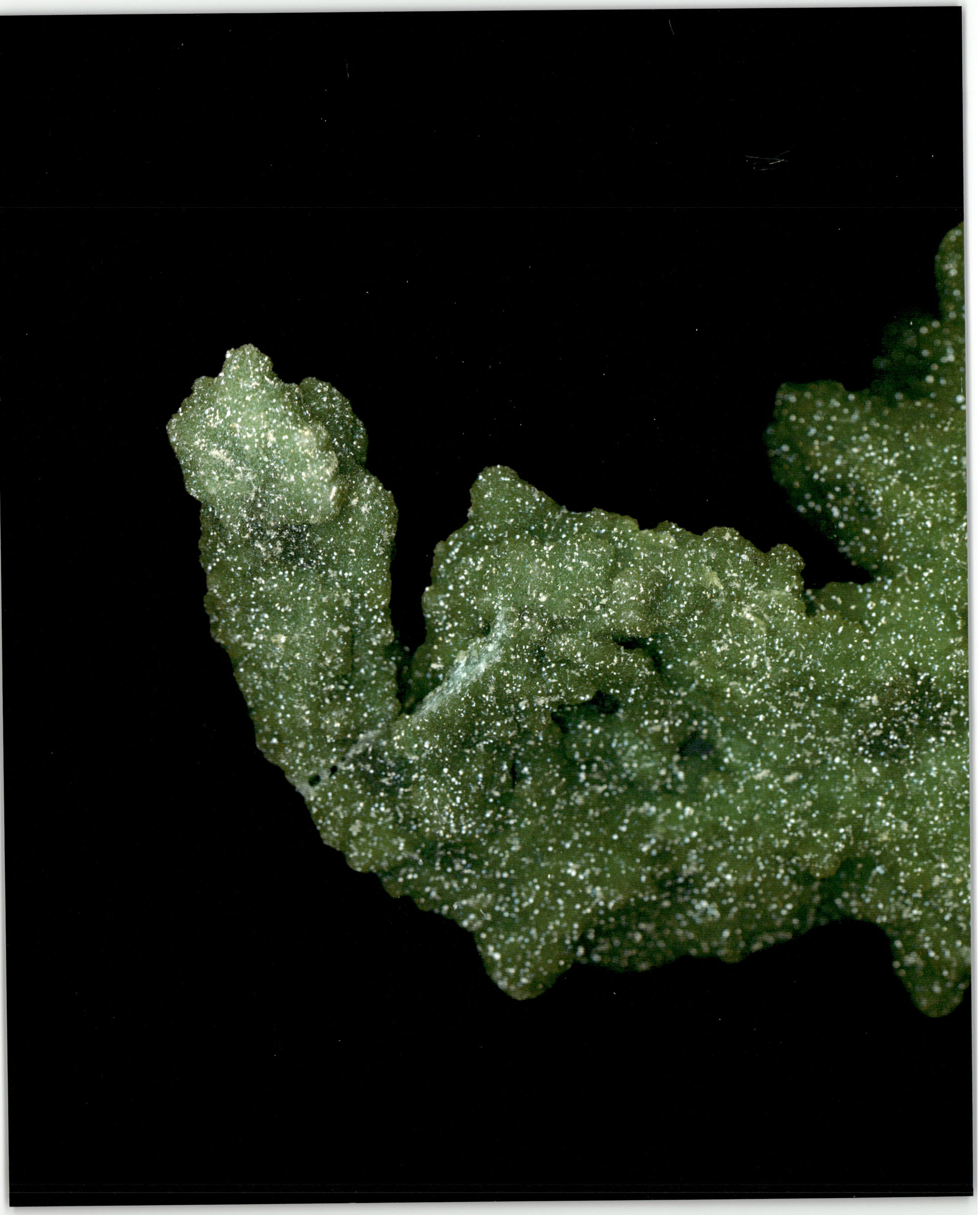

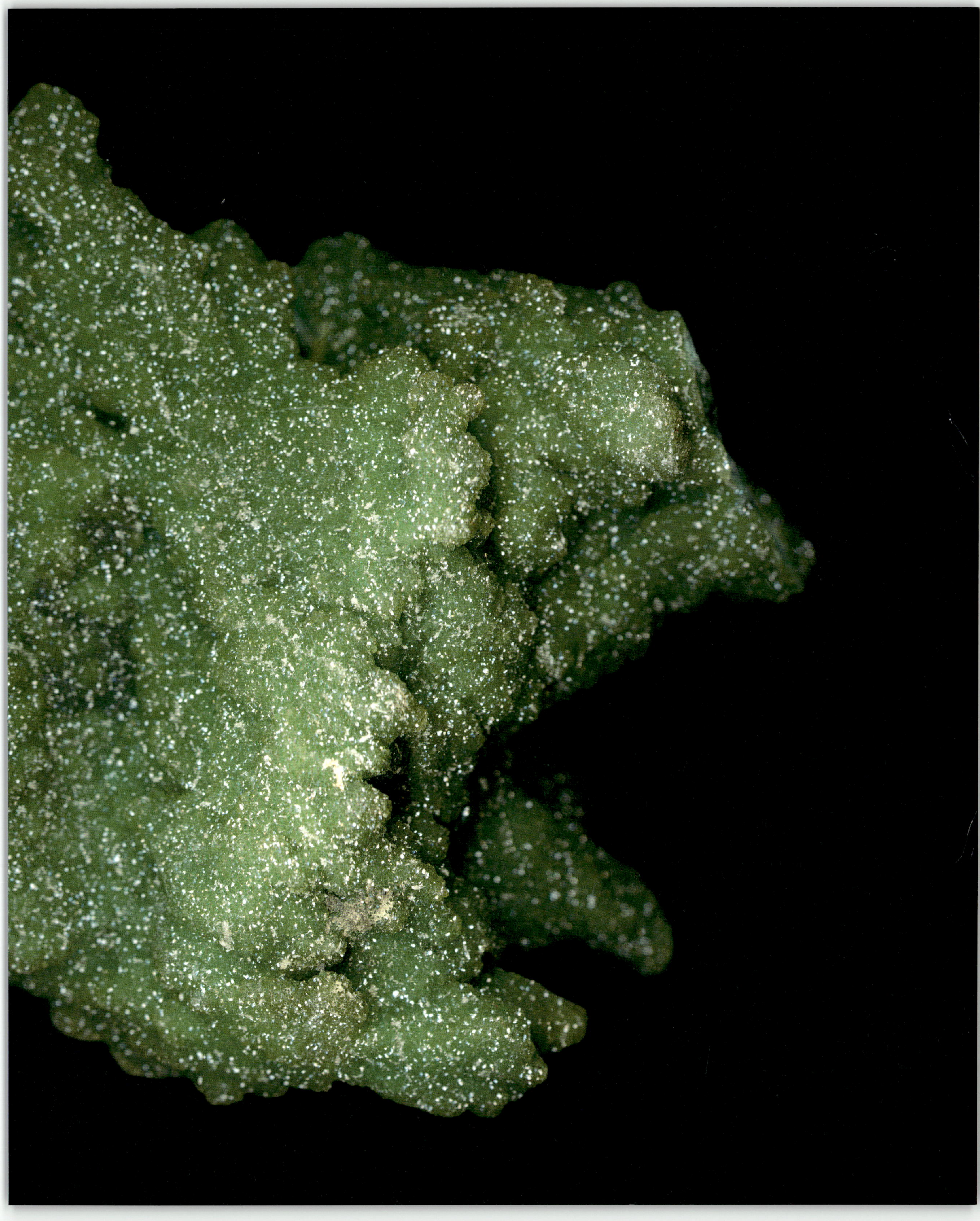

Zinkspat 5

(SMITHSONIT) , grün

Tsumeb, DSW Afrika

Nr. Sammlung Sammlung Duisberg 885

Mineralogisch-Petrologisches
Museum der Universität Bonn

Smithsonit
(Zinkspat), grün

Tsumeb, SW-Afrika

Nr. 50215 Sammlung: Duisberg (885)

50215 Inventar-Nr.:	Mineralogisches Museum der Universität Bonn	V-Karbonat Klasse n. H.Strunz

Mineralname: Smithsonit

Fundort: Tsumeb, Namibia

Schrank / Schublade: Vitrine / Position: 30, oben

Gewicht [g] Stückzahl: 1 EDV ☒

Bemerkungen: Sammlung Duisberg

Acknowledgements

I'd like to start by remembering my late grandfather Arno Eichhorn (1928–2004), who after leaving his home village in the mountains of East Germany as a teenager, was a coal miner for over three decades in West Germany—a life's work he was proud of, and which fundamentally impacted the course of our family's life in many ways. My interest in the underground is rooted in his experience. This project ballooned from a short email. When in 2018 I reached out to the director of the Mineralogical Museum of the University of Bonn, Anne Zacke, requesting a meeting, I had no idea that the one-year artist-in-residence I proposed would eventually grow into a five-year project that became *Transpositional Geologies*. I did not expect that it would bring about this publication, to which 44 people from around the world would contribute, and that it would nurture long-term collaborations in Namibia, the UK, and Germany.

I'm greatly indebted to Anne Zacke, who opened the doors of the museum to me, and for allowing the institutional space for *Transpositional Geologies* to develop beyond its initially intended scope. Thank you for your openness, all your administrative support, for our conversations on the structure and form of the Bonn collection, for giving me carte blanche for the exhibition I realised at the Museum in 2023, and especially for supporting this publication financially to meet its budgetary challenges. My personal gratitude also goes to the indefatigable Alma Hannig, the Main Coordinator at the University Museums, Collections and Art, for all her support and advice during my time in Bonn and our conversations. The cooperation with universities will continue from 2025 to 2027 with an artist-in-residence at the Mineralogical Collections of the Technical University of Berlin in collaboration with its director R. Johannes Giebel—who throughout the past years has become a dear conversation partner, a collaborator and friend: thank you Johannes for everything! I'm looking forward to what is to come.

In Namibia, particularly to the Tsumeb Municipality, I extend my deepest appreciation to the local administrator Elli Stella Udda Imalwa, for your openness, kindness, support, friendship, your presence in the field, and invaluable conversations and insights into Tsumeb, now and then. Without our encounter in 2022 and all that unfolded naturally from there this project would not be what it has become—Okunene okuhepa!

Fig. 134

Fig. 135

Fig. 136

Transpositional Geologies has been supported by individuals too numerous to mention comprehensively. Some were instrumental in the beginning; others contributed important insights near the end. I'm grateful to Alexandra Heimes for introducing me to Kathryn Yusoff's formative writing at the beginning of the project. Jamie Bradbury for extensive conversations on art, coloniality, and geology. Josh Baum for our continuous conversation, the inception of Peripatetic Thinking, and your magnanimity to dedicate a year of reading together through Sylvia Winter's and Édouard Glissant's rich and important writings. Bastian Herbst for accompanying me to Namibia and for our discussions on history, museums, and artistic strategy. Irakli Kiziria for all extra hours in conceptualising this publication's design with me; from researching hundreds of books to fine-tuning the last details—it was delightful working together, as it always is! Amanda Holmes and Manuel Pandalis for their excellent photography during our research visits in Namibia, and particularly Manuel for his photographic project documentation, and additional fine-tuning.

I thank the contributors for their excellent texts, and for trusting me with them. Especially Paul Basu, for providing a brilliant introduction, for many conversations, and his ongoing support. R. Johannes Giebel, Elizabeth Grosz—also for our conversations; Selby Hearth, also for coming to Bonn; Bastian Herbst, Chris Hill—who it has been a serendipitous pleasure to meet and engage in ongoing exchange ever since; and Herbert Jauch, the late Juuso Ndinolune David Kaluvi, Malina Lauterbach—also for thinking together; Helmut Maier, also for coming to Bonn; Okunene okuhepa to Prince Kamaazengi Marenga I, also for ongoing conversations and for introducing me to Charmaine//Gamxamûs, Maria-Oo Haihambo, Veripuami Nandee Kangumine, Jermaine Solunga, and Kuhepa Tjondu—thanks to you all for trusting me with the contextualisation of your poetic work; Hidipo Nangolo, Saima Nakuti Ndahangwapo, and Paul O'Kane—who witnessed and helped the inception of my thinking on scanned object-images together with Shahin Afrassiabi and the wonderful late Joanna Greenhill at Central Saint Martins; Peripatetic Thinking, Ellison Tjirera, and Samo Tomšič—for his generosity in exchange; and Kathryn Yusoff.

Thanks for your artistic and institutional collaborations and our conversations along the way: Paul Philipp Heinze, Elli Stella Udda Imalwa, August Kapolo (with Festus Alushiya, David David, Rakker Festus) and Richardlee Shoombe—for working together at such short notice; Irakli Kiziria, Helke Mocke and Anna Williams, Jolam Neumann, Manuel Pandalis, and Holger Risse—for putting in more hours than expected; Francesca Steinbach, and—of course—Medala's Car Wash who introduced herself with the express wish to be part of this project.

Foundations that supported the project (see p.34), especially: Kunststiftung NRW who supported the project with multiple fundings while keeping administrative constraints at a minimum—especially thanks to Dorothee Mosters, Jasmina Merz, and their team. The Institut für Auslandsbeziehungen (IFA) for making two visits to Namibia possible, which became formative to my work—thank you to Ingrid Klenner, Jochen Hetterich, and their team. My gratitude goes to the Embassy of the Federal Republic of Germany in Windhoek—here, Clarissa Judmann—and the Goethe-Institut Namibia—to Dennis Schroeder and team—as well for their ongoing multifaceted practical and admin-istrative support in Namibia. Dorothée Coßmann at Sparkassen-Kulturstiftung Rheinland for all her efforts. Thanks to the team at the Mineralogical Museum of the University of Bonn—particularly Dana Vlcek. Also to Jens Frautschy for logistical support and conversations in Tsumeb, and 711rent Hamburg for equipping our 2023 Namibia field trip. Also for their support, which is much more than just technical, since 2015: LaserSoft Imaging, especially Philipp Haarländer.

Personally, I would dearly like to thank Veronika and Engelbert. I'm indebted also to my partner Regina Peldszus, without whose resourcefulness, rigour, and patience, this book would most likely not exist (and for just telling me otherwise). Last but not least, a big thank you to everyone at Kerber Publishers who I worked with along the way—for going the whole distance together, and for allowing the editorial time the publication needed to grow: Jens Bartneck, Marius Bolduan, Lydia Fuchs, Birte Kreft, Verena Simon, and Anne Levke Vorbeck—it was a pleasure working with you! And Brendon Hooper, Sally Parker, and Helen Taylor for great copy editing, proofreading, and indexing, and Dirk Beushausen for his poignant feedback regarding the design.

Biographies

Sascha Mikloweit, MA

Artist and researcher: Berlin, DE; Brussels, BE

Sascha Mikloweit is an artist and researcher based in Berlin and Brussels. His work investigates geological and cultural entanglements through transdisciplinary methodologies, focusing on the materiality, historiography, and socio-political dimensions of geoscientific archives and extraction sites. Between 2019 and 2024, Mikloweit was a research artist-in-residence at the Mineralogical Museum of the University of Bonn, where he initiated the *Transpositional Geologies* research project. This project examines colonial heritage, histories of racism, and extraction in relation to geological science, with a particular focus on epistemic violence and ontology in geo-scientific collections. From 2025 to 2027, Mikloweit will continue developing this project during his research artist-in-residence at the Mineralogical Collection of the Technical University of Berlin. Mikloweit holds a Master of Arts (with Distinction) from Central Saint Martins, University of the Arts London, UK, and studied at the Academy of Fine Arts in Düsseldorf and Münster, DE, as well as philosophy and history at the University of Münster, DE. His research has been presented at invited lectures, symposia, and workshops, including the Summer School on Decolonising Socio-Ecological Futures organised by the University of Ghana and Bonn (2023), the Humanities Research and Innovation Group at the University of South Wales, UK (2023), and the Goethe-Institut Namibia (2022). Mikloweit's work has been supported by major German cultural and research funding bodies, including Kunststiftung NRW, the Senate Department for Culture and Europe Berlin, the Goethe-Institut Namibia, Stiftung Kunstfonds, the German Embassy in Namibia, the German Federal Government Commissioner for Culture and Media (BKM), and the Institut für Auslandsbeziehungen (ifa). His work has been developed and shown internationally at venues such as: Copernicus Science Centre, Warsaw, Poland; European Space Agency, ESOC, Darmstadt, Germany; Tsumeb Municipality, Namibia; the Mineralogical Museum of the University of Bonn, Germany; Museum of Applied Arts, Frankfurt, Germany; Museum of Applied Arts, Berlin, Germany; the 7th Liverpool Biennale, Static Gallery, Liverpool, UK; Associação Brasileira de Empresários de Diversões, São Paulo, Brazil; Platoon Kunsthalle, Seoul, South Korea; the 13th and 14th Istanbul Biennale, Istanbul, Turkey; Goethe-Institut, Warsaw, Poland; Bargehouse, London, UK; and Central Saint Martins Window Gallery, London, UK. His works are held in public and private collections across Europe, Africa, and North America. He is an Honorary Associate at the Global Heritage Lab, University of Bonn, DE, and a founding member of the German Society for Artistic Research (GKFD). *Transpositional Geologies—Spectres of Coloniality* is Mikloweit's first major publication, a culmination of over five years of research, development, and collaboration with 43 international contributors and eight institutions.

Tsumeb Municipality

Private Bag 2012, Tsumeb, Namibia; Tel: +264 67 221056

General Information: Tsumeb, deriving its name from the Haiǁom word 'Tsomsoub' and the Herero word 'Otjisume', both referring to the area's colourful, mineral-rich terrain, has a storied history. Known for its significant volcanic pipe, Tsumeb's rich mineral deposits have been exploited since prehistoric times. Vision and Mission: Our vision is to establish Tsumeb as a modern, well-managed city, attracting investors and visitors. Our mission focuses on delivering accessible and sustainable quality services through strengthened partnerships. Municipal Status: Tsumeb was established in 1905, gaining full municipal status in 1968. Since then, it has grown to be one of Namibia's leading local authorities, promoting initiatives such as the introduction of TV in northern towns and establishing the Northern Tourism Association. Economic Drivers: Key economic sectors include mining, agriculture, and tourism. Tsumeb is renowned for its mining heritage, agricultural productivity, and proximity to notable tourist sites like Etosha National Park. Infrastructure: The town boasts a robust infrastructure, including modern transport links, educational institutions, healthcare facilities, and various social amenities. Tsumeb's infrastructure supports its industrial and residential needs, contributing to its reputation as the industrial capital of the Oshikoto Region. 'Glück Auf' is our motto.

Festus Alushiya

Carpenter: Tsumeb, NA

I'm a 24 year-old man, born and raised in Omusati Region. I completed my Secondary Education at Niilo Taapopi Senior Secondary School. After completing my high school, I developed a desire to further my studies and become an Artisan, which is how I ended up enrolling at the Community Skills Development Centre (COSDEC TSUMEB) where I'm currently studying for my level 2 training in Technical Trade of Welding and Metal Fabrication. Participating in the field of Vocational and Technical Training has helped me gain skills, knowledge and attitude to enter the working Industry. In the long term, it's my dream to become a qualified Welding and Metal Fabrication Artisan.

Prof. Dr. Paul Basu

School of Anthropology and Museum Ethnography, University of Oxford, UK

Paul Basu is Professor of Anthropology and a Curator at the Pitt Rivers Museum at the University of Oxford. He is an anthropologist specialising in critical heritage, museum and material culture studies in transcultural contexts. He draws upon a wide range of ethnographic, historical and participatory methods to explore how pasts are differently materialised and mediated in the present, and how they shape futures. Paul's research examines the complex ways in which natural as well as cultural heritage is entangled in shifting regimes of value and geopolitical configurations. His work has often involved re-engagements with colonial archives and collections relating to West Africa, exploring their ambiguous status as both sites of epistemic violence and, potentially, resources for communities to recover cultural histories, memories and alternative ways of knowing and being in the world.

Josh Baum, MA

Artist: Bristol, UK; Mitzpe Ramon, IL

Josh Baum is a British-born artist and poet living and working in Israel. He has an MA in fine art from Central Saint Martins, UAL, London, where he was awarded the Future Map prize in 2010. His practice includes drawing, writing, sculpture and performance. Throughout his life and work he has sought to understand the poetic impulse of the artist: the way in which symbols and myths are constantly emerging and vanishing and the ways in which a physical object or artwork can fill with meaning and the poem becomes palpable. This has led most recently to his collaboration with Sascha Mikloweit in developing the art practice of peripatetic thinking.

Jamie Bradbury, MA, BEd

Artist: Newfoundland, CA

Jamie Bradbury is a multidisciplinary artist based in Newfoundland. His practice spans visual arts, music, and independent research-creation. Bradbury holds bachelor's degrees from OCAD University and Memorial University of Newfoundland and a Master's from Central Saint Martins in London. His work has garnered international recognition, with exhibitions across North America and Europe, including the Royal Cambrian Academy of Art, Wales, the Overture Centre for the Arts in the U.S, and Arte Fiera, Bologna. Bradbury's art has been featured in prominent publications such as Canadian Art, Create Magazine and CBC News. His pieces are held in private and public collections, notably Canada's National Collection. Collaborating with choreographer Marissa Wong, Bradbury explores the intersection of visual art, dance, and sound, at events like TEDx Soma San Francisco, US, Sound Symposium XXI and Dance Victoria, CA. He was recently a recipient of a Canadian Council of the Arts grant for Movement and Sound Creation.

David David

Carpenter: Tsumeb, NA

I'm a 26-year-old man, born and raised in Omusati Region. I completed my Secondary Education at Dr. Romanus Kumunoko Senior Secondary School. After completing my high school, I developed a desire to further my studies and become an Artisan, that's how I ended up enrolling at the Community Skills Development Centre (COSDEC TSUMEB) where am currently doing my level 2 Training in a Technical Trade of Joinery and Cabinet making (Wood Work). Participating in this field of Vocational and Technical Training has helped me gain skills, knowledge and attitude to enter the working Industry. In the Long term it's my dream to become a qualified Wood Work Artisan and open a wood work company.

Rakker Festus

Carpenter; Tsumeb, NA

I'm a 28 year old woman, born and raised in Ohangwena Region. I completed my Secondary Education at Engela Senior Secondary School. After completing my high school, I develop a desire of furthering my studies and become an Artisan, that's how I ended up enrolling at the Community Skills Development Centre (COSDEC TSUMEB) where I am currently doing my level 2 Training in a Technical Trade of Joinery and Cabinet making (Wood Work). Participating in this field of Vocational and Technical Training has helped me gain skills, knowledge and attitude to enter working Industry. In the Long term it's my dream to become a qualified Wood Work Artisan and open a wood work company.

Charmaine //Gamxamûs, BSc

Poet; Windhoek, NA

Charmaine //Gamxamûs is a Namibian poet and writer. She is an avid lover of language, expression, and storytelling through poetry, literature and music. She refines her writing acumen by reading, attending poetry and short story writing workshops and writing persistently. Moved by the power of stories, Charmaine finds that words, when expressed truthfully and simply, can spark meaningful change within the reader, whether that be contemplation, a shift in perspective, developing a deeper sense of empathy, or making a major life change. Often, authors are not privy to these changes within the reader, but the writer persists nonetheless. Charmaine placed second in the Goethe Institute's short story competition, 'In Times of Pandemic', for her story, 'Dehisce'. She is a member of the 2022-2023 Doek Collective, and a fellow of the Narrating Namibia, Narrating Africa Doek Emerging Writers Program. Charmaine took part in the inaugural Book Sprint Namibia, where she published a children's book in 12 hours, titled *Myra's Friends*.

Dr. R. Johannes Giebel

Curator of the Mineralogical Collections: Technische Universität Berlin, Berlin, DE

Research Fellow: Department of Geology, University of the Free State, Bloemfontein, ZA Johannes Giebel is a researcher and curator of the Mineralogical Collections at the

TU Berlin. He obtained his PhD in Mineralogy from the University of Tübingen (Germany) and the University of the Free State (Bloemfontein, South Africa) and completed a Master's degree in Economic Geology in Bloemfontein and a Diploma in Geology and Paleontology at the Martin-Luther University Halle-Wittenberg (Germany). Johannes worked at the State Museum of Natural History in Karlsruhe, where he was responsible for the reorganisation and digitalisation of the petrographic collection. His research interests encompass interdisciplinary fields of geology, with long-standing research activities in southern Africa. Through his work with the Mineralogical Collections, which contain a significant proportion of Namibian specimens, he is concerned with the historical background and the appropriate handling of minerals extracted under colonial conditions.

Dr. Noam Gramlich
Curator and researcher: Museum at Rothenbaum Cultures and Art of the World, Hamburg, DE
Co-curator of exhibitions on topics of coloniality, queer and trans* archives. Academic work in the field of media studies, intersectional and queer approaches, de-colonial practice, and visual art. In 2022 they submitted their dissertation with the title 'Colonial Copper. A situated media geology' (campus, forthcoming 2024). They are Co-Editor of the anthology *Feministisches Spekulieren. Genealogien, Narrationen, Zeitlichkeiten*, Berlin (Kadmos) 2020. They have recently held presentations at the University of Cologne (2023), the University of Halle (2022), and the Humboldt Forum (2022), and joined discussions on film archives and intersectionality at proQuote Film/Berlin Biennale (2021).

Prof. Dr. Elizabeth Grosz
Jean Fox O'Barr Women's Studies Distinguished Professor Emerita: Gender, Sexuality & Feminist Studies; Duke University, Durham, US
Elizabeth Grosz is Professor Emerita at Duke University, where she worked in the Programs in Literature and Gender, Sexuality and Feminist Studies. Her writings have explored the writings of Freud, Bergson, Darwin, Foucault, Deleuze and Irigaray. Her most recent book is *The Incorporeal. Ontology, Ethics and the Limits of Materialism* (Columbia University Press), which examines the place of meaning, representation and ideals in the 'minor philosophies' of the Stoics, Spinoza, Nietzsche, Deleuze, Simondon and Raymond Ruyer, in order to address the question of the excess over matter, matter's orientation to and capacity for meaning or incorporeal resonances.

Maria-Oo Haihambo
Poet: Windhoek, NA
Maria-Oo Haihambo is a Namibian poet who describes poetry as a vessel to liberate one's thoughts. She has performed at Spoken Word Namibia, open mics, and the Township Production bi-monthly poetry session in Windhoek. Her poem 'Perm Blues' was published in the *My Heart in Your Hands* anthology by UNAM Press in 2020. Maria-Oo has also contributed to several anthologies, namely *Naming the Tomb* (Township Production, 2021) and *The Griots of Ubuntu: An Anthology* (Kristrech Theatre International, 2022), which contain her two poems 'Answer' and 'Room Left'. In a society that has seen an increase in the violations of women and children, she believes that it is important to educate boys and men about the value and treatment of women. At the same time, it is important to create safe spaces for men and boys to express themselves freely without condemnation. She has also contributed several poems to the anthology *A Conversation: The Africa in ME* (Township Productions, 2023).

Dr. Alma Hannig
Director University History Museum: Main Coordinator University Museums, Collections and Art, University of Bonn, DE
Dr Alma Hannig studied Medieval and Modern History, Political Sciences, Psychology and Spanish at the Universities of Bonn, Bilbao and Salamanca. Her major fields of research interest include Austro-Hungarian and German political and diplomatic history in the 19th and 20th centuries the history of peace movements, the First World War, the history of Germanic aristocracies, the history of populism and antisemitism, and the history of university collections. She has curated numerous exhibitions in Germany and Austria, and is currently preparing an exhibition on provenance research at the University of Bonn (including 25 museums and collections). Her recent publications include Alma Hannig, Christian Meierhofer, Georg Mölich, eds., 2024. *1870/71. Der Deutsch-Französische Krieg in transnationaler, regionaler und interdisziplinärer Perspektive.* Bonn: Bonn University Press.

Prof. Dr. Selby Hearth
Associate Professor of Geology: Department of Geology, Bryn Mawr College, Bryn Mawr, US
Slby Hearth is a U.S. settler and associate professor of mineralogy at Bryn Mawr College, a small college near Philadelphia. She earned her PhD from Washington University in St. Louis, and her Master's degree from the Massachusetts Institute of Technology. She now studies the minerals that form in land impacted by coal mining, and the historical relationships between colonialism and geologic knowledge production. In addition to teaching and research, she co-curates the Bryn Mawr Mineral Collection with Dr. Carrie Robbins, the curator of Art & Artefacts at Bryn Mawr. Her interest in Tsumeb stems from the many Tsumeb minerals that live in the Bryn Mawr College Mineral Collection, and the ongoing inquiry into what ethical management of such specimens entails.

Paul Philipp Heinze, MFA
Artist, Coder, Audio Engineer: Berlin, DE
Paul Philipp Heinze (PPH) is a post-conceptual artist working across many disciplines including installation, sculpture, performance, painting, video, sound and software. He studied at the Berlin University of the Arts, the École supérieure des beaux-arts de Nantes, and the Academy of Fine Arts Leipzig, where he earned a Master of Fine Arts (summa cum laude) in New Media Arts & Fine Arts. Born (1981) in Leipzig, East-Germany, PPH refers to the immersive transition of the political and societal structures as a fundamental influence on his artistic practice, emphasising the sensual qualities of concepts. He equally works in collaborations or independently on his own initiative. His work explores topics such as 'identity-construction', 'post-humanism' and 'cognitive-capital' and has been featured in numerous international exhibitions, in Amsterdam, Athens, Basel, Bogotá, Chicago, Copenhagen, Istanbul, London, Moscow, NYC, Paris, Priština, Rīga, Vienna, and Warsaw. He is an affiliate of the Society for Artistic Research Germany.

Bastian Herbst, MA
Curator for Outreach: Humboldt Lab, Humboldt-Universität zu Berlin, Berlin, DE
Bastian Herbst holds an M.A. in Modern and Contemporary History, Political Science and Public Law from the University of Freiburg, for which he also studied in Spain, France and Switzerland. For his PhD research, he specialised in the history of mass media in the British and French Empires, particularly in Egypt, North Africa, and the wider Arab world, and researched at different archives in the UK, France, Germany, and Switzerland. Having worked at the German Historical Museum in Berlin for two years, he is now a curator at the Humboldt Lab, the exhibition and event space of Humboldt-Universität zu Berlin, located in the controversial Humboldt Forum.

Prof. Dr. Christopher Hill
Associate Professor (Research and Development): Faculty of Business and Creative Industries, Humanities Research and Innovation Group, University of South Wales, UK
Chris Hill is an Associate Professor in History at the University of South Wales. His research interests sit at the intersection of imperial history, environmental history, and the history of science and technology. His recent research, funded by the Arts and Humanities Research Council (AHRC), has explored the multifaceted role of imperialism across the cycle of nuclear development in Britain, from uranium extraction in Namibia and South Africa to nuclear weapons tests in Australia and the Pacific. Chris is Principal Investigator for the Office for Veterans' Affairs project: 'An Oral History of British Nuclear Test Veterans'.

Amanda Holmes, BA
Photographer: Berlin, DE; Mexico City, MX
Amanda Holmes studied fine arts at Bennington College, US. She grew up in Spain, the US, Chile, and Mexico. As a freelance photographer, she has travelled extensively to Asia and Europe. In addition to her commercial work for industrial clients and institutions, her focus is architecture and documentary photography. Her work is showcased in magazines and book projects about Mexico; a new one is currently in progress. Through her photos, Amanda tells stories of people and infrastructures, of the abstract connections that lie hidden and details that often go unnoticed. In her photography, Amanda brings these multifaceted nuances to light. She lives and works in Berlin and Mexico City.

Elli Stella Udda Imalwa
Spokesperson for the Municipality of Tsumeb: Tsumeb, NA
As a Public Relations practitioner, my career has been defined by my ability to strategically manage and enhance the reputation of the Municipality of Tsumeb. Having worked in the field of journalism and now communications, I have been exposed to different sectors of our societies, and I learned how to deal with individuals from different backgrounds and cultures. I have honed my skills in communication, media relations, crisis management, and strategic planning to become a trusted advisor to executives and councillors. I was drawn to the field because of its dynamic nature and the ability to have a direct impact on the success of residents and the municipality. My personal and professional experiences have shaped me to be a strong and bold character, able to initiate new ideas, and improve on existing ones through a strategic approach.

Herbert Jauch, PH, MA
Founding Director: Labour Resource and Research Institute (LaRRI), Windhoek, NA
Herbert Jauch holds an M.A.in political science from the University of the Western Cape, South Africa, and has been with the labour movement in Southern Africa for many years. He served as an executive member of the Namibia National Teachers Union (NANTU), as well as on various committees of the National Union of Namibian Workers (NUNW). For over 20 years, Herbert worked as a labour researcher, carrying out research projects for various Namibian and Southern African trade unions. Herbert was instrumental in developing a labour diploma course for Namibian trade unions and was the founding director of the trade union-based Labour Resource and Research Institute (LaRRI) in Katutura, where he worked from 1998 until 2010. He is also an active member of Namibia's Basic Income Grant Coalition since its inception in 2004, and serves as chairperson of the Economic and Social Justice Trust.

Juuso Ndinolune David Kaluvi
Time Keeper & Paymaster; Tsumeb Corporation Limited (TCL), Tsumeb, NA
Juuso Ndinolune David Kaluvi was born in 1939 in Onengali, Ohangwena region. His given name at birth was Ndinolune, a name bestowed upon him before his baptism at Edundja. Growing up, he took on responsibilities such as tending cattle and ploughing. Kaluvi started his education at Onengali and continued at Odibo St. Mary's Mission School in the Ohangwena region, returning as an adult to complete his studies. He worked for the contract labour system, 'Okaholo' in Ondangwa, before securing a supervisory position at TCL Head Office in Tsumeb. Okaholo refers to a copper necklace given to recruits as identification. Kaluvi also served as a preacher, at the Anglican church. Upon retiring, Kaluvi returned to his Onalunike village, contributing to his community until his passing. Known for his magnetic personality, peacemaking, and non-discriminatory nature, he was a cherished community member. Kaluvi married Toini Aron Shiweda from Engela, Ohangwena region. He had 22 children, 19 of whom survived him at his death. He passed away on 21st October 2023 in Onalunike, Oshikoto region.

Philemon Sheya Kaluwapa
Honorary Chairman: Lukopane-Namibia Culture Society, Berlin, DE
Philemon Sheya Kaluwapa is currently the Honorary Chairman of the Lukopane-Namibia Culture Society. He studied photography in East Berlin, and was the Secretary of Culture of the Namibian student organisation in the German Democratic Republic. In West Berlin, Kaluwapa was one of the founding members of the African Writers Association (AWA-Finnaba). Their activities included the centenary of the Berlin Africa Conference at the Hebbel Theatre in 1984 to commemorate the events which led to the division of Africa into colonies. He has recently held presentations at the Photographic Gallery Hippolyte, Helsinki; Albertinum, Dresden; the Humboldt Forum, Berlin (all 2022); and the Changing Room, Berlin (2021).

Veripuami Nandeekua Kangumine
Poet: Windhoek, NA
Veripuami Nandee Kangumine is an a Namibian poet and a writer. She has garnered success in both the local and international poetry scene with her rich imagery and feverish storytelling, her unique style has heralded her in becoming a prominent poetic voice in the African and global literary landscape. Her work wades into abstract means of storytelling inspired by African folklore she heard told around the fire. She is the Poetry winner of the 2023 Bank Windhoek Doek Literary Award. In 2022 she was named as one of the 'Top 10 Poets to Watch' by *Isele Magazine*. Her collection of poems titled 'The Loneliness of Shadows' garnered her a top 6 finalist spot for The Kinsman Quarterly's First Diaspora Award 2023. Her poems are forthcoming in *Doek!* Literary Magazine, *Isele Magazine*, The Black Diaspora Anthology, The Namibian anthology My Heart in Your Hands, and more.

August Kapolo
Artist: Tsumeb, NA
I am a 36 year-old woman, born and raised in Tsumeb. I am an Advanced Computer Certificate holder and I have 7 years of experience in Administration departments. From a very young age, I developed a keen interest in creative art. This saw me entering different art platforms. In this regard, participating in the Transpositional Geology exhibition gave me a better understanding and passion for art, whilst being aware of my natural environment, as I become more aware of nature and recycling. I use recycled materials in a creative and an artistic manner, and my future aspiration is to ensure that my art studio will not just create memories for my clients, but also increase their interest and awareness in the importance of recycling, in order to conserve and protect nature.

Irakli Kiziria, Dipl.-Ing. Arch., Dipl.-Des.
Designer, Musician, Architect: Berlin, DE; Tbilisi, GE
Irakli Kiziria is a Georgian architect, designer, musician, and cultural worker based in Berlin. He is the founder (2013) and artistic director of STAUB, a renowned Berlin-based music and cultural event series. In 2017, he established the Intergalactic Research Institute for Sound, collaborating with fellow artists to explore the experimental and abstract realms of electronic music. Kiziria's musical endeavours, characterised by their emotional depth and eclectic influences, have earned him recognition across various platforms, showcasing his versatility and creative spirit. As an architect and designer, he contributes significantly to the vibrant creative landscapes of both Berlin and Tbilisi. For *Transpositional Geologies*, Kiziria produced the soundtrack of the experimental film *transposition 001 (Happy Banality of Everything)*, and produced the exhibition and publication designs. Through his diverse projects, Kiziria continues to impact the multicultural artistic scene, seamlessly integrating his architectural expertise with his passion for music and community engagement.

Malina Lauterbach, MA
Research Assistant: Heisenberg Professorship for Contemporary Art History – Institute for Art History, Goethe-University, Frankfurt am Main, DE
Malina Lauterbach is a PhD student in art history and teaches at the Goethe University in Frankfurt am Main. Her research interests include minerals and new materialism in contemporary art, feminist and eco-sensitive art practices, and the relationship between artistic research and museum critique. Prior to this, she has worked at the Technical University of Berlin within the BUA-funded projects 'Oxford-Berlin Initiative on Museums as Spaces of Social Cohesion and Conflict', and 'Colonial Collections in Berlin Universities', where she engaged with issues of colonial traces left in the representation and materiality of mineralogical collections in Germany. Lauterbach studied Art History and Curatorial Studies in Berlin and Frankfurt, with a focus on museum studies, postcolonial theory and contemporary artistic and curatorial practices. She completed her Master's degree with a thesis on Gabriele Stötzer's body-related work with women in the context of the German Democratic Republic.

Prof. Dr. Dr Helmut Maier
Senior Researcher and Project Leader: Interdisciplinary Centre for Science and Technology Studies, University of Wuppertal, DE
Helmut Maier is a historian of science, technology and environmental history at the Interdisciplinary Centre for Science and Technology Studies, University of Wuppertal. He conducts research on the history of mining and energy, historical materials research, and the history of scientific societies and technical universities. In his research group on the history of governmental research in the 20th century, the Prussian Geological Survey and geological research form a thematic focus with two staff members. He is co-editor of the anthology *Mining and Environment in the GDR and the FRG* (in German, 2022), which examines the 'Practices of Environmental Policy and Recultivation'.

Prince Kamaazengi Marenga I
Poet: Windhoek, NA
A poem-tree farmer. I am born of very old people. My father was 63, my mother 50. I am the last born of 13. At 20, in 1939, my father was one of the 11,000 soldiers who were recruited from the former British colonies to take part in the Second World War. I always grew up with something of a double age; the youngest youth and the oldest elder. My durability is in that duality. My currency is the now and then. I don't come from a certain place. I come from a Herero people who survived the first genocide of the 21st century. I come from their history, their takes, aches and mistakes. My people would say 'okunene okuhepa' for a thank you [that a person of humble means, is a greater human]. That's my background as a human-being in the world. As a full-time poet, I perform regularly at the bi-monthly poetry sessions at the Goethe Institute Namibia.

Dr. Helke Mocke
Chief Geoscientist: National Earth Science Museum, Geological Survey of Namibia, Ministry of Mines and Energy, Windhoek, NA
Helke Mocke was born in Windhoek, Namibia. She is a chief geoscientist in the National Earth Science Museum of the Geological Survey of Namibia in Windhoek, managing the collections of minerals, rocks, meteorites and fossils, exhibitions, tours, and day-to-day running of the museum. She completed both her Honours and Masters degrees in Palaeontology at the Evolutionary Studies Institute, a Centre of Excellence at the University of the Witwatersrand in South Africa, and began working in the museum in 2004. Her research interests include the use of fossils as indicators of climate change, ancient environments, dinosaur footprints, the beginnings of life, Ediacara fossils, the earliest mammal ancestors, meteorites, and minerals from Tsumeb. She has published numerous articles on these subjects.

Hidipo Nangolo
Independent Journalist and Film Director: Tsumeb, NA
Hidipo Nangolo is a veteran journalist, independent filmmaker and activist, who started out in 1988 as a radio announcer and producer. Later, he became senior producer, executive producer and assistant news editor for Namibian national radio and TV broadcaster NBC. After 15 years he switched to cinematography, directing and print media. He also worked as audio-visual researcher and as editor-in-chief of *The Namibian Worker* and *Informante* newspapers, respectively. He worked as Music Co-Ordinator, Audio-Visual Researcher and as Second Assistant Director on set on the movie 'Namibia – The Struggle for Liberation', filmed in 2005. Between 2006-2007, Nangolo produced two township-jazz music CDs, 'Dakotas' and 'Benguela Breeze', by the Original Jazz Masters (OJM), a six-piece band of elderly people who played township-jazz in Windhoek more than half a century ago. His 2008 docu-film 'Skymaster – Township Jazz from the Old Location' about OJM was awarded a Namibia Film Award for best director and editing.

Dr. Saima Nakuti Ndahangwapo (Ashipala)
Postdoctoral Fellow: International Studies Group, University of the Free State, Bloemfontein, ZA
Saima Nakuti Ndahangwapo is a Postdoctoral Fellow with the International Studies Group at the University of the Free State and affiliated to the Faculty of Humanities and Social Sciences at the University of Namibia. She teaches history modules on the German and South African colonial periods and the process of decolonisation in Namibia. Her research interests include diplomatic, mining and labour histories in southern Africa. Her recent publications are: *Defending the Investment: Rössing Uranium and the Business of Decolonization in Namibia* (Basler Afrika Bibliographien Basel, Switzerland, forthcoming). 'Follow the Yellowcake Road': Historical Geographies of Namibian Uranium from the Rössing Mine. *Historical Social Research* 49 (1) (2024): 32-54. doi: 10.12759/hsr.49.2024.02 (with Christopher R. Hill). Technical and Vocational Education and the Place of Indigenous Labour in the Mining Industry of Namibia, 1970–1990, *Journal of Southern African Studies*, 47:1 (2021): pp.127-142, doi: 10.1080/03057070.2021.1849502

Prof. Dr. Paul O'Kane
Senior Lecturer Critical Studies Fine Art: Central Saint Martins, University of the Arts London, UK
Paul O'Kane is an artist, writer and senior lecturer on art and culture, mainly at Central Saint Martins college, University of the Arts, London. Paul has published extensively and broadly, producing numerous professional publications, including peer-reviewed articles and books, reviews, essays, and blogs, as well as stories, songs, poems and experiments in art writing. Paul's research, writing and lecturing in Fine Art is strongly informed by a historical model. In 2023, Paul published a book for Routledge titled *History in Contemporary Art & Culture*. Prior to that, he published two books titled *Technologies of Romance (Parts 1 & 2)* with eeodo. His Fine Art practices have focused on the historical and anachronistic qualities of the photographic and video image. Paul completed a PhD in 2009 on the theme of 'hesitation' while writing as an artist in the history department of Goldsmiths College, University of London. Paul is a founder member of eeodo artists' publishers, with whom he has published several artists' books. He is also a long-standing member of the International Association of Art Critics (AICA), for whom he has contributed conference papers and written reports on congresses, internationally.

Manuel Pandalis
Photographer: Hamburg, DE
Manuel Pandalis, born in Germany in 1972, is a Hamburg-based artist specialising in photography. After studying at the Hamburg University of Applied Sciences (HAW) and training as a media designer at AEP, he has worked as a freelance photographer since 2004. Pandalis undertakes photo trips and assignments across Europe and the US. In addition to his commercial work, he focuses on portrait and documentary photography, including projects in Namibia and Germany for Transpositional Geoloiesies. He has exhibited globally and collaborated withprestigious partners like Leica. His works are featured in numerous publications and book projects, including collector's editions in the German National Library. A new publication is expected at the end of 2024. Pandalis' photography emphasises looking behind facades and capturing subtle nuances. His portrait series 'Pure' offers profound insights into the expressive possibilities of his subjects, revealing hidden yet palpable depths.

Holger Risse
Visual Artist, Art Director, Motion & Graphic Designer: Cologne, DE.
Holger Risse is a distinguished visual artist, motion, and graphic designer based in Cologne, Germany. Renowned for his innovative and immersive visual concepts, Risse has been a pivotal figure in the music and festival industry for the last 25 years. He was a member of the art collective *Lichtsport*, known for their inventive live visuals. From 1998 to 2010, Risse delved deeply into live visuals, performing in collaboration with musicians like The Notwist, Aphex Twin, Étienne de Crécy, and Mouse on Mars. Holger Risse's passion for visual storytelling and his commitment to pushing creative boundaries have established him as a freelance art director for music magazines, exhibitions, and festivals; including the European Media Art Festival, Melt, Cologne Jazzweek, Lollapalooza Berlin, and Film Festival Cologne. Risse's diverse practice distils emotions and stories into fundamental forms, movements, and colours.

Jermaine Solunga
Poet: Windhoek, NA
I am Jermaine Solunga, a growing poem – stemming from a variety of Namibian gardens. I have been learning how to write poetry since the age of 15. I am 24 years of age. I am passionate about analysing symbolic language. I believe it is spoken when no one else is listening. I have conducted spoken word performances at multiple stages including, bi-monthly poetry session at the Goethe Institute's open mic night in Windhoek, at the warehouse (brewers market), as well as at the Maun International Arts Festival. To me, poetry is love, it is pain, it is good, and sometimes it is bad...

Richardlee Shoombe, MDes
Jewellery designer: Tsumeb, NA
At the age of 12, Richardlee Shoombe enrolled at the Art Performance Centre (APC) in Tsumeb. After completing his primary Visual Art course, he pursued and completed Level 8 in Visual Art at the APC (2011-2014). His exceptional jewellery designs earned him the De Beers Group Jewellery Design Shining Light Overall Winner Award in 2012, 2017, and 2018. In 2019, he received a Diploma in Accessory Design from Poli.Design, Politecnico di Milano, Italy. Since 2015, Shoombe has been a lecturer of Visual Art at the APC. He believes that teaching art and design to young generations can inspire hope and bring positive change to their lives. Richardlee draws inspiration from the culture and behaviour of others, his general impressions of life, and even the weather, all of which influence his artistic work.

Francesca Steinbach
Owner; Steinbach Beergarden, Tsuemb, NA
Franca Steinbach (ne Ardito) is resident of the copper town Tsumeb since 25 years. From the 1960s onwards, she grew up in Johannesburg, South Africa, as a child of Italian immigrants. In 1986 she met her late husband Bernd Erich Karl Steinbach, a third-generation baker, and in 1987 they married in Johannesburg. In 1998, the couple emigrated to Namibia where they opened the Steinbach Bäckerei in Tsumeb. The building they operated in was constructed in 1928 to serve as a miners' house, and from the 1950s onward, it became a bakery, until it finally closed down in 2011. Later that year, Francesca and Bernd re-opened the location as the Steinbach Beergarden & Pizzeria, and Francesca is still running the business today.

Prof. Dr. Ellison Tjirera
Senior Lecturer: Department of Sociology, University of Namibia, Windhoek, NA
Ellison Tjirera teaches Sociology at the University of Namibia, Windhoek. For over a decade as a researcher, Tjirera has been associated with the Windhoek based think tank, the Institute for Public Policy Research (IPPR) as well as the Ministry of Gender Equality and Child Welfare. He holds a PhD. in Sociology from the University of the Witwatersrand, Johannesburg. His academic interests include urban studies, social movements, gender studies, governance and post-apartheid societies.

Kuhepa Tjondu, MSc, ACCA
Poet: Windhoek, NA
Kuhepa Tjondu is a Namibian poet and a lecturer at the Namibia University of Science and Technolo and y (NUST). His poetry is influenced by the language and the culture of the people of Koakoland who raised him and the indomitable spirit of the African people all over the continent. His first published poem titled 'Truth' was published in England in an anthology entitled *Whispers on the Breeze*, in 2009. He is a regular performer at the Township Production bi-monthly poetry sessions at the Goethe Institute in Windhoek. His poems have been published in a University of Namibia (UNAM) Press Anthology titled *My Heart in Your Hands* and a number of Township Production Anthologies. In 2020 his poems were published in an African anthology titled *The Griots of Ubuntu*. He also contributed eight poems to an English/German Anthology of Namibian poetry Like a Whisper in the Wind (*Wie ein Wispern im Wind)*, which was published in 2022.

Prof. Dr. Samo Tomšič
Professor of Philosophy/Aesthetics: University of Fine Arts Dresden (HFBK), Dresden, DE
Samo Tomšič is a professor of Philosophy/Aesthetics at the University of Fine Arts Dresden (HFBK) and a senior researcher at the Humboldt University Berlin. He works in the field of continental philosophy, epistemology, history and theory of psychoanalysis, and critical theory. He is the author of numerous publications, including two monographs on the libidinal and affective politics of capitalism: *The Capitalist Unconscious: Marx and Lacan* (Verso, 2015) and *The Labour of Enjoyment: Toward a Critique of Libidinal Economy* (August Verlag, 2019).

Anna Williams
Administrative Officer: Geological Survey of Namibia, Ministry of Mines and Energy, Windhoek, NA
Anna Williams was born in Windhoek, Namibia and completed her schooling in Cuba during the struggle for Namibian Independence. She has been working at the Ministry of Mines and Energy since 1999 in various capacities, including at Human Resources, the switchboard and sales office. In the sales office she assists clients to purchase various geological maps, publications on Namibia's geology and other products. She likes to welcome visitors to the ministry with a smile and often assists with outreach activities in the National Earth Science Museum in the Geological Survey of Namibia.

Prof. Dr. Kathryn Yusoff
Professor of Inhuman Geography: School of Geography, Queen Mary University London, London, UK
Kathryn Yusoff is Professor of Inhuman Geography in the School of Geography at Queen Mary University of London. Her research examines inhuman materialisms in the context of environmental change, race, and subjectivity. She is author of *A Billion Black Anthropocenes or None* (University of Minnesota Press, 2018) and *Geologic Life: Inhuman Intimacies and the Geophysics of Race* (Duke University Press, 2024), which addresses the racial geologies of rocks and the nomenclature of natural resources. She is recipient of the Association of American Geographers Award for Creativity in Geography (2022).

Dr. Anne Zacke
Director: Mineralogical Museum, Institute of Geosciences, University of Bonn, Bonn, DE
Anne Zacke studied geology at the University of Greifswald and gained her PhD in Natural Sciences at the University of Cologne. Afterwards, she worked as research fellow at the State Museum of Natural History in Karlsruhe in the Department of Museums Education and Public Relations. Then, she took the opportunity, moved to Muscat, and worked as one of the first staff members the Geoscience Department at the newly founded German University of Technology in Oman. After six years of challenging and exciting work she moved back to Germany and became the museum director of the Mineralogical Museum of the University of Bonn. Anne Zacke works on the interface of science and public, her focus is on museums' didactics and education as well as historical scientific collections. Additionally, she teaches different courses in geosciences as well as in the master program museum studies.

Table of Figures

Author's note: Throughout the figure text the use of the first person 'I', outside of quotes, refers to the author, editor, and artist Sascha Mikloweit.

Fig. 0 [cover]
I—O_MMDUB_029-sm, Sascha Mikloweit, 2025 (*O—I*, 2019), offset print on Fedrigoni Splendorlux Mirror Argento 320g. Based on the Mineralogical Museum of the University of Bonn, no collection number. Tourmaline, dark green embedded in rock crystal; St. Anne's Mine, Mwami, Karoi, Mashonaland West, Zimbabwe, Africa. The collection's glass label states the British colony (1890–1980) 'Southern Rhodesia, Africa' as origin. The mine operated primarily during the 1960s and 1970s. In the early 1970s, Peter A. Bancroft (1916–2015) (see further text Fig.132) recalled a visit to the site: 'Just the week before, a land mine blew up a truck near St. Anne's and killed six people. Few pilots wished to jeopardise their lives or aircraft by flying in the region. Insurgents used heat-seeking surface-to-air rockets; reportedly, two aircraft had been downed in the previous weeks.' This occurred during the Zimbabwean War of Independence (1964–1979), in which the Zimbabwe African National Union (ZANU) and the Zimbabwe African People's Union (ZAPU) fought the Rhodesian white-minority government, the latter supported by the British South Africa Police (BSAP) and the Rhodesian Army. The conflict's origins trace back to the British South Africa Company's colonial actions in the region during the late-19th century. Image: Mikloweit / TG

Fig. 1 [pp.2–3]
I—O_MMdUB-70363, Sascha Mikloweit, 2025 (*O—I*, 2021), offset print on Fedrigoni Symbol Tatami 150g. Based on the Mineralogical Museum of the University of Bonn, collection object #70363. Cornetite; Kalukuluku Mine (L'Etoile du Congo Mine), Lubumbashi, Haut-Katanga, DR Congo. The mineral entered the institutional collection in 2019 as part of the donated collection Husmann. The Kalukuluku Mine was renamed by the Belgians to L'Etoile du Congo Mine (Star of the Congo Mine), and was the first mine to produce cobalt ore in 1907, opes'ted by the Belgian Union Minière du Haut Katanga. Cornetite was the first 'new' mineral 'discovered' in the Congo by the Belgian colonial geologist Jules Cornet (1865–1929) and named after him. He had a key role in identifying the rich mineral deposits of the Katanga region and facilitating its colonial exploitation. Image: Mikloweit / TG

Fig. 2 [pp.4–5]
I—O_MMdUB-60271, Sascha Mikloweit, 2025 (*O—I*, 2019), offset print on Fedrigoni Symbol Tatami 150g. Based on the Mineralogical Museum of the University of Bonn, collection object #60271. Chalcanthite; El Teniente Mine, El Teniente, Rancagua, Cachapoal Province, O'Higgins, Chile. Local mining activities are assumed to have started in 1819; the industrial-scale operations of the El Teniente Mine in Chile began in 1904, initiated by Braden Copper Company (US). This marked the beginning of capitalist mining activities at what is now the largest underground copper mine in the world. Braden Copper later became a subsidiary of the Kennecott Copper Corporation (US). The mine remained under foreign ownership until the Chilean government nationalised its copper industry in 1971, transferring ownership to the state-owned company CODELCO (Corporación Nacional del Cobre de Chile), which continues to manage it today. For the period from 1973 onwards, George Monbiot and Peter Hutchinson's 2024 publication *The Invisible Doctrine* (pp.30, 43-44) points out that 'Pinochet's violent overthrow of Salvador Allende in 1973 is widely seen as inaugurating the global neoliberal era.' His 'economic programme was devised and overseen by neoliberal economists from the University of Chicago, and enthusiastically supported by Milton Friedman [1912–2006], Friedrich Hayek [1899–1992], and the neoliberal international. Released from the constraints of democracy, Pinochet's economists were able to implement the entire neoliberal package: Chileans who resisted were imprisoned, tortured, or murdered. The nation's resources were systematically plundered, especially its principal asset, its copper mines.' Image: Mikloweit / TG

Fig. 3 [pp.6–7]
I—O_MMdUB-094-sm, Sascha Mikloweit, 2025 (*O—I*, 2021), offset print on Fedrigoni Symbol Tatami 150g. Based on the Mineralogical Museum of the University of Bonn, no collection number. Bayldonite with veins of azurite and malachite; Tsumeb Mine, Tsumeb, Oshikoto, Namibia. Entered the MMdUB collection in the 1950s. The object label references the collections Lag and Kohl, archive #SWA185 as origin. Regarding detailed information on the Tsumeb Mine, see this volume, p.303. Image: Mikloweit / TG

Fig. 4 [pp.8–9]
I—O_MMdUB-10010, Sascha Mikloweit, 2025 (*O—I*, 2021), offset print on Fedrigoni Symbol Tatami 150g. Based on the Mineralogical Museum of the University of Bonn, collection object #10010. Sulphur with calcite; Racalmuto Sulphur Mines, Racalmuto, Agrigento, Sicily, Italy. The object label references the collection C. Daniele Catania, 1962, archive #V.59.34. as origin. Industrial-scale sulphur mining in Sicily, particularly in areas like Racalmuto, began in the mid-19th century, with significant expansion in the 1850s and 1860s. Before industrialisation, sulphur was mined using artisanal methods, but the demand for sulphur—especially for use in gunpowder during the Napoleonic Wars (1803–1815) and later for sulphuric acid (1850s–1950s)—led to the development of more mechanised and large-scale operations. By the late-19th century, Sicily was producing 90% of the world's sulphur. Image: Mikloweit / TG

Fig. 5 [pp.10–11]
I—O_MMdUB-70187, Sascha Mikloweit, 2025 (*O—I*, 2019), offset print on Fedrigoni Symbol Tatami 150g. Based on the Mineralogical Museum of the University of Bonn, collection object #70187. Mimetite, campylite, and wulfenite (reverse); Chah Milleh Mine, Anarak, Nain, Isfahan, Iran. The object label references the collection R. Möckel, 1970, archive #V.74.3. as origin. Industrial-scale mining in Chah Milleh ceased in 1959 after less than a decade of operation. No marked entry date for the institutional collection. Image: Mikloweit / TG

Fig. 6 [pp.12–13]
I—O_MMdUB-60227, Sascha Mikloweit, 2025 (*O—I*, 2019), offset print on Fedrigoni Symbol Tatami 150g. Based on the Mineralogical Museum of the University of Bonn, collection object #60227. Crocoite; West Coast Range, Dundas, Tasmania, Australia (this is most likely the Adelaide Mine). The object label references the collection Friedrich Carl Duisberg (this volume, p.207), archive #1305 as origin. Institution's collection entry 1954. The industrial-scale operation of the Adelaide Mine began in the 1890s. By this time, under British colonial rule (1856–1901), the Aboriginal population of Tasmania had been devastatingly reduced due to violence, disease, and forced relocation, with the last person considered solely of Tasmanian descent—Fanny Cochrane Smith (*1834)—passing in 1905. By 1900, crocoite specimens from Tasmania were being shipped to museums and collectors worldwide. Image: Mikloweit / TG

Fig. 7 [pp.14–15]
I—O_MMdUB-10363, Sascha Mikloweit, 2025 (*O—I*, 2021), offset print on Fedrigoni Symbol Tatami 150g. Based on the Mineralogical Museum of the University of Bonn, collection object #10363, date of collection entry unknown. Copper (native); Tsumeb Mine, Tsumeb, Oshikoto, Namibia. Weight 20kg (see also text Fig.35). Regarding detailed information on the Tsumeb Mine, see this volume, p.303. Image: Mikloweit / TG

Fig. 8 [pp.16–17]
I—O_MMdUB-00937, Sascha Mikloweit, 2025 (*O—I*, 2019), offset print on Fedrigoni Symbol Tatami 150g. Based on the Mineralogical Museum of the University of Bonn's collection object #00937, entered into the collection in 1961. Wolframite; Tamboras, Sitabamba, Santiago de Chuco, La Libertad, Peru. The object label references the Stappenbeck Collection as the origin (archive #937). About 150 items from Stappenbeck's collection were added to the museum's mineral deposit collection, but the total number of specimens remains unknown due to incomplete institutional records. Richard Stappenbeck (1880–1963) was a German geologist renowned for his contributions to the study of South America's geology. He served as a State Geologist in various South American countries from 1906 to 1921 and 1932 to 1933, before becoming a professor at the Technical University of Berlin in 1936. His 1916 study on South American mineral resources highlighted their strategic importance for German industry, providing essential data on mineral deposits to support Germany's efforts to secure raw materials amid global competition and wartime needs. Colonial mining in this region predates Stappenbeck, beginning in the 16th century with the Spanish conquest and primarily targeting silver and gold. The Spanish imposed the mita system from 1572 to 1812, annually forcing over 200 Indigenous communities to send one-seventh of their adult male population to labour in mines under harsh and exploitative conditions. Image: Mikloweit / TG

Fig. 9 [pp.18–19]
I—O_MMdUB-70255, Sascha Mikloweit, 2025 (*O—I*, 2019), offset print on Fedrigoni Symbol Tatami 150g. Based on the Mineralogical Museum of the University of Bonn, collection object #70255. Vivianite; from a clay deposit near the village of Andoua, 40km from N'Gaoundéré, Cameroon. The object label references the donation by Dr Puffe in 1967. Cameroon was a German colony from 1884 to 1916. After Germany's defeat in World War I, the colony was divided between Britain and France under League of Nations' mandates, with the French controlling the larger portion. The German colonial presence ended in 1916, and this was formalised by the Treaty of Versailles in 1919. Image: Mikloweit / TG

Fig. 10 [pp.20–21]
I—O_MMdUB-70131, Sascha Mikloweit, 2025 (*O—I*, 2020), offset print on Fedrigoni Symbol Tatami 150g. Based on the Mineralogical Museum of the University of Bonn, collection object #70131. Descloizite; Tsumeb Mine, Tsumeb, Oshikoto, Namibia. The object label references the collection Stappenbeck, 12 July 1929 as origin (archive #32) (regarding Stappenbeck see text Fig. 8). Named in 1854 by Augustin Alexis Damour (1808–1902) after Alfred Louis Olivier Le Grand Des Cloizeaux (1817–1897), Professor of Mineralogy at the University of Paris, who first described the mineral in an academic context. Regarding detailed information on the Tsumeb Mine see this volume, p.303. Image: Mikloweit / TG

Fig. 11 [pp.22–23]
I—O_MMdUB-00954, Sascha Mikloweit, 2025 (*O—I*, 2019), offset print on Fedrigoni Symbol Tatami 150g. Based on the Mineralogical Museum of the University of

Bonn, collection object #00954. Wolframite interspersed with quartz; Tamboras Mine, Mundo Nuevo, Huamachuco, Sánchez Carrión, La Libertad, Peru. The object label references the collection Stappenbeck as origin (archive #954) (regarding Stappenbeck see text Fig. 8). Image: Mikloweit / TG

Fig. 12 [pp.24–25]
I—O_MMDUB_029-sm, Sascha Mikloweit, 2025 (*O—I*, 2019), offset print on Fedrigoni Symbol Tatami 150g. See text Fig. 0. Image: Mikloweit / TG

Fig. 13 [pp.26–27]
I—O_MMdUB-60048, Sascha Mikloweit, 2025 (*O—I*, 2020), offset print on Fedrigoni Symbol Tatami 150g. Based on the Mineralogical Museum of the University of Bonn, collection object #60048. Inderite; Boron Mine (now Rio Tinto Borax Mine), Boron, Kern County, California, USA. The object label references the collection Dr Shilly, 1966 as origin. Industrial-scale mining began in 1927, managed by U.S. Borax, which became part of Rio Tinto Group in 1967. It is one of the world's largest borate sources, providing ~ 30% of global demand. The Boron mine continues to hold significant international relevance due to borate's versatile applications—from agriculture to high-tech materials, and renewable energy technologies. Image: Mikloweit / TG

Fig. 14 [pp.28–29]
I—O_MMdUB-40082, Sascha Mikloweit, 2025 (*O—I*, 2019), offset print on Fedrigoni Symbol Tatami 150g. Based on the Mineralogical Museum of the University of Bonn, collection object #40082. Pyrolusite; radiating on psilomelane, Mariensberg, Westerwald, Germany. The object label references the collection Friedrich Carl Duisberg (this volume, p.207), archive #859 as origin. Institutional collection entry 1954. Image: Mikloweit / TG

Fig. 15 [pp.30–31]
I—O_MMdUB-20192, Sascha Mikloweit, 2025 (*O—I*, 2020), offset print on Fedrigoni Symbol Tatami 150g. Based on the Mineralogical Museum of the University of Bonn, collection object #20192. Stibnite (antimonite); Ichinokawa Mine, Saijō City, Ehime Prefecture, Shikoku Island, Japan. Operational from 1679 until its closure in 1957. Antimonite played a crucial role in Japan's industrialisation and imperial expansion during the Meiji and Taisho periods (1890–1930). It was especially important for military applications, including the production of alloys used in ammunition, explosives, and various industrial technologies. Image: Mikloweit / TG

Fig. 16 [p.38]
Archived image of the former 'Green Hill' in Tsumeb, Namibia, approximately 1905 (see this volume, pp.72, 124, 197, 239). Image inscription: 'Arthur Koppel Aktiengesellschaft. Otavi-Bahn: Kupferberber in Tsumeb, km 567. Kupferdruck O. Felsing. Berlin S.W. (Arthur Koppel Joint-stock Company. Otavi Railway: Copper hill in Tsumeb, km 567. Copper engraving by O. Felsig. Berlin S.W.)'. National Archives of Namibia, 2023, archive #29924. Image: National Archives of Namibia

Fig. 17 [p.47]
Presencing the coloniality of mining in the Mineral Gallery of the Natural History Museum, London. Specimens from 'Broken Hill, Northern Rhodesia' and 'Broken Hill, New South Wales' juxtaposed. Author and Photo: Paul Basu

Fig. 18 [p.52]
transposition 712 (No Geology is Neutral?), Sascha Mikloweit, 2020. A 1500 x 200 cm banner with the proposition 'No Geology Is Neutral?' was intended as an intervention in institutional and public spaces. Kathryn usoff's dictum was rephrased as a question, addressing the students and staff of the Poppelsdorfer Schloss in Bonn, Germany, which houses the Mineralogical useum of the University of Bonn, as well as the Institute of Earth Sciences with its laboratories, library, lecture hall, and student association. The proposal was rejected by Mineralogical Museum in 2020. Image: Mikloweit / TG

Fig. 19 [p.57]
We Build Our Languages Out Of Rocks—Transpositional Geologies, Sascha Mikloweit et al., 2023, Mineralogical Museum of the University of Bonn, exhibition. Floor plan of the central exhibition and list of all four exhibition locations. Illustration: Irakli Kiziria / TG

Fig. 20 [p.67]
Excerpt of SWAC letter transcript. Rogers to the SWAC Board of Directors, 6 October 1893.[20] Image: German Federal Archives

Fig. 21 [p.78]
Drive down to the former Otavi Mining and Railways Company (OMEG) / Tsumeb Copper Ltd. (TCL) Mine, Tsumeb, Namibia. Photographed during the *Transpositional Geologies (TG)* research trip in October 2022. Photo: Amanda Holmes / TG

Fig. 22 [p.82]
I—O_MSdTUB-87-1297_L_1-1, Sascha Mikloweit, 2025 (*O—I*, 2024), offset print on Munken Print White 115g. Based on the Mineralogical Collections of the Technical University of Berlin, collection object #87-1297. Object label text: 'Institute of Mineralogy, Technical University of Berlin. Boltwoodite monoclinic, yellow on calcite. Rössing, Namib, south of the Swakop, Namibia.' For the corresponding object see this volume, p.86. Image: Mikloweit / TG

Fig. 23 [p.85]
I—O_MSdTUB-1989-130b, Sascha Mikloweit, 2025 (*O—I*, 2024), offset print on Munken Print White 115g. Based on the Mineralogical Collections of the Technical University of Berlin, collection object #1989-130b. Uranophane (uranotil), monoclinic-sphenoidal yellow needles on calcite; Rössing Uranium Mine, Arandis, Erongo, Namibia. Image: Mikloweit / TG

Fig. 24 [p.86]
I—O_MSdTUB-87-1297, Sascha Mikloweit, 2025 (*O—I*, 2024), offset print on Munken Print White 115g. Based on the Mineralogical Collections of the Technical University of Berlin, collection object #87-1297. Boltwoodite, monoclinic, yellow on calcite; Rössing Uranium Mine, Arandis, Erongo, Namibia. Image: Mikloweit / TG

Fig. 25 [p.89]
I—O_MMdUB-X0022, Sascha Mikloweit, 2025 (*O—I*, 2024), offset print on Munken Print White 115g. Based on the Mineralogical Museum of the University of Bonn, collection object #X0022. Uranium; Heisenberg Cube, 4.9 x 4.9 cm, 2070g (as of 17/06/2023). Entered the collection on 26 October 1954. Obtained from the Foreign Service of the United States of America, U. S. Element, Scientific Research Division (SRD) of the Military Security Board, Reginald Bragonier (Acting Chief SRD). Image: Mikloweit / TG

Fig. 26 [p.97, top]
TG research field trip 2022 image archive: On 21 July 2022, a train carrying sulphuric acid—originating from the Dundee Precious Metals Inc. Tsumeb Specialty Smelter—derailed near Otavi, Namibia, spilling 700 tonnes of sulphuric acid from 13 tankers. Concerns over potential groundwater contamination of the sensitive karst aquifers of the region arose immediately. Nevertheless, when our research team (Mikloweit, Herbst, Holmes) encountered the site by chance on our way to Tsumeb on 20 October 2022, the site had still not been decontaminated. In fact, visible ground contamination was still spread across the area, and we had to leave after about 15 minutes as the acid contamination started to affect our respiratory systems. Photo: Amanda Holmes / TG

Fig. 27 [p.97, bottom]
TG research field trip 2023 image archive: sulphuric acid tanks awaiting filling at the Dundee Precious Metals Inc. (DPM) Tsumeb Specialty Smelter, photographed during our research team's (Mikloweit, Imalwa, Pandalis, Ngavee) tour of the smelter on 27 October 2023. The smelter's acid plant has the capacity to produce between 230,000 and 320,000 tonnes of sulphuric acid annually. In the year of the incident (Fig. 26), annual production amounted to 198,386 tonnes. DPM is Namibia's primary producer of sulphuric acid. (See this volume, pp.181, FN12; 226). Photo: Manuel Pandalis / TG

Fig. 28 [p.106]
TG research field trip 2022 image archive: File 'NAN, ZBU 1547, R.VII, L.10, Vol. 2.' from the National Archives of Namibia, Windhoek. Photo: Mikloweit / TG

Fig. 29 [p.113]
I—O_MMdUB-50239, Sascha Mikloweit, 2025 (*O—I*, 2021), offset print on Fedrigoni Symbol Tatami 150g. Based on the Mineralogical Museum of the University of Bonn, collection object #50239. Smithsonite; Tsumeb Mine, Tsumeb, Oshikoto, Namibia. The label attributes the specimen to the collection of Prof. Neuhaus (archive #V.76.7.). The reverse of the museum label states: 'Paid from GEFFRUB [editor's note: Gesellschaft der Freunde der Uni Bonn / Society of Friends of the University of Bonn] donation 1971. Original material for the work by A. Neuhaus On the Chemistry and Optics of Some Zinc Minerals from Tsumeb/SWA under III. in XX. Report of the Freiberg Geological Society 29–44.' [editor's translation]. The specimen was likely mined during the apartheid era, in the late 1960s or early 1970s, by the Tsumeb Corporation Limited (TCL) (for labour conditions, see this volume, p.77ff.). Prof Alfred Oscar Wilhelm Neuhaus (1903–1975), a renowned German geochemist and mineralogist, became a professor at Rheinische Friedrich-Wilhelms University of Bonn in 1951. As director of the Institute and Museum of Mineralogy and Petrography, he modernised the institute at Poppelsdorf Palace, leading it until his retirement in 1972. Neuhaus is regarded to have significantly advanced mineralogical research in mid-20th century Germany and received numerous honours. Image: Mikloweit / TG

Fig. 30 [p.114]
I—O_MSdTUB_57-1187_7, Sascha Mikloweit, 2025 (*O—I*, 2024), offset print on Fedrigoni Symbol Tatami 150g. Based on the Mineralogical Collections of the Technical University of Berlin, collection object #1187, entered into the collection in 1957 as part of the Karl Keller Collection. Smithsonite; Tsumeb Mine, Tsumeb, Oshikoto, Namibia. This object was acquired as part of the Karl Keller Collection in 1957. The Keller Collection, the largest contingent of Tsumeb minerals within the TU collections, was purchased by Prof. H. Strunz for 10,000 DM (this would be comparable to

a buying power of approximately 43,000 € in 2023) directly after Keller's death in April 1957. Karl Keller (1892–1957) was a significant figure at the Tsumeb mine, holding a senior position at the Otavi Mining and Railways Company (OMEG) from 1922 to 1932 under the direction of mining operations by Friedrich Wilhelm Kegel (1874–1948) (Kegel had previously worked in mining operations in Chile, the Philippines, North Korea [twice], Kiautschou, Jehol [China], and Serbia, prepared mining reports in South West Africa for Professor Friedrich Rinne [1863–1933] of the University of Leipzig, and led an expedition into the gold districts of Colombia). Keller's managerial role at the Tsumeb mine was central to the operation's success. His responsibilities likely included overseeing daily operations, managing workers, and ensuring efficient mining processes. Further, this position enabled him to amass a private collection of 1751 specimens from the Tsumeb mine. In addition to his position with OMEG, Keller was a designated member of the National Socialist German Workers' Party (NSDAP). In a handwritten CV dated 18 September 1937, Keller elaborates: 'In 1931, I founded a local chapter of the NSDAP in Tsumeb. [...] I managed this chapter [...] until my return to Germany in September 1932. Upon my arrival in Germany, the District Leader [Kreisleiter], Party Member Haus, appointed me as Base Leader [Stützpunktleiter] in Rodheim. Later, the Regional Leader [Gauleiter] confirmed me as Local Group Leader [Ortsgruppenleiter]. I continue to hold this position, and my two children from my first marriage actively serve in leadership roles within the Hitler Youth (HJ).' [editor's translation] Image: Mikloweit / TG

Fig. 31 [p.117]
I—O_MSdTUB_57-1187_2, Sascha Mikloweit, 2025 (*O—I*, 2024), offset print on Fedrigoni Symbol Tatami 150 g. Based on the Mineralogical Collections of the Technical University of Berlin, collection object #1187. Smithsonite; Tsumeb Mine, Tsumeb, Oshikoto, Namibia. Text see Fig. 30. Image: Mikloweit / TG

Fig. 32 [p.118]
I—O_MMDUB_029-sm, Sascha Mikloweit, 2025, offset print on Fedrigoni Symbol Tatami 150 g. See text Fig. 0. Image: Mikloweit / TG

Fig. 33 [p.121]
I—O_MMdUB_019-sm_3, Sascha Mikloweit, 2025 (*O—I*, 2019), offset print on Fedrigoni Symbol Tatami 150 g. Based on the Mineralogical Museum of the University of Bonn, no collection number, entered into the collection in 1987. Kutuahorite; Wessels Mine, Joe Morolong Local Municipality, John Taolo Gaetsewe District Municipality, Northern Cape, South Africa. The label states the origin as the Gebhard Collection, 1987. Wessels Mine operates in a region historically shaped by colonial extraction: the region's economy was significantly influenced by mining industries, which have been linked to exploitative labour systems during both the colonial and apartheid periods. The District Municipality was the richest mining region in the Northern Cape until the decline of asbestos mining in the 1980s. Today, the mining of manganese, iron, and tiger's eye continues. The Sishen iron-ore mine, located just 83km from Wessels, is one of the largest open-cast mines in the world. The region, home to a large Tswana-speaking population, experienced significant social and economic marginalisation due to these industries, contributing to persistent poverty and inequality. Image: Mikloweit / TG

Fig. 34 [p.122]
I—O_MMdUB_60113_2, Sascha Mikloweit, 2025 (*O—I*, 2020), offset print on Fedrigoni Symbol Tatami 150g. Based on the Mineralogical Museum of the University of Bonn, collection object #60113, entered into the collection in 1958. Hanksite; California, USA [editor's note: possibly Searles Lake area]. The collection's label states the Collection O. Wild, Idar (collection# V.55.32.), as origin. A note on the back of the institution's museum label remarks: 'Acquired for DM 4.00 from Georg O. Wild, Idar-Oberstein 2, on 13 December 1958.' [editor's translation]. Image: Mikloweit / TG

Fig. 35 [p.125]
I—O_MMdUB_10363_1, Sascha Mikloweit, 2025 (*O—I*, 2021), offset print on Fedrigoni Symbol Tatami 150g. Based on the Mineralogical Museum of the University of Bonn, collection object #10363, date of collection entry unknown. Copper (native); Tsumeb Mine, Tsumeb, Oshikoto, Namibia. Weight 20kg. The MMdUB's collection label states the collection of 'Dr Pohl, Nd.-Ramstadt' as origin. A note on the back of the museum label remarks: 'Exchange value [DM] 350.-' [editor's translation]. Image: Mikloweit / TG

Fig. 36 [p.126]
I—O_MMdUB_40082_1, Sascha Mikloweit, 2025 (*O—I*, 2019), offset print on Fedrigoni Symbol Tatami 150g. Based on the Mineralogical Museum of the University of Bonn, collection object #40082, no collection entry date. Archival records provide no additional information. Image: Mikloweit / TG

Fig. 37 [p.144]
Minerals organised according to the Strunz classification system in the display cases of Technical University of Berlin. Photo: Manuel Pandalis / TG

Fig. 38 [p.145]
Minerals organised according to the Strunz classification system in a display case of Technical University of Berlin, detail. Photo: Manuel Pandalis / TG

Fig. 39 [p.145]
O—I_MSdTUB-19-132_L_3, Sascha Mikloweit, 2024, digital file. Based on historical mineral label 'Dt. S. W. Africa' (German Southwest Africa) of the Mineralogical Collections of the Technical University of Berlin, collection object #19-132. Apatite from pegmatite; Donkerhoek, German South West Africa. Image: Mikloweit / TG

Fig. 40 [p.146]
O—I_MSdTUB-19-132_L_1-1, Sascha Mikloweit, 2024, digital file. Based on historical mineral label 'D. S. W. Afrika' (German Southwest Africa) of the Mineralogical Collections of the Technical University of Berlin, collection object #19-132. Apatite from pegmatite; Donkerhoek, German South West Africa. Image: Mikloweit / TG

Fig. 41 [p.146]
Page from Reinhard Anton Brauns' (1861–1937) 1912 English translation of *The Mineral Kingdom*. Image: Mikloweit

Fig. 42 [p.148]
transposition 015 (re—existence), Sascha Mikloweit et al., 2021–2024, performance detail. Tsumeb, Namibia, 2023 (see this volume pp.228 ff.). Photo: Manuel Pandalis / TG

Fig. 43 [p.149]
Germanite in the Mineralogical Collections of the Technical University of Berlin; drawer from the collection's archive in the basement. Photo: Manuel Pandalis / TG

Fig. 44 [p.153]
transposition 931 (Geoscientific Specimen Refusing its Further Objectification), Sascha Mikloweit, 2021 (*O—I*, 2019). Image: Mikloweit / TG

Fig. 45 [p.155]
transposition 884 (22°40'56.1"S 14°32'15.1"E •<—>• KM334), Sascha Mikloweit, 2023, sculptural intervention. Otavi, Namibia. Photo: Mikloweit / TG

Fig. 46 [pp. 156–157]
transposition 064 (22°40'56.1"S 14°32'15.1"E •<-->• 50°43'29.9"N 7°05'31.2"E), Sascha Mikloweit, 2023, sculptural intervention, detail. Photo: Manuel Pandalis / TG

Fig. 47 [p.158]
transposition 064 (22°40'56.1"S 14°32'15.1"E •<-->• 50°43'29.9"N 7°05'31.2"E), Sascha Mikloweit, 2023, installation view. Photo: Manuel Pandalis / TG

Fig. 48 [p.159]
TG research field trip 2022 image archive: *Nama & Ovaherero Genocide Monument 1904–1908*, Swakopmund, Namibia; West view (detail). Photo: Mikloweit / TG

Fig. 49–54 [p.160]
transposition 904 (22°40'56.1"S 14°32'15.1"E •<-->• 50°43'05.1"N 7°04'57.6"E), Sascha Mikloweit, 2023, details performance intervention. Mikloweit performing the intervention at Lothar von Trotha's grave, Poppelsdorfer Cemetery, Bonn, Germany, in April 2023. Photo: Manuel Pandalis / TG

Fig. 55 [p.161]
transposition 904 (22°40'56.1"S 14°32'15.1"E •<-->• 50°43'05.1"N 7°04'57.6"E), Sascha Mikloweit, 2023, detail performance intervention. The grave of Lothar von Trotha, Poppelsdorfer Cemetery, Bonn, Germany, covered by the nameless grave photographed during the TG research trip at the *Nama & Ovaherero Genocide Monument 1904–1908*, Swakopmund, in 2022—weighed down with the four unclassified Tsumeb minerals. Photo: Manuel Pandalis / TG

Fig. 56 [pp.162–163]
transposition 904 (22°40'56.1"S 14°32'15.1"E •<-->• 50°43'05.1"N 7°04'57.6"E), Sascha Mikloweit, 2023, details performance-box. Interior of the the performance kit's archival box, containing a folded piece of pure silk, printed with an image of the sandy heap of a nameless grave that I photographed during our visit to the *Nama & Ovaherero Genocide Monument 1904–1908* in Swakopmund, Namibia (2022). Underneath the silk are four unclassified Tsumeb minerals, extracted in collaboration with Elli Stella Udda Imalwa, spokesperson for the Tsumeb Municipality. Two minerals extracted from the OMEG open-pit mine—the former 'Green Hill'—and two were extracted from Level 1, where we carefully descended together, stepping slowly along the pitch-black drive that spirals down for 42 levels. This winding descent leads into the humid, slippery Tsumeb mine, the extractive operation initiated by Imperial Germany's OMEG in 1907 (see in this volume, pp. 78, 189). Photo: Manuel Pandalis / TG

Fig. 57 [p.164]
transposition 904 (22°40’56.1”S 14°32’15.1”E •<-->• 50°43’05.1”N 7°04’57.6”E), Sascha Mikloweit, 2023, detail. The sculptural intervention installed inside the Systemic Collection, Hall 1, the Mineralogical Museum of the University of Bonn. Photo: Manuel Pandalis / TG

Fig. 58 [p.165]
transposition 904 (22°40’56.1”S 14°32’15.1”E •<-->• 50°43’05.1”N 7°04’57.6”E), Sascha Mikloweit, 2023, installation view detail. Performance box and script. Photo: Manuel Pandalis / TG

Fig. 59 [pp.166–175]
transposition 904 (22°40’56.1”S 14°32’15.1”E •<-->• 50°43’05.1”N 7°04’57.6”E), Sascha Mikloweit, 2023, performance script. Images: Mikloweit

Fig. 60 [p.177]
TG research field trip 2022 image archive: *Nama & Ovaherero Genocide Monument 1904–1908*, Swakopmund, Namibia; south-east view with the Namib Desert in the background. Photo: Amanda Holmes / TG

Fig. 61 [p.178]
TG research field trip 2022 image archive: *Nama & Ovaherero Genocide Monument 1904–1908*, Swakopmund, Namibia. View of the Riverside Avenue’s affluent housing to the north. Photo: Amanda Holmes / TG

Fig. 62 [pp.184–185]
Detail of the Mineralogical Museum of the University of Bonn, Systemic Collection (Exhibition Hall 1); Manuel Pandalis, 2023. Photo: Manuel Pandalis / TG

Fig. 63 [p.189]
TG research field trip 2022 photo archive: Location of the former ‘Green Hill’ (see this volume p.38) which became the open-pit mine of the Otavi Mining and Railway Company (OMEG) in Tsumeb, Namibia. The picture was taken in October 2022. Photo: Amanda Holmes / TG

Fig. 64 [pp.192–193]
transposition 441 (Black No More, MMdUB Version), Sascha Mikloweit, 2023, installation view. The title references George S. Schuyler’s 1931 novel ‘Black No More: Being an Account of the Strange and Wonderful Workings of Science in the Land of the Free, A.D. 1933–1940’. Photo: Manuel Pandalis / TG

Fig. 65 [p.195]
transposition 441 (Black No More, MMdUB Version), Sascha Mikloweit, 2023, installation view, detail. Photo: Manuel Pandalis / TG

Fig. 66 [pp.196–197]
transposition 086 (Tsumeb Relay), Sascha Mikloweit et al., 2023, detail view. Performance intervention inside the former OMEG open pit mine, once the location of the ‘Green Hill’, Tsumeb, Namibia on 28 October 2023. Photo: Manuel Pandalis / TG

Fig. 67 [pp. 198–199]
transposition 086 (Tsumeb Relay), Sascha Mikloweit et al., 2023. Performance inside the former OMEG open pit mine, once the location of the ‘Green Hill’, Tsumeb, Namibia on 28 October 2023. Photo: Manuel Pandalis / TG

Fig. 68 [p.200, top]
transposition 086 (Tsumeb Relay) as installed at the end of the performance intervention on 28 October 2023, at the edge of the OMEG open-pit mine. Photo: Mikloweit / TG

Fig. 69 [p.200, bottom]
transposition 086 (Tsumeb Relay), Sascha Mikloweit et al., 2023. View of the permanent installation on 29 October 2023. Photo: Manuel Pandalis / TG

Fig. 70 [p.201, top]
(*transposition 086 (Tsumeb Relay)*, Sascha Mikloweit et al., 2023. Vandalised state of the permanent installation on 19 November 2023. Photo: Lukas Hoye / TG

Fig. 71 [p.201, bottom]
transposition 086 (Tsumeb Relay), Sascha Mikloweit et al., 2023. The installation further smashed into pieces on 12 March 2024. Photo: Johannes Giebel / TG

Fig. 72 [p.202]
transposition 756 (Just Another Unexpected Guest), Sascha Mikloweit, 2023, installation view, detail. The distorted reflection of *transposition 504*’s (see this volume p.204) representation of the Mineralogical Museum of the University of Bonn’s collection object #50215. Photo: Manuel Pandalis / TG

Fig. 73 [p.203]
transposition 756 (Just Another Unexpected Guest), Sascha Mikloweit, 2023, installation view, detail. Photo: Manuel Pandalis / TG

Fig. 74 [p.205]
transposition 504 (I—O_MMdUB-50215), Sascha Mikloweit, 2023, installation view. Photo: Manuel Pandalis / TG

Fig. 75 [p.206]
transposition 995 (I—O_MMdUB-50215@TM-NA), Sascha Mikloweit, 2023. View of the permanent installation inside the entrance hall of Tsumeb Municipality. Photo: Manuel Pandalis / TG

Fig. 76 [p.207]
transposition 995 (I—O_MMdUB-50215@TM-NA), Sascha Mikloweit, 2023. Photo greeting from Tsumeb Municipality sent to Mikloweit by Elli Stella Udda Imalwa on 15 July 2024. Photo: Elli Stella Udda Imalwa

Fig. 77 [pp.208–209]
transposition 741 (We Come and Die as Numbers), Sascha Mikloweit, 2023, installation view, detail. Photo: Manuel Pandalis / TG

Fig. 78 [pp. 210-211]
transposition 741 (We Come and Die as Numbers), Sascha Mikloweit, 2023, installation view, detail. Simon Zhu Mbako’s poem scratched into the glass of the emptied out vitrine. Photo: Manuel Pandalis / TG

Fig. 79 [pp. 212–213]
transposition 368 (The Future Sound of Destruction), Sascha Mikloweit and Paul Philipp Heinze, 2023, installation view, detail. Photo: Manuel Pandalis / TG

Fig. 80 [p.215]
transposition 368 (The Future Sound of Destruction), Sascha Mikloweit and Paul Philipp Heinze, 2023, installation view, detail. Photo: Manuel Pandalis / TG

Fig. 81 [p.216]
transposition 665 (I—O_OMEG-01870), Sascha Mikloweit, 2023, installation view. Photo: Manuel Pandalis / TG

Fig. 82 [p.217]
O—I_OMEG-01870, Sascha Mikloweit, 2020, digital file. Based on Otavi Mining and Railways Company (OMEG) share No 01870, Series 1, Berlin, 1921. Image: Mikloweit / TG

Fig. 83 [pp. 218–219]
transposition 610 (US / THEM), Sascha Mikloweit, Richardlee Shoombe, and Manuel Pandalis, 2023. Photo collaboration. Photo: Manuel Pandalis / TG

Fig. 84 [pp.220–221]
transposition 610 (US / THEM), Sascha Mikloweit, Richardlee Shoombe, 2023, detail. Jewellery collaboration. Image: Mikloweit/ TG

Fig. 85–90 [p.222]
transposition 521 (Does Not Contain B2Gold), Sascha Mikloweit, Medala’s Car Wash, and Manuel Pandalis, 2023. Details of the performance collaboration with Medala’s Car Wash, Tsumeb, Namibia. Photos: Manuel Pandalis / TG

Fig. 91 [p.223]
transposition 521 (Does Not Contain B2Gold), Sascha Mikloweit, Medala’s Car Wash, and Manuel Pandalis, 2023. Detail of the performance collaboration with Medala’s Car Wash, Tsumeb, Namibia. Photo: Manuel Pandalis / TG

Fig. 92 [pp.224–225]
transposition 521 (Does Not Contain B2Gold), Sascha Mikloweit, Medala’s Car Wash, and Manuel Pandalis, 2023. Photo collaboration with Manuel Pandalis: North end of the southern pit of the B2Gold Otjikoto Mine; captured on 21 October 2023 at 1230h. Photo: Manuel Pandalis / TG

Fig. 93 [p.226]
transposition 554 (Before Thought Hardens Into Language II), Sascha Mikloweit, 2023, installation view Steinbach Beergarden Tsumeb, detail. ‘Africa’ shopping bag filled with minerals collected inside the former Otavi Mining and Railways Company (OMEG) / TCL open-pit mine (see fig. 63), Tsumeb. Photos: Mikloweit/ TG

Fig. 94 [p.227]
transposition 554 (Before Thought Hardens Into Language II), Sascha Mikloweit, 2023, installation view Steinbach Beergarden Tsumeb, detail. Tailings I extracted from the Dundee Precious Metals Inc. Tsumeb Custom Smelter, placed on one of the beer garden’s tables. Photos: Mikloweit/ TG

Fig. 95 [pp.228 /231]
transposition 015 (re—existence), Sascha Mikloweit et al., 2021–2024, installation view, detail 2023. The imported Tsumeb sediment without the pulverised scientific collection object #50592, as installed at the Mineralogical Museum of the University of Bonn, Hall 2, Special Exhibitions, from 20 April to 5 June 2023 (exhibition phase 1) . Photos: Manuel Pandalis / TG

Fig. 96 [pp. 232–233]
transposition 015 (re—existence), Sascha Mikloweit et al., 2021–2024, installation view, detail 2023. The imported Tsumeb sediment topped with the powder of the destroyed specimen, collection object #50592 from Tsumeb, as installed at the Mineralogical Museum of the University of Bonn, Hall 2, Special Exhibitions, from 5 June to 11 June 2023 (exhibition phase 2). Photos: Mikloweit / TG

Fig. 97 [pp.234–235]
transposition 015 (re—existence), Sascha Mikloweit et al., 2021–2024, detail 2023. A new epistemic object—the original specimen box of the Mineralogical Museum of the University of Bonn's collection object #50592 from Tsumeb with its original archive labels, and an additionally produced TG archive label before reintegration into the institution's archive cabinet 92, drawer 8. Photos: Mikloweit / TG

Fig. 98 [p.236]
transposition 015 (re—existence), Sascha Mikloweit et al., 2021–2024, detail 2023. The Mineralogical Museum of the University of Bonn's collection object #50592 from Tsumeb as it was integrated into the MMdUB's Systemic Collection, Hall 1, from 20 April to 5 June 2023, before its destruction. Photo: Manuel Pandalis / TG

Fig. 99 [p.237]
O—I_MMdUB-50592, Sascha Mikloweit, 2023, digital file. Based on the Mineralogical Museum of the University of Bonn, collection object #50592. Cerussite; OMEG Tsumeb Mine, Tsumeb, Oshikoto, Namibia. Assumed to be extracted during the early 1900s. Image: Mikloweit / TG

Fig. 100-102 [p.238]
transposition 015 (re—existence), Sascha Mikloweit et al., 2021–2024, detail 2023. The absent epistemic object as epistemic object at the Mineralogical Museum of the University of Bonn—the original specimen box of collection object #50592 from colonial Tsumeb with its original archive labels, and an additionally produced TG archive label after their reintegration into the institution's archive cabinet 92, drawer 8, where it remains permanently as part of the archive since 5 June 2023. Photos: Mikloweit / TG

Fig. 103 [p.239]
transposition 015 (re—existence), Sascha Mikloweit et al., 2021–2024, detail 2023. Remainder of the destroyed Mineralogical Museum of the University of Bonn's collection object #50592 from Tsumeb as it was integrated into the institution's Systemic Collection, Hall 1 for the remaining duration of the exhibition; 5 June to 11 June 2023. Photos: Mikloweit / TG

Fig. 104–105 [pp.240–241]
transposition 015 (re—existence), Sascha Mikloweit et al., 2021–2024, detail 2022. Collaborative extraction performance—Elli Stella Udda Imalwa of Tsumeb Municipality and Sascha Mikloweit extracting the sediment in Tsumeb, Namibia, 22 October 2022. Photos: Amanda Holmes / TG

Fig. 106–107 [pp.242–243]
transposition 015 (re—existence), Sascha Mikloweit et al., detail. Giving back to the land and its people: the re-imported sediment, enriched with the destroyed specimen (collection object #50592) during the exhibition *We Build Our Languages Out Of Rocks—Transpositional Geologies* at the Mineralogical Museum of the University of Bonn in June 2023, is returned to its place of origin (p.242). The collective planting of the Marula tree on 1 November 2023 in this very sediment (p.243). Photos: Manuel P andalis / TG

Fig. 108 [p.244]
transposition 015 (re—existence), Sascha Mikloweit et al., 2021–2024, detail 2023. Photo greeting via email from Elli Stella Udda Imalwa of Tsumeb Municipality showing the Marula tree on 10 November 2023. Stella remarked: 'Good day Sascha, I hope this email finds you well. I managed to visit the Marula tree on Friday. It looks beautiful and resilient. It is still struggling to rain, but only time will tell. I was happy to see it and will try to visit it weekly to provide some water. All the best.' Photo: Elli Stella Udda Imalwa

Fig. 109 [p.245]
transposition 015 (re—existence), Sascha Mikloweit et al., 2021–2024, detail 2024. The absent Marula tree on 12 March 2024. Photo: Johannes Giebel /TG

Fig. 110 [pp.246–247]
transposition 241 (Before Thought Hardens into Language I), Sascha Mikloweit, 2023, installation view. The Mineralogical Museum of the University of Bonn, Special Exhibitions, Hall 2 in its transformed state, with the Systemic Collection in Hall 1 visible in the background. Photo: Manuel Pandalis / TG

Fig. 111 [p.248]
transposition 241 (Before Thought Hardens into Language I), Sascha Mikloweit, 2023, installation view. Photo: Manuel Pandalis / TG

Fig. 112 [p.249]
transposition 241 (Before Thought Hardens into Language I), Sascha Mikloweit, 2023, installation view. Photo: Manuel Pandalis / TG

Fig. 113 [pp.250–251]
transposition 241 (Before Thought Hardens into Language I), Sascha Mikloweit, 2023, installation view. Photo: Manuel Pandalis / TG

Fig. 114 [p.252]
transposition 001 (Happy Banality of Everything), Sascha Mikloweit, Irakli Kiziria, and Holger Risse, 2022. Premiere at the Kunstmuseum Bonn on 28 September 2022. The film was presented as a live image-sound performance with a projection size of 14,000 x 8,000 cm. In this context, the discourse of the cinematic work evolved into a trialogue between *transposition 001*, the neo-colonial minerality of the Kunstmuseum Bonn's architecture, and its inherent power structures. Production of the premiere event in collaboration with Förderverein Filmkultur Bonn e.V. and the Mineralogical Museum of the University of Bonn. Photo: Julian Kokott

Fig. 115 [pp.254–255]
transposition 001 (Happy Banality of Everything), Sascha Mikloweit, Irakli Kiziria, and Holger Risse, 2022, film still. Image: Mikloweit, Risse / TG

Fig. 116 [pp.256–257]
transposition 001 (Happy Banality of Everything), Sascha Mikloweit, Irakli Kiziria, and Holger Risse, 2022, film still. Image: Mikloweit, Risse / TG

Fig. 117 [pp.258–259]
transposition 001 (Happy Banality of Everything), Sascha Mikloweit, Irakli Kiziria, and Holger Risse, 2022, film still. Image: Mikloweit, Risse / TG

Fig. 118 [pp.260–261]
transposition 001 (Happy Banality of Everything), Sascha Mikloweit, Irakli Kiziria, and Holger Risse, 2022, film still. Image: Mikloweit, Risse / TG

Fig. 119 [pp. 262–263]
transposition 001 (Happy Banality of Everything), Sascha Mikloweit, Irakli Kiziria, and Holger Risse, 2022, film still. Image: Mikloweit, Risse / TG

Fig. 120 [pp.264–265]
transposition 001 (Happy Banality of Everything), Sascha Mikloweit, Irakli Kiziria, and Holger Risse, 2022, film still. Image: Mikloweit, Risse / TG

Fig. 121 [p.266]
Mikloweit's imaging laboratory setup with custom-made dark chamber and modified scanner at the Mineralogical Museum of the University of Bonn in 2021. Photo: Mikloweit / TG

Fig. 122 [p.269]
If You're Frightened We'll Protect You; IYF WWPY 09-0810-1 [ScaN 09-1302 757-833-10], Sascha Mikloweit, 2010. Series of Series of 27 images showing an inebriated, lovesick, successful, young, lawyer's briefcase that he left unattended. Subsequently it was regarded as a potential terrorist threat, and eventually—after the station was sealed off during rush hour—detonated by the British Transport Police at Waterloo Station in London at 0757h on 13 February 2009. Image: Mikloweit

Fig. 123 [p.270]
All Parts Of The Subsystem Are Present (APOTSAP); CLUSTR1_KOU-ELA-3_1996DOY156, Archiv #: CLU_ID1-2-6, Sascha Mikloweit, 2016. A visual taxonomy of Cluster I satellite debris resulting from the Ariane 5 Flight 501 explosion in 1996. Created during a residency at the European Space Operations Centre (ESOC), Darmstadt, Germany. Series of 38 images. Image: Mikloweit

Fig. 124 [p.273]
Some Parts of the Subsystem Are Present (SPOTSAP), Sascha Mikloweit, 2017, installation view, Copernicus Science Centre, Warsaw, Poland. Site-specific immersive audio-visual installation based on the *CLUSTR1_KOU-ELA3_1996DOY156* image archive, featuring a moving image loop projection with a 5-channel soundscape, 30.0 x 5.5 x 5.0 metres. Photo: Maciej Krüger

Fig. 125 [p.274]
Some Parts of the Subsystem Are Present (SPOTSAP), Sascha Mikloweit, 2017, installation view, Copernicus

Science Centre, Warsaw, Poland. Site-specific immersive audio-visual installation based on the *CLUSTR1_KOU-ELA3_1996DOY156* image archive, featuring a moving image loop projection with a 5-channel soundscape, 30.0 x 5.5 x 5.0 metres. Photo: Maciej Krüger

Fig. 126 [p.277]
I—O_MMdUB-M6209-6210, Sascha Mikloweit, 2025 (*O—I*, 2020), offset print on Fedrigoni Symbol Tatami 150g. Based on the Mineralogical Museum of the University of Bonn, Meteorite Collection object #M6209-6210. Bronzit-Chondrite (H5); Nuevo Mercurio, Zacatecas, Mexico, fall: 15 December 1978, 18:45, 3 stones, 14.1g, 12.9g, 7.7g. The Meteorites entered the institutional collection in 1980, as noted 'in exchange for 45.5 g Knyahinya *[editor's note: Ukraine, fell in 1866]* (No. 247)', presumably from the JM DuPont Collection, which stated on their included object label 'only 2kg found to date, 5 January 1979'. Image: Mikloweit / TG

Fig. 127 [pp.278/281]
I—O_MSdTUB_62-105_1, Sascha Mikloweit, 2025 (*O—I*, 2024), offset print on Fedrigoni Symbol Tatami 150g. Based on the Mineralogical Collections of the Technical University of Berlin, collection object #62-105, entered into the collection in 1962. Becquerelite; Shinkolobwe, Katanga, Congo via the Scott Williams Mineral Company, Scottsdale, Arizona. The Shinkolobwe mine was operated by the Belgian colonial mining company Union Minière du Haut-Katanga (UMHK) from 1921 to 1960. Uranium from Shinkolobwe was shipped to Belgium for processing before World War II. During Germany's occupation of Belgium in 1940, an unspecified portion of this uranium stockpile, held by Union Minière, was seized by the German Wehrmacht. Some of this material was subsequently used in Germany's unsuccessful wartime nuclear research programme (see this volume, p. 97, Fig. 25). Shinkolobwe uranium also supplied the US Manhattan Project, including the Trinity test (16 July 1945), the first nuclear weapon detonation in human history. It also fuelled the bombs dropped on Hiroshima (6 August 1945) and Nagasaki (9 August 1945), which, by the end of 1945, caused the deaths of up to 246,000 people and untold destruction to other forms of life. Subsequently, uranium from colonial Belgian Shinkolobwe became a key resource for the US nuclear weapons programme during the Cold War. In other words, Belgian colonial atrocities, economic interests, and mining know-how constituted the raw mineral foundation of these historical developments. Image: Mikloweit / TG

Fig. 128 [pp.279/280]
I—O_MSdTUB_2025-1_3, Sascha Mikloweit, 2025 (*O—I*, 2024), offset print on Fedrigoni Symbol Tatami 150g. Based on the Mineralogical Collections of the Technical University of Berlin, collection object #2025-1, entered into the collection in January 2025, donated by *Transpositional Geologies* / Sascha Mikloweit. Trinitite; Trinity test site, White Sands Missile Range (WSMR), Alamogordo, New Mexico, United States. Purchased from United Nuclear, Sandia Park, NM, USA, 2008. Trinitite is a glassy residue left on the desert floor after the Trinity nuclear bomb test of the Manhattan Project on 16 July 1945, 0529h, near Alamogordo, New Mexico, USA. The glass is primarily composed of arkosic sand that was melted by the atomic blast, along with melted fragments of this first atomic bomb, its support structures, and various radionuclides formed during this detonation. The 'success' of this test contributed to the subsequent bombings of Hiroshima (6 August 1945) and Nagasaki (9 August 1945), which caused the deaths of up to 246,000 people by the end of 1945, and triggered the Cold War nuclear arms race. Image: Mikloweit / TG

Fig. 129 [p.282]
I—O_MSdTUB_86-61_1-2, Sascha Mikloweit, 2025 (*O—I*, 2024), offset print on Fedrigoni Symbol Tatami 150g. Based on the Mineralogical Collections of the Technical University of Berlin, collection object #86-61, entered into the collection in 1986, donated by Prof H. Strunz. Stottite on germanium ore; Tsumeb Mine, Tsumeb, Oshikoto, Namibia. This specimen was mined in Tsumeb and entered the TU collections during the period of South African rule over Namibia (1915–1990), a time when apartheid policies were actively enforced. Image: Mikloweit / TG

Fig. 130 [p.283]
I—O_MMdUB_80844_1, Sascha Mikloweit, 2025 (*O—I*, 2019), offset print on Fedrigoni Symbol Tatami 150g. Based on the Mineralogical Museum of the University of Bonn, collection object #80844, with no recorded collection entry date. Tourmaline (fibrous); Minas Gerais, Brazil. During its colonial period, from 1690 to 1822, Minas Gerais was a crucial region for Portugal, which exploited its vast gold and diamond deposits. The 18th-century gold rush led to the rapid growth of towns, establishing the region as a central economic hub for the Portuguese Empire. Resistance to colonial rule began early, with protests against heavy taxation and forced or slave labour; by the late 1700s, the region was home to the Inconfidência Mineira (1789), a movement that sought to expel the Portuguese and establish an independent Brazilian republic. Image: Mikloweit / TG

Fig. 131 [p.284]
I—O_MMdUB_50020_1, Sascha Mikloweit, 2025 (*O—I*, 2021), offset print on Fedrigoni Symbol Tatami 150g. Based on the Mineralogical Museum of the University of Bonn, collection object #50020, entered into the collection in 1963. Cerussite; Tsumeb Mine, Tsumeb, Oshikoto, Namibia. The MMdUB's collection label attributes the origin to the Hellinghaus Collection, Herdorf Sieg, 1963 (collection# V.69.47.) as origin. A note on the back of the museum label remarks: 'DM 80.00, X-ray and chemical analysis determined' [editor's translation]. Image: Mikloweit / TG

Fig. 132 [p.285]
I—O_MMdUB-70291_2, Sascha Mikloweit, 2025 (*O—I*, 2023), offset print on Fedrigoni Symbol Tatami 150g. Based on the Mineralogical Museum of the University of Bonn's collection object #70291, entered into the collection in 1980. Phosphophyllite; Unificada Mine, Cerro Rico, Potosí, Tomás Frías, Bolivia. The object label references the Bancroft Collection as the origin, with the front noting in red ink: 'Rare and valuable.' The reverse side states: 'Acquired in exchange from Dr. Peter Bancroft, Ramona, California, 19.08.1970' [editor's translation]. Peter A. Bancroft (1916–2015) was a prominent American geologist and mineralogist whose collection is widely regarded for its scientific and historical value and is thought to have made significant contributions to mineralogical research and education worldwide (see also text Fig. 0). At Potosí, mining operations beginning in 1545 were among the earliest and most significant colonial operations in the Americas. By the 16th and 17th centuries, Potosí became the world's largest source of silver, with the mine accounting for up to 60-70% of global silver production during its peak in the mid-1600s. This made it a key economic engine for Spain and a major contributor to its empire's wealth. Potosí's output and its economic importance during the colonial period solidified its place as one of the most influential mining operations in history. More than 2,880,000 Indigenous people were subject to the mita conscription in Potosí alone (1572–1812), with an additional death toll of African slave labourers thought to be in the hundreds of thousands. Further, it is estimated that upwards of eight million lives were lost in the mines of Peru due to the mita system and related harsh conditions. Image: Mikloweit / TG

Fig. 133 [pp.286–287]
I—O_MMdUB-50215, Sascha Mikloweit, 2025 (*O—I*, 2019), offset print on Fedrigoni Symbol Tatami 150g. Based on the Mineralogical Museum of the University of Bonn, collection object #50215. Smithsonite; Tsumeb Mine, Tsumeb, Oshikoto, Namibia. Mined from the German colonial Otavi Mining and Railways Company (OMEG) Mine (later Tsumeb Copper Ltd. [TCL] Mine) in Tsumeb, approximately 1900–1905. Subsequently, it came into the possession of the German industrialist and *I.G. Farbenindustrie AG / IG Farben* founder Friedrich Carl Duisberg (1861–1935). In 1954, *Bayer AG* (one of the successor companies of IG Farben) outsourced Carl Duisberg's collection under a permanent loan contract to the Mineralogical Museum of the University of Bonn, where it is still on display today. This was initially the collection object I proposed for destruction in *transposition 015 (re—existence)*—the proposal was rejected by the institution in January 2023. Image: Mikloweit / TG

Fig. 134–136 [p.288]
I—O_MMdUB-50215_L_1,*I—O_MMdUB-50215_L_2*, and *I—O_MMdUB-50215_L_3*, Sascha Mikloweit, 2025 (all *O—I's*, 2019), offset print on Fedrigoni Symbol Tatami 150g. Based on the object labels of the Mineralogical Museum of the University of Bonn's collection object #50215—Friedrich Carl Duisberg's smithsonite—(this volume, pp.286–287) as transposed in *transposition 441* (this volume, p.194), *086* (p.196), *504* (p.204), and *995* (p.207). Shifting frameworks of ownership and classification are silencing the specimen's original modes of dispossession: the colonial-period label (Fig.134, top), assumed to date from the early 1900s, is attributed to the private Collection Duisberg; the mid-century institutional entry label (Fig.135, centre) is dated approximately to the mid-1950s–1954 the Collection Duisberg was incorporated into the Mineralogical Museum of the University of Bonn; and the contemporary digital database registration label (Fig.136, bottom) is from the early 2020s. Image: Mikloweit / TG

Index

Index note: page numbers in *italics* refer to images.